Audio and
Hi-Fi Handbook

Audio and
Hi-Fi Handbook
Revised edition

Editor
IAN R. SINCLAIR

Newnes

OXFORD AUCKLAND BOSTON JOHANNESBURG MELBOURNE NEW DELHI

Newnes
An imprint of Butterworth-Heinemann
Linacre House, Jordan Hill, Oxford OX2 8DP
225 Wildwood Avenue, Woburn, MA 01801-2041
A division of Reed Educational and Professional Publishing Ltd

A member of the Reed Elsevier plc group

First published as *Audio Electronics Reference Book*
by BSP Professional Books 1989
Second edition published by Butterworth-Heinemann 1993
Paperback edition 1995
Third edition 1998
Revised third edition (paperback) 2000

British Library Cataloguing in Publication Data
A catalogue record for this book is available from the British Library

ISBN 0 7506 4975 5

Typeset by Jayvee, Trivandrum, India
Printed and bound in Great Britain by Scotprint

This book is dedicated to the memory of
Fritz Langford-Smith,
Mentor and Friend

Contents

Preface   xi

Chapter 1   Sound Waves   1
Dr W. Tempest

Pure tones and complex waveforms   1
Random noise   2
Decibels   2
Sound in rooms   3
Reverberation   4
Reverberation, intelligibility and music   4
Studio and listening room acoustics   4
The ear and hearing   5
Perception of intensity and frequency   6
Pitch perception   7
Discrimination and masking   7
Binaural hearing   8
The Haas effect   8
Distortion   9
Electronic noise absorbers   12
References   12

Chapter 2   Microphones   14
John Borwick

Introduction   14
Microphone characteristics   14
Microphone types   16
The microphone as a voltage generator   18
Microphones for stereo   24
Surround sound   27
References   27

Chapter 3   Studio and Control Room Acoustics   28
Peter Mapp

Introduction   28
Noise control   28

Studio and control room acoustics   33

Chapter 4   Principles of Digital Audio   41
Allen Mornington-West

Introduction   41
Analogue and digital   41
Elementary logic processes   43
The significance of bits and bobs   44
Transmitting digital signals   46
The analogue audio waveform   47
Arithmetic   51
Digital filtering   54
Other binary operations   58
Sampling and quantising   58
Transform and masking coders   65
Bibliography   65

Chapter 5   Compact Disc Technology   67
Ken Clements

Introduction   67
The compact disc . . . some basic facts   67
The compact disc . . . what information it contains   68
Quantisation errors   69
Aliasing noise   69
Error correction   71
How are the errors corrected?   71
Interleaving   72
Control word   73
Eight to fourteen modulation   74
Compact disc construction   74
The eight to fourteen modulation process   77
Coupling bits   77
Pit lengths   77
Sync. word   78
Optical assembly   80
Servo circuits   84
The decoder   86
Digital filtering and digital to analogue conversion   87
Bibliography   92

Chapter 6   Digital Audio Recording   93
John Watkinson

Types of media   93
Recording media compared   96
Some digital audio processes outlined   97
Hard disk recorders   104
The PCM adaptor   105
An open reel digital recorder   106
Rotary head digital recorders   107
Digital compact cassette   110
Editing digital audio tape   110
Bibliography   111

Chapter 7   Tape Recording   112
John Linsley Hood

The basic system   112
The magnetic tape   112
The recording process   113
Sources of non-uniformity in frequency response 114
Record/replay equalisation   116
Head design   117
Recording track dimensions   120
HF bias   120
The tape transport mechanism   123
Transient performance   123
Tape noise   124
Electronic circuit design   125
Replay equalisation   127
The bias oscillator   129
The record amplifier   130
Recording level indication   131
Tape drive control   131
Professional tape recording equipment   131
General description   132
Multi-track machines   133
Digital recording systems   134
Recommended further reading   138

Chapter 8   Noise Reduction Systems   139
David Fisher

Introduction   139
Non-complementary systems   140
Complementary systems   142
Emphasis   142
Companding systems   143
The Dolby A system   147
Telecom C4   148
dbx   148
Dolby B   149
Dolby C   150

Dolby SR   152
Dolby S   155
Bibliography   156

Chapter 9   The Vinyl Disc   157
Alvin Gold and *Don Aldous*

Introduction   157
Background   157
Summary of major steps and processes   157
The lathe   158
Cutting the acetate   158
In pursuit of quality   160
The influence of digital processing   161
Disc cutting – problems and solutions   161
Disc pressing   162
Disc reproduction   163
Drive systems   163
Pick-up arms and cartridges   165
The cartridge/arm combination   165
Styli   167
Specifications   168
Measurement methods   169
Maintaining old recordings   169
References   170

Chapter 10   Valve Amplifiers   171
Morgan Jones

Who uses valves and why?   171
Subjectivism and objectivism   172
Fixed pattern noise   172
What is a valve?   172
Valve models and AC parameters   174
Practical circuit examples   176
Other circuits and sources of information   183

Chapter 11   Tuners and Radio Receivers   186
John Linsley Hood

Background   186
Basic requirements for radio reception   186
The influence of the ionosphere   187
Why VHF transmissions?   188
AM or FM?   189
FM broadcast standards   190
Stereo encoding and decoding   190
The Zenith-GE 'pilot tone' stereophonic system   190
The BBC pulse code modulation (PCM) programme
    distribution system   192
Supplementary broadcast signals   195
Alternative transmission methods   195

Radio receiver design 196
Circuit design 212
New developments 213
Appendix 11.1 BBC transmitted MF and VHF
   signal parameters 214
Appendix 11.2 The 57 KHz sub-carrier radio
   data system (RDS) 214

Chapter 12 Pre-amps and Inputs 215
John Linsley Hood

Requirements 215
Signal voltage and impedance levels 215
Gramophone pick-up inputs 216
Input circuitry 217
Moving coil PU head amplifier design 219
Circuit arrangements 219
Input connections 223
Input switching 223

Chapter 13 Voltage Amplifiers and Controls 226
John Linsley Hood

Preamplifier stages 226
Linearity 226
Noise levels 230
Output voltage characteristics 230
Voltage amplifier design techniques 231
Constant-current sources and 'current mirrors' 232
Performance standards 235
Audibility of distortion components 237
General design considerations 239
Controls 240

Chapter 14 Power Output Stages 252
John Linsley Hood

Valve-operated amplifier designs 252
Early transistor power amplifier designs 253
Listener fatigue and crossover distortion 253
Improved transistor output stage design 255
Power MOSFET output devices 255
Output transistor protection 258
Power output and power dissipation 259
General power amplifier design considerations 261
Slew-rate limiting and transient intermodulation
   distortion 262
Advanced amplifier designs 263
Alternative design approaches 269
Contemporary amplifier design practice 272
Sound quality and specifications 274

Chapter 15 Loudspeakers 276
Stan Kelly

Radiation of sound 276
Characteristic impedance 277
Radiation impedance 277
Radiation from a piston 277
Directivity 277
Sound pressure produced at distance *r* 277
Electrical analogue 279
Diaphragm/suspension assembly 280
Diaphragm size 280
Diaphragm profile 281
Straight-sided cones 282
Material 283
Soft domes 284
Suspensions 284
Voice coil 285
Moving coil loudspeaker 285
Motional impedance 286
References 289

Chapter 16 Loudspeaker Enclosures 290
Stan Kelly

Fundamentals 290
Infinite baffle 290
Reflex cabinets 292
Labyrinth enclosures 295
Professional systems 296
Networks 296
Components 298
Ribbon loudspeaker 298
Wide range ribbon systems 299
Pressure drive units 300
Electrostatic loudspeakers (ESL) 303

Chapter 17 Headphones 310
Dave Berriman

A brief history 310
Pros and cons of headphone listening 310
Headphone types 311
Basic headphone types 314
Measuring headphones 316
The future 317

**Chapter 18 Public Address and Sound
   Reinforcement 319**
Peter Mapp

Introduction 319
Signal distribution 319

Loudspeakers for public address and sound
  reinforcement   322
Cone driver units/cabinet loudspeakers   322
Loudspeaker systems and coverage   325
Speech intelligibility   328
Signal (time) delay systems   330
Equalisers and sound system equalisation   332
Compressor-limiters and other signal processing
  equipment   333
Amplifiers and mixers   334
Cinema systems and miscellaneous applications   335
References and bibliography   336

Chapter 19   In-Car Audio   **337**
Dave Berriman

Modern car audio   337
FM car reception   337
Power amplifiers   338
Separate power amps   339
Multi-speaker replay   340
Ambisonics   340
Cassette players   341
Compact disc   343
Digital audio tape   344
Loudspeakers   345
Installation   352
The future for in-car audio   360

Chapter 20   Sound Synthesis   **362**
Mark Jenkins

Electronic sound sources   362
Synthesizers, simple and complex   362
Radiophonics and sound workshops   363
Problems of working with totally artificial
  waveforms   366
Computers and synthesizers (MIDI and MSX)   368
Mode messages   373
Real time   376
References   377

Chapter 21   Interconnections   **378**
Allen Mornington-West

Target and scope of the chapter   378
Basic physical background   378
Resistance and electrical effects of current   381
Capacitive effects   383
Magnetic effects   384
Characteristic impedance   387
Reactive components   388
Interconnection techniques   390
Connectors   397

**Chapter 22   NICAM Stereo and Satellite
  Radio Systems**   **404**
Geoff Lewis

The signal structure of the NICAM-728
  system   404
The NICAM-728 receiver   406
The DQPSK docoder   407
Satellite-delivered digital radio (ASTRA digital
  radio ADR)   407
Coded orthogonal frequency division multiplex
  (CODFM)   411
The JPL digital system   413
Reality of digital sound broadcasting   414

**Chapter 23   Modern Audio and Hi-Fi
  Servicing**   **415**
Nick Beer

Mechanism trends   415
Circuit trends   417
Tuners   418
Power supplies   418
System control   419
Microprocessors   419
Amplifiers   419
Discrete output stage failures   422
Digital signal processing   423
Mini-disc   423
Test modes   424
Surface mounted and VLSI devices   424
Obsolete formats   425
Software problems   425
Good servicing practice   426
Test equipment   426
Conclusion   426

Chapter 24   Other Digital Audio Devices   **427**
Ian Sinclair

Video Recorders   427
HDCD   427
CD Writers   427
MPEG Systems   432
MP3   434
Transcribing a Recording by Computer   435
WAV onwards   436
DAM CD   437
DVD and Audio   437

Index   441

Preface

At one time in the 1980s, it seemed that audio had reached the limits of technology and that achieving noticeably better sound reproduction was a matter of colossal expenditure. Despite shrill claims of astonishing discoveries, many of which could never be substantiated, there seemed little to separate one set of equipment from any other at the same price, and the interest in audio technology which had fuelled the whole market seemed to be dying out.

The arrival of the compact cassette from Philips had an effect on high-quality sound reproduction that was surprising, not least to its developers. The compact cassette had been intended as a low-quality medium for distributing recordings, with the advantages of small size and easy use that set it well apart from open-reel tape and even from the predominant LP vinyl records of the day. Development of the cassette recorder, however, combined with intensive work on tape media, eventually produced a standard of quality that could stand alongside the LP, at a time when LP quality started to deteriorate because of the difficulties in finding good-quality vinyl. By the end of the 1980s, the two media were in direct competition as methods for distribution of music and the spoken word.

The whole audio scene has now been rejuvenated, led, as it always was in the past, by new technology. The first of the developments that was to change the face of audio irrevocably was the compact disc, a totally fresh approach to the problems of recording and replaying music. It is hard to remember how short the life of the compact disc has been when we read that the distribution of LP recordings is now no longer being handled by some large retail chains.

The hardest part about the swing to compact disc has been to understand even the basis of the technology. Modern trends in hi-fi up to that time could have been understood by anyone who had experience of audio engineering, particularly in the cinema, from the early 1930s onward. The compact disc, using digital rather than analogue methods, was a concept as revolutionary as the transistor and the integrated circuit, and required complete rethinking of fundamental principles by all engaged in design, construction, servicing and selling the new equipment.

The most remarkable contribution of the compact disc was to show how much the record deck and pickup had contributed to the degradation of music. Even low-priced equipment could be rejuvenated by adding a compact disc player, and the whole audio market suddenly became active again.

This book deals with compact disc technology in considerable detail, but does not neglect the more traditional parts of the audio system which are now under more intense scrutiny. The sound wave is dealt with as a physical concept, and then at each stage in the recording process until it meets the ear – which brings us back to principles discussed at the beginning.

Since the first edition, the *Audio Electronics Reference Book*, a new chapter has been added on microphones, making the chapters on recording more complete. There is now an introduction to digital principles, for the benefit of the many readers whose knowledge of analogue circuits and methods will be considerably stronger than that on digital devices and methods. Compact disc technology is now described in full technical detail and this is followed by a discussion and description relating to the newer digital devices that are now following the lead carved out by the compact disc. These include digital audio tape (DAT), NICAM (near instantaneous companding audio multiplex) stereo sound for television, digital compact cassette (DCC) and the Sony mini-disc. A new section on noise reduction systems is now included to show that the last gasp of analogue methods may well be prolonged into the next century. Filling in a gap in the previous text, there is a short section on cabling and interconnections.

The aim has been to present as wide a perspective as possible of high-quality sound reproduction, including reproduction under adverse circumstances (PA and in-car), from less conventional sources (such as synthesizers) and with regards to the whole technology from studio to ear.

1 Sound Waves

Dr W. Tempest

Audio technology is concerned with sound in all of its aspects, yet many books dealing with audio neglect the fundamentals of the sound wave, the basis of any understanding of audio. In this chapter, Dr Tempest sets the scene for all that is to follow with a clear description of the sound wave and its effects on the ear.

Energy in the form of sound is generated when a moving (in practice a vibrating) surface is in contact with the air. The energy travels through the air as a fluctuation in pressure, and when this pressure fluctuation reaches the ear it is perceived as sound. The simplest case is that of a plane surface vibrating at a single frequency, where the frequency is defined as the number of complete cycles of vibration per second, and the unit of frequency is the Hertz (Hz). When the vibrating surface moves 'outward', it compresses the air close to the surface. This compression means that the molecules of the air become closer together and the molecules then exert pressure on the air further from the vibrating surface and in this way a region of higher pressure begins to travel away from the source. In the next part of the cycle of vibration the plane surface moves back, creating a region of lower pressure, which again travels out from the source. Thus a vibrating source sets up a train of 'waves', these being regions of alternate high and low pressure. The actual pressure fluctuations are very small compared with the static pressure of the atmosphere; a pressure fluctuation of one millionth of one atmosphere would be a sound at the level of fairly loud speech.

The speed of sound in air is independent of the frequency of the sound waves and is 340 m/s at 14°C. It varies as the square root of the absolute temperature (absolute temperature is equal to Celsius temperature +273). The distance, in the travelling sound wave, between successive regions of compression, will depend on frequency. If, for instance, the source is vibrating at 100 Hz, then it will vibrate once per one hundredth of a second. In the time between one vibration and the next, the sound will travel $340/1 \times 1/100 = 3.4$ m. This distance is therefore the wavelength of the sound (λ).

$$\text{Wavelength} = \frac{\text{Sound velocity}}{\text{Frequency}} \text{ or } \lambda = \frac{c}{f}$$

A plane surface (a theoretical infinite plane) will produce a plane wave, but in practice most sound sources are quite small, and therefore the sound is produced in the form of a spherical wave, in which sound waves travel out from the source in every direction. In this case the sound energy from the source is spread out over a larger and larger area as the waves expand out around the source, and the intensity (defined as the energy per unit area of the sound wave) will diminish with distance from the source. Since the area of the spherical wave is proportional to the square of the distance from the source, the energy will decrease inversely as the square of the distance. This is known as the inverse square law.

The range of frequencies which can be detected as tones by the ear is from about 16 Hz to about 20 000 Hz. Frequencies below 16 Hz can be detected, certainly down to 1 Hz, but do not sound tonal, and cannot be described as having a pitch. The upper limit depends on the individual and decreases with increasing age (at about 1 Hz per day!)

Pure Tones and Complex Waveforms

When the frequency of a sound is mentioned, it is normally taken to refer to a sinusoidal waveform, as in Fig. 1.1(a). However, many other waveforms are possible e.g., square, triangular etc. (see Fig. 1.1(b), (c)). The choice of the sine wave as the most basic of the waveforms is not arbitrary, but it arises because all other repetitive waveforms can be produced from a combination of sine waves of different frequencies. For example, a square wave can be built up from a series of odd harmonics (f, $3f$, $5f$, $7f$, etc.) of the appropriate amplitudes (see Fig. 1.2). The series to generate the square wave is

$$\sin 2\pi ft + \frac{\sin 6\pi ft}{3} + \frac{\sin 10\pi ft}{5} + \ldots \ldots \frac{\sin 2n\pi ft}{n}$$

where f is the fundamental frequency and t is time. Similar series can be produced for other wave shapes. Conversely, a complex waveform, such as a square wave, can be analysed into its components by means of a frequency analyser, which uses a system of frequency selective filters to separate out the individual frequencies.

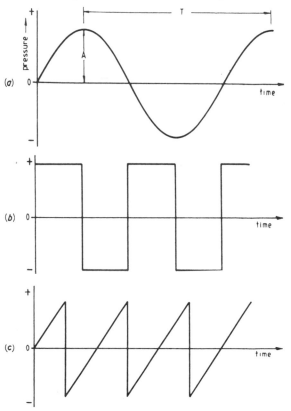

Figure 1.1   Waveforms (a) sine wave, (b) square wave, (c) triangular wave.

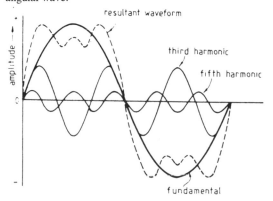

Figure 1.2   Synthesis of a square wave from its components.

Random Noise

While noise is generally regarded as an unwanted feature of a system, random noise signals have great value in analysing the behaviour of the ear and the performance of electronic systems. A random signal is one in which it is not possible to predict the future value of the signal from its past behaviour (unlike a sine wave, where the waveform simply repeats itself). Fig. 1.3 illustrates a noise waveform. Although random, a noise (voltage) for example is a measurable quantity, and has an RMS (root mean square) level which is defined in the same way as the RMS value of an alternating (sine wave) voltage, but, because of its random variability the rms value must be measured as the average over a period of time. A random noise can be regarded as a random combination of an infinite number of sine wave components, and thus it does not have a single frequency (in Hz) but covers a range of frequencies (a bandwidth). 'White' noise has, in theory, all frequencies from zero to infinity, with equal energy throughout the range. Noise can be passed through filters to produce band-limited noise. For example, a filter which passes only a narrow range of frequencies between 950 Hz and 1050 Hz will convert 'white' noise into 'narrow-band' noise with a band-width of 100 Hz (1050–950) and centre-frequency of 1000 Hz.

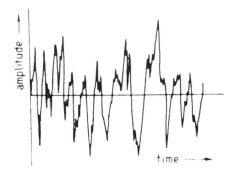

Figure 1.3   Random noise waveform.

Decibels

The pressure of a sound wave is normally quoted in Pascals (Pa). One Pascal is equal to a pressure of one Newton per square metre, and the range of pressure to which the ear responds is from about 2×10^{-5} Pa ($= 20\,\mu$Pa) to about 120 Pa, a range of six million to one. These pressure levels are the RMS values of sinusoidal waves, $20\,\mu$Pa corresponds approximately to the smallest sound that can be heard, while 120 Pa is the level above which there is a risk of damage to the ears, even from a brief exposure. Because of the very wide range of pressures involved, a logarithmic unit, the decibel, was introduced. The decibel is a unit of relative level and sound pressures are defined in relation to a reference level, normally of 20 µPa. Thus any level P (in Pascals) is expressed in decibels by the following formula:

$$dB = 20 \log \frac{P}{P_0}$$

where $P_0 = 20$ µPa.

Table 1.1 shows how the decibel and pressure levels are related.

Table 1.1

dB	P	Comment
−6	10 µPa	Inaudible
0	20 µPa	Threshold of hearing
40	2000 µPa	Very quiet speech
80	0.2 Pa	Loud speech
100	2 Pa	Damaging noise level†
120	20 Pa	Becoming painful

† Sound levels above 90 dB can damage hearing.

Sound in Rooms

Sound in 'free-space' is radiated outward from the source, and becomes weaker as the distance from the source increases. Ultimately the sound will become negligibly small.

When sound is confined to a room, it behaves quite differently since at each occasion on which the sound encounters an obstruction (i.e. a wall) some sound is absorbed, some is transmitted through the wall and some is reflected. In practice, for consideration of sound inside a room, the transmitted element is negligibly small.

When sound is reflected from a plane rigid, smooth surface, then it behaves rather like light. The reflected ray behaves as if it comes from a 'new' source, this new source being an image of the original source. The reflected rays will then strike the other walls, being further reflected and forming further images. Thus it is clear that after two reflections only, there will be numerous images in existence, and any point in the room will be 'surrounded' by these images. Thus the sound field will become 'random' with sound waves travelling in all directions. Obviously this 'random' sound field will only arise in a room where the walls reflect most of the sound falling on them, and would not apply if the walls were

highly absorbent. A further condition for the existence of a random sound field is that the wavelength of the sound is considerably less than the room dimensions. If the sound wavelength is comparable with the room size, then it is possible for 'standing waves' to be set up. A standing wave is simply a wave which travels to and fro along a particular path, say between two opposite walls, and therefore resonates between them. Standing waves can occur if the wavelength is equal to the room length (or width, or height), and also if it is some fraction such as half or one-third etc. of the room dimension. Thus if the wavelength is just half the room length, then two wavelengths will just fit into the length of the room and it will resonate accordingly. For a rectangular room of dimensions L (length) W (width) and H (height), the following formula will give the frequencies of the possible standing waves.

$$f = \frac{c}{2} \sqrt{\left(\frac{p}{L}\right)^2 + \left(\frac{q}{W}\right)^2 + \left(\frac{r}{H}\right)^2}$$

where c is the velocity of sound (340 m/s approx) and $p, q,$ & r take the integral values 0, 1, 2, etc.

For example, in a room 5 m × 4 m × 3 m, then the lowest frequency is given by $p = 1, q = 0, r = 0$, and is

$$\left(f = \frac{340}{2} \sqrt{\left(\frac{1}{5}\right)^2} = 34 \text{ Hz}\right)$$

At the lowest frequencies (given by the lowest values of $p, q,$ and r) there will be a few widely spaced frequencies (modes), but at higher values of p, q and r the frequencies become closer and closer together. At the lower frequencies these modes have a strong influence on sounds in the room, and sound energy tends to resolve itself into the nearest available mode. This may cause the reverberent sound to have a different pitch from the sound source. A simple calculation shows that a typical living room, with dimensions of say 12 × 15 ft (3.7 × 4.6 m) has a lowest mode at 37 Hz and has only two normal modes below 60 Hz. This explains why to achieve good reproduction of bass frequencies, one needs both a good loudspeaker *and* an adequately large room and bass notes heard 'live' in a concert hall have a quality which is not found in domestically reproduced sound.

At the higher frequencies, where there are very many normal modes of vibration, it becomes possible to develop a theory of sound behaviour in rooms by considering the sound field to be random, and making calculations on this basis.

Reverberation

When sound energy is introduced into a room, the sound level builds up to a steady level over a period of time (usually between about 0.25 s and, say, 15 s). When the sound source ceases then the sound gradually decays away over a similar period of time. This 'reverberation time' is defined as the time required for the sound to decay by 60 dB. This 60 dB decay is roughly equal to the time taken for a fairly loud voice level (about 80 dB) to decay until it is lost in the background of a quiet room (about 20 dB). The reverberation time depends on the size of the room, and on the extent to which sound is absorbed by the walls, furnishings etc. Calculation of the reverberation time can be made by means of the Sabine formula

$$RT = \frac{0.16\,V}{A}$$

where RT = reverberation time in seconds,
  V = room volume in cubic metres
  A = total room absorption in Sabins (= m²)
The total absorption is computed by adding together the contributions of all the absorbing surfaces.

$$A = S_1\alpha_1 + S_2\alpha_2 + S_3\alpha_3 + \ldots$$

where S_1 is the area of the surface and α_1 is its absorption coefficient.

The value of α depends on the frequency and on the nature of the surface, the maximum possible being unity, corresponding to an open window (which reflects no sound). Table 1.2 gives values of α for some commonly encountered surfaces.

The Sabine formula is valuable and is adequate for most practical situations, but modified versions have been developed to deal with very 'dead' rooms, where the absorption is exceptionally high, and very large rooms (e.g. concert halls) where the absorption of sound in the air becomes a significant factor.

Table 1.2

Material	Frequency Hz					
	125	250	500	1 k	2 k	4 k
Carpet, pile and thick felt	0.07	0.25	0.5	0.5	0.6	0.65
Board on joist floor	0.15	0.2	0.1	0.1	0.1	0.1
Concrete floor	0.02	0.02	0.02	0.04	0.05	0.05
Wood block/lino floor	0.02	0.04	0.05	0.05	0.1	0.05
Brickwork, painted	0.05	0.04	0.02	0.04	0.05	0.05
Plaster on solid backing	0.03	0.03	0.02	0.03	0.04	0.05
Curtains in folds	0.05	0.15	0.35	0.55	0.65	0.65
Glass 24–32 oz	0.2	0.15	0.1	0.07	0.05	0.05

Reverberation, Intelligibility and Music

The reverberation time of a room has important effects on the intelligibility of speech, and on the sound quality of music. In the case of speech, a short reverberation time, implying high absorption, means that it is difficult for a speaker to project his voice at a sufficient level to reach the rearmost seats. However, too long a reverberation time means that the sound of each syllable is heard against the reverberant sound of previous syllables, and intelligibility suffers accordingly. In practice, maximum intelligibility requires a reverberation time of no more than 1 s, and times in excess of 2 s lead to a rapid fall in the ability of listeners to perceive accurately every syllable. Large concert halls, by comparison, require more reverberation if they are not to sound too 'thin'. Fig. 1.4 shows how the range of reverberation times recommended for good listening conditions varies with the size of the room and the purpose for which it is to be used.

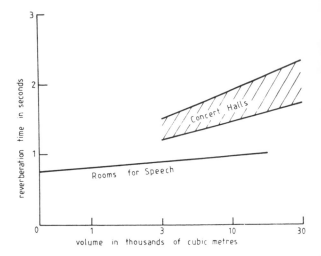

Figure 1.4   Recommended reverberation times.

Studio and Listening Room Acoustics

The recording studio and the listening room both contribute their acoustic characteristics to the sound which reaches the listener's ears. For example, both rooms add reverberation to the sound, thus if each has a reverberation time of 0.5 s then the resulting effective reverberation time will be about 0.61 s. The effective overall reverberation time can never be less than the longer time of the two rooms.

For domestic listening to reproduced sound it is usual to assume that the signal source will provide the appropriate level of reverberant sound, and therefore the listening room should be fairly 'dead', with adequate sound absorption provided by carpets, curtains and upholstered furniture. As mentioned above, the size of the room is relevant, in that it is difficult to reproduce the lower frequencies if the room is too small. In order to obtain the best effect from a stereo loudspeaker system, a symmetrical arrangement of speakers in the room is advantageous, since the stereo effect depends very largely on the relative sound levels heard at the two ears. A non-symmetrical arrangement of the room and/or speakers will alter the balance between left and right channels.

Studio design is a specialised topic which can only be briefly mentioned here. Basic requirements include a high level of insulation against external noise, and clear acoustics with a carefully controlled reverberation time. A drama studio, for radio plays, might have included a general area with a medium reverberation time to simulate a normal room, a highly reverberant 'bathroom' and a small 'dead' room, which had virtually no reverberant sound to simulate outdoor conditions. Current sound recording techniques demand clear sound but make extensive use of multiple microphones, so that the final recording is effectively 'constructed' at a sound mixing stage at which various special effects (including reverberation) can be added.

The Ear and Hearing

The human auditory system, can be divided into four sections, as follows, (see Fig. 1.5).

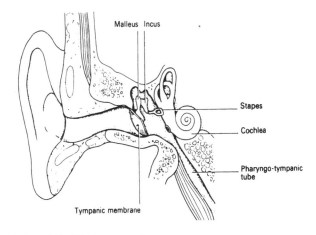

Figure 1.5 The human auditory system.

(a) the pinna, or outer ear – to 'collect the sound'
(b) the auditory canal – to conduct the sound to the eardrum (tympanic membrane)
(c) the middle ear – to transmit the movement of the eardrum to the inner ear – consisting of three bones,

the malleus, the incus and the staples, also known as the anvil, hammer and stirrup ossicles respectively
(d) The inner ear – to 'perceive' the sound and send information to the brain.

The outer ear

In man, the function of the outer ear is fairly limited, and is not big enough to act as a horn to collect much sound energy, but it does play a part in perception. It contributes to the ability to determine whether a sound source is in front of or directly behind the head.

The auditory canal

The human ear canal is about 35 mm long and serves as a passage for sound energy to reach the eardrum. Since it is a tube, open at one end and closed at the other, it acts like a resonant pipe, which resonates at 3–4 kHz. This resonance increases the transmission of sound energy substantially in this frequency range and is responsible for the fact that hearing is most sensitive to frequencies around 3.5 kHz.

The middle ear, and eardrum

Sound waves travelling down the ear canal strike the eardrum, causing it to vibrate. This vibration is then transferred by the bones of the middle ear to the inner ear, where the sound energy reaches the cochlea.

Air is a medium of low density, and therefore has a low acoustic impedance (acoustic impedance = sound velocity × density), while the fluid in the cochlea (mainly water) has a much higher impedance. If sound waves fell directly on the cochlea a very large proportion of the energy would be reflected, and the hearing process would be much less sensitive than it is. The function of the middle ear is to 'match' the low impedance of the air to the high impedance of the cochlea fluid by a system of levers. Thus the eardrum, which is light, is easily moved by sound waves, and the middle ear system feeds sound energy through to the fluid in the inner ear.

In addition to its impedance matching function, the middle ear has an important effect on the hearing threshold at different frequencies. It is broadly resonant at a frequency around 1.5 kHz, and the ear becomes progressively less sensitive at lower frequencies, (see Fig. 1.6). This reduction in sensitivity is perhaps fortunate, since man-made and natural sources (e.g. traffic noise and wind) produce much noise at low frequencies, which would be very disturbing if it were all audible. At high frequencies the bones of the middle ear, and the tissues joining them, form a filter which effectively prevents the transmission of sound at frequencies above about 20 kHz. Research into the audibility of bone-conducted sound, obtained by applying a vibrator to the head, have shown that the response of the inner ear extends to at least 200 kHz.

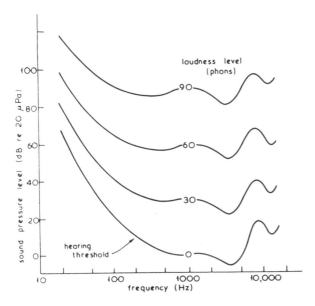

Figure 1.6   The hearing threshold and equal loudness contours.

The inner ear

The inner ear is the site of the perceptive process, and is often compared to a snail shell in its form. It consists of a spirally coiled tube, divided along its length by the basilar membrane. This membrane carries the 'hair cells' which detect sound. The structure of the cochlea is such that for any particular sound frequency, the fluid in it vibrates in a particular pattern, with a peak at one point on the basilar membrane. In this way the frequency of a sound is converted to a point of maximum stimulation on the membrane. This process provides the basis of the perception of pitch.

Perception of Intensity and Frequency

Since the sinusoid represents the simplest, and most fundamental, repetitive waveform, it is appropriate to base much of our understanding of the ear's behaviour on its response to sounds of this type.

At the simplest level, intensity relates to loudness, and frequency relates to pitch. Thus, a loud sound is one of high intensity (corresponding to a substantial flow of energy), while a sound of high pitch is one of high frequency. In practice however, the two factors of frequency and intensity interact and the loudness of a sound depends on both.

Loudness is a subjective quantity, and therefore cannot be measured directly. However, in practice, it is useful to be able to assign numerical values to the experience of loudness. This has led to a number of methods being used to achieve this objective. One of the oldest is to define 'loudness level'. Loudness level is defined as the level (in dB SPL) of a 1000 Hz tone, judged to be as loud as the sound under examination. Thus, if a tone of 100 Hz is considered, then a listener is asked to adjust the level of a 1000 Hz tone until it sounds equally loud. The level of the 1000 Hz tone (in dB) is then called the loudness level, in phons, of the 100 Hz tone. The virtue of the phon as a unit, is that it depends only upon a judgement of equality between two sounds, and it is found that the average phon value, for a group of listeners, is a consistent measure of loudness level. The phon level can be found, in this way, for any continuous sound, sine wave, or complex, but, as a unit, it only makes possible **comparisons**, it does not, in itself, tell us anything about the loudness of the sound, except that more phons means louder. For example 80 phons is louder than 40 phons, but it is **not** twice as loud.

Loudness level comparisons have been made over the normal range of audible frequencies (20 Hz to about 15 000 Hz), and at various sound pressure levels, leading to the production of 'equal loudness contours'. Fig. 1.6 shows these contours for various levels. All points on a given contour have equal loudness, thus a sound pressure level of 86 dB at 20 Hz will sound equally as loud as 40 dB at 1000 Hz. The main features of the equal loudness contours are that they rise steeply at low frequency, less steeply at high frequencies, and that they become flatter as the level rises. This flattening with increasing level has consequences for the reproduction of sound. If a sound is reproduced at a higher level than that at which it was recorded, then the low frequencies will become relatively louder (e.g. speech will sound boomy). If it is reproduced at a lower level then it will sound 'thin' and lack bass (e.g. an orchestra reproduced at a moderate domestic level). Some amplifiers include a loudness control which attempts a degree of compensation by boosting bass and possibly treble, at low listening levels.

To obtain values for 'loudness', where the numbers will represent the magnitude of the sensation, it is necessary to carry out experiments where listeners make such judgements as 'how many times louder is sound A than sound B?' While this may appear straightforward it is found that there are difficulties in obtaining self-consistent results. As an example, experiments involving the judging of a doubling of loudness do not yield the same interval (in dB) as experiments on halving. In practice, however, there is now an established unit of loudness, the sone, where a pure (sinusoidal) tone of 40 dB SPL at 1000 Hz has a loudness of one sone. The sensation of loudness is directly proportional to the number of sones, e.g. 80 sones is twice as loud as 40 sones. Having established a scale of loudness in the form of sones, it is possible to relate this to the phon scale and it is found that every addition of 10 phons corresponds to a doubling of loudness, so 50 phons is twice as loud as 40 phons.

Pitch Perception

It is well established that, for pure tones (sine waves) the basis of the perception of pitch is in the inner ear, where the basilar membrane is stimulated in a particular pattern according to the frequency of the tone, and the sensation of pitch is associated with the point along the length of the membrane where the stimulation is the greatest. However, this theory (which is supported by ample experimental evidence) does not explain all aspects of pitch perception. The first difficulty arises over the ear's ability to distinguish between two tones only slightly different in frequency. At 1000 Hz a difference of only 3 Hz can be detected, yet the response of the basilar membrane is relatively broad, and nowhere near sharp enough to explain this very high level of frequency discrimination. A great deal of research effort has been expended on this problem of how the response is 'sharpened' to make frequency discrimination possible.

The 'place theory' that perceived pitch depends on the point of maximum stimulation of the basilar membrane does not explain all aspects of pitch perception. The ear has the ability to extract pitch information from the overall envelope shape of a complex wave form. For example, when two closely spaced frequencies are presented together (say 1000 Hz and 1100 Hz) a subjective component corresponding to 100 Hz (the difference between the two tones) is heard. While the combination of the two tones does not contain a 100 Hz component, the combination does have an envelope shape corresponding to 100 Hz (see Fig. 1.7).

Discrimination and Masking

The ear – discrimination

The human ear has enormous powers of discrimination, the ability to extract wanted information from unwanted background noise and signals. However, there are limits to these discriminatory powers, particularly with respect to signals that are close either in frequency or in time.

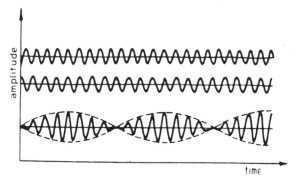

Figure 1.7 The combination of two differing frequencies to produce beats.

Masking

When two sounds, of different pitch, are presented to a listener, there is usually no difficulty in discriminating between them, and reporting that sounds of two different pitches are present. This facility of the ear, however, only extends to sounds that are fairly widely separated in frequency, and becomes less effective if the frequencies are close. This phenomenon is more conveniently looked at as 'masking', i.e. the ability of one sound to mask another, and render it completely inaudible. The extent of the masking depends on the frequency and level of the masking signal required, but as might be expected, the higher the signal level, the greater the effect. For instance, a narrow band of noise, centred on 410 Hz and at a high sound pressure level (80 dB) will interfere with perception at all frequencies from 100 Hz to 4000 Hz, the degree of masking being greatest at around 410 Hz (see Fig. 1.8). By comparison, at a 30 dB level, the effects will only extend from 200 Hz to about 700 Hz. The 'upward spread of masking', i.e. the fact that masking spreads further up the frequency scale than downwards is always present. An everyday example of the effect of masking is the reduced intelligibility of speech when it is reproduced at a high level, where the low frequencies can mask mid and high frequency components which carry important information.

Much research has been carried out into masking, and leads to the general conclusion that it is connected with the process of frequency analysis which occurs in the basilar membrane. It appears that masking is a situation where the louder sound 'takes over' or 'pre-empts', a section of the basilar membrane, and prevents it from detecting other stimuli at, or close to, the masking frequency. At higher sound levels a larger portion of the basilar membrane is 'taken over' by the masking signal.

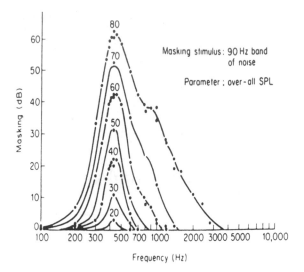

Figure 1.8 Masking by a narrow band of noise centred on 410 Hz. Each curve shows the extent to which the threshold is raised for a particular level of masking noise. (From Egan and Hake 1950.)

Temporal masking

While masking is usually considered in relation to two stimuli presented at the same time, it can occur between stimuli which are close in time, but do not overlap. A brief tone pulse presented just after a loud burst of tone or noise can be masked, the ear behaves as if it needs a 'recovery' period from a powerful stimulus. There is also a phenomenon of 'pre-stimulatory masking', where a very brief stimulus, audible when presented alone, cannot be detected if it is followed immediately by a much louder tone or noise burst. This apparently unlikely event seems to arise from the way in which information from the ear travels to the brain. A small response, from a short, quiet signal can be 'overtaken' by a larger response to a bigger stimulus, and therefore the first stimulus becomes inaudible.

Binaural Hearing

The ability of humans (and animals) to localise sources of sound is of considerable importance. Man's hearing evolved long before speech and music, and would be of value both in locating prey and avoiding predators. The term 'localisation' refers to judgements of the direction of a sound source and, in some cases its distance.

When a sound is heard by a listener, he only receives similar auditory information at both ears if the sound source is somewhere on the vertical plane through his head, i.e. directly in front, directly behind, or overhead.

If the sound source is to one side, then the shadowing effect of the head will reduce the sound intensity on the side away from the source. Furthermore, the extra path length means that the sound will arrive slightly later at the distant ear. Both intensity and arrival time differences between the ears contribute to the ability to locate the source direction.

The maximum time delay occurs when the sound source is directly to one side of the head, and is about 700 µs. Delays up to this magnitude cause a difference in the phase of the sound at the two ears. The human auditory system is surprisingly sensitive to time (or phase) differences between the two ears, and, for some types of signal, can detect differences as small as 6 µs. This is astonishingly small, since the neural processess which must be used to compare information from the two ears are much slower.

It has been found that, while for frequencies up to about 1500 Hz, the main directional location ability depends on interaural time delay, at higher frequencies differences in intensity become the dominant factor. These differences can be as great as 20 dB at the highest frequencies.

Stereophonic sound reproduction does not attempt to produce its effects by recreating, at the ears, sound fields which accurately simulate the interaural time delays and level differences. The information is conveyed by the relative levels of sound from the two loud speakers, and any time differences are, as far as possible, avoided. Thus the sound appears to come simply from the louder channel, if both are equal it seems to come from the middle.

The Haas Effect

When a loudspeaker system is used for sound reinforcement in, say, a large lecture theatre, the sound from the speaker travels through the air at about 340 ms, while the electrical signal travels to loudspeakers, set further back in the hall, practically instantaneously. A listener in the rear portion of the hall will therefore hear the sound from the loudspeaker first and will be conscious of the fact that he is hearing a loudspeaker, rather than the lecturer (or entertainer) on the platform. If, however, the sound from the loudspeaker is delayed until a short time after the direct sound from the lecture, then the listeners will gain the impression that the sound source is at the lecturer, even though most of the sound energy they receive is coming from the sound reinforcement system. This effect is usually referred to as the Haas effect, because Haas was the first to quantitatively describe the role of a 'delayed echo' in perception.

It is not feasible here to discuss details of the work by Haas (and others), but the main conclusions are that, if the amplified sound reaches the listener some 5–25 ms after the direct sound, then it can be at a level up to 10 dB higher than the direct sound while the illusion of listening to the lecturer is preserved. Thus a loudspeaker in a large hall, and placed 15 m from the platform will need a delay which allows for the fact that it will take 15/340 s = 44 ms plus say 10 ms for the Haas effect, making a total delay of about 54 ms. The system can obviously be extended to further loudspeakers placed at greater distances, and with greater delays. Due to the magnitude of the time delays required these are usually provided by a magnetic drum recorder, with pick up heads spaced round the drum. Presumably, this feature will, in due course, be taken over by a digital delay device. A useful account of the Haas effect can be found in Parkin and Humphreys (1971).

Distortion

The term 'distortion' can be most broadly used to describe (unwanted) audible differences between reproduced sound and the original sound source. It arises from a number of interrelated causes, but for practical purposes it is necessary to have some form of categorisation in order to discuss the various aspects. The classifications to be used here as follows:

(a) Frequency distortion, i.e. the reproduction of different frequencies at relative levels which differ from the relative levels in the original sound.
(b) Non-linearity. The departure of the input/output characteristic of the system from a straight line; resulting in the generation of harmonic and intermodulation products.
(c) Transient distortion. The distortion (i.e. the change in the shape) of transient signals and additionally, transient intermodulation distortion, where the occurrence of a transient gives rise to a short term distortion of other components present at the same time.
(d) Frequency modulation distortion – i.e. 'wow' and 'flutter'.

Non-linearity

A perfectly linear system will perfectly reproduce the shape of any input waveform without alteration. In practice all systems involve some degree of non-linearity, i.e. curvature, and will therefore modify any waveform passing through the system. Fig. 1.9 and 1.10 illustrate the behaviour of linear and non-linear systems for a sinusoidal input. For the case of a sine wave the change in wave shape means that the output waveform now con-

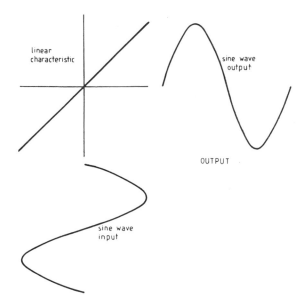

Figure 1.9 Transmission of a sine wave through a linear system.

sists of the original sine wave, together with one or more harmonic components. When a complex signal consisting of, for example, two sine waves of different frequencies undergoes non-linear distortion, intermodulation occurs. In this situation the output includes the two input frequencies, harmonics of the input frequencies together with sum and difference frequencies. These sum and difference frequencies include $f_1 f_2$ and $f_1 - f_2$ (where f_1 and f_2 are the two fundamentals), second order terms $2f_1 + f_2$, $2f_1 - f_2$, $f_1 + 2f_2$, $f_1 - 2f_2$ and higher order beats. Thus the intermodulation products may include a large number of tones. None of these is harmonically related to the original components in the signal, except by accident, and therefore if audible will be unpleasantly discordant.

In order to quantify harmonic distortion the most widely accepted procedure is to define the total harmonic distortion (THD) as the ratio of the total rms value of all the harmonics to the total rms value of the signal (fundamental plus harmonics). In practice the equation

$$d = \sqrt{(h_2)^2 + (h_3)^2 + (h_4)^2 \ldots}$$

can be used where d is percentage total harmonic distortion, h_2 = second harmonic percentage etc.

Although the use of percentage THD to describe the performance of amplifiers, pick-up cartridges etc. is widely used, it has been known for many years (since the 1940s) that it is not a satisfactory method, since THD figures do not correlate at all satisfactorily with listening

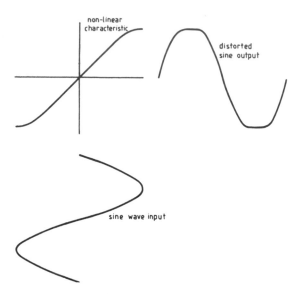

Figure 1.10   Transmission of a sine wave through a non-linear system.

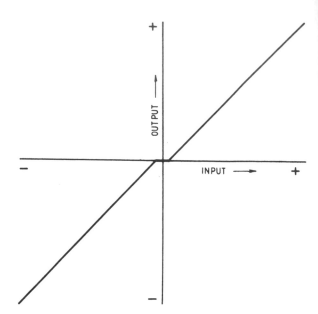

Figure 1.11   Input-output characteristic with 'cross-over' distortion.

tests. The reason for this stems from the different audibility of different harmonics, for example a smoothly curved characteristic (such as Fig. 1.10) will produce mainly third harmonic which is not particularly objectionable. By comparison the characteristic of Fig. 1.11 with a 'kink' due to 'crossover' distortion will sound harsher and less acceptable. Thus two amplifiers, with different characteristics, but the same THD may sound distinctly different in quality. Several schemes have been proposed to calculate a 'weighted distortion factor' which would more accurately represent the audible level of distortion. None of these has found much favour amongst equipment manufacturers, perhaps because 'weighted' figures are invariably higher than THD figures (see Langford-Smith, 1954).

Intermodulation testing involves applying two signals simultaneously to the system and then examining the output for sum and difference components. Various procedures are employed and it is argued (quite reasonably) that the results should be more closely related to audible distortion than are THD figures. There are however, difficulties in interpretation, which are not helped by the different test methods in use. In many cases intermodulation distortion figures, in percentage terms, are some 3–4 times higher than THD.

Any discussion of distortion must consider the question of what is acceptable for satisfactory sound reproduction. Historically the first 'high-fidelity' amplifier designs, produced in the 1945–50 period, used valves and gave THD levels of less than 0.1% at nominal maximum power levels. These amplifiers, with a smoothly curving input-output characteristic, tended mainly to produce

third harmonic distortion, and were, at the time of their development, adjudged to be highly satisfactory. These valve amplifiers, operating in class A, also had distortion levels which fell progressively lower as the output power level was reduced. The advent of transistors produced new amplifiers, with similar THD levels, but comments from users that they sounded 'different'. This difference is explicable in that class B transistor amplifiers (in which each transistor in the output stage conducts for only part of the cycle) produced a quite different **type** of distortion, tending to generate higher harmonics than the third, due to crossover effects. These designs also had THD levels which did not necessarily decrease at lower power outputs, some having roughly constant THD at all levels of output. It must therefore be concluded, that, if distortion is to be evaluated by percentage THD, then the figure of 0.1% is probably not good enough for modern amplifiers, and a design goal of 0.02% is more likely to provide a fully satisfactory performance.

Other parts of the system than amplifiers all contribute to distortion. Amplifiers distort at **all** frequencies, roughly to the same extent. Loudspeakers, by comparison, show much greater distortion at low frequencies due to large cone excursions which may either bring the cone up against the limits of the suspension, or take the coil outside the range of uniform magnetic field in the magnet. Under the worst possible conditions up to 3–5% harmonic distortion can be generated at frequencies below 100 Hz, but the situation improves rapidly at higher frequencies.

Pick-up cartridges, like loudspeakers, produce distortion, particularly under conditions of maximum amplitude, and THD levels of around 1% are common in high quality units. By comparison, compact disc systems are highly linear, with distortion levels well below 0.1% at maximum output. Due to the digital nature of the system, the actual percentage distortion may increase at lower levels.

Frequency distortion

Frequency distortion in a sound reproducing system is the variation of amplification with the frequency of the input signal. An ideal would be a completely 'flat' response from 20 Hz to 20 kHz. In practice this is possible for all the elements in the chain except the loudspeaker, where some irregularity of response is unavoidable. Furthermore, the maintenance of response down to 20 Hz tends to require a large (and expensive) loudspeaker system. In practice the human ear is fairly tolerant of minor irregularities in frequency response, and in any case the listening room, due to its natural resonances and sound absorption characteristics, can modify the response of the system considerably.

Transient distortion

Transients occur at the beginning (and end) of sounds, and contribute to the subjective quality to a considerable extent. Transient behaviour of a system can, in theory, be calculated from a knowledge of the frequency and phase response, although this may not be practicable if the frequency and phase responses are complex and irregular. Good transient response requires a wide frequency range, a flat frequency response, and no phase distortion. In practice most significant transient distortion occurs in loudspeakers due to 'hang-over'. Hang-over is the production of some form of damped oscillation, which continues after the end of the transient input signal. This is due to inadequately damped resonance at some point, and can be minimised by good design.

Transient intermodulation distortion

Current amplifier design relies heavily on the use of negative feedback to reduce distortion and to improve stability. A particular problem can arise when a transient signal with a short rise-time is applied to the amplifier. In this situation the input stage(s) of the amplifier can overload for a brief period of time, until the transient reaches the output and the correction signal is fed back to the

input. For a simple transient, such as a step function, the result is merely a slowing down of the step at the output. If, however, the input consists of a continuous tone, plus a transient, then the momentary overload will cause a loss of the continuous tone during the overload period (see Fig. 1.12).

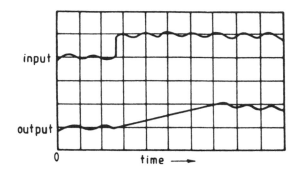

Figure 1.12 Transient inter-modulation distortion.

This brief loss of signal, while not obvious as such to a listener, can result in a loss of quality. Some designers now hold the view that in current amplifier designs harmonic and intermodulation distortion levels are so low that transient effects are the main cause of audible differences between designs and the area in which improvements can be made.

Frequency modulation distortion

When sound is recorded on a tape or disc, then any variation in speed will vary the frequency (and hence the pitch) of the reproduced sound. In the case of discs this seems to arise mainly from records with out-of-centre holes, while the compact disc has a built in speed control to eliminate this problem. The ear can detect, at 1000 Hz, a frequency change of about 3 Hz, although some individuals are more sensitive. This might suggest that up to 0.3% variations are permissible in a tape recording system. However, when listening to music in a room with even modest reverberation, a further complication arises, since a sustained note (from say a piano or organ) will be heard simultaneously with the reverberant sound from the initial period of the note. In this situation any frequency changes will produce audible beats in the form of variations in intensity and 'wow' and 'flutter' levels well below 0.3% can became clearly audible.

Phase distortion

If an audio signal is to pass through a linear system without distortion due to phase effects, then the phase response (i.e. the difference between the phase of output and input) must be proportional to frequency. This simply means that all components in a complex waveform must be delayed by the **same** time. If all components are delayed identically, then for a system with a flat frequency response, the output waveform shape will be identical with the input. If phase distortion is present, then different components of the waveform are delayed by differing times, and the result is to change the shape of the waveform both for complex tones and for transients.

All elements in the recording/reproducing chain may introduce phase distortion, but by far the largest contributions come from two elements, analogue tape recorders and most loudspeaker systems involving multiple speakers and crossover networks. Research into the audibility of phase distortion has, in many cases, used sound pulses rather than musical material, and has shown that phase distortion can be detected. Phase distortion at the recording stage is virtually eliminated by the use of digital techniques.

duced, which can provide 15–20 dB reduction in noise over a frequency range from about 50–2000 Hz. This operates on a similar principle to Olsen and May's absorber, but includes an adaptive gain control, which maintains optimum noise reduction performance, despite any changes in operating conditions.

A rather different application of an adaptive system has been developed to reduce diesel engine exhaust noise. In this case a microprocessor, triggered by a synchronising signal from the engine, generates a noise cancelling waveform, which is injected by means of a loudspeaker into the exhaust noise. A microphone picks up the result of this process, and feeds a signal to the microprocessor, which in turn adjusts the noise cancelling waveform to minimize the overall output. The whole process takes a few seconds, and can give a reduction of about 20 dB. While this adaptive system can only operate on a repetitive type of noise, other systems have been developed which can reduce random, as well as repetitive, waveforms.

Electronic Noise Absorbers

The idea of a device which could absorb noise, thus creating a 'zone of silence', was put forward in the 1930s in patent applications by Lueg (1933/4). The ideas were, at the time, in advance of the available technology, but in 1953 Olsen and May described a working system consisting of a microphone, an amplifier and a loudspeaker, which could reduce sound levels close to the speaker by as much as 20 dB over a fairly narrow range of frequencies (40–100 Hz).

The principles involved in their system are simple. The microphone picks up the sound which is then amplified and reproduced by the loudspeaker in antiphase. The sound from the speaker therefore 'cancels out' the original unwanted noise. Despite the simplicity of the principle, it is, in practice, difficult to operate such a system over a wide range of frequencies, and at the same time, over any substantial spatial volume. Olsen and May's absorber gave its best performance at a distance 8–10 cm from the loudspeaker cone, and could only achieve 7 dB attenuation at 60 cm. This type of absorber has a fundamental limitation due to the need to maintain stability in what is essentially a feedback loop of microphone, amplifier and loudspeaker. With practical transducers it is not possible to combine high loop-gain with a wide frequency response.

Olsen and May's work appears to have been confined to the laboratory, but more recent research has now begun to produce worthwhile applications. A noise reduction system for air-crew helmets has been pro-

References

Langford-Smith, F., *Radio Designer's Handbook*, (Chapter 14 Fidelity and Distortion). Illiffe (1954).

Parkin, P.H. and Humphries, H.R., *Acoustics, Noise and Buildings*. Faber and Faber, London (1971).

Rumsey, F. and McCormick, T., *Sound and Recording: An Introduction*. Focal Press, Butterworth-Heinemann (1992).

Tobias, J.V., *Foundations of Modern Auditory Theory*, vols I and II. Academic Press (1970/72).

Recommended further reading

Blauert, J., *Spatial Hearing*. Translated by J. S. Allen, MIT Press (1983).

Eargle, J., (ed). *Stereophonic Techniques – An Anthology*. Audio Engineering Society (1986).

Eargle, J., *Music, Sound, Technology*. Van Nostrand Rheinhold (1990).

Moore, B.C.J., *An Introduction to the Psychology of Hearing*. Academic Press (1989).

Rossing, T.D., *The Science of Sound*, 2nd edition. Addison-Wesley (1989).

Tobias, J., *Foundations of Modern Auditory Theory*. Academic Press (1970).

Architectural acoustics

Egan, M.D., *Architectural Acoustics*. McGraw-Hill (1988).

Parkin, P.H. and Humphries, H.R., *Acoustics, Noise and Buildings*, Faber and Faber, London (1971).

Rettinger, M., *Handbook of Architectural Acoustics and Noise Control*. TAB Books (1988).

Templeton, D. and Saunders, D., *Acoustic Design*. Butterworth Architecture (1987).

Musical acoustics

Benade, A.H., *Fundamentals of Musical Acoustics*. Oxford University Press (1976).

Campbell, M. and Greated, C., *The Musician's Guide to Acoustics*. Dent (1987).

Hall, D.E., *Musical Acoustics*, 2nd edition. Brooks/Cole Publishing (1991).

Recommended listening

Auditory Demonstrations (Compact Disc). Philips Cat. No. 1126–061. Available from the Acoustical Society of America.

2 Microphones

John Borwick

Almost all sound recording needs to make use of microphones, so that this technology is, for audio systems, the most fundamental of all. In this chapter, John Borwick explains microphone types, technology and uses.

Introduction

Microphones act as the first link in the chain of equipment used to transmit sounds over long distances, as in broadcasting and telephony. They are also used for short-distance communication in public address, sound reinforcement and intercom applications, and they supply the signals which are used to cross the barrier of time as well as distance in the field of sound recording.

Basically a microphone (Fig. 2.1) is a device which converts acoustical energy (received as vibratory motion of air particles) into electrical energy (sent along the microphone cable as vibratory motion of elementary electrical particles called electrons). All devices which convert one form of energy into another are called transducers. Clearly, whilst a microphone is an acoustical-to-electrical transducer, the last link in any audio transmission or playback system, a loudspeaker or a headphone earpiece, is a reverse electrical-to-acoustical transducer. Indeed some loudspeakers can be connected to act as microphones and vice versa.

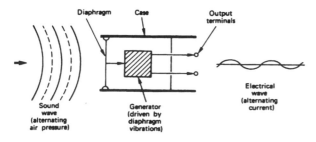

Figure 2.1 A microphone converts acoustical energy into electrical energy.

Microphone Characteristics

Microphones come in all shapes and sizes. When choosing a microphone for any particular application, some or all of the following features need to be considered.

Frequency response on axis

The microphone should respond equally to sounds over the whole frequency range of interest. Thus in high quality systems the graph of signal output voltage plotted against frequency for a constant acoustic level input over the range 20–20000 Hz (the nominal limits of human hearing) should be a straight line. Early microphones certainly failed this test, but modern microphones can come very close to it so far as the simple response on-axis is concerned.

Yet the full range may be unnecessary or even undesirable in some applications. A narrower range may be specified for microphones to be used in vehicles or aircraft to optimize speech intelligibility in noisy surroundings. Some vocalists may choose a particular microphone because it emphasizes some desired vocal quality. Lavalier or clip-on microphones need an equalised response to correct for diffraction effects, and so on.

Directivity

In most situations, of course, a microphone does not merely receive sound waves on-axis. Other sources may be located in other directions, and in addition there will be numerous reflected sound waves from walls and obstacles, all contributing in large or small measure to the microphone's total output signal. A microphone's directivity, i.e. its ability either to respond equally to sounds arriving from all directions or to discriminate against sounds from particular directions, is therefore an important characteristic.

Directivity is most easily illustrated by plotting on circular or polar graph paper the output signal level for a fixed sound pressure level at all angles in a particular plane. The most common such directivity patterns are illustrated in Fig. 2.2. They include the circle (denoting a non-directional or 'omnidirectional' microphone which responds equally at all angles), the figure-of-eight (denoting a bidirectional microphone which attenuates sounds arriving at the sides) and the cardioid or heart-shape (denoting a unidirectional microphone which attenuates sounds arriving at the back). As microphones normally operate in three-dimensional space, a clearer idea of their directivity can be obtained by rotating these polar diagrams about the main axis as illustrated in Fig. 2.3.

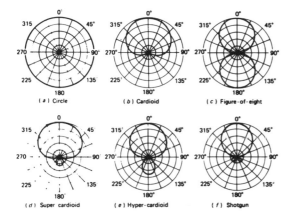

Figure 2.2 The principal microphone directivity patterns.

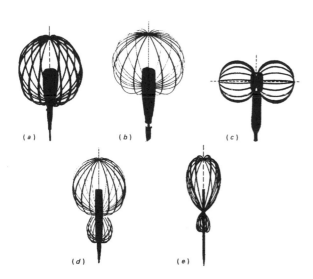

Figure 2.3 Artist's impression of three-dimensional directivity patterns: (a) circle; (b) cardioid; (c) figure-of-eight; (d) hypercardioid; (e) shotgun (courtesy Sennheiser).

Microphones exist having any one of these basic directivity patterns, and some more versatile models offer switched or continuously variable patterns. There are also microphones designed as a series in which a common body unit can be fitted with various modular capsules to provide a choice of directional types.

Frequency response off-axis

Ideally any high quality microphone, whatever its directivity pattern, should maintain the same frequency response at all angles; i.e. there is a need for polar pattern uniformity. Unfortunately this criterion is seldom met, the most common fault being irregular and falling response at oblique angles for frequencies above about 5 kHz.

The main problem is that the microphone itself begins to act as a physical obstacle to sound waves for shorter wavelength (higher frequency) sounds. The resulting diffraction and reflection produces peaks and dips in response which vary with angle of incidence. Reducing the microphone dimensions helps by pushing the problem further up the frequency scale.

Sensitivity

The conversion efficiency of a microphone, i.e. the output voltage produced by a given incident sound pressure level, should be as high as possible. This boosts the signals in relation to noise and interference along the signal path.

The most common method for specifying sensitivity is that recommended by the IEC and British Standards. This specifies the open circuit (unloaded) voltage output for an input sound pressure of 1 Pascal (10 microbars). This corresponds to a fairly loud sound since 1 Pa equals 94 dB above the threshold of hearing. An alternative rating quotes the output voltage for 0.1 Pa (74 dB SPL), or the level of normal speech at 20 cm.

Self-noise

The inherent noise level of a microphone, and this includes any built-in amplifier, should be as low as possible. This requirement has been high-lighted by the advent of digital recording and consumer media such as the compact disc where signal-to-noise ratios in excess of 90 dB have become commonplace.

Microphone self-noise is usually given as the equivalent SPL which would give the same output voltage.

Typical values are down around 15–20 dBA SPL which is about the ambient noise level in the best sound studios and corresponds to a thermal noise value of about –129 dBm. Alternatively, some manufacturers quote a 'signal-to-noise ratio' in which the equivalent noise level is measured with reference to 1 Pa (94 dB SPL).

Distortion

For the waveform of the electrical signal to be a faithful representation of the original sound wave, non-linear distortion must be as low as possible. Such distortions are mainly associated with high signal levels and the onset of overload or saturation effects in the transducer mechanism. The distortion rating for microphones is usually quoted as the maximum SPL which can be handled for a given value of total harmonic distortion. A typical value might be 130 dB for 0.5% THD.

Other criteria

A well designed microphone should be immune to induced hum or noise from electrical wiring and radio frequency interference. This is specially important in the vicinity of TV monitors, cameras, lighting regulators or transmitters. For outdoor work, or close vocals, the microphone should be proof against wind noise. Hand-held microphones and microphones mounted on television booms also need to be rugged and able to withstand mechanical shocks and vibrations.

Microphone Types

Except for a few exotic types, all microphones convert acoustical energy to electrical energy through the mechanical vibrations in response to sound waves of a thin lightweight diaphragm. The conversion may therefore be regarded as happening in two stages, though of course they occur simultaneously: (a) the varying air pressure sets the diaphragm into mechanical vibration; (b) the diaphragm vibrations generate an electric voltage.

This second stage may use any of the common electrical generator principles and microphones tend to be categorized accordingly, e.g. moving-coil, condenser, etc. However, before describing these transducer categories in detail, it is necessary to distinguish between the two basic ways in which microphones of all types first extract energy from the sound wave. These are respectively called pressure operation and pressure-gradient (or velocity) operation.

Pressure operation

In pressure operated microphones the rear surface of the diaphragm is enclosed so that the actuating force is simply that of the instantaneous air pressure at the front (Fig. 2.4). A small vent hole in the casing equalises the long-term external and internal pressures, whilst audio frequency swings above and below normal atmospheric pressure due to the incident sound waves will cause the diaphragm to move outwards and inwards. The force on the diaphragm is equal to the product of sound pressure (per unit area) and the area of the diaphragm, and is essentially independent of frequency.

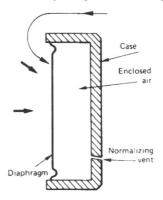

Figure 2.4 Pressure operated microphone, showing enclosed back with small vent hole.

Also, at least at low frequencies where the sound wavelength is small compared with the dimensions of the microphone, the response is the same for all angles of incidence. Therefore pressure operated microphones are generally omni-directional, having the circle and sphere polar diagrams shown earlier in Figs. 2.2(a) and 2.3(a).

At higher frequencies, however, where the microphone dimensions are equal to or greater than the wavelength, the obstacle effect comes into play. This sets up standing waves for sounds arriving on-axis, with a so-called 'pressure doubling' or treble emphasis, and explains why it is often better to speak or sing across the microphone rather than straight into it. In addition, oblique incidence of high-frequency waves causes phase differences across the face of the diaphragm and reduced output due to partial cancellations. These effects are progressively more pronounced at shorter wavelengths and so pressure operated microphones become more narrowly directional with increasing frequency (Fig. 2.5).

The effects are purely physical and relate strictly to the ratio of microphone diameter to wavelength. Therefore halving the diameter, for example, will double the frequency at which a given narrowing of the directivity pattern occurs.

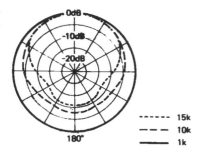

Figure 2.5 A pressure operated microphone is essentially omnidirectional but becomes more narrowly unidirectional at high frequencies.

Pressure gradient operation

Many modern microphones, including the ribbon type to be discussed later, are designed with both faces of the diaphragm open to the air. The resulting force on the diaphragm is then not simply due to the pressure on the front but to the instantaneous difference in pressure, or pressure gradient (PG) between front and back (Fig. 2.6(a)).

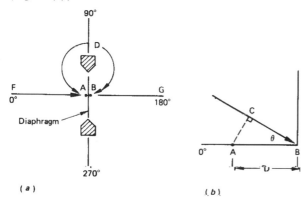

Figure 2.6 Pressure-gradient microphone: (a) for on-axis sounds the path length difference D is a maximum; (b) at oblique angles D reduces in proportion to cos θ.

At frequencies low enough for diffraction effects to be ignored, the wave arriving on-axis will produce alternating pressures on the front and back faces of the diaphragm A and B which do not coincide exactly in time. Their time of arrival or phase will be separated by an interval which is directly proportional to the extra distance travelled to reach the more remote face. This extra distance D will be a maximum for axial sounds arriving at either 0° or 180° but will steadily diminish and eventually fall to zero as the angle of incidence increases to 90° or 270°. The resultant pressure difference and therefore

force on the diaphragm may be taken to be proportional to this effective distance. As shown in Fig. 2.6(b), this shorter distance is CB and, since the ratio of this to D is CB/AB = cos θ, the microphone output at any angle θ is given by the expression Y = X cos θ, where Y is the output voltage at θ and X is the maximum output voltage on-axis (θ = 0). The value of cos θ for any angle can be obtained from cosine look-up tables, and plotting cos θ on a polar graph produces the familiar figure-of-eight diagram (Fig. 2.7). Note that there is a phase reversal of the force acting on the diaphragm for sounds arriving earlier at the back than the front.

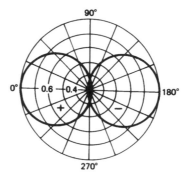

Figure 2.7 A pressure-gradient microphone has a figure-of-eight directivity pattern.

As with pressure operation, the above simple description of PG operation breaks down at higher frequencies where the wavelength becomes comparable with the physical dimensions, and in particular with the distance D. In practice the designer generally takes account of this by shaping the microphone so that pressure operation takes over above the critical frequency.

An incidental feature of all PG operated microphones is the so-called 'proximity effect'. This is the pronounced boost in low frequency output which occurs when the microphone is placed close to the sound source. It is caused by the additional pressure differences between points A and B introduced by the inverse square law increase in sound intensity at closer distances. The effect is most obvious in pure PG (figure-of-eight) microphones. However, it is also present in the combination pressure-plus-PG microphones discussed in the next section, though the bass tip-up with closer positioning is less steep.

Combination microphones

Recording engineers and other users gave an immediate welcome to the bidirectional PG microphone when it appeared in the late 1930s. Its ability to discriminate against sounds arriving at the sides gave greater flexibility in layout and balance for music and speech productions. A few years later, the choice of directivity patterns became still wider when microphones were produced which combined the characteristics of pressure and PG operation.

At first, separate pressure and PG capsules were assembled in a single casing. Then, making the axial sensitivities of the two units equal, say A, and adding their electrical outputs together, produced a combined pattern as shown diagrammatically in Fig. 2.8. At the front (left-hand side of the diagram) the outputs of the two elements OA and OB are in phase and will add to give a combined output OC which reaches a maximum value of 2A at 0°. At 90° the output of the PG element OB has fallen to zero and the combined output is reduced to A (–6 dB). At 180° (the back) the PG signal is in reverse phase and will cancel the pressure element's contribution to reduce the combined output to zero. The result is the heart-shaped or cardioid pattern which has become very popular in many applications. By simply adjusting the relative axial sensitivities of the two elements, various intermediate polar patterns can be obtained including those shown earlier in Figs. 2.2 and 2.3.

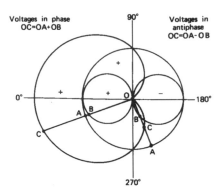

Figure 2.8 Derivation of cardioid directivity pattern by combining pressure and PG elements of equal axial sensitivity.

Instead of physically building two microphone elements into a single case, which meant that early cardioids were bulky and of variable accuracy due to the offset positions of the two diaphragms, later designs invariably use a single twin-diaphragm transducer or just one diaphragm with acoustic delay networks to establish the required PG operation.

By way of example, Fig. 2.9 outlines the geometry of a dual-diaphragm condenser (capacitor) microphone. It comprises identical diaphragms on each side of a central block having holes bored all the way through to provide for PG operation and some holes cut only part of the way through to act as 'acoustic stiffness' chambers. Put simply, with the rear diaphragm not electrically connected, the output at 0° is due to a combination of pressure oper-

ation (which tends to move both diaphragms inwards and outwards on alternate half-cycles) and PG operation (in which only the front diaphragm moves whilst the rear diaphragm remains stationary due to equal and opposite pressure and PG forces). At 90° the PG forces on both diaphragms fall to zero, reducing the output to half, and at 180° the front diaphragm experiences equal antiphase forces and the output falls to zero thus producing the cardioid pattern.

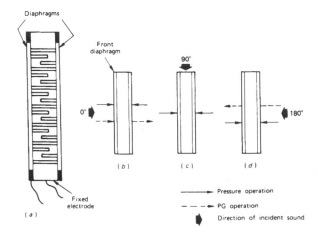

Figure 2.9 Dual-diaphragm cardioid condenser microphone, showing (a) the holes and cavities in the central fixed plate, and how the pressure and PG forces combine for sounds arriving at (b) 0°; (c) 90°; and (d) 180°.

In later versions both diaphragms are electrically connected and, with suitable switching of polarity and relative sensitivity of the back-to-back cardioids so formed, a range of directivity patterns can be selected at will.

The Microphone as a Voltage Generator

Having examined the various ways in which acoustic forces drive the diaphragm into vibrations which are analogous to the incident sound waveform, we now describe ways in which this oscillatory mechanical energy can be used to generate a corresponding electrical signal. Only the most common transducer types will be outlined in rough order of importance.

The moving-coil (dynamic) microphone

This category of microphone relies on the principle that motion of a conductor in a magnetic field generates an electromotive force (emf) causing a flow of current in the conductor (as in a dynamo). In the usual configuration the diaphragm is shaped as shown in Fig. 2.10 and carries a circular insulated coil of copper or aluminium wire centred in the annular air gap between N and S pole-pieces of a permanent magnet. The magnitude of the induced emf is given by $E = Blu$, where B is the flux density, l is the effective length of the coil in the gap and u is its velocity. Clearly B and l should be made as large as possible to give increased sensitivity. The mechanical resonance of the system is set at some mid-frequency, say 800 Hz, and smoothed as far as possible by introducing acoustic damping (resistance).

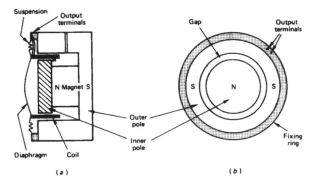

Figure 2.10   Moving-coil microphone in (a) side; and (b) plan view.

The microphone presents a low impedance source and may sometimes incorporate a step-up transformer. Most modern designs use many turns of very fine wire to produce a coil with a DC resistance of about 200 ohms which allows for long connecting cables and direct matching to standard amplifier inputs at around 1000 ohms. The majority of moving-coil microphones are omnidirectional but cardioid versions are also available.

The ribbon microphone

The electromagnetic principle is also used in the ribbon microphone. The magnet has specially shaped pole-pieces to provide a concentrated field across an elongated gap in which is suspended a thin metal ribbon (Fig. 2.11). This ribbon acts both as diaphragm and conductor and is typically made of aluminium foil about 0.1 mm thick and 2–4 mm wide. It is mildly corrugated to prevent curling and allow accurate tensioning to set the resonance to the required low frequency of around 2–4 Hz. DC resistance of the ribbon is very low and so a built-in step-up transformer is essential to provide an output at a standard matching impedance.

The low mass gives the ribbon microphone excellent transient response but it is sensitive to wind noise and mechanical movement. Ribbon microphones are therefore unsuitable for most outdoor applications or TV

boom operation. Since both ribbon surfaces are exposed to the air, except in a few designs, PG operation applies and the directivity pattern is a figure-of-eight.

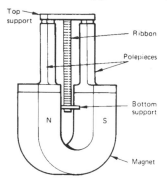

Figure 2.11   Ribbon microphone: basic construction.

The condenser (capacitor or electrostatic) microphone

This type of microphone is so-called because the transducer element is a simple condenser or capacitor whose two electrodes or plates are formed by the thin conductive diaphragm and a fixed back plate separated by a narrow air-gap (Fig. 2.12). A polarising DC voltage is applied across the two electrodes via a very high resistance R. This establishes a quasi-constant charge Q on the capacitor and the capacitance of the system is given by $C = kA/x$, where A is the effective diaphragm area, x is the electrode spacing and k is the dielectric constant (permittivity) of air.

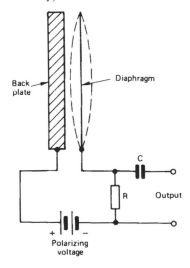

Figure 2.12   Condenser microphone: basic principle.

Since the voltage across the capacitor is $E = Q/C$, we can write $E = (Q/kA)x$ which shows that E is proportional to the spacing x. Therefore, when the diaphragm vibrates in response to a sound pressure wave, the applied DC voltage will be modulated by an AC component in linear relationship to the diaphragm vibrations. This AC component is taken through the DC-blocking capacitor C to provide the microphone's output signal.

The transducer mechanism is essentially simple and makes the condenser microphone potentially a very high quality device. However, a very narrow gap is needed to provide adequate sensitivity and impedance is very high. This calls for precision manufacture and a built-in amplifier located as close to the electrodes as possible. This adds to costs, complexity in terms of connecting cables and extremely low-noise circuitry. The DC polarizing voltage can be sent down the balanced signal cable (two live wires with an earthed screen) either in an A–B manner with the positive pole connected to one conductor and the negative to the other, or by 'phantom powering' in which the positive pole is connected to both live conductors via the junction between two identical resistors and the negative pole to the screen. Alternatively, some microphones have a compartment for the insertion of batteries to give very low-noise operation.

Condenser microphones lend themselves to designs with variable directivity by combining pressure and PG operation as previously described. Figure 2.13(a) shows a dual-diaphragm type effectively forming two cardioid

elements back-to-back, whose signal output is of course proportional to the polarizing voltage applied. The right-hand diaphragm has a fixed +60 V potential with respect to the common back plate, to form a forward facing cardioid. The voltage to the rear diaphragm is taken via a network allowing the potential to be switched from –60 V through 0 V to +60 V. Since the audio signal outputs of the two elements are connected through capacitor C, the five directivity patterns shown in Fig. 2.13(b) are obtained. Remote controlled switching gives the user a degree of flexibility in balancing musical instruments or voices and, of course, additional switch positions are possible or a continuously variable potentiometer may be fitted.

The electret microphone

As we have seen, only a low voltage DC supply is required to power the built-in amplifier of a condenser microphone. Yet there remains the practical disadvantage that, although virtually no current is drawn, a relatively high (60–120 V) DC bias voltage has also to be provided to polarize the plates. This factor has been eliminated in recent years, with a simultaneous simplification in construction, by the introduction of the electret microphone (Fig. 2.14).

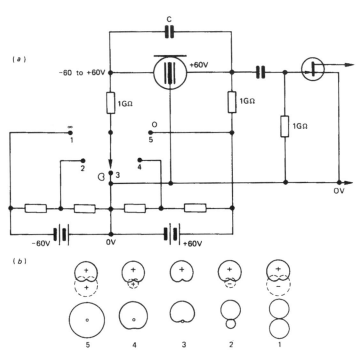

Figure 2.13   Variable directivity dual-diaphragm condenser microphone: (a) basic circuit; (b) polar patterns produced for the five switch positions.

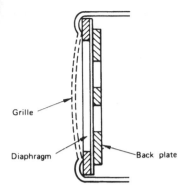

Figure 2.14 Electret microphone: basic construction.

This uses a permanently polarised material which can be regarded as the electrostatic equivalent of a permanent magnet. The electret-foil material, an insulator, is given its permanent charge, positive or negative, whilst in a strong electric field and heated either by corona discharge or electron bombardment. The design may use either a polarized diaphragm or a neutral diaphragm with the fixed plate coated with electret material ('back polarised'). With the need for a high DC voltage supply eliminated, electret microphones can also be made very small and rugged. Push–pull designs are possible; able to handle high SPL signals with low distortion. Variable directivity models using switched voltages are not practicable.

The piezoelectric (crystal) microphone

Very robust and inexpensive microphones can be produced using crystalline or ceramic materials which possess piezoelectric properties. Wafers or thin slabs of these materials generate a potential difference between opposite faces when subjected to torsional or bending stresses. The usual piezoelectric microphone capsule comprises a sandwich or 'bimorph' of oppositely polarized slabs joined to form a single unit with metal electrodes attached (Fig. 2.15). Vibrations of a conventional diaphragm are transmitted to the bimorph by a connecting rod thus setting up an alternating voltage at the output terminals which is proportional to the effective displacement.

The carbon (loose contact) microphone

Carbon microphones have been around for more than a century and are still found in millions of telephones around the world, despite the fact that they are limited in terms of frequency response, distortion and self-noise. The principle relies on the variation in DC resistance of a

package of conductive granules when they are subjected to variations in external pressure which cause their areas of contact to increase or decrease (Fig. 2.16).

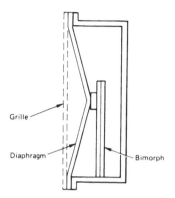

Figure 2.15 Piezoelectric (crystal) microphone: basic construction.

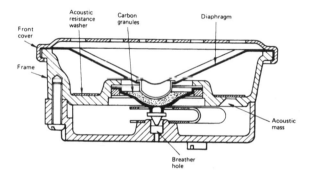

Figure 2.16 Carbon (loose contact) microphone: a modern telephone mouthpiece design.

The assembly acts like an amplifier with about 40–60 dB gain but has a number of disadvantages. The capsule generates high self-noise, is liable to non-linear distortion and becomes unreliable and less sensitive if the granules pack together, requiring a sharp knock to shake them loose again. Later push–pull dual-button transducers give improved quality but do not generally compete in studio or public address applications.

Other types of microphone

Some other types of microphone designed for specific applications will now be outlined for the sake of completeness. The *boundary microphone*, also referred to as *PZM* (*pressure zone microphone*), does not use a new type of transducer but represents a novel approach to microphone placement. It comprises a small capsule element, often an electret, housed in a flat panel in such a way that the diaphragm is located very close to any flat surface on which the unit is placed (Fig. 2.17). The main aim is to eliminate the irregular dips in frequency response (the comb filter effect) which often occurs in normal microphone placement due to interference between the direct wave and that reflected from the floor or walls. The two situations are illustrated in Fig. 2.18: (a) the microphone receives two sound waves, one direct and one reflected, so that alternate addition and subtraction will occur at frequencies for which the path length difference results in a phase shift of 0° or 180° respectively, giving (b) the familiar comb filter effect; (c) with a boundary microphone, the extreme proximity of the floor produces no significant interference effect and (d) the response is uniform.

idea is to respond strongly to on-axis sounds whilst discriminating against sounds arriving from other directions. Two basic approaches have been used, the parabolic reflector and the shotgun microphone.

A *parabolic reflector* can be used to collect sound energy over a wide area of the incident (plane) wavefront and reflect it back on to a microphone element located at the focal point (Fig. 2.19). The geometry ensures that all sounds arriving in line with the axis are focused towards the microphone, giving a considerable gain in axial signal level. At the same time, waves arriving off-axis are scattered and do not contribute much to the microphone output.

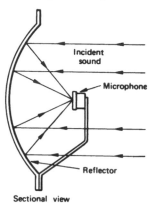

Figure 2.19 Parabolic reflector microphone: basic arrangement.

Figure 2.17 Boundary (PZM) microphone: basic construction.

Highly directional microphones are useful for picking up distant sound sources at sporting events, birdsong, audience and dance-routine sound in television, etc. The

Of course the effect works only at middle and high

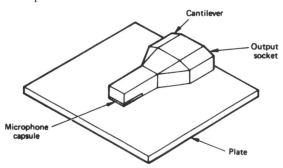

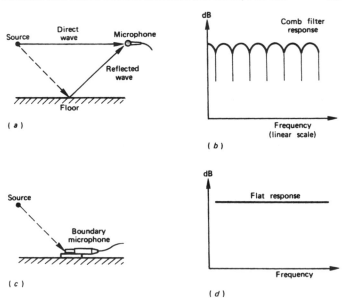

Figure 2.18 How the boundary microphone avoids comb filter distortion: (a) floor reflection produces a delayed wave; (b) uneven frequency response results; (c) boundary microphone receives only direct wave; (d) flat response results.

frequencies for which the reflector diaphragm is large compared with the wavelength, so that a 50 cm diameter parabola, for example, will be more effective than one measuring only 25 cm. A bass cut filter is usually introduced to attenuate random low-frequency signals. A gunsight is often fitted to assist precise aiming and the microphone is slightly offset from the true focal point to avoid too fine tuning of the highest frequencies.

The *shotgun* or *line microphone* consists of an acoustic line or pipe with equally spaced holes or a slit along its length and an omni or cardioid microphone element fitted at the rear end (Fig. 2.20). High directivity results from the fact that only sound waves arriving on or near the main axis produce additive pressures at the microphone diaphragm, whereas off-axis waves suffer varying phase-shift delays and tend to cancel. Again the effect is limited to middle and high frequencies, any bass extension being a function of the tube length.

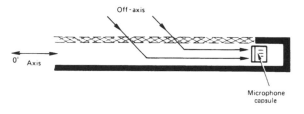

Figure 2.20   Shotgun (line) microphone: basic construction.

By a development of this idea, a *zoom microphone* can be produced which can, for example, be mounted on a video camera. The polar pattern can be varied from omni to highly directional as the TV scene demands. It uses two cardioid elements pointing forwards and spaced so as to give second-order pressure gradient performance, plus a third cardioid pointing backwards which can be added in or out of phase at various relative levels and equalization settings. The zoom effect can even be controlled automatically by linking the variable attenuators to the camera lens zoom mechanism.

Noise cancelling microphones have been developed for close-talking applications in noisy surroundings such as sports commentaries, paging and special types of telephone. One classic hand-held *lip-ribbon* design dates back to the 1930s but is still popular with radio reporters and sports commentators. It uses a ribbon transducer with a frontal guard-bar to ensure that the speaker's mouth is held at a fixed working distance of 63 mm. The proximity effect at this distance is naturally severe and must be equalized by means of a steep bass-cut filter which can be switched to three settings to suit the individual voice. Thus filtered, the response to the voice approximates to a level curve whilst external noise is drastically reduced. Indeed the noise is so far reduced that the engineer will often rig one or more separate

'atmosphere' microphones to restore the ambience of the event.

Radio (wireless) microphones use FM radio transmission instead of the traditional microphone cable and are being employed more and more for stage and televisions shows, and other applications where complete freedom of movement is demanded. The radio microphone itself is usually very small and attached to or concealed in the clothing. Larger handheld versions are also available having the transmitter built into the handgrip to which a short aerial is attached. In the smaller models the transmitter is housed in a separate unit about the size of a pocket calculator and connected to the microphone by about a metre of cable which may also act as the aerial. The microphone audio signal is frequency modulated on to the radio carrier. Frequency band allocations and licensing regulations vary from country to country, with both the VHF and UHF bands used and elaborate design measures taken to overcome the practical problems of maintaining consistent signal levels and sound quality as artists move anywhere on stage or throughout the TV studio. The 8 metre and 2 metre VHF bands, with relatively long wavelengths, suffer less from the obstacle effect but require a proportionately longer aerial (up to 75 cm). The UHF band needs only a short aerial (10–20 cm) but is easily blocked by obstacles in its path. The final choice of waveband may also be influenced by local regulations and the risk of radio interference from known sources such as taxis or ambulances in the given locality. Very large musical shows may call for a system made up of 30 or more microphones with constant switching between the available transmission channels as the show proceeds.

Of much older vintage than the true radio microphones just described is the *RF condenser microphone*. This was designed in the days before high quality head amplifiers were available and is still quite common. The microphone incorporates a built-in RF oscillator (8 Mz) plus FM modulator and demodulator circuits. Changing capacitance due to the diaphragm vibrations produces the required audio signal which modulates the RF frequency as in FM broadcasting, and the resulting signal is demodulated to provide a standard output for cable connection. High stability and robustness are claimed and the usual high polarizing voltage is reduced to about 10 volts.

Two-way microphones contain separate diaphragm/transducer elements to cover the high and low frequencies, combined via a frequency selective crossover network as in two-way loudspeakers. The main objective is to produce a more consistent directivity pattern and high-frequency response than can normally be obtained when a single diaphragm has to cover the full audio bandwidth.

Voice-activated (automatic) microphones are used in such varied situations as conference rooms, churches and security intruder monitoring. In the basic design the microphone output is muted by a level-sensitive gating circuit until a predetermined threshold sound pressure level is exceeded. In a round-table discussion, for example, each talker can address a separate microphone, yet the confusion normally introduced by background noise and multiple phase errors with all microphones permanently open is avoided. Each microphone will become live only when spoken to directly from within its frontal acceptance angle ('window'). A further development of this scheme uses two cardioid microphone capsules back-to-back to avoid the need for exact setting of the threshold level. The rear capsule continuously monitors the ambient sound level whilst the main front capsule is switched on whenever some differential level (say 10 dB) is exceeded by direct speech.

Contact microphones, as their name implies, are physically attached to the sound source and respond to mechanical vibrations rather than airborne pressure waves. They are therefore not strictly speaking microphones, but belong to the family of mechanical/electrical transducers which includes strain gauges, guitar pickups throat microphones and accelerometers used for vibration measurements. The transducer principles most commonly used are the moving-coil and condenser/electret. The unit can be fastened or taped to the soundboard or body of any acoustical musical instrument – string, wind or percussion, and of course should be lightweight.

The advantages are complete freedom of movement for the musician, good separation from the sounds of other instruments or voices and rejection of ambient (e.g. audience) noise. In one design, the electret principle is used in a strip form only about 1 mm thick, 25 mm wide and some 75 to 200 mm long.

Underwater microphones used to be improvised by suspending a standard microphone, typically moving-coil, inside a waterproof cage. Modern versions are often piezoelectric types supplied as a corrosion-free assembly with built-in amplifier and special waterproof cables capable of use in water depths of up to 1000 metres.

Microphones for Stereo

Following a rather hesitant start in the 1950s, two-channel stereophony has largely taken over as the standard method for music recording and sound broadcasting. Single-channel mono sound has certainly not disappeared completely, for example in AM radio and most television programmes, but genuine stereo and even surround sound are now becoming more common in the cinema and on video.

The stereo illusion

Stereophony, from the Greek word for 'solid', is an attempt to reproduce sounds in such a way that the listener has the same feeling of left–right spread in the frontal area, and the same ability to locate the direction of individual voices and instruments, as he has at a live performance. Important differences should be noted between natural directional hearing and the illusion of directionality created in stereophonic recording and reproduction. A listener to a stereo system of two spaced loudspeakers actually hears everything twice, since the sounds from both loudspeakers reach both ears (Fig. 2.21). This non-real duplication of signals sets a limit to the extent to which two-channel stereo can faithfully recreate the live sound experience. The stereo effect is obtained by sending to the two loudspeakers left/right signals which differ from each other in some way which creates an illusion of an extended sound stage across the imaginary arc joining the two loudspeakers. Figure 2.21 shows the standard layout for stereo listening which puts the listener and the two loudspeakers at the corners of an equilateral triangle. This implies a reproduction angle of 60°.

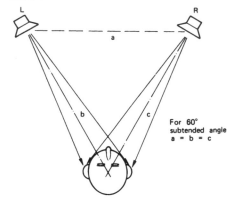

Figure 2.21 Loudspeaker stereo: in the conventional layout, a = b = c and is generally between 2 and 4 metres. Note that the sound from each loudspeaker reaches both ears, with a slight time delay to the more remote ear.

In practice, the required arc of phantom sound images can be established by using microphones in pairs (or passing a single mono signal through a panpot, to be described below) so as to make the signals from the loudspeakers differ by small amounts in intensity, time or both. By tradition, 'intensity stereo' is regarded as most effective. This is illustrated in Fig. 2.22 where the signal level fed to one loudspeaker can be attenuated. The apparent location of the source will then move progressively from the central position (when the levels are equal) towards the unattenuated loudspeaker. The same principle forms the basis of a panpot, or panoramic potentiometer, which uses a pair of variable attenuators working respectively in the clockwise and anticlockwise direction to shift a mono source to any required point between the loudspeakers.

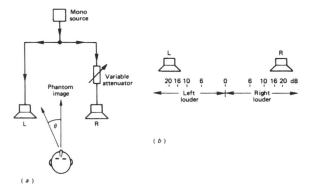

Figure 2.22 Intensity stereo: (a) attenuating one loudspeaker causes the phantom image to move towards the other; (b) typical attenuation (in decibels) for image positions across the stereo stage.

Substituting a variable time delay device for the attenuator in Fig. 2.22 will produce a similar movement of the sound image away from the central position. The source will appear to be located in one loudspeaker or the other when the time differential has reached about 2 milliseconds. Since both ears hear both loudspeakers, however, there will be some confusion due to the physical pattern of level and time differences, due to ear spacing, interfering with those from the loudspeakers themselves. In intensity stereo these fixed dimensional differences reinforce the desired effect, but in 'time delay' stereo they can introduce odd phase effects. Naturally it is possible to introduce both sorts of left/right signal differences deliberately, when the effects will be additive.

Basic Blumlein

One very common arrangement of a pair of microphones has become associated with the name of A.D. Blumlein whose comprehensive Patent No. 394325 (14.6.1933) covered nearly every aspect of two-channel stereo as it exists today. He realized that simple intensity stereo would provide all the directional clues needed by the human ear and described several pairings of microphones to produce the desired effect, including spaced omnidirectional microphones with a so-called 'shuffling'

network and the M–S (middle-and-side) arrangement, both of which will be discussed later.

The particular pairing now regarded as basic Blumlein consists of two identical figure-of-eight microphones arranged at ±45° to the front axis (Fig. 2.23). The two diaphragms are placed as nearly as possible at the same point in space, at least so far as the horizontal plane is concerned, to eliminate any time-of-arrival differences. Using separate microphones will inevitably give only approximate coincidence (Fig. 2.23(b)) but special stereo microphones with twin left/right capsules can improve matters and provide for altering the included angle by rotating the upper capsule (Fig. 2.24).

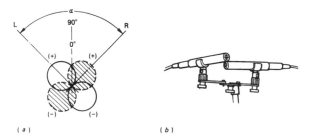

Figure 2.23 *Coincident pair (intensity) stereo: (a) basic Blumlein; (b) locating the two diaphragms for coincidence in the horizontal plane.*

This included angle is traditionally 90° and produces excellent imaging across the front quadrant which is independent of the distance from the source, as well as a uniform spread of reverberant sound between the two loudspeakers. It does suffer from the fact that the outputs of the two microphones are in antiphase in the two side quadrants, giving problems with side-wall reflections for example. Also sound sources located in the rear-right quadrant will be picked up by the rear lobe of the left microphone and vice versa, causing these sounds to be radiated by the wrong loudspeaker (cross-channelled). Modern small-diaphragm microphones are usually preferred because they can be placed nearer together. They also score by having a more consistent response off-axis than the larger types. This is particularly important since neither microphone is aimed directly at a centre soloist, for example, and much of the source may be located off-axis.

The 90° combined acceptance angle produced by the basic Blumlein arrangement may be too narrow for proper balance of some spread-out sources. It can be altered by changing the mutual angle between the microphones or made much wider, perhaps to allow a closer microphone position, by replacing the bidirectional microphones with cardioids or hypercardioids. These give a basic included angle of up to 180° and 120° respectively.

Figure 2.24 Examples of stereo microphone design allowing the mutual angle between the two capsules to be adjusted by rotating the upper capsule (AKG C422 and C34).

M–S stereo

Another arrangement of two coincident microphones covered in Blumlein's patent consists of one microphone, which can be of any directivity pattern, pointing to the front (middle) to provide a (L + R) signal, and the other a bidirectional type at right angles (side) providing a (L – R) signal. This middle-and-side or M – S pairing has a number of advantages but it does require a matrixing sum-and-difference network to recreate the conventional left and right signals. In effect this network sends the sum signal:

$$(M + S) = (L + R) + (L - R) = 2L$$

to the left channel and the difference signal:

$$(M - S) = (L + R) - (L - R) = 2R$$

to the right.

The choice of M component directivity pattern gives a range of stereo perspectives which can be further extended by adjusting the relative sensitivities of the M and S channels. It also has the advantage that a simple addition of the derived signals provides a fully compatible mono signal: i.e. (M + S) + (M–S) = 2M. This is important for broadcasters who generally have both mono and stereo listeners to consider. The mono signal can be of the highest quality, not subject to the off-axis colorations and phase ambiguities present with other X-Y methods.

Non-coincident (spaced) microphones

The use of spaced microphones for stereo has been around for just as long as Blumlein's coincident pair method. It will be realised, however, that placing spaced microphones in front of a sound source will mean that the waves reaching the left and right microphones differ in time-of-arrival or phase as well as intensity. This can be used to produce a pleasing open effect but may give less consistent imaging of more distant sources, vague centre imaging and unpredictable mono results. If the spacing is too great, a 'hole in the middle' effect can result in which the musicians on either side of centre become crowded together in the two loudspeakers. This can be cured by adding a third microphone at the centre, whose relative level setting can then act as a 'width' control.

Near-coincident microphones

In recent years, engineers in the fields of broadcasting and recording have developed various stereo microphone configurations which avoid the main limitations of both the coincident and the widely spaced methods. In general these near-coincident arrays use a pair of angled cardioid microphones 16–30 cm apart. They give clear unclouded directional clues at high frequencies and behave like an ideal coincident pair at low frequencies.

A special type of near-coincident array is used for binaural recording in which a pair of microphones are arranged at the same spacing as the human ears, usually mounted on a dummy head or baffle (Fig. 2.25). The result can be uncannily realistic, when listened to on headphones and each ear receives exclusively the signal from one channel. It has been used successfully for radio drama and documentaries but has not been developed more widely because it does not translate well to loudspeakers unless an adaptor network is inserted.

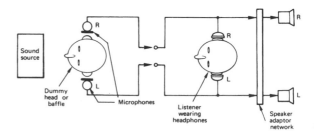

Figure 2.25 Binaural recording: using a dummy head or baffle can produce realistic stereo on headphones, but need an adaptor network for loudspeaker listening.

Surround Sound

While stereo reproduction can provide a realistic spread of sound over a standard 60° arc in the horizontal plane, much research has gone into extending this to full-scale three-dimensional 'surround sound', perhaps with height information as well as a 360° horizontal spread. Only then, it must be admitted, will the real-life situation of receiving sounds from all possible directions be recreated.

Early so-called quadraphonic systems in the 1970s relied on the use of four microphone channels, with matrix networks to encode these on to two channels for recording or transmission on normal stereo systems. Quadraphony did not succeed, partly because the public found the results inconsistent and not worth the expense of extra loudspeakers and amplifiers, but mainly because there was no true sense of the timing, phase dispersal and direction of the reverberant sound waves.

The British system known as Ambisonics solves most of these problems and offers a 'hierarchy' of encoding schemes from a stereo-compatible UHJ format, which some record companies are already using, all the way to 'periphony' giving full-sphere playback from a suitable array of loudspeakers. At the heart of the Ambisonics system is the use of a special cluster of four microphone capsules as found in the Calrec Soundfield microphone (Fig. 2.26). The four diaphragms are arranged as a regular tetrahedron array, sampling the soundfield as if on the surface of a sphere and reproducing the sound pressure at the effective centre. The four output channels represent one omni (pressure) element and three bidirectional (pressure gradient) signals in the left/right, front/back and up/down planes.

Surround sound has become popular in the cinema and is transferred to television broadcasts and the related video tapes and discs. The Dolby Surround cinema system is heard at its best when played back through three full-range behind-the-screen loudspeakers plus a U-shaped array of 'surround' loudspeakers positioned around the rear half of the theatre and sometimes additional bass loudspeakers. Microphone techniques for surround sound films are basically the same as those used for mono and stereo, using quadraphonic panpots to steer voices, music and sound effects as necessary. Dialogue is concentrated at the centre front and surround tracks are delayed slightly to maintain the impression that most sounds are coming from the area of the screen.

Figure 2.26 Soundfield microphone (courtesy Calrec/AMS).

References

AES Anthology, Microphones, New York: Audio Engineering Society (1979).

Bartlett, B., *Stereo Microphone Techniques*. Oxford: Focal Press (1991).

Borwick, J., *Microphones: Technology and Technique*. Oxford: Focal Press (1990).

Clifford, M., *Microphones*, 3rd Edition. TAB (1986).

Eargle, J., *The Microphone Handbook*. New York: Elar Publishing (1981).

Huber, D.M., *Microphone Manual*. Boston: Focal Press (1992).

Nisbett, A., *The Use of Microphones*, 4th edn. Oxford: Focal Press (1979).

Woram, J., *Sound Recording Handbook*. Indianapolis: Howard W. Sams (1989).

3 Studio and Control Room Acoustics

Peter Mapp

The reproduction of natural instrumental sound begins with microphones, but the behaviour of microphones cannot be separated from the acoustics of the studio. Peter Mapp shows here the principles and practices of studio acoustics as used today.

Introduction

Over the past few years the performance, sophistication and quality of the recording medium and ancillary hardware and technology have advanced at a considerable rate. The 50 to 60 dB dynamic range capability of conventional recording and disc technology suddenly has become 90 to 100 dB with the introduction of digital recording and the domestic digital compact disc (CD). In the foreseeable future a dynamic range of 110 dB could well become commonplace for digital mastering.

The increased dynamic range coupled with other advances in loudspeaker and amplifier technology now mean that audio, broadcast and hi-fi systems can today offer a degree of resolution and transparency to the domestic market unachievable only a few years ago even with the best professional equipment.

The acoustic environments of the studios in which the majority of the recordings or broadcasts originate from have become correspondingly more critical and important. No longer can recordings be made in substandard environments. Control rooms and studios exhibiting an uneven or coloured acoustic response or too high a level of ambient noise, which previously could be lost or masked by traditional analogue recording process, can no longer be tolerated. The transparency of the digital or FM broadcast medium immediately highlights such deficiencies.

To many, the subject of studio acoustics is considered a black art, often surrounded by considerable hype and incomprehensible terminology. However, this is no longer the case. Today, there certainly is an element of art in achieving a desirable and a predictable acoustic environment, but it is very much based on well-established scientific principles and a comprehensive understanding of the underlying physics of room acoustics and noise control.

Studios and control rooms create a number of acoustic problems which need to be overcome. Essentially they can be divided into two basic categories – noise control and room acoustics with the latter including the interfacing of the monitor loudspeakers to the control room.

Noise Control

The increased dynamic range of the hi-fi medium has led to corresponding decreases in the levels of permissible background noise in studios and control rooms.

A single figure rating (e.g. 25 dBA) is not generally used to describe the background noise requirement as it is too loose a criterion. Instead a set of curves which take account of the spectral (frequency) content of the noise are used. The curves are based on an octave or 1/3 octave band analysis of the noise and also take account of the ear's reduced sensitivity to lower frequency sounds.

The curves most frequently used are the NC (noise criterion) and NR (noise rating) criteria. Figures 3.1 and 3.2 graphically present the two sets of criteria in terms of the octave band noise level and frequency. (Although as Figs. 3.1 and 3.2 show, the criteria are not exactly the same, the corresponding target criteria e.g. NC20 or NR20 are frequently interchanged.)

The NC system is primarily intended for rating air conditioning noise, whilst the NR system is more commonplace in Europe and can be used to rate noises other than air conditioning. (An approximate idea of the equivalent dBA value can be obtained by adding 5 to 7 dB to the NC/NR level).

The following Table 3.1 presents typical design targets for various studio and recording formats. Many organisations e.g. the BBC demand even more stringent criteria particularly at low frequencies. From this table and curves shown in Figs. 3.1 and 3.2, it can be seen that the background noise level requirements are pretty stringent typically being 25 dBA or less which subjectively is very quiet. (The majority of domestic living rooms, considered by their occupants to be quiet, generally measure 30–40 dBA.)

The background noise level in a studio is made up from a number of different sources and components and for control purposes these may be split up into four basic categories:

- external airborne noise
- external structure borne noise
- internally and locally generated noise
- internal noise transfer.

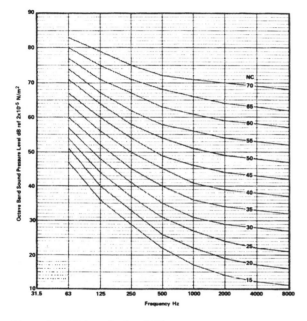

Figure 3.1 Noise criterion (NC) curves.

Each source of noise requires a slightly different approach in terms of its containment or control.

Achieving low noise levels in modern studios is a complex and generally costly problem but any short cuts or short circuits of the noise control measures will destroy the integrity of the low noise countermeasures and allow noise into the studio.

External airborne noise

Here we are typically concerned with noise from local road traffic, aircraft and railways. Any of these sources can generate levels of over 100 dB at low frequencies at the external façade of a building. Controlling such high levels of low-frequency noise is extremely difficult and costly. Therefore, if possible when planning a sensitive acoustic area such as a studio, it is obviously preferable to avoid locations or buildings exposed to high external

levels of noise. An extensive noise survey of the proposed site is therefore essential.

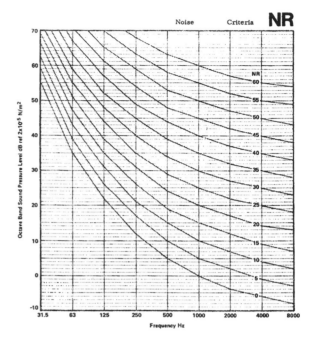

Figure 3.2 Noise criterion (NC) curves.

Wherever possible the walls to a studio or control room should not form part of the external envelope of the building. Instead a studio should be built within the main body of the building itself so a buffer area between the studio and external noise source can be created e.g. by corridors or offices or other non critical areas. Similarly it is not good practice to locate a studio on the top floor of a building. However, if it is not possible to create suitable buffer areas, a one metre minimum separation should be created between the external and studio walls or roof and ceiling etc.

Table 3.1

Studio/recording environment	NC/NR requirement
Studio with close mic technique (300 mm)	
(Bass cut may be required)	30
TV studio control rooms	25
TV studios and sound control rooms	20
Non-broadcast – pre-production	25
Broadcast studio and listening rooms	15–20
Music sound studio	15–20
Concert halls	15–20
Drama studios (broadcast)	15
Sound effects and post production (preferred)	5
(max.)	15
Commercial recording and pop studios	
(depending on type and budget)	15–25

Internally and locally generated noise

Noise apart from being transmitted by sound pressure waves within the air, can also be transmitted by the building structure itself. Structural noise can be induced either directly by the vibration of the structure – e.g. by locating the building over or adjacent to a railway or underground train line – or sound pressure waves (particularly at low frequencies) can cause the structure itself to vibrate. Again, the best method of control is to separate the sensitive areas from such potential noise sources as structural isolation is required. One solution is to spend the available budget on the equipment of rubber pads, or literally steel springs if really effective low frequency isolation is required. However, the best solution is to choose the site carefully in the first place and spend the available budget on the equipment and acoustic finishes rather than on expensive structural isolation.

If the studio is to be built in a building with other occupants/activities, again a thorough survey should be conducted to establish any likely sources of noise which will require more than usual treatment (e.g. machinery coupled into the structure on floors above or below the studio. This includes lifts and other normal 'building services' machinery such as air conditioning plant etc.).

Internal noise transfer

The biggest noise problem in most studios is noise generated or transmitted by the air conditioning equipment. Studios must provide a pleasant atmosphere to work in, which apart from providing suitable lighting and acoustics also means that the air temperature and humidity need to be carefully controlled. Sessions in pop studios may easily last 12 to 15 hours. The studio/control room must therefore have an adequate number of air changes and sufficient fresh air supply in order for the atmosphere not to become 'stuffy' or filled with cigarette smoke etc.

It should be remembered that the acoustic treatments also tend to provide additional thermal insulation and in studios with high lighting capacity (e.g. TV or film) or where a large number of people may be working, the heat cannot easily escape, but is contained within the studio, so requiring additional cooling capacity over that normally expected.

Large air conditioning ducts with low air flow velocities are therefore generally employed in studios with special care being taken to ensure that the air flow through the room diffuser grilles is low enough not to cause air turbulence noise (e.g. duct velocities should be kept below 500 ft^3/min, and terminal velocities below 250 ft^3/min). Studios should be fed from their own air

conditioning plant, which is not shared with other parts of the building.

Ducts carrying air to and from the studios are fitted with attenuators or silencers to control the noise of the intake or extract fans. Crosstalk attenuators are also fitted in ducts which serve more than one area, e.g. studio and control room to prevent the air conditioning ducts acting as large speaking tubes or easy noise transmission paths.

Studio air conditioning units should not be mounted adjacent to the studio but at some distance away to ensure that the air conditioning plant itself does not become a potential noise source. Duct and pipework within the studio complex will need to be mounted with resilient mountings or hangers, to ensure that noise does not enter the structure or is not picked up by the ductwork and transmitted to the studio.

Waterpipes and other mechanical services ducts or pipes should not be attached to studio or control room walls – as again noise can then be readily transmitted into the studio.

Impact noise from footfall or other movement within the building must also be dealt with if it is not be transmitted through the structure and into the studio. A measure as simple as carpeting adjacent corridors or office floors etc can generally overcome the majority of such problems. Wherever possible, corridors adjacent to studios should not be used at critical times or their use should be strictly controlled. Main thoroughfares should not pass through studio areas.

Sound insulation

Having created a generally suitable environment for a studio, it is then up to the studio walls themselves to provide the final degree of insulation from the surrounding internal areas.

The degree of insulation required will very much depend on:

(1) the uses of the studio e.g. music, speech, drama, etc.
(2) the internal noise level target e.g. NC20
(3) the noise levels of the surrounding areas
(4) the protection required to the surrounding area (e.g. in a pop music studio, the insulation is just as much there to keep the noise in so that it will not affect adjoining areas or studios as it is to stop noise getting into the studio itself).

Appropriate insulation or separation between the control room and studio is also another very important aspect which must be considered here. 55–60 dB is probably about the minimum value of sound insulation required between control room and studio. Monitoring levels are generally considerably higher in studio control rooms than in the home. For example, 85 dBA is quite typical for TV or radio sound monitoring, whereas pop studio control rooms will operate in the high 90s and above.

Sound insulation can only be achieved with mass or with mass and separation. Sound absorbing materials are not sound insulators although there is much folklore which would have you believe otherwise. Table 3.2 sets out the figures, showing that only very high values of sound absorption produce usable sound reductions.

Table 3.2 *Sound absorption*

Absorption coefficient	Percentage absorbed (%)	Insulation produced (dB)	Incident sound energy transmitted (%)
0.1	10	0.5	90
0.5	50	3	50
0.7	70	5	30
0.9	90	10	10
0.99	99	20	1
0.999	99.9	30	0.1
0.9999	99.99	40	0.01

To put the above figures in perspective a single 4.5 in. (112 mm) brick wall has a typical sound insulation value of 45 dB, and a typical domestic brick cavity wall 50–53 dB – and you know how much sound is transmitted between adjoining houses! Furthermore, typical sound absorbing materials have absorption coefficients in the region of 0.7 to 0.9 which is equivalent to a noise reduction of only 5–10 dB.

Figure 3.3 shows the so called mass law of sound insulation – which shows that sound insulation increases by approximately 5 dB for every doubling of the mass. A number of typical constructions and building elements have been drawn on for comparison purposes.

It should be noted that the sound insulating properties of a material are also affected by the frequency of the sound in question. In fact the insulation generally increases by 5 dB for every doubling of frequency. This is good news as far as the insulation of high frequency sound is concerned, but bad news for the lower frequencies.

Generally, a single value of sound insulation is often quoted – this normally refers to the average sound insulation achieved over the range 100 Hz to 3.15 KHz and is referred to as the sound reduction index (SRI). (STC in USA based on the average between 125 Hz and 4 KHz.)

An approximate guide to the performance of a material can be obtained from the 500 Hz value of insulation as this is often equivalent to the equated SRI. For example, our 112 mm brick wall will tend to have an SRI of 45 dB and have a sound insulation of 45 dB at 500 Hz, 40 dB at 250 Hz and only 35 dB at 125 Hz – which begins to

explain why it is always the bass sound or beat which seems to be transmitted. In practice, therefore, the studio designer or acoustic consultant first considers controlling break-in (or break-out) of the low frequency sound components as overcoming this problem will invariably automatically sort out any high frequency problems.

Table 3.3 *Typical sound absorption coefficients*

Material	Hz 125	250	500	1K	2K	4K
Drapes						
• Light velour	0.07	0.37	0.49	0.31	0.65	0.54
• Medium velour	0.07	0.31	0.49	0.75	0.70	0.60
• Heavy velour	0.14	0.35	0.55	0.72	0.70	0.65
Carpet						
• Heavy on concrete	0.02	0.06	0.14	0.37	0.60	0.65
• Heavy on underlay or felt	0.08	0.24	0.57	0.70	0.70	0.73
• Thin	0.01	0.05	0.10	0.20	0.45	0.65
Cork floor tiles and linoleum, plastic flooring	0.02	0.04	0.05	0.05	0.10	0.05
Glass (windows)						
• 4 mm	0.30	0.20	0.10	0.07	0.05	0.02
• 6 mm	0.10	0.08	0.04	0.03	0.02	0.02
Glass, tile and marble	0.01	0.01	0.01	0.01	0.02	0.02
Glass fibre mat						
• 80 kg/m³ – 25 mm	0.10	0.30	0.55	0.65	0.75	0.80
• 80 kg/m³ – 50 mm	0.20	0.45	0.70	0.80	0.80	0.80
Wood fibre ceiling tile	0.15	0.40	0.55	0.70	0.80	0.70
Plasterboard over deep air space	0.20	0.15	0.10	0.05	0.05	0.05

From the graph in Fig. 3.3, it can be seen that some pretty heavy constructions are required in order to achieve good sound separation/insulation. However, it is possible to also achieve good sound insulation using lightweight materials by using multi-layered construction techniques.

Combinations of layers of plasterboard and softboard, separated by an air space are frequently used. Figure 3.4 shows a typical construction. The softboard is used to damp out the natural resonances of the plasterboard which would reduce its sound insulation. The airspace may also be fitted with loose acoustic quilting (e.g. fibreglass or mineral wool, to dampout any cavity resonances).

Note that an airtight seal has to be created for optimum sound insulation efficiency. Also note that different sized air spaces are used.

A double-leaf construction as shown in Fig. 3.4 can achieve a very good level of sound insulation particularly at mid and high frequencies.

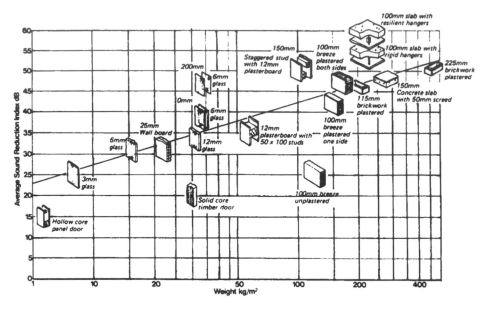

Figure 3.3   Sound insulation performance of building materials compared with mass law.

Typically such a partition might achieve the following insulation values.

Frequency (Hz)	63	125	250	500	1 K	2 K	4 K
Insulation (dB)	40	50	60	65	70	75	80

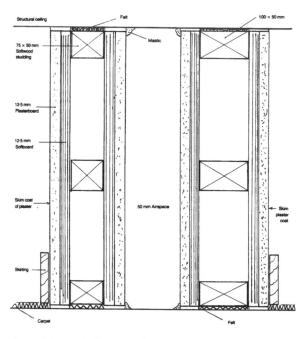

Figure 3.4   Typical construction of double-skin lightweight compartment.

However, the final value will very much depend on the quality of the construction and also on the presence of flanking transmission paths, i.e. other sound paths such as ducting or trunking runs etc. which short circuit or bypass the main barrier.

Triple layer constructions and multiple layer heavy masonry constructions can be employed to achieve improved levels of insulation – particularly at the lower frequencies.

Studios can also be built from modular construction assemblies based on steel panels/cavities lined with sound absorptive material.

The following table compares a double leaf steel panel construction with an equivalent thickness (300 mm) of masonry wall construction.

Frequency (Hz)	63	125	250	500	1 K	2 K	4 K
Masonry	28	34	34	40	56	73	76
Steel Panelling	40	50	62	71	80	83	88

Again it must be remembered that when it comes to sound insulation, large differences frequently occur between the theoretical value (i.e. what can be achieved) and the actual on-site value (i.e. what is actually achieved). Great care has to be taken to the sealing of all joints and to avoid short-circuiting panels by the use of wall ties or debris within the cavities. Furthermore, where doors or observation windows etc. penetrate a wall, particular care must be taken to ensure that these do not degrade the overall performance. For example, the door or window should have the same overall sound insulation capabilities as the basic studio construction.

Studio doors should therefore be solid core types, and are frequently fitted with additional lead or steel sheet linings. Proprietory seals and door closers must be fitted to form an airtight seal round the edge of the door. Soft rubber compression or magnetic sealing strips are commonly employed whilst doors are rebated to help form a better joint. (Wherever possible a sound lock should be employed i.e. two doors with an acoustically treated space between, such that one door is closed before the other opened.) Other methods of improving door insulation include the use of two separate doors on a single frame, each fitted with appropriate seals etc.

Control room windows may either be double or triple glazed depending on the performance required. Large air spaces, e.g. 200–300 mm plus, are necessary in order to achieve appropriate insulation. Tilting one of the panes not only helps to cut down visual reflections, but also breaks up single frequency cavity resonances, which reduce the potential insulation value significantly. The reveals to the windows should also be lined with sound absorbing material to help damp out cavity resonances.

Window glass needs to be much thicker/heavier than typical domestic glazing. For example, a typical 0.25 in., 6 mm, domestic window would have a sound insulation performance of around 25–27 dB (21–23 dB at 125 Hz), whereas a 200 mm void fitted with 12 mm and 8 mm panes could produce an overall SRI of 52 dB (31 dB at 125 Hz).

Special laminated windows and panes can also be used to improve performance.

Studios are frequently built on the box within a box principle, whereby the second skin is completely isolated from the outer. This is achieved by building a 'floating' construction. For example, the floor of the studio is isolated from the main structure of the building by constructing it on resilient pads or on a continuous resilient mat. The inner walls are then built up off the isolated floor with minimal connection to the outer leaf. The ceiling is supported from the inner leaves alone, and so does not bridge onto the outer walls which would short-circuit the isolation and negate the whole effect. With such constructions particular care has to be taken with the connection of the various services into the studio. Again attention to detail is of tantamount importance if optimum performance is to be achieved. Note that the sealing of any small gaps or cracks is extremely important as shown by the plastered and unplastered walls in Table 3.4.

Studio and Control Room Acoustics

Today, the acoustic response of the control room is recognised as being as equally important, if not more important, than that of the studio itself. This is partly due to improvements in monitor loudspeakers, recent advances in the understanding of the underlying

Table 3.4 *Summary of some typical building materials*

Materials	SRI (dB)
Walls	
Lightweight block work (not plastered)	35
(plastered)	40
200 mm lightweight concrete slabs	40
100 mm solid brickwork (unplastered)	42
(plastered 12 mm)	45
110 mm dense concrete (sealed)	45
150 mm dense concrete (sealed)	47
230 mm solid brick (unplastered)	48
(plastered)	49
200 mm dense concrete (blocks well sealed/plastered)	50
250 mm cavity brick wall (i.e. 2 × 110)	50
340 mm brickwork (plastered both sides 12 mm)	53
450 mm brick/stone plastered	55
112 mm brick + 50 mm partition of one layer plasterboard and softboard per side	56
225 brick – 50 mm cavity – 225 brick	65
325 brick – 230 mm cavity – 225 brick (floated)	75
12 mm plasterboard on each side 50 mm stud frame	33
12 mm plasterboard on each side 50 mm stud frame with quilt in cavity	36
Two × 12 mm plasterboard on 50 mm studs with quilt in cavity	41
As above with 75 mm cavity	45
Three layers 12 mm plasterboard on separate timberframe with 225 mm air gap with quilt	49
Floors	
21 mm + t & g boards or 19 mm chipboard	35
110 mm concrete and screed	42
21 mm + t & g boards or 19 mm chipboard with plasterboard below and 50 mm sand pugging	45
125 mm reinforced concrete and 150 mm screed	45
200 mm reinforced concrete and 50 mm screed	47
125 mm reinforced concrete and 50 mm screed on glass fibre or mineral wool quilt	50
21 mm + t & g boards or 19 mm chipboard in form of raft on min fibre quilt with plasterboard and 50 mm sand pugging	50
150 mm concrete on specialist floating raft	55–60
Windows	
4 mm glass well sealed	
23–25	
6 mm glass well sealed	27
6 mm glass – 12 mm gap – 6 mm glass	28
12 mm glass well sealed	31
12 mm glass laminated	35
10 mm glass – 80 mm gap – 6 mm glass	37
4 mm glass – 200 mm gap – 4 mm glass	39
10 mm glass – 200 mm gap – 6 mm glass	44
10 mm glass – 150 mm gap – 8 mm glass	44
10 mm glass – 200 mm gap – 8 mm glass	52

psychoacoustics and the recent trend of using the control room as a recording space, e.g. for synthesisers, whereby better contact can be maintained between musician, producer and recording engineer.

For orchestral or classical music recording, the studios need to be relatively large and lively, with reverberation times lying somewhere between 1.2 and 2 seconds depending on their size and use. Pop studios tend to be much more damped than this, perhaps typically varying from 0.4 to 0.7 seconds.

Broadcast studios would typically have reverberation times as follows:

- Radio talks   0.25 s   (60 m³)
- Radio drama   0.4 s   (400 m³)
- Radio music
  - small   0.9 s   (600 m³)
  - large   2.0 s   (10000 m³)
- Television   0.7 s   (7000 m³)
- Control rooms   0.25 s

Whilst it has been recognised for some time now that reverberation time alone is not a good descriptor of studio acoustics, it is difficult to determine any other easily predictable or measurable parameters which are. Reverberation time therefore continues to be used.

Reverberation time is effectively a measure of the sound absorption within a room or studio, etc. Its importance therefore really lies within the way it alters with frequency rather than on its absolute value. It is therefore generally recognised that reverberation time should be maintained essentially constant within tight limits (e.g. 10%) across the audio band of interest with a slight rise at the lower bass frequencies generally being permissible. By maintaining the reverberation time constant, the room/studio is essentially affecting all frequencies equally – at least on a statistical basis. However, in practice, just because a studio or control room has a perfectly flat reverberation time characteristic this does not necessarily guarantee that it will sound all right, but it is a good baseline from which to begin.

Studios are generally designed so that they can either be divided up, e.g. by use of portable screens or by having their acoustic characteristics altered/adjusted using hinged or fold over panels, for example, with one side being reflective and the other absorptive or by the use of adjustable drapes etc. Isolation booths or voice-over booths are also frequently incorporated in commercial studios to increase separation. Drama studios frequently have a live end and a dead end to cater for different dramatic requirements and scenes. Recent advances in digital reverberation, room simulation and effects have, however, added a new dimension to such productions, enabling a wide variety of environments to be electronic-

ally simulated and added in at the mixing stage. The same is the case with musical performances/recordings.

It is important not to make studios and control rooms too dead, as this can be particularly fatiguing and even oppressive.

TV studios have to cater for a particularly wide range of possible acoustic environments, ranging from the totally reverberation free outdoors to the high reverberant characteristics of sets depicting caves, cellars or tunnels etc., with light entertainment and orchestral music and singing somewhere in between the two.

As the floor of the TV studio has to be ruler flat and smooth for camera operations and the roof is a veritable forest of lighting, the side walls are the only acoustically useful areas left. In general these are covered with wide band acoustic absorption, e.g. 50 mm mineral wool/glass fibre over a 150 mm partioned air-space and covered in wire mesh to protect it. Modular absorbers are also used and allow a greater degree of final tuning to be carried out should this be desired.

Absorbers

Proprietary modular absorbers are frequently used in commercial broadcast/radio studios. These typically consist of 600 mm square boxes of various depths in which cavities are created by metal or cardboard dividers over which a layer of mineral wool is fixed together with a specially perforated faceboard.

By altering the percentage perforations, thickness/density of the mineral wool and depth of the cavities, the absorption characteristics can be adjusted and a range of absorbers created, varying from wideband mid and high frequency units to specifically tuned low frequency modules which can be selected to deal with any particularly strong or difficult room resonances or colorations – particularly those associated with small rooms where the modal spacing is low or where coincident modes occur.

Resonances

Room modes or Eigentones are a series of room resonances formed by groups of reflections which travel back and forth in phase within a space, e.g. between the walls or floor and ceiling.

At certain specific frequencies, the reflections are perfectly in phase with each other, and cause a resonance peak in the room frequency response. The resonant frequencies are directly related to the dimensions of the room, the first resonance occurring at a frequency corresponding to the half wavelength of sound equal to the spacing between the walls or floor etc. (see Fig. 3.5). Resonances also occur at the harmonics of the first (fundamental) frequencies. At frequencies related to the quarter wavelength dimension, the incident and reflected waves are out of phase with each other and by destructive wave interference attempt to cancel each other out, resulting in a series of nulls or notches in the room frequency response. Room modes therefore give

rise to a harmonic series of peaks and dips within the room frequency response.

At different positions throughout the room, completely different frequency responses can occur depending on whether a peak or null occurs at a given location for a given frequency. Variations of 10–15 dB within the room response are quite common unless these modal resonances are brought under control. Figure 3.6 presents a frequency response trace of a monitor loudspeaker in a poorly treated control room.

There are three types of room mode, the axial, oblique and tangential, but in practice it is the axial modes which generally cause the most severe problems. Axial modes are caused by in and out of phase reflections occurring directly, between the major axes of the room, i.e. floor to ceiling, side wall to side wall and end wall to end wall.

Figure 3.7 shows the variation in sound pressure level (loudness) which occurred within a poorly treated control room at the fundamental and first harmonic modal frequencies. Subjectively the room was criticised as suffering from a gross bass imbalance, having no bass at the centre but excess at the boundaries – which is exactly what Fig. 3.7 shows, there being a 22 dB variation in the level of the musical pitch corresponding to the fundamental modal frequency and 14 dB at the first harmonic.

When designing a control room or studio, it is vitally important to ensure that the room dimensions are chosen such as not to be similar or multiples of each other, as this will cause the modal frequencies in each dimension to coincide, resulting in very strong resonance patterns and an extremely uneven frequency response anywhere within the room.

The room (studio) dimensions should therefore be selected to given as even as possible distribution of the Eigentones. A number of ratios have been formulated

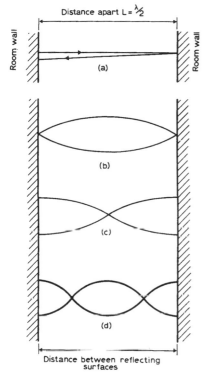

Figure 3.5   Formation of room modes.

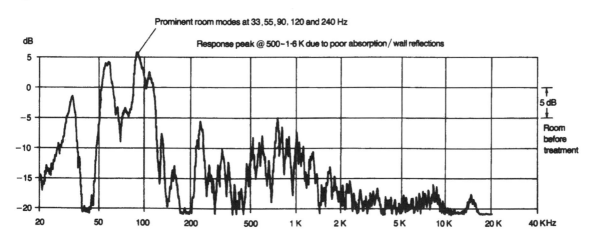

Figure 3.6   Frequency response trace of a monitor loudspeaker in a poorly treated control room.

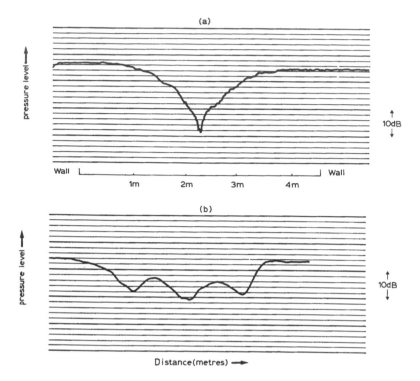

Figure 3.7   Sound pressure level (loudness) variation of first and second modes measured in poorly treated control room.

which give a good spread. These are sometimes referred to as Golden Ratios. Typical values include:

$$
\begin{array}{ccc}
H & W & L \\
1 : & 1.14 : & 1.39 \\
1 : & 1.28 : & 1.54 \\
1 : & 1.60 : & 2.33
\end{array}
$$

Figure 3.8 presents a graphical check method for ensuring an optimal spread of room resonances.

The frequencies at which main axial room resonances occur can be simply found using the following formula:

$$
F_{res} = \frac{172(n)}{L}
$$

where L is the room dimension (metres)
   $n = 1, 2, 3, 4$ etc.

The graph presented in Fig. 3.9 provides a quick method of estimating room mode frequencies.

The complete formula for calculating the complete series of modes is as follows:

$$
F = 172 \left[\left(\frac{nx}{Lx} \right)^2 + \left(\frac{ny}{Ly} \right)^2 + \left(\frac{nz}{Lz} \right)^2 \right]^{\frac{1}{2}}
$$

At low frequencies, the density of room modes is low, causing each to stand out and consequently be more audible. However, at higher frequencies, the density becomes

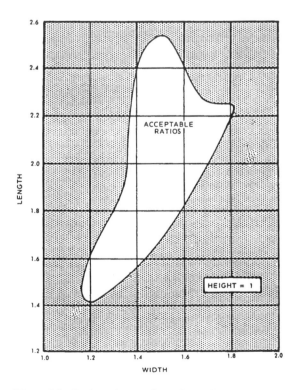

Figure 3.8   Preferred room dimension ratios.

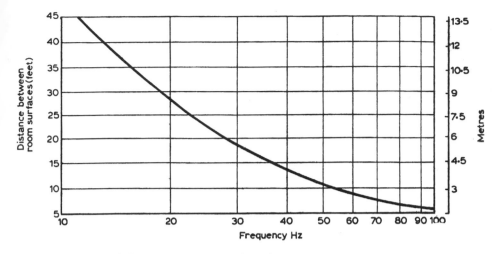

Figure 3.9 Axial mode frequency versus room dimension.

very much greater, forming a continuous spectrum, i.e. at any given frequency, a number of modes will occur, which counteract each other. Room modes therefore generally cause problems at low or lower mid frequencies (typically up to 250–500 Hz). Whilst this is primarily due to the low modal density at these frequencies it is also exacerbated by the general lack of sound absorption which occurs at low frequencies in most rooms.

After determining that the room/studio dimensions are appropriate to ensure low coincidence of modal frequencies, control of room modes is brought about by providing appropriate absorption to damp down the resonances.

Absorber performance

Sound absorbers effectively fall into four categories.

(1) high and medium frequency dissipative porous absorbers
(2) low frequency panel or membrane absorbers
(3) Helmholtz tuned frequency resonator/absorber
(4) quadratic residue/phase interference absorbers/diffusers.

Porous absorbers include many common materials including such items as drapes, fibreglass/mineral wool, foam, carpet, acoustic tile etc.

Figure 3.10 illustrates the frequency range over which porous absorbers typically operate. The figure clearly shows the performance of these types of absorber to fall off at medium to low frequencies – unless the material is very thick (comparable in fact to 1/4 wavelength of the sound frequency to be absorbed).

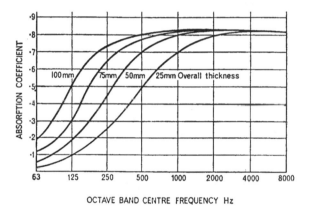

Figure 3.10 Typical absorption characteristics of porous material.

Panel or membrane absorbers (Fig. 3.11) therefore tend to be used for low frequency absorption. The frequency of absorption can be tuned by adjusting the mass of the panel and the airspace behind it. By introducing some dissipative absorption into the cavity to act as a damping element, the absorption curve is broadened. Membrane absorption occurs naturally in many studios and control rooms where relatively lightweight structures are used to form the basic shell, e.g. plasterboard and stud partitions, plasterboard or acoustic tile ceilings with an airspace above etc.

The frequency at which maximum absorption occurs can be calculated from the following formula

$$f = \frac{60}{\sqrt{md}}$$

where m = mass of panel in kg/m^2
d = airspace in metres

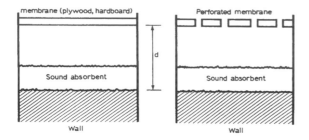

Figure 3.11   Typical panel or membrane absorber construction.

The action of porous and membrane type absorbers is often combined to form a single wide frequency band absorber or to extend the low frequency characteristics of porous materials.

Bass traps are generally formed by creating an absorbing cavity or layer of porous absorption equivalent to a quarter wavelength of the sound frequency or bass note in question. Table 3.5 presents the dimensions required for the range 30–120 Hz.

Figure 3.12 shows basic principles and typical absorption characteristics of a Helmholtz or cavity absorber.

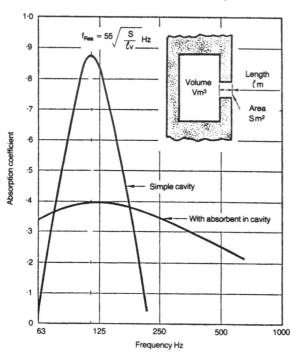

Figure 3.12   Typical absorption characteristics of a resonant absorber.

The frequency of maximum absorption occurs at the natural resonant frequency of the cavity which is given by the following formula:

$$F_{res} = 55 \sqrt{\frac{S}{lV}} \quad \text{Hz}$$

where S is the cross sectional area of the neck (m²)
l is the length of the neck (m)
V is the volume of air in the main cavity (m³)

Table 3.5   *Frequency 1/4 wavelength dimension*

Hz	m	ft
30	2.86	11.38
40	2.14	7.04
50	1.72	5.64
60	1.43	4.69
70	1.23	4.03
80	1.07	3.52
90	0.96	3.13
100	0.86	2.82
120	0.72	2.35

Often membrane absorbers are combined with cavity absorbers to extend their range. Some commercial modular absorbers also make use of such techniques and, by using a range of materials/membranes/cavities, can provide a wide frequency range of operation within a standard size format. Figure 3.13 illustrates general characteristics of membrane, cavity and dissipative/porous absorbers.

A newer type of 'absorber' is the Quadratic Residue Diffuser. This device uniformly scatters or diffuses sound striking it, so that although individual reflections are attenuated and controlled, the incident sound energy is essentially returned to the room. This process can therefore be used to provide controlled low-level reflections or reverberation enabling studios or control rooms to be designed without excessive absorption or subjective oppressiveness which often occurs when trying to control room reflections and resonances. When designing a studio or control room etc, the various absorptive mechanisms described above are taken into account and combined to produce a uniform absorption/frequency characteristic i.e. reverberation time.

Reverberation and reflection

Reverberation times can be calculated using the following simple Sabine formula:

$$RT = \frac{0.16\,lV}{A}$$

where A is the total sound absorption in m² which is computed by multiplying each room surface by its sound absorption coefficient and summing these to give the total absorption present, or

$$RT = \frac{0.16\,lV}{S\bar{a}}$$

where $\bar{a}$ is the average absorption coefficient

In acoustically dead rooms the following Eyring formula should be used:

$$RT = \frac{0.16\,lV}{-S\,log_e\,(1 - \bar{a}) + MV}$$

where V is the volume of the room in m³ and M is an air absorption constant.

Achieving a uniform room reverberation time (absorption) characteristic does not necessarily ensure good acoustics, the effect and control of specific reflections must also be fully taken into account. For example, strong reflections can strongly interfere with the recorded or perceived live sound causing both tonal colourations and large frequency response irregularities. Such reflections can be caused by poorly treated room surfaces, by large areas of glazing, off doors or by large pieces of studio equipment including the mixing console itself.

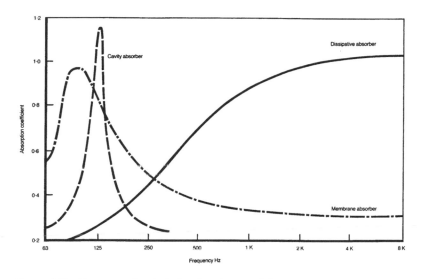

Figure 3.13 Summary of typical absorber characteristics.

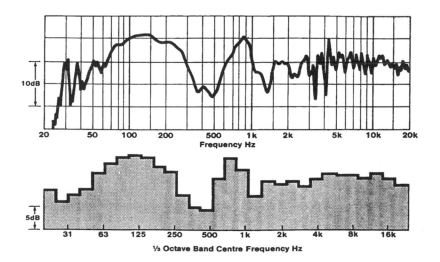

Figure 3.14 Frequency response plot of monitor loudspeaker mounted adjacent to wall.

Figure 3.14 illustrates the effect well, being a frequency response plot of a monitor loudspeaker (with a normally very flat response characteristic) mounted near to a reflective side wall.

Apart from causing gross frequency response irregularities and colorations, the side wall reflections also severely interfere with stereo imaging precision and clarity. Modern studio designs go to considerable lengths to avoid such problems by either building the monitor loudspeakers into, but decoupled from, the structure and treating adjacent areas with highly absorbing material, or by locating the monitors well away from the room walls and again treating any local surfaces. Near field monitoring overcomes many of these problems, but reflections from the mixing console itself can produce combfiltering irregularities. However, hoods or careful design of console-speaker geometry/layout can be used to help overcome this.

The so called Live End Dead End (LEDE) approach to control room design uses the above techniques to create a reflection-free zone enabling sound to be heard directly from the monitors with a distinct time gap before any room reflections arrive at the mixing or prime listening position. Furthermore, such reflections which do occur are carefully controlled to ensure that they do not cause sound colorations nor affect the stereo image, but just add an acoustic liveliness or low-level reverberant sound field which acts to reduce listening fatigue and provide better envelopment within the sound field.

4 Principles of Digital Audio

Allen Mornington-West

The impact that digital methods have made on audio has been at least as remarkable as it was on computing. Allen Mornington-West uses this chapter to introduce the digital methods which seem so alien to anyone trained in analogue systems.

Introduction

The term digital audio is used so freely by so many that you could be excused for thinking there was nothing much new to tell. It is easy in fast conversation to present the impression of immense knowledge on the subject but it is more difficult to express the ideas concisely yet readably. The range of topics and disciplines which need to be harnessed in order to cover the field of digital audio is very wide and some of the concepts may appear paradoxical at first sight. One way of covering the topics would be to go for the apparent precision of the mathematical statement but, although this has its just place, a simpler physical understanding of the principles is of greater importance here. Thus in writing this chapter we steer between excessive arithmetic precision and ambiguous over-simplified description.

Analogue and Digital

Many of the physical things which we can sense in our environment appear to us to be part of a continuous range of sensation. For example, throughout the day much of coastal England is subject to tides. The cycle of tidal height can be plotted throughout the day. Imagine a pen plotter marking the height on a drum in much the same way as a barograph is arranged, Fig. 4.1. The continuous line which is plotted is a feature of analogue signals in which the information is carried as a continuous infinitely fine variation of a voltage, current or, as in this case, height of the sea level.

When we attempt to take a measurement from this plot we will need to recognise the effects of limited measurement accuracy and resolution. As we attempt greater resolution we will find that we approach a limit described by the noise or random errors in the measurement technique. You should appreciate the difference between resolution and accuracy since inaccuracy gives rise to distortion in the measurement due to some non-linearity in the measurement process. This facility of measurement is useful. Suppose, for example, that we wished to send the information regarding the tidal heights we had measured to a colleague in another part of the country. One, admittedly crude, method might involve turning the drum as we traced out the plotted shape whilst, at the far end an electrically driven pen wrote the same shape onto a second drum, Fig. 4.2(a). In this method we would be subject to the non-linearity of both the reading pen and the writing pen at the far end. We would also have to come to terms with the noise which the line, and any amplifiers, between us would add to the signal describing the plot. This additive property of noise and distortion is characteristic of handling a signal in its analogue form and, if an analogue signal has to travel through many such links then it can be appreciated that the quality of the analogue signal is abraded irretrievably.

As a contrast consider describing the shape of the curve to your colleague by measuring the height of the curve at frequent intervals around the drum, Fig. 4.2(b). You'll need to agree first that you will make the measurement at each ten-minute mark on the drum, for example, and you will need to agree on the units of the measurement. Your colleague will now receive a string of numbers from you. The noise of the line and its associated amplifiers will not affect the accuracy of the received information since the received information should be a recognisable number. The distortion and noise performance of the line must be gross for the spoken numbers to be garbled and thus you are very well assured of correctly conveying the information requested. At the receiving end the numbers are plotted on to the chart and, in the simplest approach, they can be simply joined up with straight lines. The result will be a curve looking very much like the original.47

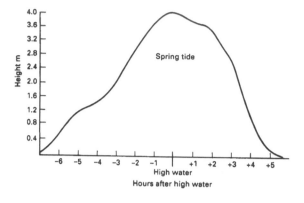

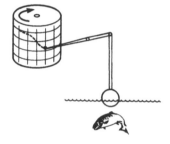

Figure 4.1   The plot of tidal height versus diurnal time for Portsmouth UK, time in hours. Mariners will note the characteristically distorted shape of the tidal curve for the Solent. We could mark the height as a continuous function of time using the crude arrangement shown.

Let's look at this analogy a little more closely. We have already recognised that we have had to agree on the time interval between each measurement and on the meaning of the units we will use. The optimum choice for this rate is determined by the fastest rate at which the tidal height changes. If, within the ten minute interval chosen, the tidal height could have ebbed and flowed then we would find that this nuance in the change of tidal height would not be reflected in our set of readings. At this stage we would need to recognise the need to decrease the interval between readings. We will have to agree on the resolution of the measurement since, if an arbitrarily fine resolution is requested it will take a much longer time for all of the information to be conveyed or transmitted. We will also need to recognise the effect of inaccuracies in marking off the time intervals at both the transmit or coding end and at the receiving end since this is a source of error which affects each end independently.

In this simple example of digitising a simple wave-shape we have turned over a few ideas. We note that the method is robust and relatively immune to noise and distortion in the transmission and we note also that

provided we agree on what the time interval between readings should represent then small amounts of error in the timing of the reception of each piece of data will be completely removed when the data is plotted. We also note that greater resolution requires a longer time and that the choice of time interval affects our ability to resolve the shape of the curve. All of these concepts have their own special terms and we will meet them slightly more formally.

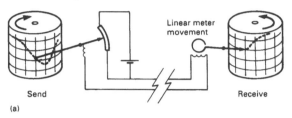

Figure 4.2   Sending the tidal height data to a colleague in two ways: (a) by tracing out the curve shape using a pen attached to a variable resistor and using a meter driven pen at the far end; (b) by calling out numbers, having agreed what the scale and resolution of the numbers will be.

In the example above we used implicitly the usual decimal base for counting. In the decimal base there are ten digits (0 through 9). As we count beyond 9 we adopt the convention that we increment our count of the number of tens by one and recommence counting in the units column from 0. The process is repeated for the count of hundreds, thousands and so on. Each column thus represents the number of powers of ten ($10 = 10^1$, $100 = 10^2$, $1000 = 10^3$ and so on). We are not restricted to using the number base of ten for counting. Amongst the bases which are in common use these days is the base 16 (known more commonly as the hexadecimal base) the base 8 (known as octal) and the simplest of them all, base 2 (known as binary). Some of these scales have been, and continue to be, in common use. We recognise the old coinage system in the UK used the base of 12 for pennies as, indeed, the old way of marking distance still uses the unit of 12 inches to a foot.

The binary counting scale has many useful properties. Counting in the base of 2 means that there can only be two unique digits, 1 and 0. Thus each column must represent a power of 2 ($2 = 2^1$, $4 = 2^2$, $8 = 2^3$, $16 = 2^4$ and so on) and, by convention, we use a 1 to mark the presence of a power of two in a given column. We can represent any number by adding up an appropriate collection of powers of 2 and, if you try it, remember that 2^0 is equal to 1. We refer to each symbol as a bit (actually a contraction of the words **bi**nary dig**it**). The bit which appears in the units column is referred to as the least significant bit, or LSB, and the bit position which carries the most weight is referred to as the most significant bit, or MSB.

Binary arithmetic is relatively easy to perform since the result of any arithmetic operation on a single bit can only be either 1 or 0.

We have two small puzzles at this stage. The first concerns how we represent numbers which are smaller than unity and the second is how are negative numbers represented. In the everyday decimal (base of 10) system we have adopted the convention that numbers which appear to the right of the decimal point indicate successively smaller values. This is in exactly the opposite way in which numbers appearing to the left of the decimal point indicated the presence of increasing powers of 10. Thus successive columns represent $0.1 = 1/10 = 10^{-1}$, $0.01 = 1/100 = 10^{-2}$, $0.001 = 1/1000 = 10^{-3}$ and so on. We follow the same idea for binary numbers and thus the successive columns represent $0.5 = 1/2 = 2^{-1}$, $0.25 = 1/4 = 2^{-2}$, $0.125 = 1/8 = 2^{-3}$ and so on.

One of the most useful properties of binary numbers is the ease with which arithmetic operations can be carried out by simple binary logic. For this to be viable there has to be a way of including some sign in the number itself since we have only the two symbols 0 and 1. Here are two ways it can be done. We can add a 1 at the beginning of the number to indicate that it was negative, or we can use a more flexible technique known as two's complement. Here the positive numbers appear as we would expect but the negative numbering is formed by subtracting the value of the intended negative number from the largest possible positive number incremented by 1. Table 4.1 shows both of these approaches. The use of a sign bit is only possible because we will arrange that we will use the same numbering and marking convention. We will thus know the size of the largest positive or negative number we can count to. The simple use of a sign bit leads to two values for zero which is not elegant or useful. One of the advantages of two's complement coding is that it makes subtraction simply a matter of addition. Arithmetic processes are at the heart of digital signal processing (DSP) and thus hold the key to handling digitised audio signals.

There are many advantages to be gained by handling analogue signals in digitised form and, in no particular order, they include:

- great immunity from noise since the digitised signal can only be 1 or 0;
- exactly repeatable behaviour;
- ability to correct for errors when they do occur;
- simple arithmetic operations, very easy for computers;

- more flexible processing possible and easy programmability;
- low cost potential;
- processing can be independent of real time.

Table 4.1 *Four-bit binary number coding methods*

| Decimal number | Binary number representation | | |
	Sign plus magnitude	Two's complement	Offset binary
7	0111	0111	1111
6	0110	0110	1110
5	0101	0101	1101
4	0100	0100	1100
3	0011	0011	1011
2	0010	0010	1010
1	0001	0001	1001
0	0000	0000	1000
–0	1000	(0000)	(1000)
–1	1001	1111	0111
–2	1010	1110	0110
–3	1011	1101	0101
–4	1100	1100	0100
–5	1101	1011	0011
–6	1110	1010	0010
–7	1111	1001	0001
–8		1000	0000

Elementary Logical Processes

We have described an outline of a binary counting scale and shown how we may implement a count using it but some physical method of performing this is needed. We can represent two states, a 1 and an 0 state, by using switches since their contacts will be either open or closed and there is no half-way state. Relay contacts also share this property but there are many advantages in representing the 1 and 0 states by the polarity or existence of a voltage or a current not least of which is the facility of handling such signals at very high speed in integrated circuitry. The manipulation of the 1 and 0 signals is referred to as logic and, in practice, is usually implemented by simple logic circuits called gates. Digital integrated circuits comprise collections of various gates which can number from a single gate (as in the eight input NAND gate exemplified by the 74LS30 part number) to many millions (as can be found in some microprocessors).

All logic operations can be implemented by the appropriate combination of just three operations:

- the AND gate, circuit symbol &, arithmetic symbol '.'
- the OR gate, circuit symbol |, arithmetic symbol '+'
- the inverter or NOT gate, circuit and arithmetic symbol '–'.

From this primitive trio we can derive the NAND, NOR and EXOR (exclusive-OR gate). Gates are characterised by the relationship of their output to combinations of their inputs, Fig. 4.3. Note how the NAND (literally negated AND gate) performs the same logical function as OR gate fed with inverted signals and, similarly note the equivalent duality in the NOR function. This particular set of dualities is known as De Morgan's Theorem.

Practical logic systems are formed by grouping many gates together and naturally there are formal tools available to help with the design, the most simple and common of which is known as Boolean algebra. This is an algebra which allows logic problems to be expressed in symbolic terms. These can then be manipulated and the resulting expression can be directly interpreted as a logic circuit diagram. Boolean expressions cope best with logic which has no timing or memory associated with it: for such systems other techniques, such as state machine analysis, are better used instead.

The simplest arrangement of gates which exhibit memory, at least whilst power is still applied, is the cross coupled NAND (or NOR) gate. Fig. 4.4. More complex arrangements produce the wide range of flip-flop (FF) gates including the set-reset latch, the D type FF which is edge triggered by a clock pulse, the JK FF and a wide range of counters (or dividers) and shift registers, Fig. 4.5. These circuit elements and their derivatives find their way into the circuitry of digital signal handling for a wide variety of reasons. Early digital circuitry was based around standardised logic chips, but it is much more common nowadays to use application-specific ICs (ASICs).

The Significance of Bits and Bobs

Groupings of eight bits which represent a symbol or a number are usually referred to as bytes and the grouping

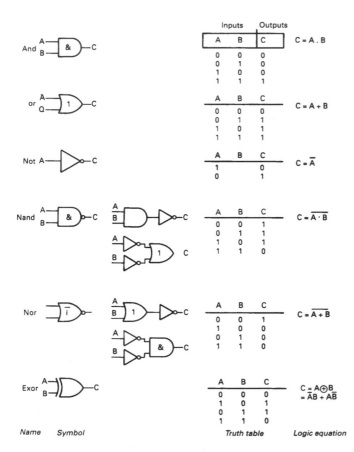

Figure 4.3 The symbols and truth tables for the common basic gates. For larger arrays of gates it is more useful to express the overall logical function as a set of sums (the OR function) and products (the AND function), and this is the terminology used by gate array designers.

of four bits is, somewhat obviously, sometimes called a nibble (sometimes spelt nybble). Bytes can be assembled into larger structures which are referred to as words. Thus a three byte word will comprise twenty-four bits (though word is by now being used mainly to mean two bytes, and DWord to mean four bytes). Blocks are the next layer of structure perhaps comprising 512 bytes (a common size for computer hard discs). Where the arrangement of a number of bytes fits a regular structure the term frame is used. We will meet other terms which describe elements of structure in due course.

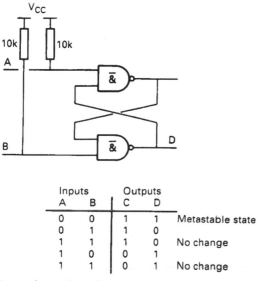

Inputs		Outputs		
A	B	C	D	
0	0	1	1	Metastable state
0	1	1	0	
1	1	1	0	No change
1	0	0	1	
1	1	0	1	No change

Convention and good practice is to use pull-up resistors on open gate inputs.

Figure 4.4 A simple latch. In this example the outputs of each of two NAND gates is cross-coupled to one of the other inputs. The unused input is held high by a resistor to the positive supply rail. The state of the gate outputs will be changed when one of the inputs is grounded and this output state will be steady until the other input is grounded or until the power is removed. This simple circuit, the R-S flip-flop, has often been used to debounce mechanical contacts and as a simple memory.

Conventionally we think of bits, and their associated patterns and structures, as being represented by one of two voltage levels. This is not mandatory and there are other ways of representing the on/off nature of the binary signal. You should not forget alternatives such as the use of mechanical or solid state switches, presence or absence of a light, polarity of a magnetic field, state of waveform phase and direction of an electric current. The most common voltage levels referred to are those which are used in the common 74×00 logic families and are often referred to as TTL levels. A logic 0 (or low) will be any voltage which is between 0 V and 0.8 V whilst a logic 1 (or high) will be any voltage between 2.0 V and the sup-

ply rail voltage which will be typically 5.0 V. In the gap between 0.8 V and 2.0 V the performance of a logic element or circuit is not reliably determinable as it is in this region where the threshold between low and high logic levels is located. Assuming that the logic elements are being correctly used the worst case output levels of the TTL families for a logic 0 is between 0 V and 0.5 V and for a logic 1 is between 2.4 V and the supply voltage. The difference between the range of acceptable input voltages for a particular logic level and the range of outputs for the same level gives the noise margin. Thus for the TTL families the noise margin is typically in the region of 0.4 V for both logic low and logic high. Signals whose logic levels lie outside these margins may cause misbehaviour or errors and it is part of the skill of the design and lay-out of such circuitry that this risk is minimised.

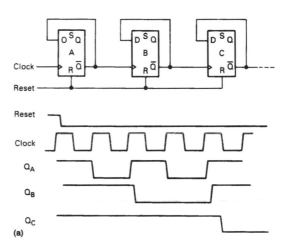

(a)

Figure 4.5 (a) The simplest counter is made up of a chain of edge triggered D type FFs. For a long counter, it can take a sizeable part of a clock cycle for all of the counter FFs to change state in turn. This ripple-through can make decoding the state of the counter difficult and can lead to transitory glitches in the decoder output, indicated in the diagram as points where the changing edges do not exactly line up. Synchronous counters in which the clock pulse is applied to all of the counting FFs at the same time, are used to reduce the overall propagation delay to that of a single stage.

Logic elements made using the CMOS technologies have better input noise margins because the threshold of a CMOS gate is approximately equal to half of the supply voltage. Thus after considering the inevitable spread of production variation and the effects of temperature, the available input range for a logic low (or 0) lies in the range 0 V to 1.5 V and for a logic high (or 1) in the range of 3.5 V to 5.0 V (assuming a 5.0 V supply). However the output impedance of CMOS gates is at least three times higher than that for simple TTL gates and thus in a 5.0 V supply system interconnections in CMOS systems are more susceptible to reactively coupled noise. CMOS systems produce their full benefit of high noise margin when they are operated at higher voltages but this is not possible for the CMOS technologies which are intended to be compatible with 74x00 logic families.

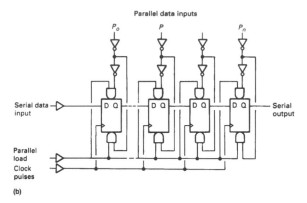

Figure 4.5   (b) This arrangement of FFs produces the shift register. In this circuit a pattern of 1s and 0s can be loaded into a register (the load pulses) and then they can be shifted out serially one bit at a time at a rate determined by the serial clock pulse. This is an example of a parallel in serial out (PISO) register. Other varieties include LIFO (last in first out), SIPO (serial in parallel out) and FILO (first in last out). The diagrams assume that unused inputs are tied to ground or to the positive supply rail as needed.

Transmitting Digital Signals

There are two ways in which you can transport bytes of information from one circuit or piece of equipment to another. Parallel transmission requires a signal line for each bit position and at least one further signal which will indicate that the byte now present on the signal lines is valid and should be accepted. Serial transmission requires that the byte be transmitted one bit at a time and in order that the receiving logic or equipment can recognise the correct beginning of each byte of information it is necessary to incorporate some form of signalling in the serial data in order to indicate (as a minimum) the start of each byte. Figure 4.6 shows an example of each type.

Parallel transmission has the advantage that, where it is possible to use a number of parallel wires, the rate at which data can be sent can be very high. However it is not easy to maintain a very high data rate on long cables using this approach and its use for digital audio is usually restricted to the internals of equipment and for external use as an interface to peripheral devices attached to a computer.

The serial link carries its own timing with it and thus it is free from errors due to skew and it clearly has benefits when the transmission medium is not copper wire but

infra-red or radio. It also uses a much simpler single circuit cable and a much simpler connector. However the data rate will be roughly ten times that for a single line of a parallel interface. Achieving this higher data rate requires that the sending and receiving impedances are accurately matched to the impedance of the connecting cable. Failure to do this will result in signal reflections which in turn will result in received data being in error. This point is of practical importance because the primary means of conveying digital audio signals between equipments is by the serial AES/EBU signal interface at a data rate approximately equal to 3 Mbits per second.

You should refer to a text on the use of transmission lines for a full discussion of this point but for guidance here is a simple way of determining whether you will benefit by considering the transmission path as a transmission line.

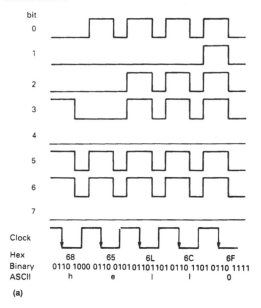

(a)

Figure 4.6   Parallel transmission: (a) a data strobe line $\overline{\text{DST}}$ (the — sign means active low) would accompany the bit pattern to clock the logic state of each data line on its falling edge, and is timed to occur some time after the data signals have been set so that any reflections, crosstalk or skew in the timing of the individual data lines will have had time to settle. After the $\overline{\text{DST}}$ signal has returned to the high state the data lines are reset to 0 (usually they would only be changed if the data in the next byte required a change).

- Look up the logic signal rise time, t_R.
- Determine the propagation velocity in the chosen cable, v. This will be typically about 0.6 of the speed of light.
- Determine the length of the signal path, l.
- Calculate the propagation delay, $\tau = l/v$.
- Calculate the ratio of t_R/τ.
- If the ratio is greater than 8 then the signal path can be

considered electrically short and you will need to consider the signal path's inductance or capacitance, whichever is dominant.

• If the ratio is less than 8 then consider the signal path in terms of a transmission line.

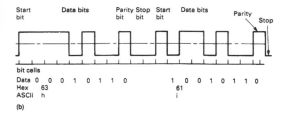

(b)

Figure 4.6 (b) serial transmission requires the sender and receiver to use and recognise the same signal format or protocol, such as RS232. For each byte, the composite signal contains a start bit, a parity bit and a stop bit using inverted logic (1 = –12 V; 0 = –12 V). The time interval between each bit of the signal (the start bit, parity bit, stop bit and data bits) is fixed and must be kept constant.

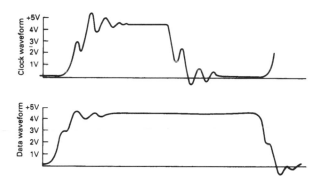

Figure 4.7 A practical problem arose where the data signal was intended to be clocked in using the rising edge of a separate clock line, but excessive ringing on the clock line caused the data line to be sampled twice, causing corruption. In addition, due to the loading of a large number of audio channels the actual logic level no longer achieved the 4.5 V target required for acceptable noise performance, increasing the susceptibility to ringing. The other point to note is that the falling edge of the logic signals took the data line voltage to a negative value, and there is no guarantee that the receiving logic element would not produce an incorrect output as a consequence.

The speed at which logic transitions take place determines the maximum rate at which information can be handled by a logic system. The rise and fall times of a logic signal are important because of the effect on integrity. The outline of the problem is shown in Fig. 4.7 which has been taken from a practical problem in which serial data was being distributed around a large digitally controlled audio mixing desk. Examples such as this illustrate the paradox that digital signals must, in fact, be considered from an analogue point of view.

The Analogue Audio Waveform

It seems appropriate to ensure that there is agreement concerning the meaning attached to words which are freely used. Part of the reason for this is in order that a clear understanding can be obtained into the meaning of phase. The analogue audio signal which we will encounter when it is viewed on an oscilloscope is a causal signal. It is considered as having zero value for negative time and it is also continuous with time. If we observe a few milliseconds of a musical signal we are very likely to observe a waveform which can be seen to have an underlying structure. Fig. 4.8. Striking a single string can produce a waveform which appears to have a relatively simple structure. The waveform resulting from striking a chord is visually more complex though, at any one time, a snapshot of it will show the evidence of structure. From a mathematical or analytical viewpoint the complicated waveform of real sounds are impossibly complex to handle and, instead, the analytical, and indeed the descriptive, process depends on us understanding the principles through the analysis of much simpler waveforms. We rely on the straightforward principle of superposition of waveforms such as the simple cosine wave.

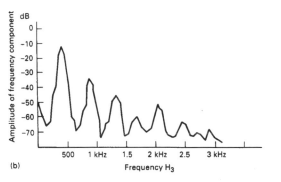

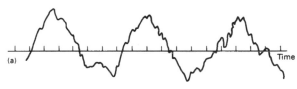

Figure 4.8 (a) An apparently simple noise, such as a single string on a guitar, produces a complicated waveform, sensed in terms of pitch. The important part of this waveform is the basic period of the waveform; its fundamental frequency. The smaller detail is due to components of higher frequency and lower level. (b) An alternative description is analysis into the major frequency components. If processing accuracy is adequate then the description in terms of amplitudes of harmonics (frequency domain) is identical to the description in terms of amplitude and time (time domain).

On its own an isolated cosine wave, or real signal, has no phase. However from a mathematical point of view the apparently simple cosine wave signal which we consider as a stimulus to an electronic system can be considered more properly as a complex wave or function which is accompanied by a similarly shaped sine wave, Fig. 4.9. It is worthwhile throwing out an equation at this point to illustrate this:

$$f(t) = \mathrm{Re}\, f(t) + \mathrm{j}\, \mathrm{Im}\, f(t)$$

where $f(t)$ is a function of time, t which is composed of Re $f(t)$, the real part of the function and j Im $f(t)$, the imaginary part of the function and j is $\sqrt{-1}$.

It is the emergence of the $\sqrt{-1}$ which is the useful part here because you may recall that the analysis of the simple analogue circuit, Fig. 4.10 involving resistors and capacitors produces an expression for the attenuation and the phase relationship between input and output of that circuit which is achieved with the help of $\sqrt{-1}$.

The process which we refer to glibly as the Fourier transform considers that all waveforms can be considered as constructed from a series of sinusoidal waves of the appropriate amplitude and phase added linearly. A continuous sine wave will need to exist for all time in order that its representation in the frequency domain will consist of only a single frequency. The reverse side, or dual, of this observation is that a singular event, for

example an isolated transient, must be composed of all frequencies. This trade-off between the resolution of an event in time and the resolution of its frequency components is fundamental. You could think of it as if it were an uncertainty principle.

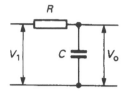

Figure 4.10   The simple resistor and capacitor attenuator can be analysed to provide us with an expression for the output voltage and the output phase with respect to the input signal.

The reason for discussing phase at this point is that the topic of digital audio uses terms such as linear phase, minimum phase, group delay and group delay error. A rigorous treatment of these topics is outside the brief for this chapter but it is necessary to describe them. A system has linear phase if the relationship between phase and frequency is a linear one. Over the range of frequencies for which this relationship may hold the system output is effectively subjected to a constant time delay with respect to its input. As a simple example consider that a linear phase system which exhibits –180 degrees of phase shift at 1 kHz will show –360 degrees of shift at 2 kHz.

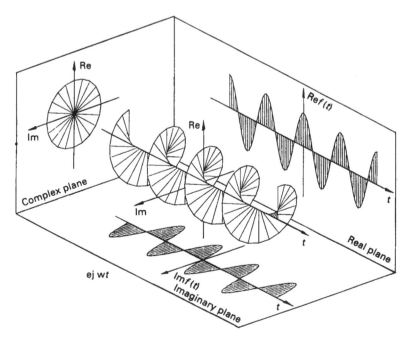

Figure 4.9   The relationship between the cosine (real) and the sine (imaginary) waveforms in the complex exponential e_{jwt}. This assists in understanding the concept of phase. Note that one property of the spiral form is that its projection onto any plane which is parallel to the time axis will produce a sinusoidal waveform.

From an auditive point of view a linear phase performance should preserve the waveform of the input and thus be benign to an audio signal.

Most of the common analogue audio processing systems such as equalisers exhibit minimum phase behaviour. Individual frequency components spend the minimum necessary time being processed within the system. Thus some frequency components of a complex signal may appear at the output at a slightly different time with respect to others. Such behaviour can produce gross waveform distortion as might be imagined if a 2 kHz component were to emerge 2 ms later than a 1 kHz signal. In most simple circuits, such as mixing desk equalisers, the output phase of a signal with respect to the input signal is usually the ineluctable consequence of the equaliser action. However, for reasons which we will come to, the process of digitising audio can require special filters whose phase response may be responsible for audible defects.

One conceptual problem remains. Up to this point we have given examples in which the output phase has been given a negative value. This is comfortable territory because such phase lag is readily converted to time delay. No causal signal can emerge from a system until it has been input otherwise our concept of the inviolable physical direction of time is broken. Thus all practical systems must exhibit delay. Systems which produce phase lead cannot actually produce an output which, in terms of time, is in advance of its input. Part of the problem is caused by the way we may measure the phase difference between input and output. This is commonly achieved using a dual channel oscilloscope and observing the input and output waveforms. The phase difference is readily observed and it can be readily shown to match calculations such as that given in Fig. 4.10. The point is that the test signal has essentially taken on the characteristics of a signal which has existed for an infinitely long time exactly as it is required to do in order that our use of the relevant arithmetic is valid. This arithmetic tacitly invokes the concept of a complex signal, that is one which for mathematical purposes is considered to have real and imaginary parts, see Fig. 4.9. This invocation of phase is intimately involved in the process of composing, or decomposing, a signal by using the Fourier series. A more physical appreciation of the response can be obtained by observing the system response to an impulse.

Since the use of the idea of phase is much abused in audio at the present time it may be worthwhile introducing a more useful concept. We have referred to linear phase systems as exhibiting simple delay. An alternative term to use would be to describe the system as exhibiting a uniform (or constant) group delay over the relevant band of audio frequencies. Potentially audible problems start to exist when the group delay is not constant but changes with frequency. The deviation from a fixed delay value is called group delay error and it can be quoted in ms.

The process of building up a signal using the Fourier series produces a few useful insights, Fig. 4.11(a). The classic example is that of the square wave and it is shown here as the sum of the fundamental, third and fifth harmonics. It is worth noting that the 'ringing' on the waveform is simply the consequence of band-limiting a square wave and that simple, minimum phase, systems will produce an output rather like Fig. 4.11(b) and there is not much evidence to show that the effect is audible. The concept of building up complex waveshapes from simple components is used in calculating the shape of tidal heights. The accuracy of the shape is dependent on the number of components which we incorporate and the process can yield a complex wave shape with only a relatively small number of components. We see here the germ of the idea which will lead to one of the methods available for achieving data reduction for digitised analogue signals.

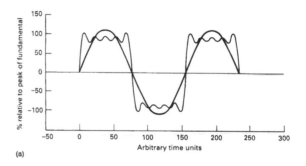

(a)

Figure 4.11 (a) Composing a square wave from the harmonics is an elementary example of a Fourier series. For the square wave of unit amplitude the series is of the form:

$$4/\pi[sin\omega + 1/3sin3\omega + 1/5sin5\omega + 1/7sin7\omega....]$$

where $\omega = 2\pi f$, the angular frequency.

The composite square wave has ripples in its shape, due to band limiting, since this example uses only the first four terms, up to the seventh harmonic. For a signal which has been limited to an audio bandwidth of approximately 21 kHz this square wave must be considered as giving an ideal response even though the fundamental is only 3 kHz. The 9% overshoot followed by a 5% undershoot, the Gibbs phenomenon, will occur whenever a Fourier series is truncated or a bandwidth is limited.

Instead of sending a stream of numbers which describe the waveshape at each regularly spaced point in time we first analyse the waveshape into its constituent frequency components and then send (or store) a description of the frequency components. At the receiving end these numbers are unravelled and, after some calculation the waveshape is re-constituted. Of course this requires that both the sender and the receiver of the information know how to process it. Thus the receiver will attempt to apply the inverse, or opposite, process to that applied during coding at the sending end. In the extreme it is possible to encode a complete Beethoven symphony in a single 8-bit byte. Firstly we must equip both ends of our communication link with the same set of raw data, in this case a collection of CD's containing recordings of Beethoven's work. We then send the number of the disc which contains the recording which we wish to 'send'. At the receiving end the decoding process uses the received byte of information, selects the disc and plays it. Perfect reproduction using only one byte to encode 64 minutes of stereo recorded music . . . and to CD quality!

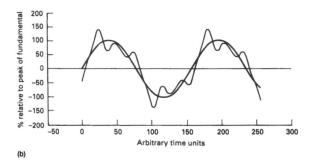

(b)

Figure 4.11 (b) In practice, a truly symmetrical shape is rare since most practical methods of limiting the audio bandwidth do not exhibit linear phase but, delay progressively the higher frequency components. Band limiting filters respond to excitation by a square wave by revealing the effect of the absence of higher harmonics and the so-called 'ringing' is thus not necessarily the result of potentially unstable filters.

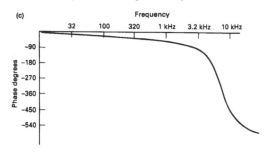

Figure 4.11 (c) The phase response shows the kind of relative phase shift which might be responsible.

A very useful signal is the impulse. Figure 4.12 shows an isolated pulse and its attendant spectrum. Of equal value is the waveform of the signal which provides a uniform spectrum. Note how similar these waveshapes are. Indeed if we had chosen to show in Fig. 4.12(a) an isolated square-edged pulse then the pictures would be identical save that the references to the time and frequency domains would need to be swapped. You will encounter these waveshapes in diverse fields such as video and in the spectral shaping of digital data waveforms. One important advantage of shaping signals in this way is that since the spectral

bandwidth is better controlled so the effect of the phase response of a bandlimited transmission path on the waveform is also limited. This will result in a waveform which is much easier to restore to clean 'square' waves at the receiving end.

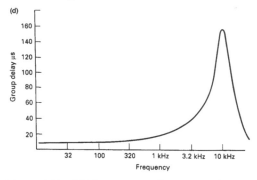

Figure 4.11 (d) The corresponding group delay curve shows a system which reaches a peak delay of around 160 µs.

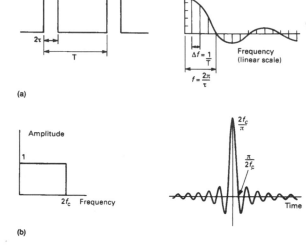

Figure 4.12 (a) A pulse with a period of 2π seconds is repeated every T seconds, producing the spectrum as shown. The spectrum appears as having negative amplitudes since alternate 'lobes' have the phase of their frequency components inverted, although it is usual to show the modulus of the amplitude as positive and to reflect the inversion by an accompanying plot of phase against frequency. The shape of the lobes is described by the simple relationship:

$$A = k(\sin x)/x$$

(b) A further example of the duality between time and frequency showing that a widely spread spectrum will be the result of a narrow pulse. The sum of the energy must be the same for each so that we would expect a narrow pulse to be of a large amplitude if it is to carry much energy. If we were to use such a pulse as a test signal we would discover that the individual amplitude of any individual frequency component would be quite small. Thus when we do use this signal for just this purpose we will usually arrange to average the results of a number of tests.

Arithmetic

We have seen how the process of counting in binary is carried out. Operations using the number base of 2 are characterised by a number of useful tricks which are often used. Simple counting demonstrates the process of addition and, at first sight, the process of subtraction would need to be simply the inverse operation. However, since we need negative numbers in order to describe the amplitude of the negative polarity of a waveform, it seems sensible to use a coding scheme in which the negative number can be used directly to perform subtraction. The appropriate coding scheme is the two's complement coding scheme. We can convert from a simple binary count to a two's complement value very simply. For positive numbers simply ensure that the MSB is a zero. To make a positive number into a negative one first invert each bit and then add one to the LSB position thus:

$$+9_{10} > 1001_2 > 01001_{2c} \text{ (using a 5 bit word length)}$$
$$-9_{10} > -01001_2 > 10110 + 1_{\text{invert and add 1}} > 10111_{2c}$$

We must recognise that since we have fixed the number of bits which we can use in each word (in this example to 5 bits) then we are naturally limited to the range of numbers we can represent (in this case from +15 through 0 to −16). Although the process of forming two's complement numbers seems lengthy it is very speedily performed in hardware. Forming a positive number from a negative one uses the identical process. If the binary numbers represent an analogue waveform then changing the sign of the numbers is identical to inverting the polarity of the signal in the analogue domain. Examples of simple arithmetic should make this a bit more clear:

decimal, base 10	binary, base 2	2's complement
addition		
12	01100	01100
+ 3	+ 00011	+ 00011
= 15	= 01111	= 01111
subtraction		
12	01100	01100
− 3	− 00011	+ 11101
= 9		= 01001

Since we have only a 5 bit word length any overflow into the column after the MSB needs to be handled. The rule is that if there is overflow when a positive and a negative number are added then it can be disregarded. When overflow results during the addition of two positive numbers or two negative numbers then the resulting answer will be incorrect if the overflowing bit is neglected. This requires special handling in signal processing, one

approach being to set the result of an overflowing sum to the appropriate largest positive or negative number. The process of adding two sequences of numbers which represent two audio waveforms is identical to that of mixing the two waveforms in the analogue domain. Thus when the addition process results in overflow the effect is identical to the resulting mixed analogue waveform being clipped.

We see here the effect of word length on the resolution of the signal and, in general, when a binary word containing n bits is added to a larger binary word comprising m bits the resulting word length will require $m + 1$ bits in order to be represented without the effects of overflow. We can recognise the equivalent of this in the analogue domain where we know that the addition of a signal with a peak to peak amplitude of 3 V to one of 7 V must result in a signal whose peak to peak value is 10 V. Don't be confused about the rms value of the resulting signal which will be:

$$\frac{\sqrt{3^2 + 7^2}}{2.82} = 2.69 \text{ V}_{\text{rms}}$$

assuming uncorrelated sinusoidal signals.

A binary adding circuit is readily constructed from the simple gates referred to above and Fig. 4.13 shows a two bit full adder. More logic is needed to be able to accommodate wider binary words and to handle the overflow (and underflow) exceptions.

If addition is the equivalent of analogue mixing then multiplication will be the equivalent of amplitude or gain change. Binary multiplication is simplified by only having 1 and 0 available since $1 \times 1 = 1$ and $1 \times 0 = 0$.

Since each bit position represents a power of 2 then shifting the pattern of bits one place to the left (and filling in the vacant space with a 0) is identical to multiplication by 2. The opposite is, of course, true of division. The process can be appreciated by an example:

decimal	binary
3	00011
× 5	00101
= 15	00011
+	00000
+	00011
+	00000
+	00000
=	000001111

and another example:

12	01100
×13	01101
= 156	= 010011100

The process of shifting and adding could be programmed in a series of program steps and executed by a microprocessor but this would take too long. Fast multipliers work by arranging that all of the available shifted combinations of one of the input numbers are made available to a large array of adders whilst the other input number is used to determine which of the shifted combinations will be added to make the final sum. The resulting word width of a multiplication equals the sum of both input word widths. Further, we will need to recognise where the binary point is intended to be and arrange to shift the output word appropriately. Quite naturally the surrounding logic circuitry will have been designed to accommodate a restricted word width. Repeated multiplication must force the output word width to be limited. However limiting the word width has a direct impact on the accuracy of the final result of the arithmetic operation. This curtailment of accuracy is cumulative since subsequent arithmetic operations can have no knowledge that the numbers being processed have been 'damaged'.

Two techniques are important in minimising the 'damage'. The first requires us to maintain the intermediate stages of any arithmetic operation at as high an accuracy as possible for as long as possible. Thus although most conversion from analogue audio to digital (and the converse digital signal conversion to an analogue audio signal) takes place using 16 bits the intervening arithmetic operations will usually involve a minimum of 24 bits.

The second technique is called **dither** and we will cover this fully later. Consider, for the present, that the output word width is simply cut (in our example above such as to produce a 5 bit answer). The need to handle the large numbers which result from multiplication without overflow means that when small values are multiplied they are likely to lie outside the range of values which can be expressed by the chosen word width. In the example above if we wish to accommodate the most significant digits of the second multiplication (156) as possible in a 5 bit word then we shall need to lose the information contained in the lower four binary places. This can be accomplished by shifting the word four places (thus effectively dividing the result by 16) to the right and losing the less significant bits. In this example the result becomes 01001 which is equivalent to decimal 9. This is clearly only approximately equal to 156/16.

When this crude process is carried out on a sequence of numbers representing an audio analogue signal the error results in an unacceptable increase in the signal to noise ratio. This loss of accuracy becomes extreme when we apply the same adjustment to the lesser product of $3 \times 5 = 15$ since, after shifting four binary places, the result is zero. Truncation is thus a rather poor way of handling the restricted output word width. A slightly better approach is to round up the output by adding a fraction to each output number just prior to truncation. If we added 00000110, then shifted four places and truncated the output would become 01010 (= 1010) which, although not absolutely accurate is actually closer to the true answer of

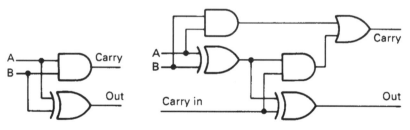

Data A	Bits B	Carry in	Out	Carry out
0	0	0	0	0
0	0	1	1	0
0	1	0	1	0
0	1	1	0	1
1	0	0	1	0
1	0	1	0	1
1	1	0	0	1
1	1	1	1	1

Figure 4.13   A two bit full adder needs to be able to handle a carry bit from an adder handling lower order bits and similarly provide a carry bit. A large adder based on this circuit would suffer from the ripple through of the carry bit as the final sum would not be stable until this had settled. Faster adding circuitry uses look-ahead carry circuitry.

9.75. This approach moves the statistical value of the error from 0 to –1 towards +/–0.5 of the value of the LSB but the error signal which this represents is still very highly correlated to the required signal. This close relationship between noise and signal produces an audibly distinct noise which is unpleasant to listen to.

An advantage is gained when the fraction which is added prior to truncation is not fixed but random. The audibility of the result is dependent on the way in which the random number is derived. At first sight it does seem daft to add a random signal (which is obviously a form of noise) to a signal which we wish to retain as clean as possible. Thus the probability and spectral density characteristics of the added noise are important. A recommended approach commonly used is to add a random signal which has a triangular probability density function (TPDF), Fig. 4.14. Where there is sufficient reserve of processing power it is possible to filter the noise before adding it in. This spectral shaping is used to modify the spectrum of the resulting noise (which you must recall is an error signal) such that it is biased to those parts of the audio spectrum where it is least audible. The mathematics of this process are beyond this text.

A special problem exists where gain controls are emulated by multiplication. A digital audio mixing desk will usually have its signal levels controlled by digitising the position of a physical analogue fader (certainly not the only way by which to do this, incidentally). Movement of the fader results in a stepwise change of the multiplier value used. When such a fader is moved any music signal being processed at the time is subjected to stepwise changes in level. Although small, the steps will result in audible interference unless the changes which they represent are themselves subjected to the addition of dither. Thus although the addition of a dither signal reduces the correlation of the error signal to the program signal it must, naturally, add to the noise of the signal. This reinforces the need to ensure that the digitised audio signal remains within the processing circuitry with as high a precision as possible for as long as possible. Each time that the audio signal has its precision reduced it inevitably must become noisier.

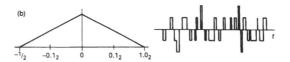

Figure 4.14 (b) The addition of two uncorrelated RPDF sequences gives rise to one with triangular distribution (TPDF). When this dither signal is added to a digitised signal it will always mark the output with some noise since there is a finite possibility that the noise will have a value greater than 0, so that as a digitised audio signal fades to zero value the noise background remains fairly constant. This behaviour should be contrasted with that of RPDF for which, when the signal fades to zero, there will come a point at which the accompanying noise also switches off. This latter effect may be audible in some circumstances and is better avoided. A wave form associated with this type of distribution will have values which range from $+1.0_2$ through 0_2 to -1.0_2.

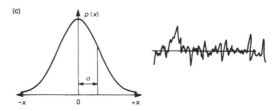

Figure 4.14 (c) Noise in the analog domain is often assumed to have a Gaussian distribution. This can be understood as the likelihood of the waveform having a particular amplitude. The probability of an amplitude x occurring can be expressed as:

$$\rho(x) = \frac{e^{\dfrac{-(\chi - \mu)^2}{2\sigma^2}}}{\sigma\sqrt{2\pi}}$$

where μ = the mean value
       σ = variance
       X = the sum of the squares of the deviations x from the mean.

In practice, a 'random' waveform which has a ratio between the peak to mean signal levels of 3 can be taken as being sufficiently Gaussian in character. The spectral balance of such a signal is a further factor which must be taken into account if a full description of a random signal is to be described.

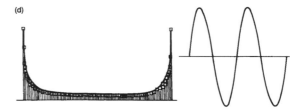

Figure 4.14 (d) The sinusoidal wave form can be described by the simple equation:

$$x(t) = A \sin(2\pi ft)$$

where $x(t)$ = the value of the sinusoidal wave at time t
       A = the peak amplitude of the waveform
       f = the frequency in Hz and
       t = the time in seconds
and its probability density function is as shown here.

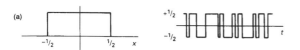

Figure 4.14 (a) The amplitude distribution characteristics of noise can be described in terms of the amplitude probability distribution characteristic. A square wave of level 0 V or +5 V can be described as having a rectangular probability distribution function (RPDF). In the case of the 5 bit example which we are using the RPDF wave form can be considered to have a value of $+0.1_2$ or -0.1_2 [(meaning +0.5 or –0.5), equal chances of being positive or negative.]

(e)

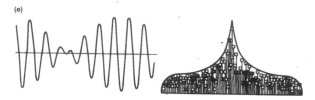

Figure 4.14 (e) A useful test signal is created when two sinusoidal waveforms of the same amplitude but unrelated in frequency are added together. The resulting signal can be used to check amplifier and system non-linearity over a wide range of frequencies. The test signal will comprise two signals to stimulate the audio system (for testing at the edge of the band 19 kHz and 20 kHz can be used) whilst the output spectrum is analysed and the amplitude of the sum and difference frequency signals is measured. This form of test is considerably more useful than a THD test.

Digital Filtering

Although it may be clear that the multiplication process controls signal level it is not immediately obvious that the multiplicative process is intrinsic to any form of filtering. Thus multipliers are at the heart of any significant digital signal processing and modern digital signal processing (DSP) would not be possible without the availability of suitable IC technology. You will need to accept, at this stage, that the process of representing an analogue audio signal in the form of a sequence of numbers is readily achieved and thus we are free to consider how the equivalent analogue processes of filtering and equalisation may be carried out on the digitised form of the signal.

In fact the processes required to perform digital filtering are performed daily by many people without giving the process much thought. Consider the waveform of the tidal height curve of Fig. 4.15. The crude method by which we obtained this curve (Fig. 4.1) contained only an approximate method for removing the effect of ripples in the water by including a simple dashpot linked to the recording mechanism. If we were to look at this trace more closely we would see that it was not perfectly smooth due to local effects such as passing boats and wind-driven waves. Of course tidal heights do not normally increase by 100 mm within a few seconds and so it is sensible to draw a line which passes through the average of these disturbances. This averaging process is filtering and, in this case, it is an example of low-pass filtering. To achieve this numerically we could measure the height indicated by the tidal plot each minute and calculate the average height for each four minute span (and this involves measuring the height at five time points).

$$h_{\text{average}} = \frac{1}{5} \sum_{\tau=t}^{\tau=t+4} h_\tau$$

Done simply this would result in a stepped curve which still lacks the smoothness of a simple line. We could reduce the stepped appearance by using a moving average in which we calculate the average height in a four minute span but we move the reference time forward by a single minute each time we perform the calculation. The inclusion of each of the height samples was made without weighting their contribution to the average and this is an example of rectangular windowing. We could go one step further by weighting the contribution which each height makes to the average each time we calculate the average of a four minute period. Shaped windows are common in the field of statistics and they are used in digital signal processing. The choice of window does affect the result, although as it happens the effect is slight in the example we have given here.

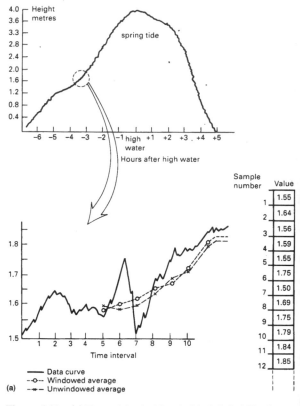

(a)

— Data curve
--o-- Windowed average
--+-- Unwindowed average

Figure 4.15 (a) To explain the ideas behind digital filtering we review the shape of the tidal height curve (Portsmouth, UK, spring tides) for its underlying detail. The pen Plotter trace would record also every passing wave, boat and breath of wind; all overlaid on the general shape of the curve.

One major practical problem with implementing practical FIR filters for digital audio signals is that controlling the response accurately or at low frequencies forces the number of stages to be very high. You can appreciate this through recognising that the FIR (finite impulse response) filter response is determined by the number and value of the coefficients which are applied to each of the taps in the delayed signal stages. The value of these coefficients is an exact copy of the filter's impulse response. Thus an impulse response which is intended to be effective at low frequencies is likely to require a great many stages. This places pressure on the hardware which has to satisfy the demand to perform the necessary large number of multiplications within the time allotted for processing each sample value. In many situations a sufficiently accurate response can be obtained with less circuitry by feeding part of a filter's output back to the input.

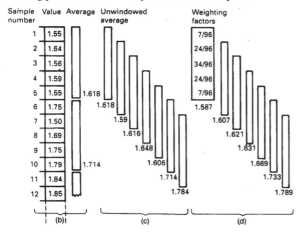

Sample number	Value	Average	Unwindowed average		Weighting factors
1	1.55				7/96
2	1.64				24/96
3	1.56				34/96
4	1.59				24/96
5	1.55	1.618			7/96
6	1.75	1.618			1.587
7	1.50		1.59		1.607
8	1.69		1.616		1.621
9	1.75		1.648		1.631
10	1.79	1.714	1.606		1.669
11	1.84		1.714		1.733
12	1.85		1.784		1.789

(b)   (c)   (d)

Figure 4.15 (b) For a small portion of the curve, make measurements at each interval. In the simplest averaging scheme we take a block of five values, average them and then repeat the process with a fresh block of five values. This yields a relatively coarse stepped waveform. (c) The next approach carries out the averaging over a block of five samples but shifts the start of each block only one sample on at a time, still allowing each of the five sample values to contribute equally to the average each time. The result is a more finely structured plot which could serve our purpose. (d) The final frame in this sequence repeats the operations of (c) except that the contribution which each sample value makes to the averaging process is weighted, using a five-element weighting filter or window for this example whose weighting values are derived by a modified form of least squares averaging. The values which it returns are naturally slightly different from those of (c).

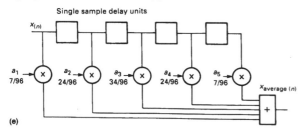

(e)

Figure 4.15 (e) A useful way of showing the process which is

being carried out in (d) is to draw a block diagram in which each time that a sample value is read it is loaded into a form of memory whilst the previous value is moved on to the next memory stage. We take the current value of the input sample and the output of each of these memory stages and multiply them by the weighting factor before summing them to produce the output average. The operation can also be expressed in an algebraic form in which the numerical values of the weighting coefficients have been replaced by an algebraic symbol:

$$x_{average}n = (a_1 x_{n-1} + a_2 x_{n-2} + a_3 x_{n-3} + a_4 x_{n-4}).$$

This is a simple form of a type of digital filter known as a finite impulse response (FIR) or transversal filter. In the form shown here it is easy to see that the delay of the filter is constant and thus the filter will show linear phase characteristics.

If the input to the filter is an impulse, the values you should obtain at the output are identical, in shape, to the profile of the weighting values used. This is a useful property which can be used in the design of filters since it illustrates the principle that the characteristics of a system can be determined by applying an impulse to it and observing the resultant output.

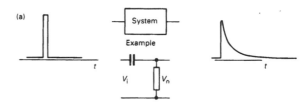

Figure 4.16 (a) An impulse is applied to a simple system whose output is a simple exponential decaying response:

$$V_o = V_i e^{-t/RC}$$

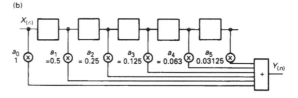

Figure 4.16 (b) A digital filter based on a FIR structure would need to be implemented as shown. The accuracy of this filter depends on just how many stages of delay and multiplication we can afford to use. For the five stages shown the filter will cease to emulate an exponential decay after only 24 dB of decay. The response to successive n samples is:

$$Y_n = 1X_n + (1/2)X_{n-1} + (1/4)X_{n-2} + (1/8)X_{n-3} + (1/16)X_{n-4}$$

We have drawn examples of digital filtering without explicit reference to their use in digital audio. The reason is that the principles of digital signal processing hold true no matter what the origin or use of the signal which is being processed. The ready accessibility of analogue audio 'cook-books' and the simplicity of the signal structure has drawn a number of less than structured practitioners into the field. For whatever inaccessible reason, these practitioners have given the audio engineer the peculiar notions of directional copper cables, specially coloured capacitors and compact discs outlined with green felt tip pens. Bless them all, they have their place as entertainment and they remain free in their pursuit of the arts of the audio witch-doctor. The world of digital processing requires more rigour in its approach and practice. Figure 4.17 shows examples of simple forms of first and second-order filter structures. Processing an audio signal in the digital domain can provide a flexibility which analogue processing denies. You may notice from the examples how readily the characteristics of a filter can be changed simply by adjustment of the coefficients which are used in the multiplication process. The equivalent analogue process would require much switching and component matching. Moreover each digital filter or process will provide exactly the same performance for a given set of coefficient values and this is a far cry from the miasma of tolerance problems which beset the analogue designer.

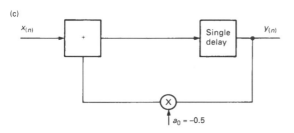

Figure 4.16   (c) This simple function can be emulated by using a single multiplier and adder element if some of the output signal is fed back and subtracted from the input. The use of a multiplier in conjunction with an adder is often referred to as a multiplier-accumulator or MAC. With the correct choice of coefficient in the feedback path the exponential decay response can be exactly emulated:

$$Y_n = X_n - 0.5Y_{n-1}$$

This form of filter will continue to produce a response for ever unless the arithmetic elements are no longer able to handle the decreasing size of the numbers involved. For this reason it is known as an infinite impulse response (IIR) filter or, because of the feedback structure, a recursive filter. Whereas the response characteristics of FIR filters can be comparatively easily gleaned by inspecting the values of the coefficients used the same is not true of IIR filters. A more complex algebra is needed in order to help in the design and analysis which we will not cover here.

The complicated actions of digital audio equalisation are an obvious candidate for implementation using IIR (infinite impulse response) filters and the field has been heavily researched in recent years. Much research has been directed towards overcoming some of the practical problems such as limited arithmetic resolution or precision and limited processing time. Practical hardware considerations force the resulting precision of any digital arithmetic operation to be limited. The limited precision also affects the choice of values for the coefficients whose value will determine the characteristics of a filter. This

limited precision is effectively a processing error and this will make its presence known through the addition of noise to the output. The limited precision also leads to the odd effects for which there is no direct analogue equivalent such as limit cycle oscillations. The details concerning the structure of a digital filter have a very strong effect on the sensitivity of the filter to noise and accuracy in addition to the varying processing resource requirement. The best structure thus depends a little on the processing task which is required to be carried out. The skill of the engineer is, as ever, in balancing the factors in order to optimise the necessary compromises.

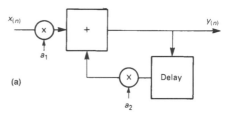

Figure 4.17   (a) The equivalent of the analogue first-order high-and low-pass filters requires a single delay element. Multipliers are used to scale the input (or output) values so that they lie within the linear range of the hardware. Digital filter characteristic are quite sensitive to the values of the coefficients used in the multipliers. The output sequence can be described as:

$$Y_n = a_1 X_n + a_2 X_{n-1}$$

If $0 > a_2 > -1$ the structure behaves as a first-order lag. If $a_2 > 0$ than the structure produces an integrator. The output can usually be kept in range by ensuring that $a_1 = 1 - a_2$

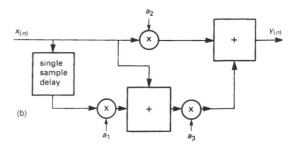

Figure 4.17   (b) The arrangement for achieving high-pass filtering and differentiation again requires a single delay element. The output sequence is given by:

$$Y_n = a_2 X_n + a_3(a_1 X_{n-1} + X_n)$$

The filter has no feedback path so it will always be stable. Note that $a_1 = -1$ and with $a_2 = 0$ and $a_3 = 1$ the structure behaves as a differentiator. These are simple examples of first-order structures and they are not necessarily the most efficient in terms of their use of multiplier or adder resources. Although a second-order system would result if two first-order structures were run in tandem full flexibility of second-order IIR structures requires recursive structures. Perhaps the most common of these emulates the analogue biquad (or twin integrator loop) filter.

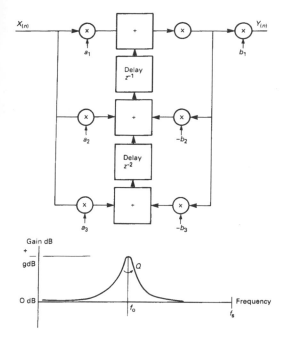

Figure 4.17 (c) To achieve the flexibility of signal control which analogue equalisers exhibit in conventional mixing desks an IIR filter can be used. Shown here it requires two single-sample value delay elements and six multiplying operations each time it is presented with an input sample value. We have symbolised the delay elements by using the z^{-1} notation which is used when digital filter structures are formally analysed. The output sequence can be expressed as:

$$b_1 Y_n = a_1 X_n + a_2 X_{n-1} - b_2 Y_{n-1} + a_3 X_{n-2} - b_3 Y_{n-2}$$

The use of z^{-1} notation allows us to express this difference or recurrence equation as:

$$b_1 Y_n z^{-0} = a_1 X_n z^{-0} + a_2 X_n z^{-1} - b_2 Y_n z^{-1} + a_3 X_n z^{-2} - b_3 Y_n z^{-2}$$

The transfer function of the structure is the ratio of the output over the input, just as it is in the case of an analogue system. In this case the input and output happen to be sequences of numbers and the transfer function is indicated by the notation H(z):

$$H(z) = \frac{Y(z)}{X(z)} = \frac{a_1 + a_2 z^{-1} + a_3 z^{-2}}{b_1 + b_2 z^{-1} + b_3 z^{-2}}$$

The value of each of the coefficients can be determined from a knowledge of the rate at which samples are being made available F_s and your requirement for the amount of cut or boost and of the Q required. One of the first operations is that of pre-warping the value of the intended centre frequency f_c in order to take account of the fact that the intended equaliser centre frequency is going to be comparable to the sampling frequency. The 'warped' frequency is given by:

$$f_w = F_s/\pi \tan \pi f_c/F_s$$

And now for the coefficients:

$$a_1 = 1 + \pi f_w (1+k)/F_s Q + (\pi f_w/F_s)^2$$
$$a_2 = b_2 = -2(1 + (\pi f_w/F_s)^2)$$
$$a_3 = 1 - \pi f_w (1+k)/F_s Q + (\pi f_w/F_s)^2$$
$$b_1 = 1 + \pi f_w/F_s Q + (\pi f_w/F_s)^2$$
$$b_3 = 1 - \pi f_w/F_s Q + (\pi f_w/F_s)^2$$

The mathematics concerned with filter design certainly appear more complicated than that which is usually associated with analogue equaliser design. The complication does not stop here though since a designer must take into consideration the various compromises brought on by limitations in cost and hardware performance.

While complicated filtering is undoubtedly used in digital audio signal processing spare a thought for the simple process of averaging. In digital signal processing terms this is usually called interpolation, Fig. 4.18. The process is used to conceal unrecoverable errors in a sequence of digital sample values and, for example, is used in the compact disc for just this reason.

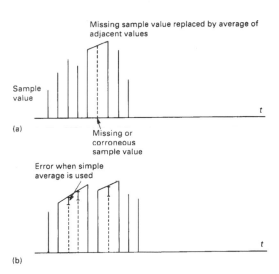

Figure 4.18 (a) Interpolation involves guessing the value of the missing sample. The fastest guess uses the average of the two adjacent good sample values, but an average based on many more sample values might provide a better answer. The use of a simple rectangular window for including the values to be sampled will not lead to as accurate a replacement value. The effect is similar to that caused by examining the spectrum of a continuous signal which has been selected using a simple rectangular window. The sharp edges of the window function will have frequency components which cannot be separated from the wanted signal. (b) A more intelligent interpolation uses a shaped window which can be implemented as a FIR, or transversal, filter with a number of delay stages each contributing a specific fraction of the sample value of the output sum. This kind of filter is less likely than the simple linear average process to create audible transitory aliases as it fills in damaged sample values.

Other Binary Operations

One useful area of digital activity is related to filtering and its activity can be described by a similar algebra. The technique uses a shift register whose output is fed back and combined with the incoming logic signal. Feedback is usually arranged to come from earlier stages as well as the final output stage. The arrangement can be considered as a form of binary division. For certain combinations of feedback the output of the shift register can be considered as a statistically dependable source of random numbers. In the simplest form the random output can be formed in the analogue domain by the simple addition of a suitable low-pass filter. Such a random noise generator has the useful property that the noise waveform is repeated and this allows the results of stimulating a system with such a signal to be averaged.

When a sequence of samples with a nominally random distribution of values is correlated with itself the result is identical to a band-limited impulse, Fig. 4.19. If such a random signal is used to stimulate a system (this could be an artificial reverberation device, an equaliser or a loudspeaker under test) and the resulting output is correlated with the input sequence the result will be the impulse response of the system under test. The virtue of using a repeatable excitation noise is that measurements can be made in the presence of other random background noise or interference and if further accuracy is required the measurement is simply repeated and the results averaged. True random background noise will average out leaving a 'cleaner' sequence of sample values which describe the impulse response. This is the basis behind practical measurement systems.

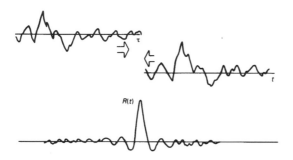

Figure 4.19   Correlation is a process in which one sequence of sample values is checked against another to see just how similar both sequences are. A sinusoidal wave correlated with itself (a process called auto correlation) will produce a similar sinusoidal wave. By comparison a sequence of random sample values will have an autocorrelation function which will be zero everywhere except at the point where the samples are exactly in phase, yielding a band-limited impulse.

Shift registers combined with feedback are also used in error detecting and correction systems and the reader is

referred to other chapters for their implementation and use.

Sampling and Quantising

It is not possible to introduce each element of this broad topic without requiring the reader to have some foreknowledge of future topics. The above text has tacitly admitted that you will wish to match the description of the processes involved to a digitised audio signal although we have pointed out that handling audio signals in the digital domain is only an example of some of the flexibility of digital signal processing.

The process of converting an analogue audio signal into a sequence of sample values requires two key operations. These are sampling and quantisation. They are not the same operation, for while sampling means that we only wish to consider the value of a signal at a fixed point in time the act of quantising collapses a group of amplitudes to one of a set of unique values. Changes in the analogue signal between sample points are ignored. For both of these processes the practical deviations from the ideal process are reflected in different ways in the errors of conversion.

Successful sampling depends on ensuring that the signal is sampled at a frequency at least twice that of the highest frequency component. This is Nyquist's sampling theorem. Figure 4.20 shows the time domain view of the operation whilst Fig. 4.21 shows the frequency domain view.

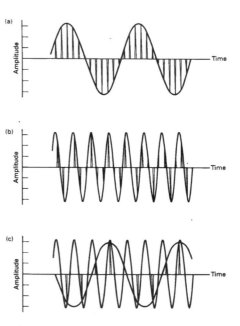

Figure 4.20   (a) In the time domain the process of sampling is like one of using a sequence of pulses, whose amplitude is either 1 or 0 and multiplying it by the value of the sinusoidal

waveform. A sample and hold circuit holds the sampled signal level steady while the amplitude is measured (b) At a higher frequency sampling is taking place approximately three times per sinusoid input cycle. Once more it is possible to see that even by simply joining the sample spikes that the frequency information is still retained. (c) This plot shows the sinusoid being under sample and on reconstituting the original signal from the spikes the best fit sinusoid is the one shown in the dashed line. This new signal will appear as a perfectly proper signal to any subsequent process and there is no method for abstracting such aliases from properly sampled signals. It is necessary to ensure that frequencies greater than half of the sampling frequency F_s are filtered out before the input signal is presented to a sampling circuit. This filter is known as an anti-aliassing filter.

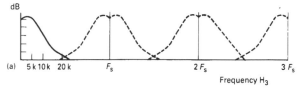

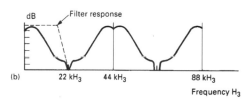

Figure 4.21 (a) The frequency domain view of the sampling operation requires us to recognise that the spectrum of a perfectly shaped sampling pulse continues for ever. In practice sampling waveforms do have finite width and practical systems do have limited bandwidth. We show here the typical spectrum of a musical signal and the repeated spectrum of the sampling pulse using an extended frequency axis. Note that even modern musical signals do not contain significant energy at high frequencies and, for example, it is exceedingly rare to find components in the 10 kHz region more than –30 dB below the peak level. (b) The act of sampling can also be appreciated as a modulation process since the incoming audio signal is being multiplied by the sampling waveform. The modulation will develop sidebands which are reflected either side of the carrier frequency (the sampling waveform), with a set of sidebands for each harmonic of the sampling frequency. The example shows the consequence of sampling an audio bandwidth signal which has frequency components beyond $F_s/2$, causing a small but potentially significant amount of the first lower sideband of the sampling frequency to be folded or aliassed into the intended audio bandwidth. The resulting distortion is not harmonically related to the originating signal and it can sound truly horrid. The use of anti-alias filter before sampling restricts the leakage of the sideband into the audio signal band. The requirement is ideally for a filter with an impossibly sharp rate of cutoff and in practice a small guard band is allowed for tolerance and finite cut off rates. Realising that the level of audio signal with a frequency around 20 kHz is typically 60 dB below the peak signal level it is possible to perform practical filtering using seventh-order filters. However even these filters are expensive to manufacture and they represent a significant design problem in their own right.

Sampling

Practical circuitry for sampling is complicated by the need to engineer ways around the various practical difficulties. The simple form of the circuit is shown in Fig. 4.22. The analogue switch is opened for a very short period t_{ac} each $1/F_s$ seconds. In this short period the capacitor must charge (or discharge) to match the value of the instantaneous input voltage. The buffer amplifier presents this voltage to the input of the quantiser or analogue to digital converter (ADC). There are several problems. The series resistance of the switch sets a limit on how large the storage capacitor can be whilst the input current requirements of the buffer amplifier set a limit on how low the capacitance can be. The imperfections of the switch mean that there can be significant energy leaking from the switching waveform as the switch is operated and there is the problem of crosstalk from the audio signal across the switch when it is opened. The buffer amplifier itself must be capable of responding to a step input and settling to the required accuracy within a small fraction of the overall sample period. The constancy or jitter of the sampling pulse must be kept within very tight tolerances and the switch itself must open and close in exactly the same way each time it is operated. Finally the choice of capacitor material is itself important because certain materials exhibit significant dielectric absorption.

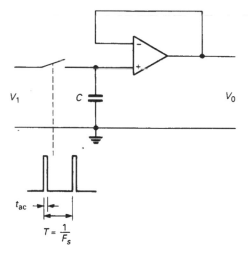

Figure 4.22 An elementary sample and hold circuit using a fast low distortion semiconductor switch which is closed for a short time to allow a small-valued storage capacitor to charge up to the input voltage. The buffer amplifier presents the output to the quantiser.

The overall requirement for accuracy depends greatly on the acceptable signal to noise ratio (SNR) for the process and this is much controlled by the resolution and accuracy of the quantiser or converter. For audio purposes we may assume that suitable values for F_s will be in the range 44 kHz to 48 kHz. The jitter or aperture uncertainty will need to be in the region of 120 picoseconds, acquisition and settling time need to be around 1 µs and the capacitor discharge rate around 1 V/s for a signal which will be quantised to 16 bits if the error due to that cause is not to exceed +/–0.5 LSB. The jitter performance is complex to visualise completely because of the varying amplitude and frequency component of the jitter itself.

At this stage you will need to be sure that you are confident that you appreciate that the combined effect of the anti-alias filtering and the proper implementation of the sampling process mean that the sampled data contains perfectly all of the detail up to the highest frequency component in the signal permitted by the action of the anti-alias filter.

Quantising

The sampled input signal must now be measured. The dynamic range which can be expressed by an n bit number is approximately proportional to 2^n and this is more usually expressed in dB terms. The converters used in test equipment such as DVMs are too slow for audio conversion but it is worthwhile considering the outline of their workings, Fig. 4.23. A much faster approach uses a successive approximation register (SAR) and a digital to analogue converter (DAC), Fig. 4.24.

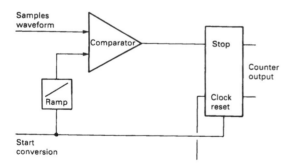

Figure 4.23 The simplest ADC uses a ramp generator which is started at the beginning of conversion. At the same time a counter is reset and the clock pulses are counted. The ramp generator output is compared with the signal from the sample and hold and when the ramp voltage equals the input signal the counter is stopped. Assuming that the ramp is perfectly linear (quite difficult to achieve at high repetition frequencies) the count will be a measure of the input signal. The problem for audio bandwidth conversion is the speed at which the counter must run in order to achieve a conversion within approximately 20 µs. This is around 3.2768 GHz and the comparator would need to be able to change state within 150 ps with, in the worst case, less than 150 µV of differential voltage. There have been many developments of this conversion technique for instrumentation purposes.

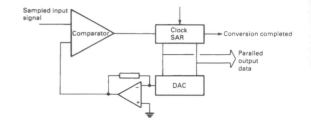

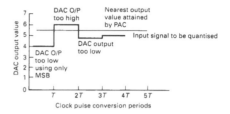

Figure 4.24 The SAR operates with a DAC and a comparator, initially reset to zero. At the first clock period the MSB is set and the resulting output of the DAC is compared to the input level. In the example given here the input level is greater than this and so the MSB value is retained and, at the next clock period, the next MSB is set. In this example the comparator output indicates that the DAC output is too high, the bit is set to 0 and the next lower bit is set. This is carried out until all of the DAC bits have been tried. Thus a 16 bit ADC would require only 17 clock periods (one is needed for reset) in which to carry out a onversion.

A very simple form of DAC is based on switching currents into a virtual earth summing point. The currents are derived from a R–2R ladder which produces binary weighted currents, Fig. 4.25. The incoming binary data directly controls a solid state switch which routes a current either into the virtual earth summing point of the output amplifier or into the local analogue ground. Since the voltage across each of the successive 2R resistors is halved at each stage the current which is switched is also halved. The currents are summed at the input to the output buffer amplifier. The limit to the ultimate resolution and accuracy is determined partly by the accuracy of adjustment and matching of the characteristics of the resistors used and also by the care with which the converter is designed into the surrounding circuitry. Implementation of a 16 bit converter requires that all of the resistors are trimmed to within 0.0007 per cent (half of an LSB) of the ideal value and, further, that they all maintain this ratio over the operational temperature of the converter. The buffer amplifier must be able to settle quickly and it must not contribute to the output noise significantly.

There are many variations on the use of this technique with one common approach being to split the conversion into two stages. Typically a 16 bit converter would have the top 8 most significant bits control a separate conversion stage which sets either the voltage or the current with which the lower 8 LSBs operate. The approach has to contend with the problem of ensuring that the changeover point between the two stages remains matched throughout the environmental range of the converter. One solution to the problem of achieving an accurate binary ratio between successive currents is to use a technique called dynamic element balancing.

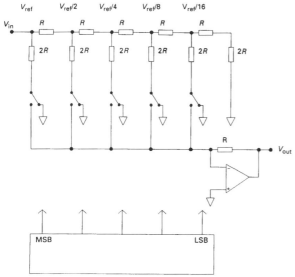

Switch control – digital value to be converted to analogue

Figure 4.25 The basic form of the R–2R digital to analogue converter is shown here implemented by ideal current switches. The reference voltage can be an audio bandwidth signal and the DAC can be used as a true four quadrant multiplying converter to implement digitally controlled analogue level and equalisation changes. Other forms of switching currents are also used and these may not offer a true multiplication facility.

Whereas sampling correctly executed loses no information quantising inevitably produces an error. The level of error is essentially dependent on the resolution with which the quantising is carried out. Figure 4.26 illustrates the point by showing a sinusoid quantised to 16 quantising levels. A comparison of the quantised output with the original has been used to create the plot of the error in the quantising. The error waveform of this example clearly shows a high degree of structure which is strongly related to the signal itself. The error can be referred to as quantising distortion, granulation noise or simply quantising noise.

One obvious non-linearity will occur when the input signal amplitude drops just below the threshold for the first quantising level. At this stage the quantising output will remain at zero and all knowledge of the size of the signal will be lost. The remedy is to add a small amount of noise to the signal prior to quantising, Fig. 4.27. This deliberate additional noise is known as dither noise. It

does reduce the dynamic range by an amount which depends on its exact characteristics and amplitude but the damage is typically 3 dB. One virtue is that as the original input signal amplitude is reduced below the ± 1 quantising level thresholds (q) the noise is still present and therefore, by virtue of the intrinsic non-linearity of the quantiser, so are the sidebands which contain vestiges of the original input signal. Thus the quantiser output must also contain information about the original input signal level even though it is buried in the noise. However, the noise is wide band and a spectral plot of the reconstituted waveform will show an undistorted signal component standing clear of the wide band noise.

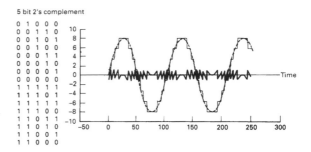

Figure 4.26 The input sinusoid is shown here prior to sampling as a dotted line superimposed on the staircase shape of the quantised input signal. The two's complement value of the level has been shown on the left hand edge. The error signal is the difference between the quantised value and the ideal value assuming a much finer resolution. The error signal, or quantising noise, lies in the range of $\pm 1q$. Consideration of the mean square error leads to the expression for the rms value of the quantising noise:

$$V_{noise} = q/\sqrt{(12)}$$

where q = the size of a quantising level.

The maximum rms signal amplitude which can be described is:

$$V_{signal} = q\, 2^{n-1}/\sqrt{2}$$

Together the expression combine to give the expression for SNR (in dB):

$$SNR_{dB} = 6.02n + 1.76$$

Unfortunately it is also quite a problem to design accurate and stable quantisers. The problems include linearity, missing codes, differential non-linearity and non-monotonicity. Missing codes should be properly considered a fault since they arise when a converter either does not respond to or produce the relevant code. A suitable test is a simple ramp. A powerful alternative test method uses a sinusoidal waveform and checks the amplitude probability density function. Missing, or misplaced codes, and non-monotonic behaviour can show up readily in such a plot.

Linearity can be assessed overall by using a sinusoidal test signal since the output signal will contain harmonics. The performance of an ADC can be carried out in the digital domain by direct use of the discrete fast Fourier transform (DFFT). The DAC can be checked by driving it with a computer-generated sequence and observing the output in the analogue domain. The trouble with using the simple harmonic distortion test is that it is not easy to check the dynamic performance over the last decade of bandwidth and for this reason the CCIR twin tone intermodulation distortion (IMD) is much to be preferred.

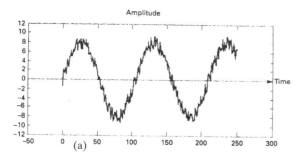

Figure 4.27   (a) Adding a small amount of random noise to the signal prior to quantising can help to disguise the otherwise highly correlated quantising noise. Aided by the binary modulation action of the quantiser the sidebands of the noise are spread across the whole audio band width and to a very great degree their correlation with the original distortion signal is broken up. In this illustration the peak to peak amplitude of the noise has been set at ±1.5q.

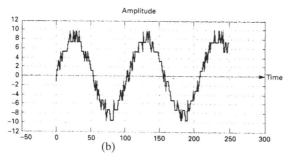

Figure 4.27   (b) The quantiser maps the noisy signal onto one of a range of unique levels as before.

Differential non-linearity is the random unevenness of each quantisation level. This defect can be assessed by measuring the noise floor in the presence of a signal. In a good DAC the rms noise floor should be approximately 95 dB below the maximum rms level (assuming a modest margin for dither). The output buffer amplifier will contribute some noise but this should be at a fixed level and not dependent on the DAC input sample values.

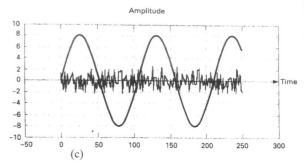

Figure 4.27   (c) You can compare the resulting quantising noise with the original signal and this time you can see that the noise waveform has lost the highly structured relationship which is shown in Figure 4.26.

The basic ADC element simply provides an output dependent on the value of the digital input. During the period whilst a fresh sample is settling its output can be indeterminate. Thus the output will usually be captured by a sample and hold circuit as soon as the DAC has stabilised. The sample and hold circuit is a zero order hold circuit which imposes its own frequency response on the output signal, Fig. 4.28, correction for which can be accommodated within the overall reconstruction filter. The final filter is used to remove the higher components and harmonics of the zero order hold.

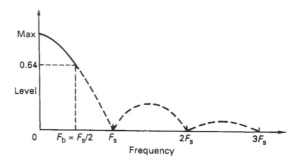

Figure 4.28   Finally the DAC output is fed to a zero-order hold circuit which performs a similar operation to the sample and hold circuit and thence to a reconstruction or output anti-aliasing filter. The plot of the spectral shape of the zero-order hold shows that there are frequency components, at decreasing level, at harmonic intervals equal to F_s.

Other forms of ADC and DAC

Flash converters, Fig. 4.29, function by using an array of comparators, each set to trigger at a particular quantising threshold. The output is available directly. These days the technique is most commonly employed directly as shown in digitising video waveforms. However, there is a use for the technique in high quality oversampling converters for audio signals.

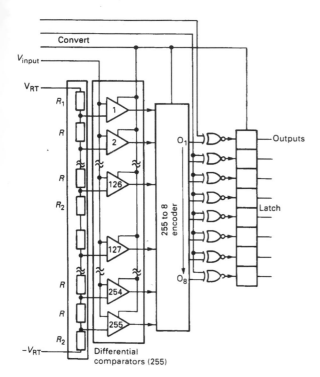

Figure 4.29 A chain of resistors provides a series of reference voltages for a set of comparators whose other input is the input signal. An 8 bit encoder will need 255 comparators. Their output will drive an encoder which maps the output state of the 255 comparators onto an 8 bit word. The NMINV control line is used to convert the output word from an offset binary count to a two's complement form. A 16 bit converter would require an impracticably large number of comparators (65536) in addition to posing serious difficulties to setting the 65536 resistor values to the required tolerance value. The technique does have one virtue in speed and in not needing a sample and hold amplifier to precede it.

One great benefit of operating with digital signals is their robustness, they are, after all, construed as either 1 or 0 irrespective of the cable or optical fibre down which they travel. Their disadvantage is that the digital signals do cover a wide bandwidth. Since bandwidth is a valuable resource there has been much effort expended in devising ways in which an apparently high quality signal can be delivered using fewer bits. The telephone companies were among the first to employ digital compression and expansion techniques but the technology has been used for non-telephonic audio purposes. In the A and μ law converters, Fig. 4.30, the quantising steps, q, do not have the same value. For low signal levels the quantising levels are closely spaced and they become more widely spaced at higher input levels. The decoder implements the matching inverse conversion.

Another approach to providing a wide coding range with the use of fewer bits than would result if a simple linear approach were to be taken is exemplified in the

flying comma or floating point type of converter. In this approach a fast converter with a limited coding range is presented with a signal which has been adjusted in level such that most of the converter's coding range is used. The full output sample value is determined by the combination of the value of the gain setting and of the sample value returned by the converter. The problem here is that the change in gain in the gain stage is accompanied by a change in background noise level and this too is coded. The result is that the noise floor which accompanies the signal is modulated by the signal level and this produces a result which does not meet the performance requirement for high quality audio. A more subtle approach is exemplified in syllabic companders. The NICAM approach, see Chapter 6, manages a modest reduction from around 1 Mbs⁻¹ to 7.04 kbs⁻¹ and we see in it an early approach to attempts to adapt the coding process to the psychoacoustics of human hearing.

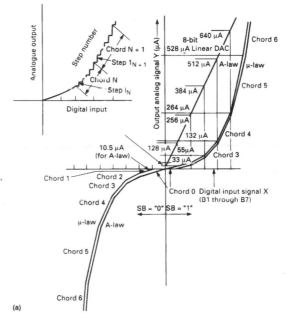

Figure 4.30 (a) The relationship between digital input word and analogue output current is not linear. The sign bit is the MSB and the next three bits are used to set the chord slope. The lower 4 bits set the output steps within each chord. The drawing shows the equivalent output for a linear 8 bit DAC.

The encoder will need to have the matching inverse characteristic in order that the net transfer characteristic is unity. The dynamic range of an 8 bit m or A law converter is around 62 dB and this can be compared to the 48 dB which a linear 8 bit converter can provide. The use of companding (compressing and then expanding the coding range) could be carried in the analogue domain prior to using a linear converter. The difficulty is then one of matching the analogue sections. This is an approach which has been taken in some consumer video equipment.

Delta sigma modulators made an early appearance in digital audio effects processors for a short while. One of the shortcomings of the plain delta modulator is the limitation in the rate at which it can track signals with high slew rates. As we have shown it, each pulse is worth one unit of charge to the integrator. To make the integrator climb faster the rate of charge can be increased so that high slew rate signals can be tracked.

Oversampling techniques can be applied to both ADCs and DACs. The oversampling ratio is usually chosen as a power of 2 in order to make computation more efficient. Figure 4.31 shows the idea behind the process for either direction of conversion. The 4 × oversampling shown is achieved by adding samples with zero value at each new sample point. At this stage of the process the spectrum of the signal will not have been altered and thus there are still the aliases at multiples of F_s. The final stage is to filter the new set of samples with a low-pass filter set to remove the components between the top of the audio band and the lower sideband of the $4 \times F_s$ component.

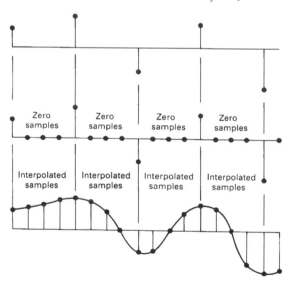

Figure 4.31   The oversampling process adds zero valued dummy samples to the straight sampled signal (a). If oversampling is being carried out in the DAC direction then the digital sequence of samples is treated as if these extra dummy samples had been added in. The sequence is then filtered using an interpolation filter and this creates useful values for the dummy samples.

The process effectively transfers the anti-alias filter from the analogue to the digital domain with the attendant advantages of digital operation. These include a near ideal transfer function, low ripple in the audio band, low group delay distortion, wide dynamic range, exactly

repeatable manufacture and freedom from a wide range of analogue component and design compromises. The four times up-sampling process spreads the quantisation noise over four times the spectrum thus only 1/4 of the noise power now resides in the audio band. If we assume that the noise spectrum is uniform and that dither has been judiciously applied this is equivalent to obtaining a 1 bit enhancement in dynamic range within the audio bandwidth in either digital or analogue domains. This performance can be further improved by the process of noise shaping.

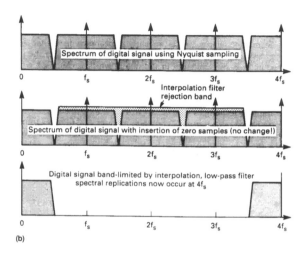

(b)

Figure 4.31   (b) at the new sample rate (here shown as $4 \times F_s$). The spectrum of the signal now extends to $4 \times F_s$ although there is only audio content up to $F_s/2$. Thus when the signal is passed to the DAC element (an element which will have to be able to work at the required oversampling speed) the resulting audio spectrum can be filtered simply from the nearest interfering frequency component which will be at $4 \times F_s$. Note that the process of interpolation does not add information.

If the oversampling is being carried out in the ADC direction the analogue audio signal itself will be sampled and quantised at the higher rate. The next stage requires the reduction of the sequence of data by a factor of four. First the data is filtered in order to remove components in the band between the top of the required audio band and the lower of the $4 \times F_s$ sideband and then the data sequence can be simply sub-sampled (only one data word out of each four is retained).

The information capacity of a communication channel is a function of the SNR and the available bandwidth. Thus there is room for trading one against the other. The higher the oversampling ratio the wider is the bandwidth in which there is no useful information. If samples were to be coarsely quantised it should be possible to place the extra quantisation noise in part of the redundant spectrum. The process of relocating the noise in the redundant spectrum and out of the audio band is known as noise shaping and it is accomplished using a recursive filter structure (the signal is fed back to the filter). Practical noise shaping filters are high order structures which incorporate integrator elements in feedback path along with necessary stabilisation.

The process of oversampling and noise shaping can be taken to an extreme and the implementation of this approach is available in certain DACs for compact disc systems. The audio has been oversampled by 256x in the Philips bitstream device, 758x in the Technic's MASH device and 1024x in Sony's device. The output is in the form of a pulse train modulated either by its density (PDM), by its width (PWM) or by its length (PLM). High oversampling ratios are also used in ADCs which are starting to appear on the market at the current time.

cient work carried out on the way in which concatenations of coders will affect the quality of the sound passing through. Though this problem affects the broadcaster more, the domestic user of such signals may also be affected. Be sure that perceptual coding techniques must remove data from the original and this is data which cannot be restored. Thus a listener who wishes to maintain the highest quality of audio reproduction may find that the use of his pre-amplifier's controls or room equaliser provide sufficient change to an encoded signal that the original assumptions concerning masking powers of the audio signal may no longer be valid. Thus the reconstituted analogue signal may well be accompanied by unwelcome noise.

Transform and Masking Coders

We indicated very early on that there may be some advantage in terms of the reduction in data rate to taking the Fourier transform of a block of audio data and transmitting the coefficient data. The use of a technique known as the discrete cosine transform (DCT) is similar in concept and is used in the AC-2 system designed by Dolby Laboratories. This system can produce a high quality audio signal with 128 kb per channel.

The MUSICAM process also relies on a model of the human ear's masking processes. The encoder receives a stream of conventionally encoded PCM samples which are then divided into 32 narrow bands by filtering. The allocation of the auditive significance of the contribution that each band can make to the overall programme is then carried out prior to arranging the encoded data in the required format. The principle in this encoder is similar to that planned for the Philips digital compact cassette (DCC) system.

The exploitation of the masking thresholds in human hearing lies behind many of the proposed methods of achieving bit rate reduction. One significant difference between them and conventional PCM converters is the delay between applying an analogue signal and the delivery of the digital sample sequence. A similar delay is involved when the digital signal is reconstituted. The minimum delay for a MUSICAM encoder is in the region of 9 ms to 24 ms depending on how it is used. These delays do not matter for a programme which is being replayed but they are of concern when the coders are being used to provide live linking programme material in a broadcast application.

A second, potentially more damaging, problem with these perceptual encoders is that there has been insuffi-

Bibliography

There are no numbered references in the text but the reader in search of more detailed reading may first begin with some of the texts listed below. One risk exists in this multidisciplinary field of engineering and that is the rate at which the state of the art of digital audio is being pushed forward. Sometimes it is simply the process of ideas which were developed for one application area (for example submarine holography) becoming applicable to high quality audio processing.

A useful overall text is that of John Watkinson (*The Art of Digital Audio*, Butterworth-Heinemann, ISBN 0-240-51270-7).

No text covers every aspect of the subject and a more general approach to many topics can be found in the oft-quoted Rabiner and Gold (*Theory and Application of Digital Signal Processing*, Rabiner and Gold, ISBN 0-13-914-101-4). Although it was initially published in 1975 the principles have not changed, indeed it is salutary to read the book and to realise that so much of what we think of as being modern was developed in principle so long ago.

Undoubtedly the best book to help the reader over the tricky aspects of understanding the meaning of transforms (Laplace, Fourier and z) is by Peter Kraniauskus (*Transforms in Signals and Systems*, Addison-Wesley, ISBN 0-201-19694-8).

Texts which describe the psychoneural, physiological and perceptive models of human hearing can be found in Brian Moore's tome (*An Introduction to the Psychology of Hearing*, Academic Press ISBN 0-12-505624-9), and in James Pickles's *An Introduction to the Physiology of Hearing* (Academic Press, ISBN 0-12-554754-4). For both of these texts a contemporary publication date is essential as developments in our basic understanding of the hearing progress are still taking place.

Almost any undergraduate text which handles signals and modulation will cover the topic of sampling, quantising, noise and errors sufficiently well. A visit to a bookshop which caters for university or college requirements should prove useful.

Without doubt the dedicated reader should avail themselves of copies of the *Journal of the Audio Engineering Society* in whose pages many of the significant processes involved in digital audio have been described. The reader can achieve this readily by becoming a member. The same society also organises two conventions each year at which a large number of papers are presented.

Additional sources of contemporary work may be found in the Research Department Reports of the BBC Research Department, Kingswood Warren, UK whilst the American IEEE ICASSP proceedings and the related IEEE journals have also held details of useful advances.

Other titles of interest

Baert *et al*, *Digital Audio and Compact Disc Technology*, 2nd Edition. Butterworth-Heinemann (1992).

Pohlmann, K.C., *Advanced Digital Audio*, Sams (1991).

Pohlmann, K.C., *Principles of Digital Audio*, 2nd Edition. Sams (1989).

Sinclair, R., *Introducing Digital Audio*, PC Publishing (1991).

5 Compact Disc Technology

Ken Clements

The compact disc has changed audio almost beyond recognition in the last few years, but clear accounts of the system have been few. Ken Clements explains in this chapter the nature of the system and the production of discs, building on the introduction to digital principles of Chapter 4.

Introduction

When the compact disc (CD) first appeared on the market in 1982, it was a result of much research that originated back to the mid 1970s. It is fair to say that the compact disc was developed as the result of proposals from a number of manufacturers for a digital audio disc, after earlier joint primary development between Philips and Sony, with Philips providing the optical disc technology, and Sony contributing from their experience in error correction techniques.

In fact prototype players were believed to have been tested around 1977, and in 1979, utilising the combined auspicious talents of Philips and Sony, these two companies agreed in principle to combine their own specialised efforts.

Agreement towards a signal format and the type of disc material were adopted after discussions in 1980, of the Digital Audio Disc Committee, a group representing over 35 manufacturers; the compact disc digital audio system of today was effectively given birth.

Further development continued with respect to the necessary semi-conductor technology that was required, to enable the joint launch of Philips and Sony compact disc players in 1982.

An interesting story with respect to a prototype Sony player at a major audio show relates to this particular prototype player being demonstrated in all its glory on an impressive plinth. The size of the player was in fact remarkably small in view of the technology involved, especially as the LSI semiconductor devices of today were not then available. Enquiries from surrounding onlookers regarding as to how this type of technology

had been achieved, were only answered by knowing nods and smiles. The one question that was not asked was 'What was in the plinth?', which apparently contained the decoding, error correction, and digital to analogue conversion circuits, in the form of a massive number of discrete devices, which were to be developed into the LSI components to finally appear in the first generation players.

As various generations of compact disc players have appeared, whether for the domestic hi-fi or the in-car markets, numerous improvements, innovations and developments have continually taken place, but whatever those various developments may have been, the compact disc itself still remains in the same original format.

The Compact Disc . . . Some Basic Facts

There are two sizes of compact disc, i.e. 8 cm and 12 cm, with the starting speed of either size of disc being in the region of 486 to 568 rpm. and the scanning of the disc commencing from the centre. As the compact disc continues to be played towards the outer edge of the disc, the actual rotation speed slows down as a result of maintaining a constant surface speed of the disc passing a laser beam which retrieves the data information from the disc.

The constant surface speed of the compact disc passing the laser beam is in the region of 1.2 to 1.4 metres per second, and this aspect of driving a compact disc is referred to as constant linear velocity (CLV), whereas the original vinyl record having a fixed speed of 33 1/3 rpm was referred to as constant angular velocity (CAV).

The important difference between the two types of drive is related to the fact that more data or information can be effectively 'packed' onto a CLV disc than a CAV disc.

Again comparison to the original vinyl record reveals that the surface speed of the record at the outer edge is much faster than at the centre indicating that the analogue audio information is effectively more 'stretched

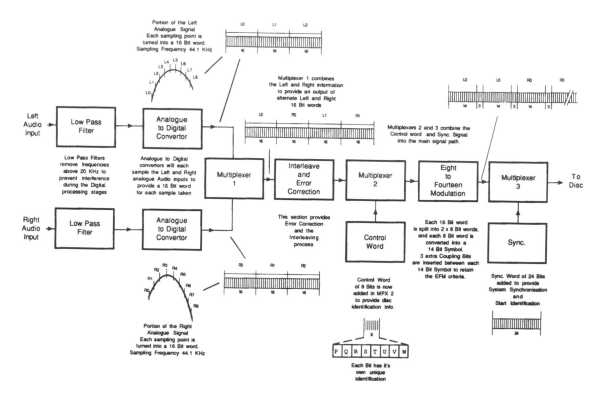

Figure 5.1   Basic compact disc system recording process.

out' at the outer edge compared to that nearer the centre. The normally accepted maximum playing time for the 12 cm compact disc is in the region of 74 minutes, whilst the maximum playing time for the 8 cm disc is around 20 minutes, but by pushing the limits of the manufacturing tolerances increased maximum playing times can become possible.

The finishing speeds of the two sizes of disc are approximately 300 to 350 rpm for the 8 cm disc, and 196 to 228 rpm for the 12 cm disc, with the actual finishing speed being related to whichever surface speed was chosen, within the range of 1.2 to 1.4 metres per second, at the time of recording the disc. The information to be placed on a compact disc is in the form of digital data, i.e. 0s and 1s, to be contained in a series of 'pits' or 'bumps' placed along a fine helical or spiral track originating from the centre of the disc. These pits or bumps vary in length, depending on when the 1s appear in the digital data stream, with the start or finish of a pit or bump being the moment when a digital 1 occurs. The 0s in the data stream are in effect not recorded onto the compact disc, but are retrieved within a decoder section of the compact disc player.

The track dimensions on the compact disc can appear somewhat mind-boggling when trying to find a suitable physical comparison with a track width of 0.5 μm and a track spacing or pitch of 1.6 μm. The thickness of a human hair could contain approximately 30 tracks of a compact disc, whilst the actual track length on a maximum play 12 cm disc could extend to more than three miles. Another interesting comparison is that if one of the pits that represents a small element of digital information, were to be enlarged to the size of a grain of rice, then the equivalent diameter of a compact disc would be in the region of half a mile.

The Compact Disc . . . What Information It Contains

The audio analogue information to be recorded onto the disc is regularly sampled at a specific rate in an analogue to digital convertor, with each sample being turned into digital data in the form of 16-bit binary data words.

There are 65 536 different combinations of the 16-bit binary code from all the 0s through to all the 1s, and the sampling frequency used for the compact disc system is 44.1 kHz.

With a stereo system of recording there are two analogue to digital convertors, with one for the left channel and one for the right. The process of sampling the analogue information is referred to as quantisation, and the overall operation can produce certain effects as quantisation errors and aliasing noise.

Quantisation Errors

Whenever an analogue signal is sampled, the sampling frequency represents the actual times of measuring the signal, whereas quantisation represents the level of the signal at the sample time. Unfortunately whatever the length of the data word that portrays or depicts a specific sample of the analogue signal, quantisation errors occur when the level of the sampled signal at the moment of sampling, lies between two quantisation levels.

As an example consider a portion of an analogue signal, which for convenience of description is converted into a 4-bit data signal.

The analogue values shown in Fig. 5.2 each relate to a specific 4-bit data word, also the sampling points indicate the moment when the waveform is sampled in order to determine the relevant 4-bit data word. The example shown indicates that at the moment of sampling, the instantaneous value of the waveform lies between two specific data words, and thus whichever data word is selected contains a quantisation error, which, in turn, reflects an analogue error when that digital information is converted back into analogue at a later stage.

This problem is related to the specific frequency of the analogue signal, and also the length of the data words.

Aliasing Noise

During the process of recording analogue information onto the compact disc, the audio frequencies are first of all passed through a low-pass filter before being applied to the analogue to digital convertor as shown in Fig. 5.3 for the left audio channel. The purpose of this filter is to remove all audio frequencies above 20 kHz. When the audio signal is converted from analogue to digital, the signal is sampled at regular intervals, and the value that is obtained at each sample interval is in turn converted into a 16-bit digital signal.

The sampling frequency for compact disc is 44.1 kHz, and is the standard determined to provide the lowest acceptable sampling frequency that enables the audio spectrum of 0–20 kHz to be reproduced as faithfully as possible, following a criterion that was stated by Nyquist, working in the Bell Telephone Laboratories in the USA in 1928. It was he who stated words to the effect that if an analogue signal is sampled, providing you sample that signal at a rate of at least twice the highest frequency in the bandwidth that is being used, it is possible faithfully to reproduce all the frequencies within that bandwidth.

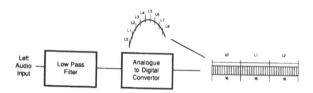

Figure 5.3 Low-pass filter and analogue to digital conversion.

Thus with CD, the audio bandwidth being from 0–20 kHz, the logical sampling frequency would be 40 kHz, but to ensure that effects such as aliasing are minimised, the standard was set at 44.1 kHz.

As previously mentioned, the audio signal is first filtered via a low-pass filter to remove frequencies in excess of 20 kHz, the reason for this process being to minimise the effect of aliasing, which can be caused when frequencies above 20 kHz are passed through to the D to A convertor. The sampling frequency of the D to A convertor can, in effect, sample frequencies above the audio range to produce resultant frequencies which can occur within the audio range (see Fig. 5.4).

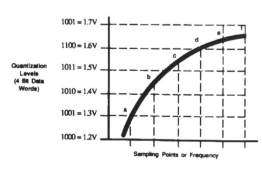

The Sampling Points a and d occur at precise quantization levels, and therefore the digital value will be an accurate representation of those specific samples, whilst those at b, c, e, and f fall between particular quantization levels.

Whichever quantization level is chosen to represent these particular samples, will not enable the original levels to be reproduced when the digital values are eventually converted back into analogue.

This illustrates Quantisation Error, which in effect is limited to one half LSB, but will improve by increasing the number of bits of the data word for each Quantization Level.

Figure 5.2 Quantisation errors.

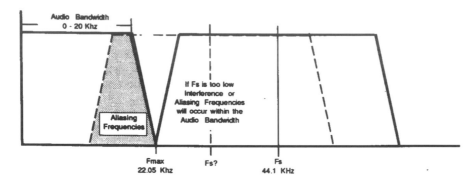

Figure 5.4   Frequency spectrum showing the effects of aliasing if the sampling frequency is too low.

Consider a sampling frequency in the D to A convertor of 44 kHz: if a frequency of, say, 33 kHz, as a result of harmonics outside of the normal listening range from a particular musical instrument, were to be present and therefore sampled, a resultant aliasing frequency of 11 kHz would become apparent. In fact as these harmonics approached the sampling frequency, a range of descending frequencies could be present.

A similar effect of creating something that was not there in the first place can be recalled in those early Western movies, when as the wagons began to roll the wheels appeared at first to go forward, and whilst the wagon increased in speed, the wheels suddenly seemed to throw themselves into reverse, with the wagon still maintaining its forward direction.

As the analogue information is processed into digital data, in the form of 16-bit data words, there is in effect a 'natural' clock frequency regarding this data information, related to the sampling frequency and the number of bits in a data word. For the analogue to digital convertor processing either channel of the stereo audio information this clock frequency is 44.1 kHz × 16 = 705.6 kHz.

During the process of preparing the analogue information for recording onto the compact disc, and after each of the left and right analogue signals have been sampled and converted into a series of 16-bit data words, there are now effectively two separate channels of data information which have to be placed onto what will become a single track on the compact disc. This is achieved by passing the separate left and right data words into a multiplexer (Fig. 5.5), the output of which is a series of alternate left and right data words, with the left data being the first data word from the output, and thus the first audio data word to appear from the compact disc.

It is a requirement of the compact disc specification that the first audio information data word to be laid on the CD track is a left data word. This is to enable the decoder within the CD player to 'know' that the first audio data word output from the decoder will be a left data word and that data word will be 'directed' towards the left audio channel for subsequent processing. Accordingly the next data word will be a right data word which will in turn be 'directed' towards the right audio channel.

As a result of this alternating procedure, the natural clock frequency of the data stream must be increased in order that the audio information can be maintained

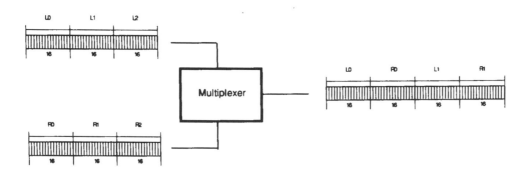

Figure 5.5   Multiplexing the left and right channels.

within the correct time scale. The bit or clock frequency is increased by two times to 705.6 kHz × 2 = 1.411200 MHz.

Error Correction

Whilst there are benefits to be gained from the process of digital audio recording as used with the compact disc, such as the improvements in the dynamic range, signal to noise ratio and also stereo separation when compared to the old-fashioned vinyl analogue recordings, problems do arise when some of the data information becomes corrupted for some reason. Corruption of the digital data can occur during the manufacturing process of a compact disc, or as a result of improper care of the disc arising in scratches and dirt in various forms becoming present on the surface of the disc; these effects can cause incorrect data words which in turn would provide an unsuitable output (see Fig. 5.6).

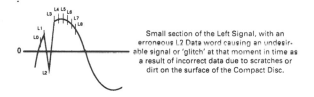

Figure 5.6 Effect of an erroneous data word.

Unfortunately the compact disc is not as indestructible as one was first led to believe, and in fact the treatment of a compact disc should be on a par to the treatment of a vinyl long playing record: those original 'raspberry jam' demonstrations when compact disc was first unveiled, have a lot to answer for.

The main physical advantage with compact disc is that it does not wear in use as there is no mechanical 'stylus' touching the surface of the disc, the data being read from the disc via a reflected beam of laser light.

So how does error correction take place? Basically extra information, such as parity bits, is added to the data steam and eventually recorded onto the compact disc.

When the disc is being played in the CD Player, the extra information such as the parity bits, assists in identifying any defective words as the data stream is being processed. The error correction stage within the CD player applies certain types of correction depending upon how much error is present. Overall this is an extremely complex process, for which there are excellent references available which can provide the enquiring mind more than enough information in that area: suffice to say, by the addition of this extra information, it is possible within the CD player to identify and endeavour to

'correct' the errors, if they are within certain criteria, or the audio output can be muted if those errors exceed those criteria (see Fig. 5.7).

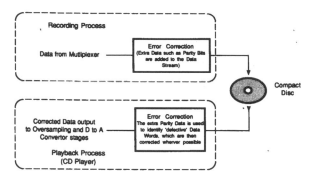

Figure 5.7 The error correction process.

If the errors prove to be too much of a 'handful' the CD player will usually 'throw in the towel' and shut down as far as the operation of that particular disc is concerned.

How are the Errors Corrected?

It is useful at this stage to consider, in a very basic form, how the errors are effectively 'corrected' within the CD player.

Consider the sampled section of the analogue signal, compared to the same sample when one of the data words has gone on 'walkabout', which could provide an incorrect output (Fig. 5.8).

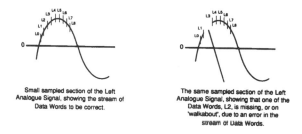

Figure 5.8 Comparison of good and faulty data word areas.

The overall processing of the digital information within compact disc technology is very much a complex mathematical operation but one of the aims of this book is to minimise the mathematical content, and endeavour to consider as much of the technology in a more 'low key' approach, and fortunately this approach can be applied to the principles of error correction. Within the compact disc player there are three main methods of overcoming errors as they occur, and depending upon how much error occurs at a specific moment in time.

Muting

It is quite a logical concept that, when an error occurs, the result of that error would, in real terms, be a different level data word, which in turn would produce an undesirable sound or 'glitch' from the output. Again it would be logical to mute the sound output when an error has been identified, so that the recipient does not hear undesirable sounds; unfortunately life rarely follows a logical path.

Consider the sampled signal and the application of muting when the error occurs. If the portion of the analogue waveform shown in Fig. 5.9 is the motion of the loudspeaker as it responds to the audio signal, then when the muting is applied, a bigger and better 'glitch' could occur that would probably be even more undesirable than the original error if that had been allowed to be reproduced. In fact history reveals that in the early design stages of compact disc, when this form of error correction was applied at high output levels, the final effect proved detrimental to loudspeakers whereby the speech coil and the cone parted company quite effectively. However muting does play its part in the error correction process, and is normally applied at the times when an extensive series of errors occur, and other methods prove unacceptable.

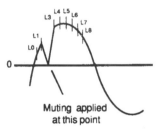

Figure 5.9   Error correction/concealment – muting.

Previous word hold

Though many of the operations that take place when manipulating the digital information, whether when putting that information onto the disc or in the digital processing within the CD player, a substantial amount of memory circuits are used, especially within the CD player itself. In fact as the audio data is being processed within the player, there is a period of time when all the digital audio information is sequentially held in memory for a brief period of time. It could therefore be logical to replace an 'iffy' data word by the nearest one that is similar in value (Fig. 5.10). When the original analogue signal is sampled, and each sample turned into a 16-bit word,

there are in fact 65 536 different combinations of the 16-bit code from all the 0s to all the 1s. Thus it is reasonable to accept that the previous data word to the error word could be approximately 1/65 000 different in level to the original correct data word, and as again it is quite logical to assume that there is no person in existence that is capable of determining those differences in sound levels, then previous word hold is an acceptable substitute.

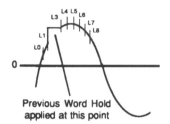

Figure 5.10   Error correction/concealment – previous word hold.

Linear interpolation

The previous word hold method of error correction can be further improved with linear interpolation, where it is possible to compare the data word before the error word and the data word after the error word and by taking the average of the sum of these two words and using this as the substitute for the error word, it is possible to achieve a more accurate assumption of the missing word (Fig. 5.11).

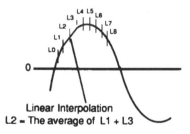

Figure 5.11   Error correction/concealment – linear interpolation.

Interleaving

Interleaving is an extremely complex process and is complementary to the error correction process, where the basic operation is in effect, to 'jumble up' the audio data words on the compact disc in a defined arrangement, and to re-arrange those data words back into their original order within the compact disc player before the process of restoring the digital data back into the original analogue information.

The interleaving process is achieved by inserting the data words into a series of delays. These delays are in multiples of one sampling period or data word in length and are in the form of random access memory (RAM), where they are initially stored and then extracted from that memory in a strictly defined sequence. This in effect actually causes data words to be delayed by varying amounts, which in turn causes the data words to be effectively 'jumbled up' on the disc, and providing these data words are 'unjumbled' within the CD player in the reverse manner the data words will re-appear in the correct sequence.

Figure 5.12 illustrates the basic concept of the interleaving process. During the recording process the serial data words are turned into parallel data words by the first serial to parallel convertor and then passed into each delay with the result that the output is the data words appearing at differing times. These are converted back into serial data by the subsequent parallel to serial convertor before eventually arriving onto the compact disc.

Within the CD player this process is reversed through a series of delays which are set in the reverse order, and it is this operation which effectively 'restructures' the data sequence back into the original order.

Control Word

After the error correction and interleaving processes, it is necessary to insert an extra 8-bit data word, which is referred to as the control word, or sub-code to give it another title, with another multiplexer performing this task at regular and specific intervals, (Fig. 5.13).

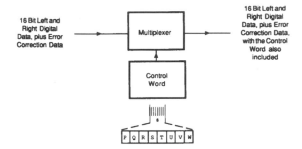

Figure 5.13 The control word.

The purpose of the control word is to provide disc identification information, and each of the 8 bits that comprise the control word have their own unique identification. The 8 bits or the control word are labelled with letters P to W.

As the data information is 'laid' along the track on the compact disc, the control word is inserted immediately before a group or block of audio data information, and immediately after a 'sync. word', the purpose of which is described later.

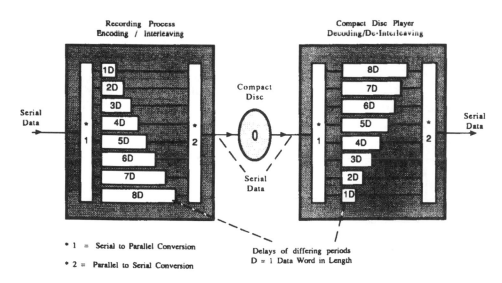

* 1 = Serial to Parallel Conversion
* 2 = Parallel to Serial Conversion

Delays of differing periods
D = 1 Data Word in Length

The basic delay concept, where the Serial Data is turned into Parallel Data, in terms of blocks of Data Words.
Each Data Word is delayed by differing amounts during the recording process, and delayed again during the Playback process.
However, providing the decoding is in effect the reverse of the encoding, the original Data Word arrangement will be restored.

Figure 5.12 Basic interleaving/de-interleaving process.

As the disc rotates within the CD player, the control word is identified within the processing circuits of the player, each bit, with its own identification, is inserted into a specific memory, and the relevant memory is analysed at regular intervals to determine the data information that has been 'built up' over a period of time, which is in effect after 98 control words have been stored.

Only the P and Q bits are in general use with CD players, whilst the R through to W bits are intended for use with computer displays, and the relevant software to display graphics, such as the words of the song being sung on the disc, a new concept of 'Singalong-a-Max', though not to be confused with Karaoke. The R through to W bits of the control word are not in general use with the domestic disc player.

The P bit is used to 'inform' the CD player that the music from the disc is about to commence, and enables the internal muting circuits on some of the early CD players to be switched off enabling the analogue information to be applied to the audio circuits. The later and more sophisticated players tend to release the mute circuits in relation to the availability of the data ready for audio processing.

The Q bit contains an extensive range of information including the following:

(a) total playing time
(b) total number of music tracks
(c) individual music track identification
(d) elapsed playing time for each music track
(e) end of playing area information to enable the player to be stopped
(f) de-emphasis information: a requirement by the recording authorities to be able to apply additional emphasis on any specific track and therefore enable the relative de-emphasis circuits to be switched in or out accordingly.

Other information can be available such as copyright, catalogue number, as well as recording date and serial number, none of which is used within most domestic players currently available.

The total playing time and the total number of music tracks comprise the table of contents (TOC) information which all CD players require to 'know' before commencing to play a compact disc. The control word is processed within the decoder section of the compact disc player.

Eight to Fourteen Modulation

After the process of interleaving the data, and after adding the control word, the next major operation is the technique of eight to fourteen modulation (EFM).

Within this stage of operation during recording processes onto compact disc, the 16-bit data words are changed into two 8-bit data words, and each 8-bit data word is then changed into a 14-bit symbol (Fig. 5.14). The term symbol is used to differentiate between a data word of 8 or 16 bits, and its transformation into a 14-bit word or symbol.

To appreciate the reasons for the technique of eight to fourteen modulation, it is useful to determine the problems that can occur when the data is actually put onto the compact disc, and therefore the basic process of disc manufacture is now worth considering.

Compact Disc Construction

When a compact disc is manufactured, a disc of glass, of extremely high precision, is coated with a photo-resist material, which is sensitive to laser light (Fig. 5.15). This glass disc is placed into the recording machinery (Fig. 5.16) at the time of manufacture to expose a fine

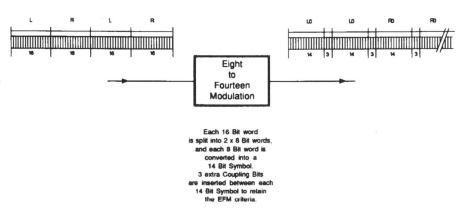

Figure 5.14 Eight to fourteen modulation.

helical track onto the photo resist material, the width of the track being 0.5 μm.

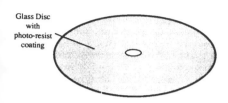

Figure 5.15 Glass mastering disk.

The laser exposes the photo-resist surface with a 0.5 μm width spiral track, as the disc rotates, commencing from the centre and working towards the outer edge. The laser beam is operated in relation to the data information, whereby the laser is switched on and off as the 1s information appears, creating a track of exposed photo-resist material which comprises a series of dashes, the length of which and the intervals between which are related to when the 1s occur. No 0s are recorded onto the disc, they are re-generated within the CD player.

Whenever a 1 occurs in the digital data, the laser beam recording the information onto the photo-resist surface alternately switches ON and OFF as each 1 occurs,

exposing the photo-resist surface of the disc and creating a series of 'dashes' 0.5 μm wide, the length being dependent on the period or distance between the 1s.

On completion of the recording process, the unexposed photo-resist material is chemically removed, leaving a helical track across the surface of the glass disc, which becomes the master to produce the injection moulding formers for the mass production of the resultant compact disc. The playing surface of a compact disc is a thin layer of a special injection moulded plastic material, which is indented from the non-playing side, with a series of 'pits', (Fig. 5.17).

The data stream shown in Fig. 5.18 to illustrate how a disc is manufactured 'conveniently' has no consecutive 1s in the data stream; of course the digital data can comprise any combination of 0s and 1s, and instances will frequently occur when there will be long series of 0s or 1s.

When a series of 1s occurs (Fig. 5.19) the following problems can arise:

(a) the frequency rate will increase;
(b) the 'pit' length becomes too short, and would be shorter than the track width;
(c) the high 1s rate can be integrated within the servo circuits to create varying DC levels which can cause the servo circuits to become unstable.

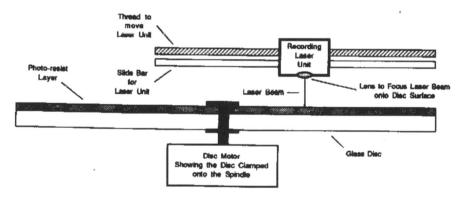

Figure 5.16 Basic mechanical arrangement for disc recording.

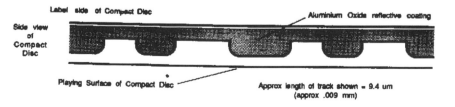

Figure 5.17 Enlarged side view of compact disc, showing pits indented into the back of the playing surface.

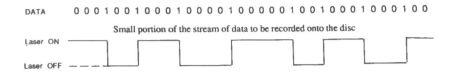

Figure 5.18   The sequence of dashes created where the laser beam has been switched on.

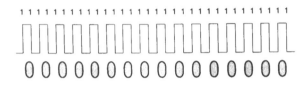

Figure 5.19   A long sequence of 1s.

When a series of 0s occurs the following problems can arise:

(a) Without any 1s occurring at regular intervals, the lock of the phase lock loop (PLL) controlling the voltage controlled oscillator (VCO) within the CD player decoder circuit can become unstable.

(b) If a long series of 0s causes long distances between the pits or bumps, the 'trackability' of the player can be affected due to long distances without any track being present (Fig. 5.20)

To overcome these problems the method of eight to fourteen modulation was devised, and is purely a process of being able to effectively transfer the digital data onto the compact disc without any of the problems that have been previously outlined.

When a 16-bit word is prepared for the eight to fourteen modulation process, the 16-bit word is split into two

8-bit words; this stage is easily achieved by taking the first 8 bits followed by the second 8 bits. Of the 8-bit sequence of data there are 256 different combinations, from all the 0s through to all the 1s.

With the 14-bit code there are 16 364 possible combinations, from all the 0s to all the 1s. However there are 267 of these combinations that satisfy the following criteria:

> No two 1s are consecutive
> A minimum of two 0s exist between two 1s
> A maximum of ten 0s exist between two 1s

Of the 267 combinations that satisfy the above criteria, 256 have been specifically selected, and put, in effect, into a 'look-up' table. Each 8-bit combination is then 'paired' with a specific 14-bit code, of which three examples are shown below:

$$00000010 = 10010000100000$$
$$01011001 = 10000000000100$$
$$11111100 = 01000000010010$$

The new 14-bit 'symbols' now represent the original data information when 'embodied' onto the compact disc, and when this information is retrieved from the disc within the CD player, another 'look-up' table enables the 14-bit data to be transferred back into the original 8-bit code,

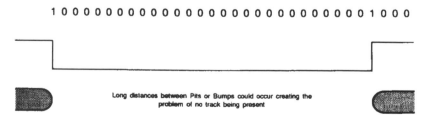

Figure 5.20   A long sequence of 0s.

and two 8-bit 'words' together now comprise the original 16-bit word that was developed in the beginning.

The Eight to Fourteen Modulation Process (Fig. 5.21)

Consider the following 16-bit word

0111101001111100

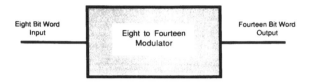

Figure 5.21 The eight to fourteen modulator.

which is split into two 8-bit words

01111010 01111100

and fed into the eight to fourteen modulator, where they are converted into two 14-bit symbols:

10010000000010 01000000000010

Again consider the following 16-bit word

0111101001111100

split into two 8-bit words

01111010 01111100

and fed into the eight to fourteen modulator (Fig. 5.22).

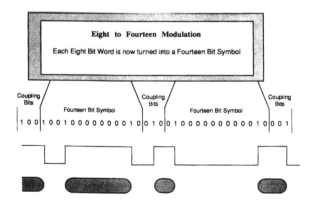

Figure 5.22 Details of eight to fourteen modulation on a typical word.

Coupling Bits

From the above process, it can be observed that three extra bits have been inserted between each 14-bit 'sym-

bol', which are referred to as coupling bits. The purpose of these extra bits is to allow for the fact that throughout the eight to fourteen modulation process, it could not be guaranteed that when one 14-bit symbol ended in a 1, that the next 14-bit symbol would not commence in a 1, therefore disrupting the achievements gained with this process.

The extra bits are inserted under 'computer control' by analysing subsequent symbols in order that the relevant extra bits enable the required criteria to be maintained. In fact these extra bits are identified within the processing circuits of the CD player and then literally 'thrown away' as they fulfil no further function.

Pit Lengths

As a result of the eight to fourteen modulation process, there are only nine different 'pit' lengths to carry all the necessary digital data and information for effective CD player operation.

Reference is made in Fig. 5.23 to the final clock frequency of 4.3218 MHz, which is the final 'clock' rate for the data on compact disc, and will be referred to later.

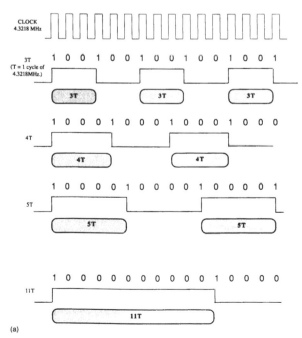

Figure 5.23 (a) The nine possible pit lengths.

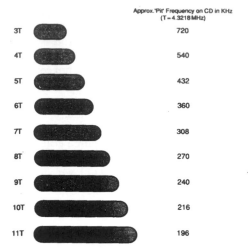

Approx. 'Pit' Frequency on CD in KHz
(T = 4.3218 MHz)

3T	720
4T	540
5T	432
6T	360
7T	308
8T	270
9T	240
10T	216
11T	196

Minimum 'Pit' length (3T) 0.833 - 0.972 μm, maximum 'Pit' length (11T) 3.054 - 3.56 μm
depending upon disc velocity (1.2 - 1.4 metres /sec).

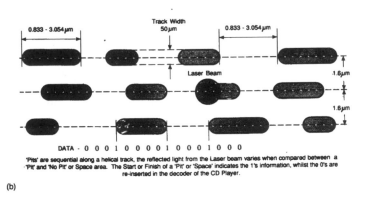

DATA - 0 0 0 1 0 0 0 0 1 0 0 0 1 0 0 0

'Pits' are sequential along a helical track, the reflected light from the Laser beam varies when compared between a 'Pit' and 'No Pit' or Space area. The Start or Finish of a 'Pit' or 'Space' indicates the 1's information, whilst the 0's are re-inserted in the decoder of the CD Player.

(b)

Figure 5.23   (b) comparison of pit lengths.

Sync. Word

The final information to be inserted in the recording sequence is the sync. word, which is used in the compact disc player, as the 'start point' for the data to commence the necessary processing, and also as a signal to be compared for the disc speed control circuits.

The sync. signal is uniquely different to all other data information on compact disc, and comprises a 24-bit word as follows (Fig. 5.24):

100000000001000000000010

The diagram showing how the track is laid down (Fig. 5.25) indicates how the data is positioned within one frame.

The path of the original analogue signal from its humble beginnings through to becoming an almost insignificant part of a 'pit' or 'bump', depending upon which way you

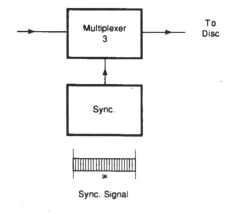

Figure 5.24   The sync. word.

are looking at the disc, or even the space between, has now been completed.

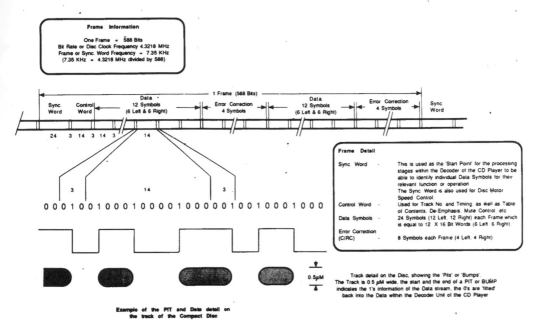

Figure 5.25 Basic compact disc system track detail.

Compact disc technology is a very complex process, and it is often a wonder that the technology works as well as it does, especially when installed in a motor vehicle bumping along some of the 'better' highways.

Many engineers have been known to almost run the other way when it comes to compact disc player scvicing, but with a reasonable amount of knowledge they do not constitute a major problem, and having considered some of the basic concepts of the compact disc, it is now worthwhile considering a typical compact disc player and its basic operation.

The block diagram of the compact disc player is shown at Fig. 5.26. The main function of the unit is to play the compact disc at the relevant speed and retrieve the digital data from the disc by means of a reflected laser beam, and reproduce that digital data back into the original analogue form as accurately as possible.

Much of a compact disc player comprises various servo

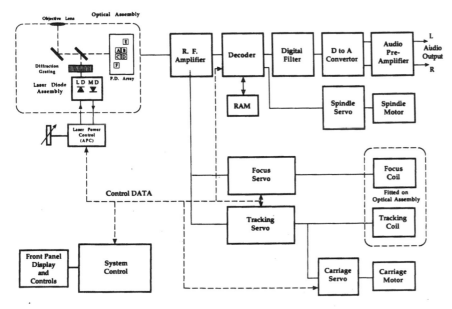

Figure 5.26 Compact disc player – basic block diagram.

systems that enable the laser beam to be accurately focused onto the surface of the disc and simultaneously track the laser beam across the surface of the disc, whilst it is rotated at the correct speed.

All compact disc players are required to operate in a specific manner, the sequence of which is controlled by some form of system control. Most of the electronic operations take place within large-scale integrated circuits, which despite their complexity, are usually extremely reliable.

Mechanical operations are relatively simple and are limited to motors to drive the disc, optical assembly and the loading and unloading mechanism, and a pair of coils which enable the lens within the optical assembly to move vertically and laterally.

Optical Assembly

A typical optical assembly arrangement is outlined in Fig. 5.27, and is of the three beam type, which is typical of optical assemblies in most CD players.

Reference to the diagram indicates that an optical assembly, or pick-up unit, can comprise a fair number of components, each of which are identified and described as follows:

Laser and photo diode assembly

The laser diode (LD) emits a single beam of low power infra-red laser light, the power of which is maintained at a stabilised power level in conjunction with the photo diode (PD) and an automatic power control (APC) circuit.

The laser output power of a typical optical assembly within the domestic range of compact disc player is usually in the region of 0.12 mW (120 microwatts).

Diffraction grating

Most optical assemblies used in current compact disc players are of the three beam type, where the main beam is used for data retrieval from the disc, as well as maintaining

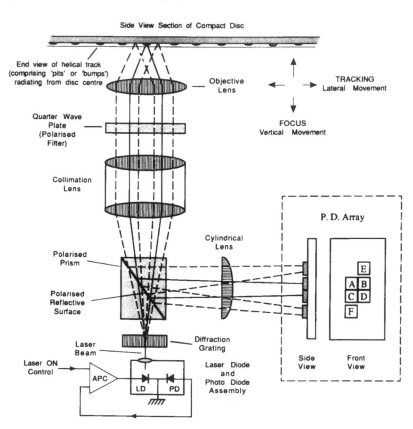

Figure 5.27   Optical assembly.

focus of the laser beam onto the compact disc, and the other two beams, side beams, provide the tracking information. The diffraction grating, a very small lens similar to the much larger 'add on' lens used in photography to provide multiple images, enables the required three beams to be obtained from the single laser beam.

Polarised prism

This prism enables the direct laser light to be passed through to the compact disc, whilst the reflected light from the disc is deflected towards the PD array. The reflective surface of the prism is polarised to ensure that the reflected light is efficiently passed to the PD array.

Collimation lens

This lens together with the objective lens ensures that the correct focal length of the optical assembly is achieved.

Quarter-wave plate (polarising filter)

This filter causes the optical polarisation of the direct laser beam compared to the reflected laser light from the disc to differ by 90 degrees and ensures that the reflected light is of the correct polarisation to be efficiently deflected by the polarised prism towards the PD array.

Objective lens

This lens can be observed on the optical assembly and is capable of moving vertically to enable focus to be achieved onto the compact disc, and moving laterally to allow the laser beam to track across the disc. To enable the lateral and vertical movements to be achieved, a pair of coils, one for each operation, are attached to the lens and operated by the relevant servo control circuits.

Photo detector (PD) array

This unit comprises six photo diodes arranged as shown in Fig. 5.28. These diodes are used to retrieve the digital information from the compact disc, to develop the signals for focusing the laser beam onto the disc and to enable the laser beam to track across the disc.

The photo diodes are connected to separate circuits to process the data retrieval, focus and tracking operations. To obtain the required signal for the focus operation the A, B, C, and D photo diodes are connected in a certain

manner to obtain the required signal in conjunction with the effect of the cylindrical lens.

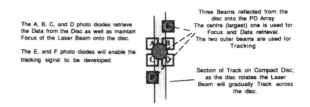

The A, B, C, and D photo diodes retrieve the Data from the Disc as well as maintain Focus of the Laser Beam onto the disc.

The E, and F photo diodes will enable the tracking signal to be developed.

Three Beams reflected from the disc onto the PD Array The centre (largest) one is used for Focus and Data retrieval. The two outer beams are used for Tracking

Section of Track on Compact Disc, as the disc rotates the Laser Beam will gradually Track across the disc.

Figure 5.28   Photo detector array.

Cylindrical lens

The purpose of this lens is to enable the focus signal to be developed. When the laser beam is in focus on the compact disc, a circular beam of laser light, the centre beam, will land equally on the surface of the centre four photo diodes A, B, C, and D. Should the laser beam go out of focus in one direction or the other (Fig. 5.29), the cylindrical lens will distort the beam into an ellipse, which will effect the diagonal photo diodes A and D or B and D, depending upon which direction the mis-focus has occurred.

Out of focus in one direction

In-focus

Out of focus in the other direction

Figure 5.29   A method of developing the focus error signal.

Other methods of determining the focus error signal have made their appearance during the development of compact disc players, but the method described above is typical of many of the optical assemblies currently in use.

Another method that made a brief appearance with manufacturers was the critical angle prism (Fig. 5.30) that made use of an optical phenomenon where by a focused beam of light moved fractionally sideways when reflected through a prism, with the angle of the reflected surface set at a specific angle. This system was usually associated with single-beam optical assemblies, but usually involved more complicated electronic circuitry to develop the separate focus and tracking error signals.

The signals from the photo diode array are processed to produce three basic signals, which comprise the data from the disc, plus the focus and the tracking signals. The basic circuit illustrated in Fig. 5.31 is normally contained within an integrated circuit, with the relevant outputs as indicated.

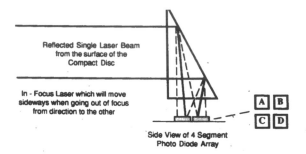

Figure 5.30 Critical angle prism (angle of prism surface to reflected laser beam = 42 degrees).

PD array outputs

The six photo diode elements provide outputs which are related to reflected information from the playing surface of the compact disc, with the four centre elements, A, B, C, and D, being connected diagonally to form two pairs of A and D as well as B and C. These two pairs are fed to the focus error amplifier and also the summation amplifier. Photo diode elements E and F are each fed to the tracking error amplifier. (See Fig. 5.31.)

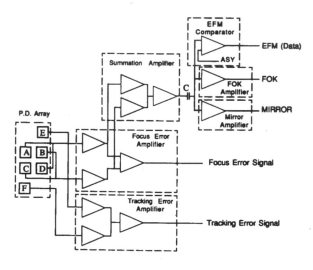

Figure 5.31 PD array signal processing.

Focus error amplifier

This circuit amplifies the two pairs of signals from the A, B, C, and D elements, and then the signals are passed to a differential amplifier to provide the focus error output, the value and polarity of which is related to the amount and direction of focus error.

When the in-focus situation is achieved, the focus error signal will be zero, but when mis-focus occurs, then, depending upon the direction that the out of focus situation arises, the polarity of the signal will be either positive or negative with respect to zero, and the amount or level of the signal will relate to the extent of the mis-focus.

Tracking error amplifier

The outputs from the E and F elements are used for tracking purposes only. As the compact disc rotates, the helical track on the disc causes different amounts of reflectivity from the surface of the disc: this, in turn, develops an output from the differential amplifier which produces the tracking error signal, a signal or voltage, the value and polarity of which will be related to the amount and direction of tracking error.

Summation amplifier

The input to this amplifier comprises the two pairs of outputs from the A, B, C, and D elements. These two pairs of signals are summed or added together within this amplifier to produce an output which is the sum of the four centre elements added together. This signal is usually referred to as the RF signal or eye pattern waveform (Fig. 5.32) and comprises the data information from the compact disc which is to be further processed in order that the original analogue audio information may be obtained.

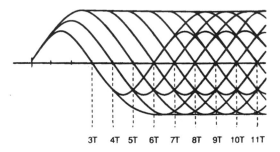

Figure 5.32 Eye pattern waveform. Resultant signal due to the variations in intensity of reflected laser beam light from the playing surface or 'pits' on the compact disc. (T = one cycle of 4.3218 MHz).

The output from this amplifier is fed via a capacitor to three more circuits. The purpose of the capacitor is twofold; firstly to isolate any DC content from the summation amplifier and to produce an RF output signal, comprising frequencies within the range of 196 to 720 kHz with the level varying either side of zero depending upon the reflectivity of the laser beam from the playing surface of the compact disc. Secondly, the waveform resulting from the reflected signals from the disc does not tend to be symmetrical due to variations during disc manufacture, as well as the actual shape of the pits or bumps on the disc. Here the capacitor tends to ensure that the waveform is reasonably symmetrical either side of the zero line, and therefore as the waveform effectively crosses the zero line, can indicate that a '1' is present within the data stream.

The RF or eye pattern waveform is characteristic of any compact disc player and virtually any type of compact disc. The period lengths are related to the differing lengths of the 'pits' or 'bumps' on the playing surface of the disc, which in turn are related to the final clock frequency as a result of all the various processing of the data that takes place prior to the recording of the data onto the compact disc.

The shortest period in the waveforms are the result of the shortest pit length or space between pits, and is equivalent to three cycles of the final clock frequency of 4.3218 MHz. Each period length thereafter is equal to the next pit length and therefore a further cycle of the final clock frequency.

EFM comparator

The RF signal from the capacitor is now passed to the EFM comparator, where it is effectively amplified to provide a square wave output with clean or steep leading and trailing edges. It is extremely important to obtain a square output as the leading and trailing edges indicate when a '1' occurs within the data stream.

During the manufacturing process of the compact disc in the injection mould or 'pressing' stage, variations in quality can occur during the lifetime of a specific mould as it begins to wear with each successive 'pressing', causing the edges of the pits or bumps to become less defined (Fig. 5.33). This can result in problems within the compact disc player of identifying when a pit starts or commences, and thus impair its ability to produce a '1' at the correct point in time.

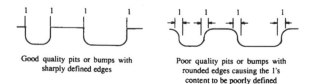

Good quality pits or bumps with sharply defined edges

Poor quality pits or bumps with rounded edges causing the 1's content to be poorly defined

Figure 5.33 How poor quality pressing can effect pit shapes.

The EFM comparator within most CD players has two inputs, one being the RF signal, and the other being the ASY (asymmetry) voltage or reference voltage, which is related to the timing of the EFM, or squared RF signal waveform, compared with a phase locked clock (PLCK) within the decoder, which after filtering, produces a reference voltage (ASY) for the EFM comparator to operate. The resultant output from the comparator is now passed to the decoder for further processing, to eventually enable the required 16-bit data words to be obtained for conversion back into their original analogue counterparts.

FOK amplifier

When a compact disc is inserted into the player, it is necessary for the focus servo initially to operate the objective lens to achieve focus onto the playing surface of the disc. This is achieved by causing the lens to elevate up and down, usually two or three times, and when focus has been achieved, which is related to a maximum RF signal from the disc, a HIGH output will be obtained from the FOK amplifier (focus OK), which in turn communicates to the relevant control circuits that focus has been achieved.

In many players the FOK signal is also used to inform the control circuits of the player that a disc is present and ready for playing.

Mirror amplifier

The tracking servo enables the laser beam to move gradually across the playing surface of the compact disc by a combination of moving the objective sideways as well as physically moving the complete optical assembly.

It is possible for the servo to allow the laser beam to track between the tracks containing the digital information, therefore it is essential for the tracking servo to 'know' that it is tracking correctly along the data track.

The mirror signal enables the control circuits to be aware that incorrect tracking may be occurring, and this is determined by the fact that this signal will be LOW when actually 'on track' and HIGH when 'off track'. If this signal stays HIGH beyond a defined period of time, the control circuit will effectively give the tracking servo a 'kick' until the mirror signal maintains a LOW level, thereby indicating an 'on track' condition.

The mirror signal is also used when track jumping takes place when a new music track on the disc is selected by the operator. The start of the required music track is computed within the player and the optical assembly traverses across the disc whilst simultaneously counting individual tracks as they are crossed.

Servo Circuits

Within most compact disc players there are usually four individual servo systems:

Focus servo – To maintain the laser beam 'in-focus' on the playing surface of the compact disc, by the vertical movement of the objective 1 lens.

Tracking servo – To enable the laser beam to track across the playing surface of the compact disc by the sideways movement of the objective lens.

Carriage servo – Fractionally to move the optical assembly when the objective lens begins to reach the limits of its operation. This servo works in conjunction with the tracking servo, and is often described as either the sled servo or slider servo.

Spindle servo – The speed of the disc is related to the position of the laser beam, and therefore the optical assembly, as it obtains the data information from the compact disc. The data rate from the compact disc is compared to an internal reference within the compact disc player, usually a crystal frequency source for stability, with a resultant signal to enable the compact disc to rotate at the correct speed, which is in the region of 500 rpm at the centre or start of the disc, decreasing to around 180 rpm at the end of a 12 cm compact disc. As the data is processed within the decoder, the 16-bit data words are briefly stored in a random access memory (RAM), and then taken out of the memory as required. This operation is linked to the crystal clock that drives the disc at the approximately correct speed, and as the data is removed from the memory, the disc speed will increase and decrease to maintain a certain amount of data within the RAM. This servo is frequently referred to as the disc servo or occasionally as the turntable servo.

Focus servo (Fig. 5.34)

The focus servo amplifies the focus error signal and applies the signal to the focus coil which is attached to the objective lens, to enable it to move vertically to maintain focus on the playing surface of the compact disc.

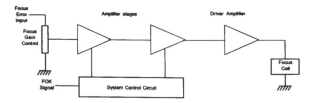

Figure 5.34   Focus servo.

The system control circuit controls the operation of the servo, ensuring that it operates at the correct time usually after the laser has been switched on. The first operation is to cause the lens to move vertically in order to search for the correct focal point; having achieved this, and when the FOK signal has been received, the servo will then be allowed to operate normally maintaining the required 'in-focus' condition. During the focus search sequence the first amplifier stage is normally switched off to prevent the developing focus error signal from counteracting the search operation.

Tracking servo (Fig. 5.35)

The tracking servo has many similarities to the focus servo, but its function is to enable the tracking error signal to be amplified and passed to the tracking coil, which in turn is attached to the objective lens to enable the relevant sideways movement to maintain the tracking requirement across the playing surface of the compact disc.

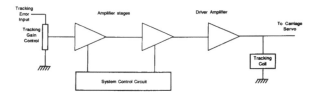

Figure 5.35   Tracking servo.

The objective lens has only a limited amount of lateral or sideways movement, approximately 2 mm, but as the movement increases laterally this can be interpreted as an increasing output from the tracking circuit which can be used to drive the carriage servo when necessary.

Again the system control circuits control the operation of the tracking servo, enabling the servo to commence the tracking function when the compact disc is rotating, which on many players commences after the laser has been switched on and the focus servo has, in turn, commenced its own operation.

Carriage servo (Fig. 5.36)

The function of the carriage servo is to move the optical assembly gradually across the surface of the compact disc by driving the carriage motor which, in turn, is mechanically connected via a long threaded drive to the optical assembly. The gearing ratio is extremely high with the effect that it can take up to 75 minutes, depending on the playing time of the compact disc, to enable the optical assembly to cover the playing area of the disc, which can be in the region of 4.5 cm.

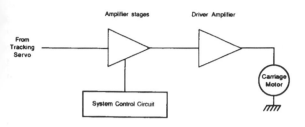

Figure 5.36 Carriage servo

As the objective lens moves laterally, up to approximately 1 mm of movement, in conjunction with the tracking servo, the increasing signal to the tracking coil is passed to the carriage servo where it is amplified. When the output voltage has reached a level sufficient to operate the carriage motor, the motor operates to move the optical assembly a very small amount. Simultaneously the objective lens centralises its position again, with the resultant decrease in the signal to the tracking coil stopping the carriage motor, until the level has increased again to repeat the process. This procedure continues until the disc has completed the playing sequence.

Some compact disc players utilise a linear drive motor system instead of a conventional motor, with a similar if not more complicated operating process. Again the servo is controlled by the system control circuit, which allows the carriage servo to come into operation when the tracking servo is functioning, but also signals can be developed from the control circuit which drives the motor during track location sequences when searching for selected music tracks on the disc, as well as returning the optical assembly back to the centre or start position on completion of playing the compact disc, in readiness for playing another.

Spindle servo

This servo is required to rotate the compact disc at the correct speed within the range of approximately 500 to 180 rpm. The data retrieved from the disc is compared to an internal reference within the compact disc player to produce a control voltage which drives the disc at the correct speed compatible with the correct data rate from the disc.

The effective clock frequency from the disc is 4.3218 MHz, with 588 bits of the digital information comprising one frame of information. Dividing 4.3218 MHz by 588 provides a frame or sync. frequency of 7.35 kHz.

The majority of the spindle servo circuitry is contained within the decoder, and despite its apparent complexity, its operation proves to be extremely reliable in the majority of compact disc players. When the compact disc is operating normally the disc runs at a speed which enables the surface speed passing the laser beam to be in the region of 1.2 to 1.4 metres per second. As the optical assembly tracks towards the outer edge of the disc the disc will appear to slow down as it maintains a constant linear velocity (CLV) of 1.2–1.4 m/s, which in effect is maintaining a constant data rate from the disc. A simplified block diagram of the spindle motor is illustrated in Fig. 5.37, with the four important sections being a crystal reference oscillator, a phase locked loop voltage controlled oscillator, together with rough and fine servos.

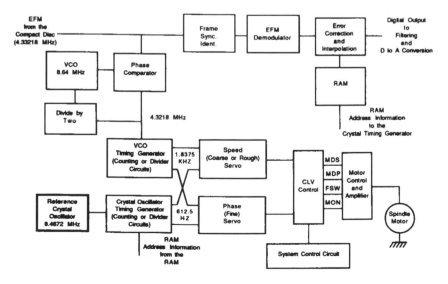

Figure 5.37 Block diagram of spindle or disc motor servo system.

Essentially the reference crystal oscillator is the 'master reference' with respect to the overall operation of the spindle servo. In most compact disc players the disc commences spinning once the laser has been switched on and focus has been achieved. The system control circuit 'kick' starts the motor, via the MON (motor ON) signal, and as the motor gathers speed a period occurs during which the effective data rate coming from the disc (the EFM signal) approaches 4.3218 MHz. The EFM information is passed to various sections but one particular path is via a phase comparator to enable a voltage controlled oscillator to produce an output frequency of 8.6436 MHz, which is twice the effective frequency from the disc, to enable improved frequency lock and stability of the VCO.

The VCO frequency is divided by two to produce the CD clock frequency which is passed to the VCO timing circuits to produce two frequencies of 612.5 Hz and 1.8375 MHz. At the same time the 'master reference' frequency of 8.4672 MHz is passed to the crystal oscillator timing circuits to produce the same two frequencies. The higher frequency of 1.8375 MHz is used to control the rough speed of the motor, whilst the lower frequency of 612.5 Hz becomes the phase or fine speed control.

The outputs from the coarse and fine servos are passed to the CLV control and then on to the motor control to provide the necessary signals to drive the spindle motor at the correct speed.

The MDS and MDP provide the required motor speed and phase signals, with the MON providing the motor on/off control. The FSW signal controls the operation of a filter circuit within the motor drive circuit to enable its sensitivity to be varied to provide improved motor control, with a lower sensitivity whilst the motor is running up to speed, and a higher sensitivity once the motor has achieved the correct speed during normal playing of the compact disc.

As the data is extracted from the disc, it is initially necessary to identify the sync. signal, in order to identify the 14-bit symbols that immediately follow. These symbols are now converted back into the original 8-bit words, and the restructured 16-bit words are fed to the random access memory (RAM) for the purpose of error correction and re-assembling the data words back into their correct order (de-interleaving).

Within the RAM there are defined upper and lower limits of acceptable data storage, identifiable by the RAM address lines, and should the storage fall below the lower acceptable level the disc is momentarily increased in speed, effectively to top up the RAM to within the accepted limits. Likewise, as the upper limit is reached then the disc speed is momentarily decreased to prevent overloading the memory, with subsequent loss of information. This fractional variation in speed is achieved by altering part of the dividing function within the crystal oscillator timing circuit to slightly alter the phase servo control to increase or decrease the speed accordingly.

Many references are made to the fact that the wow and flutter of compact disc is virtually immeasurable, whilst from the foregoing it is fairly obvious that the spindle motor suffers excessive wow and flutter, but the fact that the data is clocked through the random access memory at a rate linked to the crystal reference oscillator, the stability of which may be measured in so many parts per million, reduces wow and flutter to this same order of magnitude.

The Decoder

Much of the spindle servo circuitry is usually contained within the decoder, but despite its apparently small size as component within the compact disc player, usually about the size of a standard postage stamp, it is fair to say that the decoder does quite a lot of work, with the support of a small amount of external circuitry.

Figure 5.38 illustrates the main features of the decoder section of a compact disc player. Most of the decoder section of a compact disc player is contained within an integrated circuit, the size of which is comparable to that of a postage stamp.

Once the data from the disc has been initially processed from the original pits or bumps, the data is identified as the eight to fourteen modulation (EFM) signal and thus becomes the input to the decoder. Amongst the many processes that now takes place, this signal is used to phase lock a voltage controlled oscillator (VCO), usually operating at 8.64 MHz, twice the effective compact disc clock frequency. The sync signal is removed within the 23-bit shift register, and passed to the frame sync circuits, with the main bulk of the data now passing to the EFM demodulator. Within the EFM demodulator, the 14-bit symbols are restored to their original 8-bit counterparts, with the three coupling bits being virtually discarded. With the exception of the 8-bit control word, the remaining 8-bit data words are now paired together to form the original 16-bit data words that are in effect the digital samples of the original analogue information, but in a 'jumbled up' condition as a result of the interleaving process.

De-interleaving takes place with the use of the random access memory, with error correction being applied using interpolation and relevant substitutions from the RAM as applicable, resulting from the error correction information contained within the CRCC data from the disc. The 8-bit control word is detected by the sub-code detector, with the P and Q bits being utilised accordingly.

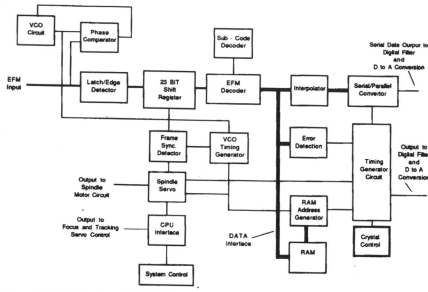

Figure 5.38 Basic decoder block diagram.

System control plays its own part in the operation of the CD player by 'kick starting' the disc and detecting that data is being retrieved from the disc. If any errors are being detected the system control will make decisions as to the type of correction to be applied, and if too many errors occur will, in due course, stop the operation of the player. An interface is also present to link up to the focus and tracking servos to ensure their correct operation.

The final digital data for processing into analogue can be either in serial or parallel form, with serial data being the normal requirement for domestic CD players, and is usually fed in modern players to a digital filter for the next stage of processing.

The complete decoder operation is strictly controlled by the crystal control circuit which acts as a master reference for the majority of the decoder functions.

Digital Filtering and Digital to Analogue Conversion

When the digital data has been converted back into its original analogue form, it is necessary to pass the analogue signals through a filter network to remove any effects of the sampling frequency, 44.1 kHz, which would appear as interference in the form of aliasing noise in the analogue output circuits. In order that the effects of the interference are minimised it is essential that the first subharmonic of the sampling frequency, 22.05 kHz is removed to ensure that any effects from 44.1 kHz will be minimal (Fig. 5.39). However the filter cut-off characteristics must operate from 20 kHz to 20.05 kHz, a cut-off

rate which is not possible with conventional inductive, capacitive, resistive filters without degrading the analogue top-end frequency response, which was a problem with the earlier compact disc players.

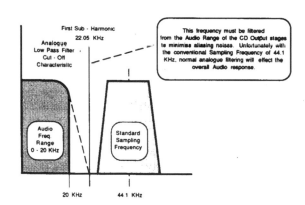

Figure 5.39 Problems with conventional sampling at Fs = 44.1 kHz.

Many developments have occurred within compact disc players with respect to improvements in the digital to analogue conversion and filtering stages, and whilst filtering is an area to be improved, it is worthwhile considering the restructuring of the analogue information from the digital data from the compact disc (Fig. 5.40).

As can be seen from Fig. 5.40 another problem occurs with respect to quantisation noise, whereby noise pulses can occur at the sampling points, which though at 44.1 kHz, can still cause effects to decrease the signal to noise ratio of the CD player.

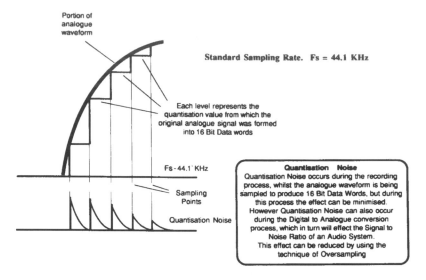

Figure 5.40   Restructing the original analogue information.

A method of overcoming both of these problems, i.e. reduction in the high frequency response in order to overcome aliasing noise and the effect of quantisation noise, is to introduce the technique of oversampling. But it must be remembered that the digital data on the compact disc is effectively 16-bit data words at a sampling frequency of 44.1 kHz and therefore cannot be altered, whereas techniques within the CD player can.

By the use of additional memory, delaying and multiplication techniques it is possible to make acceptable predictions of a data word that can be included between any two of the sampling points described in Fig. 5.40.

A simple method of describing the achievement of an additional data word which can be effectively 'fitted in between' two original samples is by interpolation, i.e. add two consecutive data words together, divide by two and fit the result in between the two original samples. Two important factors emerge from this concept:

1. The sampling frequency must double in frequency to 88.2 kHz.
2. Quantisation noise will reduce, thereby improving the signal to noise ratio.

The effective doubling of the sampling frequency ensures that a conventional filter can be used to remove the first sub-harmonic, which is now 44.1 kHz, with the effect of improving the high frequency response of the CD player.

Figures 5.41 (a) and (b) illustrate the effect of two times and four times oversampling techniques, though many players utilise even higher sampling frequencies, especially in the realms of bit stream digital to analogue methods which will be highlighted later.

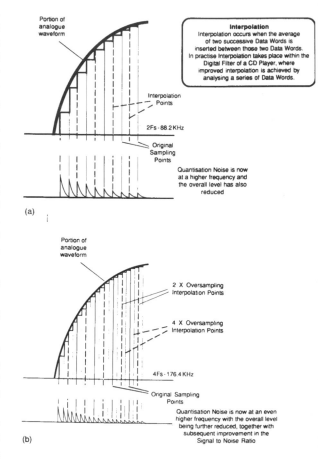

Figure 5.41   Oversampling: (a) two times oversampling. Fs = 88.2 kHz: (b) four times oversampling, Fs = 176.4 kHz.

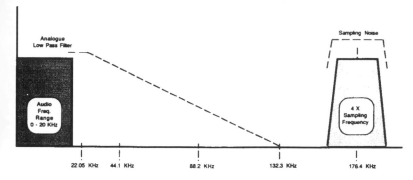

Figure 5.42 Improved filtering effects with 4 × oversampling, Fs = 176.4 kHz.

The frequency response of the filtering circuits in the output stages is illustrated in Fig. 5.42, where a new sampling frequency of four times the standard used on the compact disc, i.e. 176.4 kHz, enables a more conventional filter with less steep characteristics, to be designed.

Digital filtering

The increases in the sampling frequencies can be achieved by multiplication and delaying techniques as illustrated in Fig. 5.43.

From Fig. 5.42 it may be appreciated with the concept of four times oversampling it is possible to obtain three extra digitally produced data words, which are in fact predictions, to be inserted between any two actual samples from the compact disc.

With the technique of delaying the original 16-bit data words, and then multiplying each resultant word four times, each multiplication being a different co-efficient, a resultant series of 28-bit words are produced. By summing a complete sequence or group of these resultant 28-bit words, a weighted average of a large number of samples can be achieved, and whilst this description relates to four times oversampling, further improvements can be achieved using eight times oversampling, where the sampling frequency now becomes 352.8 kHz.

The resultant 28-bit data words are passed to noise shaping circuits where further improvements in the signal to noise ratio can be achieved.

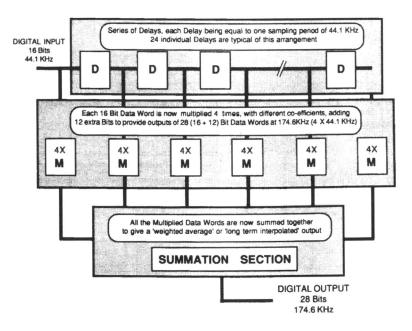

Figure 5.43 Basic illustration of digital filtering.

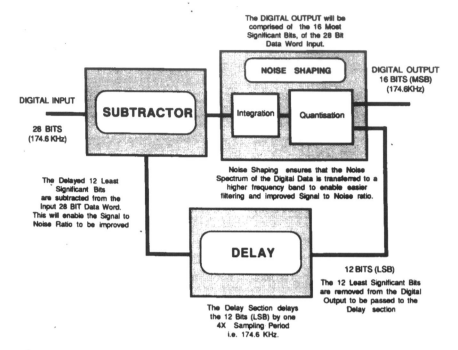

Figure 5.44   The noise shaping process.

Noise shaping

Noise shaping circuits (Fig. 5.44) can be used to improve the signal to noise ratio, by utilising the fact that the noise level of the signal is contained within the least significant bit area, and by extracting the 12-least significant bits and delaying them by one sampling period and then subtracting the resultant from the input signal, the output comprises the 16 most significant bits. Figure 5.42 provides an example of this concept, and can be identified as a single integration type of noise shaping circuit.

More complex types are available comprising double integration or even multi-stage noise shaping (MASH) circuits, all of which are progressions in the evolution, by various manufacturers, of the processing of the original 16-bit data words coming from the compact disc. The basic concept of these will be covered at a later stage.

The 16-bit data words from the noise shaping circuit can now be passed to the digital to analogue convertor for restoring back into the original analogue information.

Digital to analogue conversion

The 16-bit data words are passed to the D to A convertor where the data information in terms of 0s and 1s controls internal switches to enable a current to be produced that

ideally relates to the original analogue value of the signal when it was first recorded onto compact disc.

The simplified diagram (Fig. 5.45) illustrates a series of current generators, each of which provide a current that is equal to half of the current provided by its left-hand counterpart. The left-hand current generator will provide half of the total value of current available.

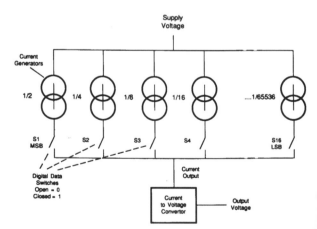

Figure 5.45   Digital to analogue conversion by current switching.

The 16-bit data word is passed into the circuit, and whenever a 1 occurs within the word, the relevant switch S1–S16 will close, and the sum of the currents related to where the 1s occur will pass the current to voltage convertor to provide an output which should be equivalent to the original analogue value at the moment of sampling or quantisation during the recording process.

Each subsequent data word will therefore enable the original analogue signal to be sequentially restructured, and therefore the speed of operation is related to the sampling frequency of the data words being applied. However D to A convertors suffer from inherent problems that can cause discrepancies in the output signal in the form of non-linearity errors, zero cross-over distortion and glitches.

Non-linearity errors

Each current generator is usually formed by a resistive network to enable a specific current to flow, and it is essential for each generator to be a specific fraction of the one next to it. Thus if the MSB current generator is designed to provide half of the total current required, then the next generator must provide half of that value and so forth. Any variations of these tolerances will not enable a faithful reproduction of the original analogue signal to be achieved.

Zero cross-over distortion

When the analogue signal is operating within its positive half cycle the 16 bit digital data will be the complement of its opposite negative value, thus as the analogue signal traverses the zero cross-over point the sequence of bits will be reversed. The MSB during the positive half cycle is always a 0, whilst during the negative half cycle the MSB will always be a 1. Therefore with low level signals and any non-linearity within the LSB area of the D to A convertor, distortion of the signal output can be caused.

Glitches

These are caused when any of the switches do not operate at the precise moment of the occurrence of each relevant bit, with the result that an incorrect output can be momentarily achieved.

To overcome some of these problems, 18-bit and 20-bit D to A convertors have been designed which enable a more faithful replication of the original signal to be achieved. Eighteen or 20-bit data words can be derived from the digital filtering or oversampling process, but whether it be a 16, 18 or even 20-bit D to A convertor, accurate manufacture of these items for the domestic compact disc player can prove extremely expensive.

Another method of digital to analogue conversion

proving to be extremely popular with modern players is the bit stream or one bit system.

As previously mentioned there has been a gradual evolution, by various manufacturers, regarding the processing of the digital data back into its original analogue form as accurately as possible, and noise shaping is an essential pre-requisite.

Multi-stage noise shaping

The digital input from the digital filter can be in the order of 16-bit data words at a sampling frequency of eight times the original sampling frequency (i.e. $8F_s = 8 \times 44.1$ kHz = 352.8 kHz). This implies that there are seven interpolated samples between any two data words from the compact disc. The multi-stage noise shaping circuit now processes the digital input at a much higher rate, in the order of 32 times the original sampling frequency (i.e. $32F_s = 44.1$ kHz $\times 32 = 1.411$ MHz); it would appear at this stage that frequencies are becoming inordinately high when compared to the original sampling frequency. But the implication now is that at $32F_s$ there must be 31 interpolated samples between any two data words from the disc, and as a result of this it is now not necessary to maintain such a high number of bits in each data word. In fact the output from the MASH circuit (Fig. 5.46) is in 4-bit form, which will provide sufficient detail to effectively indicate the analogue content in due course.

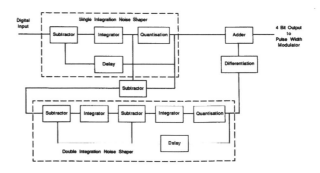

Figure 5.46 Multistage noise shaping.

The 4-bit output is passed to a pulse width modulator, in which 11 of the possible 4-bit combinations are used to provide 11 different pulse widths, the width of these pulses being related to the analogue information. By passing these varying pulse widths to a low-pass filter, or integrator, the required analogue output is obtained. (See Fig. 5.47.)

Whilst a high clock frequency is used, timing circuits will provide the relevant clock frequencies for the different circuits. The output from the pulse width modulator is described as 1-bit information insomuch that it is a series of single pulses the width of which is varying, up to eleven different widths, and is therefore not a stream of data.

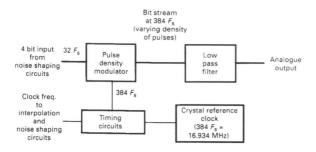

Figure 5.47   Pulse width modulation.

Passing the information through a low-pass filter enables the original analogue information to be retrieved with much greater accuracy and simplicity compared to the more conventional method of digital to analogue conversion, and therefore the problems of non-linearity, zero cross-over distortion, and glitches are resolved. There is, however, one main problem with this method and that is jitter (Fig. 5.48), which is related to the accuracy of the actual start and finishing points of each pulse width, and which can effect the final sound quality. This can be quite effectively resolved by a further development of the bit stream concept with the introduction of pulse density modulation, instead of pulse width modulation. In effect, a varying series of very high frequency pulses ($384\ F_s$ i.e. 16.934 MHz), are used, and again, passing these through the low-pass filter produces the desired analogue output.

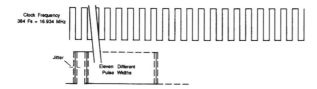

Figure 5.48   1 bit pulses, locked to the clock, but varying in width.

With the pulse density concept the circuit arrangement is virtually identical, the main difference being that instead of producing varying pulse widths, suffering from

possible jitter, a stream of $384\ F_s$ pulses is used, the number or density of which will relate to the original analogue output from the low-pass filter. (See Fig. 5.49.)

The pulse density modulation or bit stream principle is virtually the latest technology being utilised within compact disc players and it is difficult to see where further developments can occur. No doubt the Sony MiniDisc will be the next development as a source of permanent recorded material for domestic use, with the probability of digital audio tape (DAT) and/or digital compact cassette (DCC) being the main future concerning owner recordable material until the compact disc recorder makes its long awaited appearance for general domestic use.

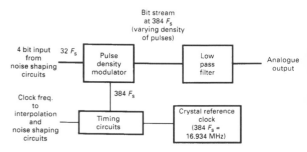

Figure 5.49   Pulse density modulation.

This compact disc technology section is intended for the reader to achieve an appreciation of most of the concepts being used with this particular aspect of recorded media, providing an overall view without emphasising any particular manufacturers methods or techniques. For any reader wishing to pursue any particular topic in greater depth there is available a wide range of publications of which a brief selection is offered below.

Bibliography

Baert, L., Theunissen, L. and Vergult, G., *Digital Audio and Compact Disc Technology*. Heinemann Newnes (1992).

Pohlman, K.C., *Principles of Digital Audio*. Howard Sams, ISBN 0-672-22634-0.

Pohlman, K.C., *Compact Disc Handbook*. Oxford University Press (1992).

Sinclair, I., *Introducing Digital Audio*. PC Publishing (1991).

Watkinson, J., *The Art of Digital Audio*. Focal Press, ISBN 0-240-51270-0.

6 Digital Audio Recording

John Watkinson

In the previous chapters the conversion from analogue signals into the digital domain was discussed. Once such a conversion has taken place, audio has become data, and a digital audio recorder is no more than a data recorder adapted to record samples from convertors. Provided that the original samples are reproduced with their numerical value unchanged and with their original timebase, a digital recorder causes no loss of information at all. The only loss of information is due to the conversion processes unless there is a design fault or the equipment needs maintenance. In this chapter John Watkinson explains the various techniques needed to record audio data.

Types of Media

There is considerably more freedom of choice of digital media than was the case for analogue signals, and digital media take advantage of the research expended in computer recording.

Digital media do not need to be linear, nor do they need to be noise-free or continuous. All they need to do is allow the player to be able to distinguish some replay event, such as the generation of a pulse, from the lack of such an event with reasonable rather than perfect reliability. In a magnetic medium, the event will be a flux change from one direction of magnetisation to another. In an optical medium, the event must cause the pickup to perceive a change in the intensity of the light falling on the sensor. In CD, the contrast is obtained by interference. In some discs it will be through selective absorption of light by dyes. In magneto-optical discs the recording itself is magnetic, but it is made and read using light.

Magnetic recording

Magnetic recording relies on the hysteresis of certain magnetic materials. After an applied magnetic field is removed, the material remains magnetised in the same direction. By definition the process is non-linear, and analogue magnetic recorders have to use bias to linearise it. Digital recorders are not concerned with the non-linearity, and HF bias is unnecessary.

Figure 6.1 shows the construction of a typical digital record head, which is just like an analogue record head. A magnetic circuit carries a coil through which the record current passes and generates flux. A non-magnetic gap forces the flux to leave the magnetic circuit of the head and penetrate the medium. The current through the head must be set to suit the coercivity of the tape, and is arranged to almost saturate the track. The amplitude of the current is constant, and recording is performed by reversing the direction of the current with respect to time. As the track passes the head, this is converted to the reversal of the magnetic field left on the tape with respect to distance. The recording is actually made just after the trailing pole of the record head where the flux strength from the gap is falling. The width of the gap is generally made quite large to ensure that the full thickness of the magnetic coating is recorded, although this cannot be done if the same head is intended to replay.

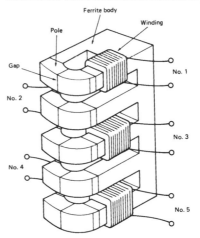

Figure 6.1 A typical ferrite head-windings are placed on alternate sides to save space, but parallel magnetic circuits have high crosstalk.

Figure 6.2 shows what happens when a conventional inductive head, i.e. one having a normal winding, is used to replay the track made by reversing the record current. The head output is proportional to the rate of change of flux and so only occurs at flux reversals. The polarity of the resultant pulses alternates as the flux changes and changes back. A circuit is necessary which locates the peaks of the pulses and outputs a signal corresponding to the original record current waveform.

The head shown in Fig. 6.2 has the frequency response shown in Fig. 6.3. At DC there is no change of flux and no output. As a result inductive heads are at a disadvantage at very low speeds. The output rises with frequency until the rise is halted by the onset of thickness loss. As the frequency rises, the recorded wavelength falls and flux from the shorter magnetic patterns cannot be picked up so far away. At some point, the wavelength becomes so short that flux from the back of the tape coating cannot reach the head and a decreasing thickness of tape contributes to the replay signal. In digital recorders using short wavelengths to obtain high density, there is no point in using thick coatings. As wavelength further reduces, the familiar gap loss occurs, where the head gap is too big to resolve detail on the track.

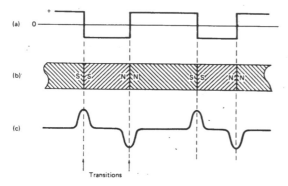

Figure 6.2   Basic digital recording. At (a) the write current in the head is reversed from time to time, leaving a binary magnetisation pattern shown at (b) When replayed, the waveform at (c) results because an output is only produced when flux in the head changes. Changes are referred to as transitions.

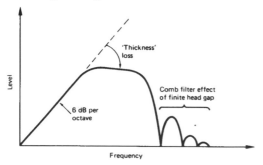

Figure 6.3   The major mechanisms defining magnetic channel bandwidth.

As can be seen, the frequency response is far from ideal, and steps must be taken to ensure that recorded data waveforms do not contain frequencies which suffer excessive losses.

A more recent development is the magneto-resistive (MR) head. This is a head which measures the flux on the tape rather than using it to generate a signal directly. Flux measurement works down to DC and so offers advantages at low tape speeds. Unfortunately flux measuring heads are not polarity conscious and if used directly they sense positive and negative flux equally, as shown in Fig. 6.4. This is overcome by using a small extra winding carrying a constant current. This creates a steady bias field which adds to the flux from the tape. The flux seen by the head now changes between two levels and a better output waveform results.

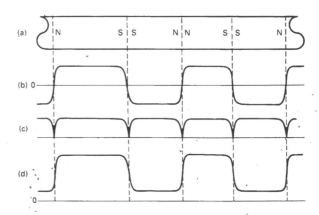

Figure 6.4   The sensing element in a magneto-resistive head is not sensitive to the polarity of the flux, only the magnitude. At (a) the track magnetisation is shown and this causes a bidirectional flux variation in the head as at (b) resulting in the magnitude output at (c) However, if the flux in the head due to the track is biased by an additional field, it can be made unipolar as at (d) and the correct output waveform is obtained.

Recorders which have low head-to-medium speed, such as DCC (Digital Compact Cassette) use MR heads, whereas recorders with high speeds, such as DASH (digital audio stationary head), RDAT (rotary head digital audio tape) and magnetic disc drives use inductive heads.

Heads designed for use with tape work in actual contact with the magnetic coating. The tape is tensioned to pull it against the head. There will be a wear mechanism and need for periodic cleaning.

In the hard disc, the rotational speed is high in order to reduce access time, and the drive must be capable of staying on line for extended periods. In this case the heads do not contact the disc surface, but are supported on a boundary layer of air. The presence of the air film causes spacing loss, which restricts the wavelengths at which the head can replay. This is the penalty of rapid access.

Digital audio recorders must operate at high density in order to offer a reasonable playing time. This implies that the shortest possible wavelengths will be used. Figure 6.5 shows that when two flux changes, or transitions, are recorded close together, they affect each other on replay. The amplitude of the composite signal is reduced, and the position of the peaks is pushed outwards. This is known as inter-symbol interference, or peak-shift distortion and it occurs in all magnetic media.

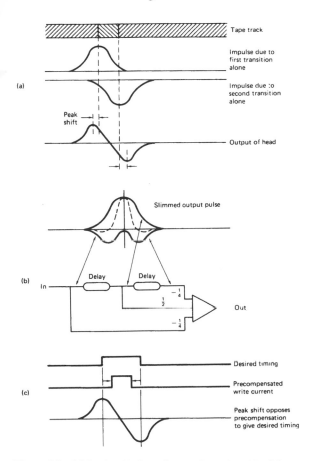

Figure 6.5 (a) Peak shift distortion can be reduced by (b) equalization in replay or (c) precompensation.

The effect is primarily due to high frequency loss and it can be reduced by equalisation on replay, as is done in most tapes, or by pre-compensation on record as is done in hard discs.

Optical discs

Optical recorders have the advantage that light can be focused at a distance whereas magnetism cannot. This means that there need be no physical contact between the pickup and the medium and no wear mechanism.

In the same way that the recorded wavelength of a magnetic recording is limited by the gap in the replay head, the density of optical recording is limited by the size of light spot which can be focused on the medium. This is controlled by the wavelength of the light used and by the aperture of the lens. When the light spot is as small as these limits allow, it is said to be diffraction limited. The recorded details on the disc are minute, and could easily be obscured by dust particles. In practice the information layer needs to be protected by a thick transparent coating. Light enters the coating well out of focus over a large area so that it can pass around dust particles, and comes to a focus within the thickness of the coating. Although the number of bits per unit area is high in optical recorders the number of bits per unit volume is not as high as that of tape because of the thickness of the coating.

Figure 6.6 shows the principle of readout of the compact disc which is a read-only disc manufactured by pressing. The track consists of raised bumps separated by flat areas. The entire surface of the disc is metallised, and the bumps are one quarter of a wavelength in height. The player spot is arranged so that half of its light falls on top of a bump, and half on the surrounding surface. Light returning from the flat surface has travelled half a wavelength further than light returning from the top of the bump, and so there is a phase reversal between the two components of the reflection. This causes destructive interference, and light cannot return to the pickup. It must reflect at angles which are outside the aperture of the lens and be lost. Conversely, when light falls on the flat surface between bumps, the majority of it is reflected back to the pickup. The pickup thus sees a disc apparently having alternately good or poor reflectivity.

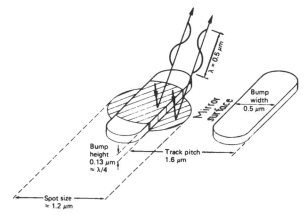

Figure 6.6 CD readout principle and dimensions. The presence of a bump causes destructive interference in the reflected light.

Some discs can be recorded once, but not subsequently erased or re-recorded. These are known as WORM (write once read mostly) discs. One type of WORM disc uses a thin metal layer which has holes punched in it on recording by heat from a laser. Others rely on the heat raising blisters in a thin metallic layer by decomposing the plastic material beneath. Yet another alternative is a layer of photo-chemical dye which darkens when struck by the high powered recording beam. Whatever the recording principle, light from the pickup is reflected more or less, or absorbed more or less, so that the pickup once more senses a change in reflectivity. Certain WORM discs can be read by conventional CD players and are thus called recordable CDs, whereas others will only work in a particular type of drive.

Magneto-optical discs

When a magnetic material is heated above its Curie temperature, it becomes demagnetised, and on cooling will assume the magnetisation of an applied field which would be too weak to influence it normally. This is the principle of magneto-optical recording used in the Sony MiniDisc. The heat is supplied by a finely focused laser, the field is supplied by a coil which is much larger.

Figure 6.7 assumes that the medium is initially magnetised in one direction only. In order to record, the coil is energised with the waveform to be recorded. This is too weak to influence the medium in its normal state, but when it is heated by the recording laser beam the heated area will take on the magnetism from the coil when it cools. Thus a magnetic recording with very small dimensions can be made.

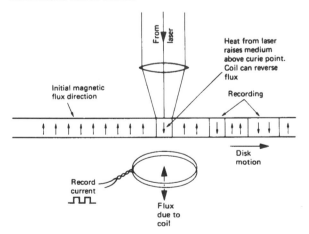

Figure 6.7   The thermomagneto-optical disk uses the heat from a laser to allow a magnetic field to record on the disk.

Readout is obtained using the Kerr effect, which is the rotation of the plane of polarisation of light by a magnetic field. The angle of rotation is very small and needs a sensitive pickup. The recording can be overwritten by reversing the current in the coil and running the laser continuously as it passes along the track.

A disadvantage of magneto-optical recording is that all materials having a Curie point low enough to be useful are highly corrodible by air and need to be kept under an effectively sealed protective layer.

All optical discs need mechanisms to keep the pickup following the track and sharply focused on it, and these will be discussed in the chapter on CD and need not be treated here.

The frequency response of an optical disc is shown in Figure 6.8. The response is best at DC and falls steadily to the optical cut-off frequency. Although the optics work down to DC, this cannot be used for the data recording. DC and low frequencies in the data would interfere with the focus and tracking servos. In practice the signal from the pickup is split by a filter. Low frequencies go to the servos, and higher frequencies go to the data circuitry. As a result the data channel has the same inability to handle DC as does a magnetic recorder, and the same techniques are needed to overcome it.

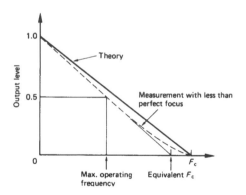

Figure 6.8   Frequency response of laser pickup. Maximum operating frequency is about half of cut-off frequency F_c.

Recording Media Compared

Of the various media discussed so far, it might be thought that one would be the best and would displace all the others. This has not happened because there is no one best medium; it depends on the application.

Random access memory (RAM) offers extremely short access time, but the volume of data generated by digital audio precludes the use of RAM for anything more than a few seconds because it would be too expensive. In addition loss of power causes the recording to be lost.

Tape has the advantage that it is thin and can be held compactly on reels. However, this slows down the access time because the tape has to be wound, or shuttled, to the appropriate place. Tape is, however, inexpensive, long lasting and is appropriate for archiving large quantities of data.

On the other hand, discs allow rapid access because their entire surface is permanently exposed and the positioner can move the heads to any location in a matter of milliseconds. The capacity is limited compared to tape because in the case of magnetic discs there is an air gap between the medium and the head. Exchangeable discs have to have a certain minimum head flying height below which the risk of contamination and a consequent head crash is too great. In Winchester technology the heads and disc are sealed inside a single assembly and contaminants can be excluded. In this case the flying height can be reduced and the packing density increased as a consequence. However, the disc is no longer exchangeable. In the case of optical discs the medium itself is extremely thick and multiple platter drives are impracticable because of the size of the optical pickup.

If the criterion is access time, discs are to be preferred. If the criterion is compact storage, tape is to be preferred. In computers, both technologies have been used in a complementary fashion for many years. In digital audio the same approach could be used, but to date the steps appear faltering.

In tape recording, the choice is between rotary and stationary heads. In a stationary head machine, the narrow tracks required by digital recordings result in heads with many parallel magnetic circuits, each of which requires its own read and write circuitry. Gaps known as guard bands must be placed between the tracks to reduce crosstalk. Guard bands represent wasted tape.

In rotary head machines, the tracks are laid down by a small number of rapidly rotating heads and less read/write circuitry is required. The space between the tracks is controlled by the linear tape speed and not by head geometry and so any spacing can be used. If azimuth recording is used, as described on page 108 no guard bands are necessary. A further advantage of rotary head recorders is that the high head to tape speed raises the frequency of the off-tape signals, and with a conventional inductive head this results in a larger playback signal compared to the thermal noise from the head and the preamplifiers.

As a result the rotary head tape recorder offers the highest storage density yet achieved, despite the fact that available formats are not yet in sight of any fundamental performance limits.

Some Digital Audio Processes Outlined

Whilst digital audio is a large subject, it is not necessarily a difficult one. Every process can be broken down into smaller steps, each of which is relatively easy to assimilate. The main difficulty with study is not following the simple step, but to appreciate where it fits in the overall picture. The next few sections illustrate various important processes in digital audio and show why they are necessary. Such processes are combined in various ways in real equipment.

The sampler

Figure 6.9 consists of an ADC which is joined to a DAC by way of a quantity of random access memory (RAM). What the device does is determined by the way in which the RAM address is controlled. If the RAM address increases by one every time a sample from the ADC is stored in the RAM, a recording can be made for a short period until the RAM is full. The recording can be played back by repeating the address sequence at the same clock rate but reading data from the memory into the DAC. The result is generally called a sampler. By running the replay clock at various rates, the pitch and duration of the reproduced sound can be altered. At a rate of one million bits per second, a megabyte of memory gives only eight seconds' worth of recording, so clearly samplers will be restricted to a fairly short playing time.

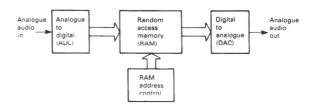

Figure 6.9 In the digital sampler, the recording medium is a random access memory (RAM). Recording time available is short compared to other media, but access to the recording is immediate and flexible as it is controlled by addressing the RAM.

Using data reduction, the playing time of a RAM based recorder can be extended. Some telephone answering machines take messages in RAM and eliminate the cassette tape. For pre-determined messages read only memory (ROM) can be used instead as it is non-volatile. Announcements in aircraft, trains and elevators are one application of such devices.

The programmable delay

If the RAM of Fig. 6.9 is used in a different way, it can be written and read at the same time. The device then becomes an audio delay. Controlling the relationship between the addresses then changes the delay. The addresses are generated by counters which overflow to zero after they have reached a maximum count. As a result the memory space appears to be circular as shown in Fig. 6.10. The read and write addresses are driven by a common clock and chase one another around the circle. If the read address follows close behind the write address, the delay is short. If it just stays ahead of the write address, the maximum delay is reached. Programmable delays are useful in TV studios where they allow audio to be aligned with video which has been delayed in various processes. They can also be used in auditoria to align the sound from various loudspeakers.

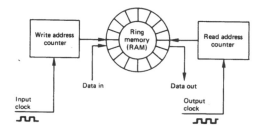

Figure 6.10   TBC memory is addressed by a counter which periodically overflows to give a ring structure. Memory allows read side to be non-synchronous with write side.

In digital audio recorders, a device with a circular memory can be used to remove irregularities from the replay data rate. The offtape data rate can fluctuate within limits but the output data rate can be held constant. A memory used in this way is called a timebase corrector. All digital recorders have timebase correctors to eliminate wow and flutter.

Time compression

When samples are converted, the ADC must run at a constant clock rate and it outputs an unbroken stream of samples. Time compression allows the sample stream to be broken into blocks for convenient handling.

Figure 6.11 shows an ADC feeding a pair of RAMs. When one is being written by the ADC, the other can be read, and vice versa. As soon as the first RAM is full, the ADC output switched to the input of the other RAM so that there is no loss of samples. The first RAM can then be read at a higher clock rate than the sampling rate. As a result the RAM is read in less time than it took to write it, and the output from the system then pauses until the second RAM is full. The samples are now time-compressed. Instead of being an unbroken stream which is difficult to handle, the samples are now arranged in blocks with convenient pauses in between them. In these pauses numerous processes can take place. A rotary head recorder might switch heads; a hard disc might move to another track. On a tape recording, the time compression of the audio samples allows time for synchronising patterns, subcode and error-correction words to be recorded.

In digital audio recorders which use video cassette recorders (VCRs) time compression allows the continuous audio samples to be placed in blocks in the unblanked parts of the video waveform, separated by synchronising pulses.

Subsequently, any time compression can be reversed by time expansion. Samples are written into a RAM at the incoming clock rate, but read out at the standard sampling

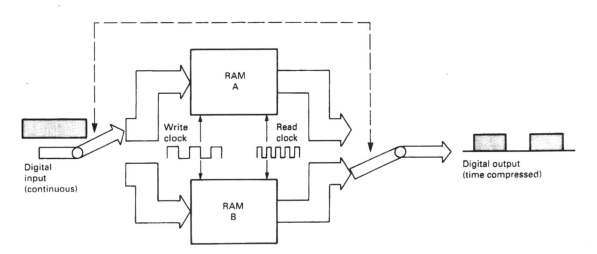

Figure 6.11   In time compression, the unbroken real-time stream of samples from an ADC is broken up into discrete blocks. This is accomplished by the configuration shown here. Samples are written into one RAM at the sampling rate by the write clock. When the first RAM is full, the switches change over, and writing continues into the second RAM whilst the first is read using a higher frequency clock. The RAM is read faster than it was written and so all of the data will be output before the other RAM is full. This opens spaces in the data flow which are used as described in the text.

rate. Unless there is a design fault, time compression is totally inaudible. In a recorder, the time-expansion stage can be combined with the timebase-correction stage so that speed variations in the medium can be eliminated at the same time. The use of time compression is universal in digital audio recording. In general the *instantaneous* data rate at the medium is not the same as the rate at the convertors, although clearly the *average* rate must be the same.

Another application of time compression is to allow more than one channel of audio to be carried on a single cable. If, for example, audio samples are time compressed by a factor of two, it is possible to carry samples from a stereo source in one cable.

In digital video recorders both audio and video data are time compressed so that they can share the same heads and tape tracks.

Synchronisation

In addition to the analogue inputs and outputs, connected to convertors, many digital recorders have digital inputs which allow the convertors to be bypassed. This mode of connection is desirable because there is no loss of quality in a digital transfer. Transfer of samples between digital audio devices is only possible if both use a common sampling rate and they are synchronised. A digital audio recorder must be able to synchronise to the sampling rate of a digital input in order to record the samples. It is frequently necessary for such a recorder to be able to play back locked to an external sampling rate reference so that it can be connected to, for example, a digital mixer. The process is already common in video systems but now extends to digital audio.

Figure 6.12 shows how the external reference locking process works. The timebase expansion is controlled by the external reference which becomes the read clock for the RAM and so determines the rate at which the RAM address changes. In the case of a digital tape deck, the write clock for the RAM would be proportional to the tape speed. If the tape is going too fast, the write address will catch up with the read address in the memory, whereas if the tape is going too slow the read address will catch up with the write address. The tape speed is controlled by subtracting the read address from the write address. The address difference is used to control the tape speed. Thus if the tape speed is too high, the memory will fill faster than it is being emptied, and the address difference will grow larger than normal. This slows down the tape.

Thus in a digital recorder the speed of the medium is constantly changing to keep the data rate correct. Clearly this is inaudible as properly engineered timebase correction totally isolates any instabilities on the medium from the data fed to the convertor.

In multitrack recorders, the various tracks can be synchronised to sample accuracy so that no timing errors can exist between the tracks. In stereo recorders image shift due to phase errors is eliminated.

In order to replay without a reference, perhaps to provide an analogue output, a digital recorder generates a

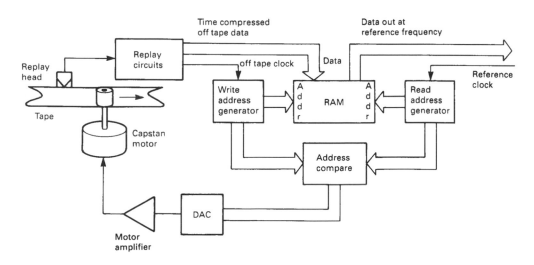

Figure 6.12 In a recorder using time compression, the samples can be returned to a continuous stream using RAM as a timebase corrector (TBC). The long-term data rate has to be the same on the input and output of the TBC or it will lose data. This is accomplished by comparing the read and write addresses and using the difference to control the tape speed. In this way the tape speed will automatically adjust to provide data as fast as the reference clock takes it from the TBC.

sampling clock locally by means of a crystal oscillator. Provision will be made on professional machines to switch between internal and external references.

Error correction and concealment

As anyone familiar with analogue recording will know, magnetic tape is an imperfect medium. It suffers from noise and dropouts, which in analogue recording are audible. In a digital recording of binary data, a bit is either correct or wrong, with no intermediate stage. Small amounts of noise are rejected, but inevitably, infrequent noise impulses cause some individual bits to be in error. Dropouts cause a larger number of bits in one place to be in error. An error of this kind is called a burst error. Whatever the medium and whatever the nature of the mechanism responsible, data are either recovered correctly, or suffer some combination of bit errors and burst errors. In compact disc, random errors can be caused by imperfections in the moulding process, whereas burst errors are due to contamination or scratching of the disc surface.

The audibility of a bit error depends upon which bit of the sample is involved. If the LSB of one sample was in error in a loud passage of music, the effect would be totally masked and no-one could detect it. Conversely, if the MSB of one sample was in error in a quiet passage, no-one could fail to notice the resulting loud transient. Clearly a means is needed to render errors from the medium inaudible. This is the purpose of error correction.

In binary, a bit has only two states. If it is wrong, it is only necessary to reverse the state and it must be right. Thus the correction process is trivial and perfect. The main difficulty is in identifying the bits which are in error. This is done by coding the data by adding redundant bits. Adding redundancy is not confined to digital technology, airliners have several engines and cars have twin braking systems. Clearly the more failures which have to be handled, the more redundancy is needed. If a four-engined airliner is designed to fly normally with one engine failed, three of the engines have enough power to reach cruise speed, and the fourth one is redundant. The amount of redundancy is equal to the amount of failure which can be handled. In the case of the failure of two engines, the plane can still fly, but it must slow down; this is graceful degradation. Clearly the chances of a two-engine failure on the same flight are remote.

In digital audio, the amount of error which can be corrected is proportional to the amount of redundancy and within this limit the samples are returned to exactly their original value. Consequently *corrected* samples are inaudible. If the amount of error exceeds the amount of redundancy, correction is not possible, and, in order to allow graceful degradation, concealment will be used. Concealment is a process where the value of a missing sample is estimated from those nearby. The estimated sample value is not necessarily exactly the same as the original, and so under some circumstances concealment can be audible, especially if it is frequent. However, in a well designed system, concealments occur with negligible frequency unless there is an actual fault or problem.

Concealment is made possible by re-arranging or shuffling the sample sequence prior to recording. This is shown in Fig. 6.13 where odd-numbered samples are separated from even-numbered samples prior to recording. The odd and even sets of samples may be recorded in different places, so that an uncorrectable burst error only affects one set. On replay, the samples are recombined into their natural sequence, and the error is now split up so that it results in every other sample being lost. The waveform is now described half as often, but can still be reproduced with some loss of accuracy. This is better than not being reproduced at all even if it is not perfect. Almost all digital recorders use such an odd/even shuffle for concealment. Clearly if any errors are fully correctable, the shuffle is a waste of time; it is only needed if correction is not possible.

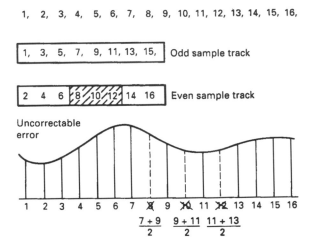

Figure 6.13 In cases where the error correction is inadequate, concealment can be used provided that the samples have been ordered appropriately in the recording. Odd and even samples are recorded in different places as shown here. As a result an uncorrectable error causes incorrect samples to occur singly, between correct samples. In the example shown, sample 8 is incorrect, but samples 7 and 9 are unaffected and an approximation to the value of sample 8 can be had by taking the average value of the two. This interpolated value is substituted for the incorrect value.

In high density recorders, more data are lost in a given sized dropout. Adding redundancy equal to the size of a dropout to every code is inefficient. Figure 6.14(a) shows that the efficiency of the system can be raised using interleaving. Sequential samples from the ADC are assembled into codes, but these are not recorded in their

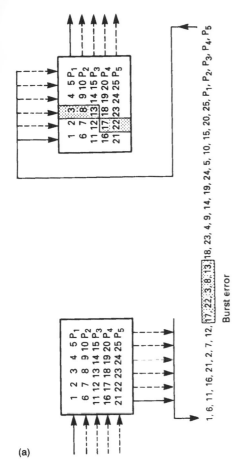

(a)

Figure 6.14 (a) Interleaving is essential to make error correction schemes more efficient. Samples written sequentially in rows into a memory have redundancy P added to each row. The memory is then read in columns and the data are sent to the recording medium. On replay the non-sequential samples from the medium are de-interleaved to return them to their normal sequence. This breaks up the burst error (shaded) into one error symbol per row in the memory, which can be corrected by the redundancy P.

natural sequence. A number of sequential codes are assembled along rows in a memory. When the memory is full, it is copied to the medium by reading down columns. On replay, the samples need to be de-interleaved to return them to their natural sequence. This is done by writing samples from tape into a memory in columns, and when it is full, the memory is read in rows. Samples read from the memory are now in their original sequence so there is no effect on the recording. However, if a burst

error occurs on the medium, it will damage sequential samples in a vertical direction in the de-interleave memory. When the memory is read, a single large error is broken down into a number of small errors whose size is exactly equal to the correcting power of the codes and the correction is performed with maximum efficiency.

An extension of the process of interleave is where the memory array has not only rows made into codewords, but also columns made into codewords by the addition of vertical redundancy. This is known as a product code. Figure 6.14(b) shows that in a product code the redundancy calculated first and checked last is called the outer code, and the redundancy calculated second and checked first is called the inner code. The inner code is formed along tracks on the medium. Random errors due to noise are corrected by the inner code and do not impair the burst correcting power of the outer code. Burst errors are declared uncorrectable by the inner code which flags the bad samples on the way into the de-interleave memory. The outer code reads the error flags in order to locate the erroneous data. As it does not have to compute the error locations, the outer code can correct more errors.

An alternative to the product block code is the convolutional cross interleave, shown in Fig. 6.14(c). In this system, the data are formed into an endless array and the code words are produced on columns and diagonals. The compact disc and DASH formats use such a system because it needs less memory than a product code.

The interleave, de-interleave, time-compression and timebase-correction processes cause delay and this is evident in the time taken before audio emerges after starting a digital machine. Confidence replay takes place later than the distance between record and replay heads would indicate. In DASH format recorders, confidence replay is about one-tenth of a second behind the input. Synchronous recording requires new techniques to overcome the effect of the delays.

The presence of an error-correction system means that the audio quality is independent of the tape/head quality within limits. There is no point in trying to assess the health of a machine by listening to it, as this will not reveal whether the error rate is normal or within a whisker of failure. The only useful procedure is to monitor the frequency with which errors are being corrected, and to compare it with normal figures. Professional digital audio equipment should have an error rate display.

Some people claim to be able to hear error correction and misguidedly conclude that the above theory is flawed. Not all digital audio machines are properly engineered, however, and if the DAC shares a common power supply with the error correction logic, a burst of errors will raise the current taken by the logic, which loads the power supply and can interfere with the

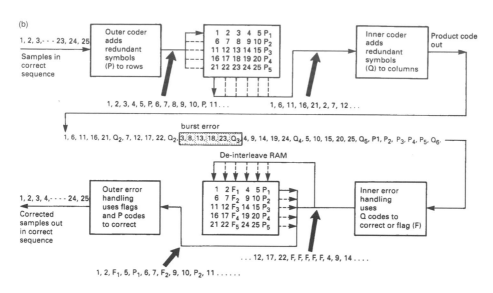

Figure 6.14   (b) In addition to the redundancy P on rows, inner redundancy Q is also generated on columns. On replay, the Q code checker will pass on flags F if it finds an error too large to handle itself. The flags pass through the de-interleave process and are used by the outer error correction to identify which symbol in the row needs correcting with P redundancy. The concept of crossing two codes in this way is called a product code.

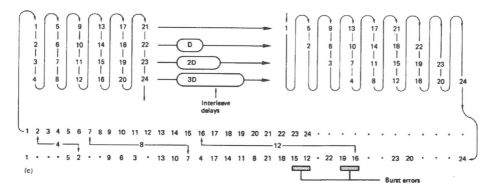

Figure 6.14   At (c) convolutional interleave is shown. Instead of assembling samples in blocks, the process is continuous and uses RAM delays. Samples are formed into columns in an endless array. Each row of the array is subject to a different delay so that after the delays, samples in a column are available simultaneously which were previously on a diagonal. Code words which cross one another at an angle can be obtained by generating redundancy before and after the delays.

operation of the DAC. The effect is harder to eliminate in small battery powered machines where space for screening and decoupling components is hard to find, but it is only a matter of design: there is no flaw in the theory.

Channel coding

In most recorders used for storing digital information, the medium carries a track which reproduces a single waveform. Clearly data words representing audio samples contain many bits and so they have to be recorded serially, a bit at a time. Some media, such as CD, only have one track, so it must be totally self contained. Other media, such as digital compact cassette (DCC) have many parallel tracks. At high recording densities, physical tolerances cause phase shifts, or timing errors, between parallel tracks and so it is not possible to read them in parallel. Each track must still be self-contained until the replayed signal has been timebase corrected.

Recording data serially is not as simple as connecting the serial output of a shift register to the head. In digital audio, a common sample value is all zeros, as this corresponds to silence. If a shift register is loaded with all zeros and shifted out serially, the output stays at a constant low level, and nothing is recorded on the track. On replay there is nothing to indicate how many zeros were present, or even how fast to move the medium. Clearly serialised raw data cannot be recorded directly, it has to be modulated in to a waveform which contains an embedded clock irrespective of the values of the bits in the samples. On replay a circuit called a data separator can lock to the embedded clock and use it to separate strings of identical bits.

The process of modulating serial data to make it self-clocking is called channel coding. Channel coding also shapes the spectrum of the serialised waveform to make it more efficient. With a good channel code, more data can be stored on a given medium. Spectrum shaping is used in CD to prevent the data from interfering with the focus and tracking servos, and in RDAT to allow re-recording without erase heads.

A self-clocking code contains a guaranteed minimum number of transitions per unit time, and these transitions must occur at multiples of some basic time period so that they can be used to synchronise a phase locked loop. Figure 6.15 shows a phase locked loop which contains an oscillator whose frequency is controlled by the phase error between input transitions and the output of a divider. If transitions on the medium are constrained to occur at multiples of a basic time period, they will have a constant phase relationship with the oscillator, which can stay in lock with them even if they are intermittent. As the damping of the loop is a low-pass filter, jitter in the incoming transitions, caused by peak-shift distortion or by speed variations in the medium will be rejected and the oscillator will run at the average frequency of the off-tape signal. The phase locked loop must be locked before data can be recovered, and to enable this, every data block is preceded by a constant frequency recording known as a preamble. The beginning of the data is identified by a unique pattern known as a sync. pattern.

Irrespective of the channel code used, transitions always occur separated by a range of time periods which are all multiples of the basic clock period. If such a replay signal is viewed on an oscilloscope a characteristic display called an eye pattern is obtained. Figure 6.16 shows an eye pattern, and in particular the regular openings in the trace. A decision point is in the centre of each opening, and the phase locked loop acts to keep it centred laterally, in order to reject the maximum amount of jitter. At each decision point along the time axis, the waveform is above or below the point, and can be returned to a binary signal.

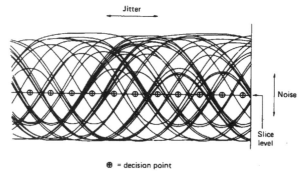

⊕ = decision point

Figure 6.16 At the decision points, the receiver must make binary decisions about the voltage of the signal, whether it is above or below the slicing level. If the eyes remain open, this will be possible in the presence of noise and jitter.

Occasionally noise or jitter will cause the waveform to pass the wrong side of a decision point, and this will result in an error which will require correction.

Figure 6.17 shows an extremely simple channel code known as FM (frequency modulation) which is used for the AES/EBU digital interface and for recording time code on tape.

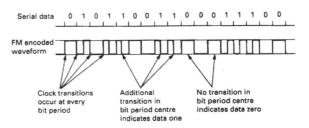

Figure 6.17 FM channel code, also known as Manchester code or bi-phase mark (BMC) is used in AES/EBU interface and for timecode recording. The waveform is encoded as shown here. See text for details.

Every bit period begins with a transition, irrespective of the value of the bit. If the bit is a one, an additional transition is placed in the centre of the bit period. If the bit is a zero, this transition is absent. As a result, the waveform is always self-clocking irrespective of the values of the data bits. Additionally, the waveform spends as much

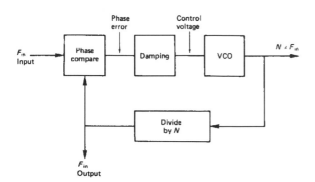

Figure 6.15 A typical phase-locked loop where the VCO is forced to run at a multiple of the input frequency. If the input ceases, the output will continue for a time at the same frequency until it drifts.

time in the low state as it does in the high state. This means that the signal has no DC component, and it will pass through capacitors, magnetic heads and transformers equally well. However simple FM may be, it is not very efficient because it requires two transitions for every bit and jitter of more than half a bit cannot be rejected.

More recent products use a family of channel codes known as group codes. In group codes, groups of bits, commonly eight, are associated together into a symbol for recording purposes. Eight-bit symbols are common in digital audio because two of them can represent a 16-bit sample. Eight-bit data have 256 possible combinations, but if the waveforms obtained by serialising them are examined, it will be seen that many combinations are unrecordable. For example all ones or all zeros cannot be recorded because they contain no transitions to lock the clock and they have excessive DC content. If a larger number of bits is considered, a greater number of combinations is available. After the unrecordable combinations have been rejected, there will still be 256 left which can each represent a different combination of eight bits. The larger number of bits are channel bits; they are not data because all combinations are not recordable. Channel bits are simply a convenient way of generating recordable waveforms. Combinations of channel bits are selected or rejected according to limits on the maximum and minimum periods between transitions. These periods are called run-length limits and as a result group codes are often called run-length-limited codes.

In RDAT, an 8/10 code is used where 8 data bits are represented by 10 channel bits. Figure 6.18 shows that this results in jitter rejection of 80% of a data bit period: rather better than FM. Jitter rejection is important in RDAT because short wavelengths are used and peak shift will occur. The maximum wavelength is also restricted in RDAT so that low frequencies do not occur.

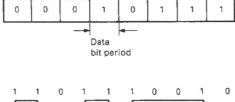

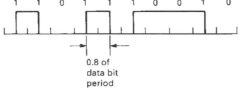

Figure 6.18   In RDAT an 8/10 code is used for recording. Each eight data bits are represented by a unique waveform generated by ten channel bits. A channel bit one causes a transition to be recorded. The transitions cannot be closer than 0.8 of a data bit, and this is the jitter resistance. This is rather better than FM which has a jitter window of only 0.5 bits.

In CD, an 8/14 code is used where 8 data bits are represented by 14 channel bits. This only has a jitter rejection of 8/14 of a data bit, but this is not an issue because the rigid CD has low jitter. However, in 14 bits there are 16K combinations, and this is enough to impose a minimum run length limit of three channel bits. In other words transitions on the disc cannot occur closer than 3 channel bits apart. This corresponds to 24/14 data bits. Thus the frequency generated is less than the bit rate and a result is that more data can be recorded on the disc than would be possible with a simple code.

Hard Disc Recorders

The hard disc recorder stores data on concentric tracks which it accesses by moving the head radially. Rapid access drives move the heads with a moving coil actuator, whereas lower cost units will use stepping motors which work more slowly. The radial position of the head is called the cylinder address, and as the disc rotates, data blocks, often called sectors, pass under the head. To increase storage capacity, many discs can be mounted on a common spindle, each with its own head. All the heads move on a common positioner. The operating surface can be selected by switching on only one of the heads. When one track is full, the drive must select another head. When every track at that cylinder is full, the drive must move to another cylinder. The drive is not forced to operate in this way; it is equally capable of obtaining data blocks in any physical sequence from the disc.

Clearly while the head is moving it cannot transfer data. Using time compression to smooth out the irregular data transfer, a hard disc drive can be made into an audio recorder with the addition of a certain amount of memory.

Figure 6.19 shows the principle. The instantaneous data rate of the disc drive is far in excess of the sampling rate at the convertor, and so a large time-compression factor can be used. The disc drive can read a block of data from disc, and place it in the timebase corrector in a fraction of the real time it represents in the audio waveform. As the timebase corrector steadily advances through the memory, the disk drive has time to move the heads to another track before the memory runs out of data. When there is sufficient space in the memory for another block, the drive is commanded to read, and fills up the space. Although the data transfer at the medium is highly discontinuous, the buffer memory provides an unbroken stream of samples to the DAC and so continuous audio is obtained.

Recording is performed by using the memory to assemble samples until the contents of one disc block are available. These are then transferred to disc at high data rate. The drive can then reposition the head before the next block is available in memory.

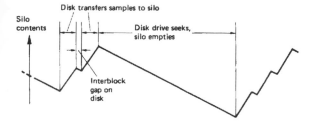

Figure 6.19 During an audio replay sequence, the silo is constantly emptied to provide samples, and is refilled in blocks by the drive.

An advantage of hard discs is that access to the audio is much quicker than with tape, as all of the data are available within the time taken to move the head. This speeds up editing considerably.

After a disc has been in use for some time, the free blocks will be scattered all over the disc surface. The random access ability of the disc drive means that a continuous audio recording can be made on physically discontinuous blocks. Each block has a physical address, known as the block address, which the drive controller can convert into cylinder and head selection codes to locate a given physical place on the medium. The size of each block on the disc is arranged to hold the number of samples which arrive during a whole number of time-code frames. It is then possible to link each disc block address used during a recording with the time code at which it occurred. The time codes and the corresponding blocks are stored in a table. The table is also recorded on the disc when the recording is completed.

In order to replay the recording, the table is retrieved from the disc, and a time code generator is started at the first code. As the generator runs, each code is generated in sequence, and the appropriate data block is read from the disc and placed in memory, where it can be fed to the convertor.

If it is desired to replay the recording from elsewhere than the beginning, the time code generator can be forced to any appropriate setting, and the recording will play from there. If an external device, such as a video-recorder, provides a time code signal, this can be used instead of the internal time code generator, and the machine will automatically synchronise to it.

The transfer rate and access time of the disc drive is such that if sufficient memory and another convertor are available, two completely independent playback processes can be supplied with data by the same drive. For the purpose of editing, two playback processes can be controlled by one time code generator. The time code generator output can be offset differently for each

process, so that they can play back with any time relationship. If it is required to join the beginning of one recording to the end of another, the operator specifies the in-point on the second recording and the out-point on the second recording. By changing the time code offsets, the machine can cause both points to occur simultaneously in the data accessed from the disc and played from memory. In the vicinity of the edit points, both processes are providing samples simultaneously and a crossfade of any desired length can be made between them.

The arrangement of data on the disc surface has a bearing on the edit process. In the worst case, if all the blocks of the first recording were located at the outside of the disc and all of the blocks of the second recording were located at the inside, the positioner would spend a lot of time moving. If the blocks for all recordings are scattered over the entire disc surface, the average distance the positioner needs to move is reduced.

The edit can be repeated with different settings as often as necessary without changing the original recordings. Once an edit is found to be correct, it is only necessary to store the handful of instructions which caused it to happen, and it can be executed at any time in the future in the same way. The operator has the choice of archiving the whole disc contents on tape, so different edits can be made in the future, or simply recording the output of the current edit so that the disc can be freed for another job.

The rapid access and editing accuracy of hard disc systems make them ideal for assembling sound effects to make the sound tracks of motion pictures.

The use of data reduction allows the recording time of a disc to be extended considerably. This technique is often used in plug-in circuit boards which are used to convert a personal computer into a digital audio recorder.

The PCM Adaptor

The PCM adaptor was an early solution to recording the wide bandwidth of PCM audio before high density digital recording developed. The video recorder offered sufficient bandwidth at moderate tape consumption. Whilst they were a breakthrough at the time of their introduction, by modern standards PCM adaptors are crude and obsolescent, offering limited editing ability and slow operation.

Figure 6.20 shows the essential components of a digital audio recorder using this technique. Input analogue audio is converted to digital and time compressed to fit into the parts of the video waveform which are not blanked. Time compressed samples are then odd-even shuffled to allow concealment. Next, redundancy is added and the data are interleaved for recording. The data are serialised and set on the active line of the video signal as black and white levels shown in Fig. 6.21. The video is sent to the recorder, where the analogue FM modulator switches between two frequencies representing the black and white levels, a system called frequency shift keying (FSK). This takes the place of the channel coder in a conventional digital recorder.

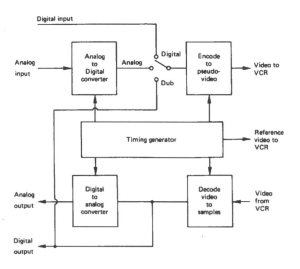

Figure 6.20   Block diagram of PCM adaptor. Note the dub connection needed for producing a digital copy between two VCRs.

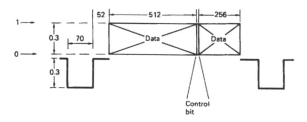

Figure 6.21   Typical line of video from PCM-1610. The control bit conveys the setting of the pre-emphasis switch or the sampling rate depending on position in the frame. The bits are separated using only the timing information in the sync pulses.

On replay the FM demodulator of the video recorder acts to return the FSK recording to the black/white video waveform which is sent to the PCM adaptor. The PCM adaptor extracts a clock from the video sync. pulses and uses it to separate the serially recorded bits. Error correction is performed after de-interleaving, unless the errors are too great, in which case concealment is used after the de-shuffle. The samples are then returned to the standard sampling rate by the timebase expansion process, which also eliminates any speed variations from the recorder. They can then be converted back to the analogue domain.

In order to synchronise playback to a reference and to simplify the circuitry, a whole number of samples is recorded on each unblanked line. The common sampling rate of 44.1 kHz is obtained by recording three samples per line on 245 active lines at 60 Hz. The sampling rate is thus locked to the video sync. frequencies and the tape is made to move at the correct speed by sending the video recorder syncs which are generated in the PCM adaptor.

An Open Reel Digital Recorder

Figure 6.22 shows the block diagram of a machine of this type. Analogue inputs are converted to the digital domain by converters. Clearly there will be one convertor for every audio channel to be recorded. Unlike an analogue machine, there is not necessarily one tape track per audio channel. In stereo machines the two channels of audio samples may be distributed over a number of tracks each in order to reduce the tape speed and extend the playing time.

The samples from the convertor will be separated into odd and even for concealment purposes, and usually one set of samples will be delayed with respect to the other before recording. The continous stream of samples from the convertor will be broken into blocks by time compression prior to recording. Time compression allows the insertion of edit gaps, addresses and redundancy into the data stream. An interleaving process is also necessary to re-order the samples prior to recording. As explained above, the subsequent de-interleaving breaks up the effects of burst errors on replay.

The result of the processes so far is still raw data, and these will need to be channel coded before they can be recorded on the medium. On replay a data separator reverses the channel coding to give the original raw data with the addition of some errors. Following de-interleave, the errors are reduced in size and are more readily correctable. The memory required for de-interleave may double as the timebase correction memory, so that variations in the speed of the tape are rendered indetectible. Any errors which are beyond the power of the correction system will be concealed after the odd-even shift is reversed. Following conversion in the DAC an analogue output emerges.

On replay a digital recorder works rather differently to an analogue recorder, which simply drives the tape at constant speed. In contrast, a digital recorder drives the tape at constant sampling rate. The timebase corrector works by reading samples out to the convertor at constant frequency. This reference frequency comes typically from a crystal oscillator. If the tape goes too fast, the memory will be written faster than it is being read,

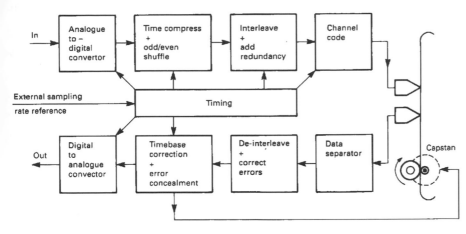

Figure 6.22 Block diagram of one channel of a stationary head digital audio recorder. See text for details of the function of each block. Note the connection from the timebase corrector to the capstan motor so that the tape is played at such a speed that the TBC memory neither underflows nor overflows.

and will eventually overflow. Conversely, if the tape goes too slow, the memory will become exhausted of data. In order to avoid these problems, the speed of the tape is controlled by the quantity of data in the memory. If the memory is filling up, the tape slows down, if the memory is becoming empty, the tape speeds up. As a result, the tape will be driven at whatever speed is necessary to obtain the correct sampling rate.

Rotary Head Digital Recorders

The rotary head recorder borrows technology from videorecorders. Rotary heads have a number of advantages over stationary heads. One of these is extremely high packing density: the number of data bits which can be recorded in a given space. In a digital audio recorder packing density directly translates into the playing time available for a given size of the medium.

In a rotary head recorder, the heads are mounted in a revolving drum and the tape is wrapped around the surface of the drum in a helix as can be seen in Fig. 6.23. The helical tape path results in the heads traversing the tape in a series of diagonal or slanting tracks. The space between the tracks is controlled not by head design but by the speed of the tape and in modern recorders this space is reduced to zero with corresponding improvement in packing density.

The added complexity of the rotating heads and the circuitry necessary to control them is offset by the improvement in density. The discontinuous tracks of the rotary head recorder are naturally compatible with time compressed data. As Fig. 6.24 illustrates, the audio

samples are time compressed into blocks each of which can be contained in one slant track.

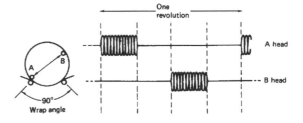

Figure 6.23 Rotary head recorder. Helical scan records long diagonal tracks.

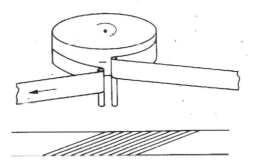

Figure 6.24 The use of time compression reduces the wrap angle necessary, at the expense of raising the frequencies in the channel.

In a machine such as RDAT (rotary head digital audio tape) there are two heads mounted on opposite sides of the drum. One rotation of the drum lays down two tracks. Effective concealment can be had by recording odd numbered samples on one track of the pair and even numbered samples on the other. Samples from the two audio channels are multiplexed into one data stream which is shared between the two heads.

As can be seen from the block diagram shown in Fig. 6.25 a rotary head recorder contains the same basic steps as any digital audio recorder. The record side needs ADCs, time compression, the addition of redundancy for error correction, and channel coding. On replay the channel coding is reversed by the data separator, errors are broken up by the de-interleave process and corrected or concealed, and the time compression and any fluctuations from the transport are removed by timebase correction. The corrected, time stable, samples are then fed to the DAC.

One of the reasons for the phenomenal recording density at which RDAT operates is the use of azimuth recording. In this technique, alternate tracks on the tape are laid down with heads having different azimuth angles. In a two-headed machine this is easily accommodated by having one head set at each angle. If the correct azimuth head subsequently reads the track there is no difficulty, but as Fig. 6.26 shows, the wrong head suffers a gross azimuth error.

Azimuth error causes phase shifts to occur across the width of the track and, at some wavelengths, this will result in cancellation except at very long wavelengths where the process is no longer effective. The use of 8/10 channel coding in RDAT ensures that no low frequencies are present in the recorded signal and so this characteristic of azimuth recording is not a problem. As a result the pickup of signals from the adjacent track is effectively prevented, and the tracks can be physically touching with no guard bands being necessary.

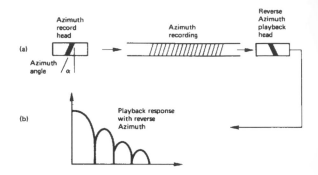

Figure 6.26 In azimuth recording (a), the head gap is tilted. If the track is played with the same head, playback is normal, but the response of the reverse azimuth head is attenuated (b).

As the azimuth system effectively isolates the tracks from one another, the replay head can usefully be made wider than the track. A typical figure is 50 per cent wider.

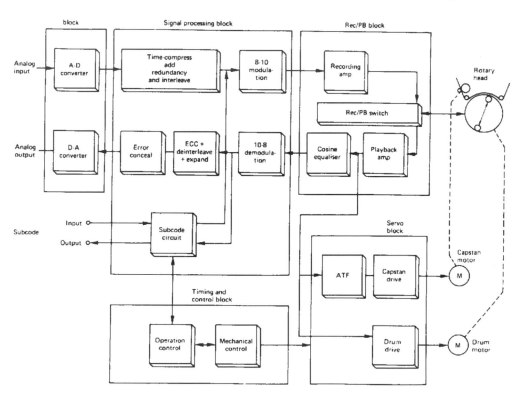

Figure 6.25   Block diagram of RDAT.

Tracking error of up to +/– 25 per cent of the track width then causes no loss of signal quality.

In practice the same heads can also be used for recording, even though they are too wide. As can be seen in Fig. 6.27, the excess track width is simply overwritten during the next head sweep. Erase heads are unnecessary, as the overlapping of the recorded tracks guarantees that the whole area of a previous recording is overwritten. A further advantage of the system is that more than one track width can be supported by the same mechanism simply by changing the linear tape speed. Pre-recorded tapes made by contact duplication have lower coercivity coatings, and to maintain the signal level the tracks are simply made wider by raising the tape speed. Any RDAT machine can play such a recording without adjustment.

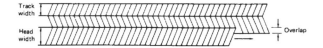

Figure 6.27 In azimuth recording, the tracks can be made narrower than the head pole by overwriting the previous track.

In any rotary head recorder, some mechanism is necessary to synchronise the linear position of the tape to the rotation of the heads, otherwise the recorded tracks cannot be replayed. In a conventional video recorder, this is the function of the control track which requires an additional, stationary, head. In RDAT the control track is dispensed with, and tracking is obtained by reading patterns in the slant tracks with the normal playback heads.

Figure 6.28 shows how the system works. The tracks are divided into five areas. The PCM audio data are in the centre, and the subcode data are at the ends. The audio and subcode data are separated by tracking patterns. The tracking patterns are recorded and played back along with the data. The tracking is measured by comparing the level of a pilot signal picked up from the tracks on each side of the wanted track. If the replay head drifts towards one side, it will overlap the next track on that side by a greater amount and cause a larger pilot signal to be picked up. Pilot pick-up from the track on the opposite side will be reduced. The difference between the pilot levels is used to change the speed of the capstan which has the effect of correcting the tracking.

Ordinarily, azimuth effect prevents the adjacent tracks being read, but the pilot tones are recorded with a wavelength much longer than that of the data. They can then be picked up by a head of the wrong azimuth.

The combination of azimuth recording, an active tracking system and high coercivity tape (1500 Oersteds compared to 200 Oersteds for analogue audio tape) allows the tracks to be incredibly narrow. Heads of 20 micrometres width, write tracks 13 micrometres wide.

About ten such tracks will fit in the groove of a vinyl disc. Although the head drum spins at 2000 rpm, the tape speed needed is only 8.15 millimetres per second.

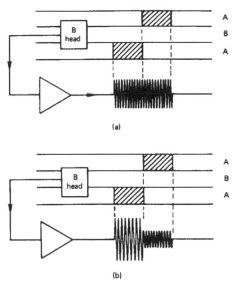

Figure 6.28 (a) A correctly tracking head produces pilot-tone bursts of identical amplitude. (b) The head is off-track, and the first pilot burst becomes larger, whereas the second becomes smaller. This produces the tracking error.

The subcode of RDAT functions in a variety of ways. In consumer devices, the subcode works in the same way as in CD, having a table of contents and flags allowing rapid access to the beginning of tracks and carrying signals to give a playing time readout.

In professional RDAT machines, the subcode is used to record timecode. A timecode format based on hours, minutes, seconds and DAT frames (where a DAT frame is one drum revolution) is recorded on the tape, but suitable machines can convert the tape code to any video, audio or film time code and operate synchronised to a timecode reference. As the heads are wider than the tracks, a useful proportion of the data can be read even when the tape is being shuttled. The subcode data are repeated many times so that they can be read at any speed. In this way an RDAT machine can chase any other machine and remain synchronised to it.

Whilst there is nothing wrong with the performance of RDAT, it ran into serious political problems because its ability to copy without loss of quality was seen as a threat by copyright organisations. Launch of RDAT as a consumer product was effectively blocked until a system called SCMS (serial copying management system) was incorporated. This allows a single generation of RDAT copying of copyright material. If an attempt is made to copy a copy, a special flag on the copy tape defeats recording on the second machine.

In the mean time, RDAT found favour in the professional audio community where it offered exceptional sound quality at a fraction of the price of professional equipment. Between them the rapid access of hard disc based recorders and the low cost of RDAT have effectively rendered ¼inch analogue recorders and stereo open reel digital recorders obsolete.

Digital Compact Cassette

Digital compact cassette (DCC) is a consumer stationary head digital audio recorder using data reduction. Although the convertors at either end of the machine work with PCM data, these data are not directly recorded, but are reduced to one quarter of their normal rate by processing. This allows a reasonable tape consumption similar to that achieved by a rotary head recorder. In a sense the complexity of the rotary head transport has been exchanged for the electronic complexity of the data reduction and subsequent expansion circuitry.

Figure 6.29 shows that DCC uses stationary heads in a conventional tape transport which can also play analogue cassettes. Data are distributed over nine parallel tracks which occupy half the width of the tape. At the end of the tape the head rotates about an axis perpendicular to the tape and plays the other nine tracks in reverse. The advantage of the conventional approach with linear tracks is that tape duplication can be carried out at high speed. This makes DCC attractive to record companies.

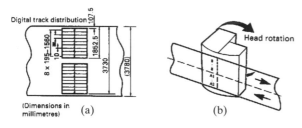

(Dimensions in millimetres)   (a)   (b)

Figure 6.29   In DCC audio and auxiliary data are recorded on nine parallel tracks along each side of the tape as shown at (a) The replay head shown at (b) carries magnetic poles which register with one set of nine tracks. At the end of the tape, the replay head rotates 180 degrees and plays a further nine tracks on the other side of thè tape. The replay head also contains a pair of analogue audio magnetic circuits which will be swung into place if an analogue cassette is to be played.

However, reducing the data rate to one quarter and then distributing it over nine tracks means that the frequency recorded on each track is only about 1/32 that of a PCM machine with a single head. At such a low frequency, conventional inductive heads which generate

a voltage from flux changes cannot be used, and DCC has to use active heads which actually measure the flux on the tape at any speed. These magneto-resistive heads are more complex than conventional inductive heads, and have only recently become economic as manufacturing techniques have been developed.

As was introduced in Chapter 4, data reduction relies on the phenomenon of auditory masking and this effectively restricts DCC to being a consumer format. It will be seen from Fig. 6.30 that the data reduction unit adjacent to the input is complemented by the expansion unit or decoder prior to the DAC. The sound quality of a DCC machine is not a function of the tape, but depends on the convertors and on the sophistication of the data reduction and expansion units.

Editing Digital Audio Tape

Digital recordings are simply data files and editing digital audio should be performed in the same way that a word processor edits text. No word processor attempts to edit on the medium, but brings blocks of data to a computer memory where it is edited before being sent back for storage.

In fact this is the only way that digital audio recordings can be edited, because of the use of interleave and error correction.

Interleave re-orders the samples on the medium, and so it is not possible to find a physical location on the medium which linearly corresponds to the time through the recording. Error correction relies on blocks of samples being coded together. If part of a block is changed, the coding will no longer operate.

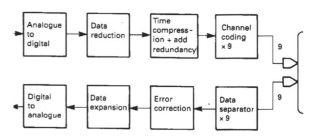

Figure 6.30   In DCC, the PCM data from the convertors is reduced to one quarter of the original rate prior to distribution over eight tape tracks (plus an auxiliary data track). This allows a slow linear tape speed which can only be read with an MR head. The data reduction unit is mirrored by the expansion unit on replay.

Figure 6.31 shows how an audio edit is performed. Samples are played back, de-interleaved and errors are corrected. Samples are now available in their natural

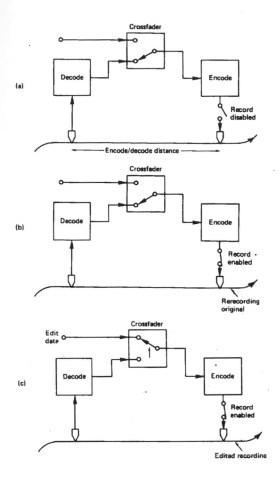

Figure 6.31 Editing a convolutionally interleaved recording. (a) Existing recording is decoded and re-encoded. After some time, record can be enabled at (b) when the existing tape pattern is being rerecorded. The crossfader can then be operated, resulting (c) in an interleaved edit on the tape.

real-time sequence and can be sent to a cross-fader where external material can be inserted. The edited sam-ples are then re-coded and interleaved before they can be re-recorded. De-interleave and interleave cause delay, and by the time these processes have been per-formed, the tape will have moved further through the machine. In simple machines, the tape will have to be reversed, and the new data recorded in a second pass. In more sophisticated machines, an edit can be made in a single pass because additional record heads are placed further down the tape path.

In a stationary head machine, these are physically dis-placed along the head block. In a rotary head machine, the extra heads are displaced along the axis of the drum.

Displaced heads also allow synchronous recording to be performed on multitrack digital audio recorders.

Some stationary head digital formats allow editing by tape cutting. This requires use of an odd-even sample shift and concealment to prevent the damaged area of the tape being audible. With electronic editing, now widely available, tape-cut editing is obsolete as it does not offer the ability to preview or trim the result and causes damage to the medium. The glue on the splicing tape tends to migrate in storage and cause errors.

Bibliography

Baert, L., Theunissen, L. and Vergult, G., *Digital Audio and Compact Disc Technology*, 2nd Edition. Butterworth-Heinemann (1992).

Pohlmann, K., *The Compact Disc*. Oxford University Press (1989).

Pohlmann, K.C., *Advanced Digital Audio*. Sams (1991).

Rumsey, F., *Tapeless Sound Recording*. Focal Press (1990).

Rumsey, F., *Digital Audio Operations*. Focal Press (1991).

Sinclair, R., *Introducing Digital Audio*. PC Publishing (1991).

Watkinson, J., *Art of Digital Audio*. Focal Press, Butterworth-Heinemann (1988).

Watkinson, J., *Coding for Digital Recording*. Focal Press (1990).

7 Tape Recording

John Linsley Hood

No matter what method is used for a mass-produced recording, virtually all records start as tape masters, and a very substantial proportion of recordings are sold in cassette form. John Linsley Hood explains here the complexity of modern tape and cassette recording just prior to the start of DAT.

The Basic System

In principle, the recording of an alternating electrical signal as a series of magnetic fluctuations on a continuous magnetisable tape would not appear to be a difficult matter, since it could be done by causing the AC signal to generate corresponding changes in the magnetic flux across the gap of an electromagnet, and these could then be impressed on the tape as it passes over the recording electromagnet head.

In practice, however, there are a number of problems, and the success of tape recording, as a technique, depends upon the solution of these, or, at least, on the attainment of some reasonable working compromise. The difficulties which exist, and the methods by which these are overcome, where possible, are considered here in respect of the various components of the system.

The Magnetic Tape

This is a thin continuous strip of some durable plastics base material, which is given a uniform coating of a magnetisable material, usually either 'gamma' ferric oxide (Fe_2O_3), chromium dioxide (CrO_2), or, in some recently introduced tapes, of a metallic alloy, normally in powder form, and held by some suitable binder material. Various 'dopants' can also be added to the coating, such as cobalt, in the case of ferric oxide tapes, to improve the magnetic characteristics.

To obtain a long playing time it is necessary that the total thickness of the tape shall be as small as practicable, but to avoid frequency distortion on playback it is essential that the tape shall not stretch in use. It is also important that the surface of the tape backing material shall be hard, smooth and free from lumps of imperfectly extruded material (known as 'pollywogs') to prevent inadvertent momentary loss of contact between the tape and the recording or play-back heads, which would cause 'drop-outs' (brief interruptions in the replayed signal). The tape backing material should also be unaffected, so far as is possible, by changes in temperature or relative humidity.

For cassette tapes, and other systems where a backup pressure pad is used, the uncoated surface is chosen to have a high gloss. In other applications a matt finish will be preferred for improved spooling.

The material normally preferred for this purpose, as the best compromise between cost and mechanical characteristics, is biaxially oriented polyethylene terephthalate film (Melinex, Mylar, or Terphan). Other materials may be used as improvements in plastics technology alter the cost/performance balance.

The term 'biaxial orientation' implies that these materials will be stretched in both the length and width directions during manufacture, to increase the surface smoothness (gloss), stiffness and dimensional stability (freedom from stretch). They will normally also be surface treated on the side to which the coating is to be applied, by an electrical 'corona discharge' process, to improve the adhesion of the oxide containing layer. This is because it is vitally important that the layer is not shed during use as it would contaminate the surface or clog up the gaps in the recorder heads, or could get into the mechanical moving parts of the recorder.

In the tape coating process the magnetic material is applied in the form of a dope, containing also a binder, a solvent and a lubricant, to give an accurately controlled coating thickness. The coated surface is subsequently polished to improve tape/head contact and lessen head wear. The preferred form of both ferric oxide and chromium dioxide crystals is needle-shaped, or 'acicular', and the best characteristic for audio tapes are given when these are aligned parallel to the surface, in the direction of magnetisation. This is accomplished during manufacture by passing the tape through a strong, unidirectional magnetic field, before the coating becomes fully dry. This aligns the needles in the longitudinal direction. The tape is then demagnetized again before sale.

Chromium dioxide and metal tapes both have superior properties, particularly in HF performance, resistance to 'print through' and deterioration during repeated playings, but they are more costly. They also require higher magnetic flux levels during recording and for bias and erase purposes, and so may not be suitable for all machines.

The extra cost of these tape formulations is normally only considered justifiable in cassette recorder systems, where reproduction of frequencies in the range 15–20 kHz, especially at higher signal levels, can present difficulties.

During the period in which patent restrictions limited the availability of chromium dioxide tape coatings, some of the manufacturers who were unable to employ these formulations for commercial reasons, put about the story that chromium dioxide tapes were more abrasive than iron oxide ones. They would, therefore, cause more rapid head wear. This was only marginally true, and now that chromium dioxide formulations are more widely available, these are used by most manufacturers for their premium quality cassette tapes.

Composite 'ferro-chrome' tapes, in which a thinner surface layer of a chromium dioxide formulation is applied on top of a base ferric oxide layer, have been made to achieve improved HF performance, but without a large increase in cost.

In 'reel-to-reel' recorders, it is conventional to relate the tape thickness to the relative playing time, as 'Standard Play', 'Double Play' and so on. The gauge of such tapes is shown in Table 7.1. In cassette tapes, a more straight forward system is employed, in which the total playing time in minutes is used, at the standard cassette playing speed. For example, a C60 tape would allow 30 minutes playing time, on each side. The total thicknesses of these tapes are listed in Table 7.2.

Table 7.1   *Tape thicknesses (reel-to-reel)*

Tape	Thickness (in)
'Standard play'	0.002
'Long play'	0.0015
'Double play'	0.001
'Triple play'	0.00075
'Quadruple play'	0.0005

For economy in manufacture, tape is normally coated in widths of up to 48 in. (1.2 m), and is then slit down to the widths in which it is used. These are 2 in. (50.8 mm) 1 in. (25.4 mm) 0.5 in. (12.7 mm) and 0.25 in. (6.35 mm) for professional uses, and 0.25 in. for domestic reel-to-reel machines. Cassette recorders employ 0.15 in. (3.81 mm) tape.

High-speed slitting machines are complex pieces of precision machinery which must be maintained in good order if the slit tapes are to have the required parallelism and constancy of width. This is particularly important in cassette machines where variations in tape width can cause bad winding, creasing, and misalignment over the heads.

For all of these reasons, it is highly desirable to employ only those tapes made by reputable manufacturers, where these are to be used on good recording equipment, or where permanence of the recording is important.

Table 7.2   *Tape thicknesses (cassette)*

Tape	Thickness (μm)
C60	18 (length 92m)
C90	12 (length 133m)
C120	9 (length 184m)

Tape base thicknesses 12μm, 8μm and 6μm respectively.

The Recording Process

The magnetic materials employed in tape coatings are chosen because they possess elemental permanent magnets on a sub-microscopic or molecular scale. These tiny magnetic elements, known as 'domains', are very much smaller than the grains of spherical or needle-shaped crystalline material from which oxide coatings are made.

Care will be taken in the manufacture of the tape to try to ensure that all of these domains will be randomly oriented, with as little 'clumping' as possible, to obtain as low a zero-signal-level noise background as practicable. Then, when the tape passes over a recording head, shown schematically in Fig. 7.1, these magnetic domains will be realigned in a direction and to an extent which depend on the magnetic polarity and field strength at the trailing edge of the recording head gap.

This is where first major snag of the system appears. Because of the magnetic inertia of the material, small applied magnetic fields at the recording head will have very little effect in changing the orientation of the domains. This leads to the kind of characteristic shown in Fig. 7.2, where the applied magnetising force (H), is related to the induced flux density in the tape material, (B).

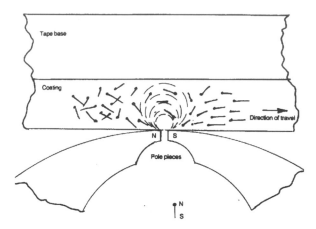

Figure 7.1 The alignment of magnetic domains as the magnetic tape passes over the recording head.

If a sinusoidal signal is applied to the head, and the flux across the recording head gap is related to the signal voltage, as shown in Fig. 7.2, the remanent magnetic flux induced in the tape – and the consequent replayed signal – would be both small in amplitude and badly distorted.

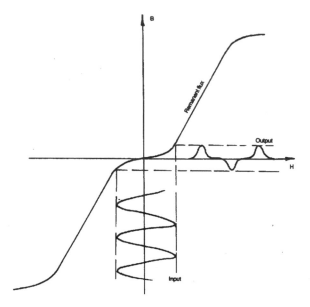

Figure 7.2 The effect of the B-H non-linearity in magnetic materials on the recording process.

This problem is removed by applying a large high-frequency signal to the recording head, simultaneously with the desired signal. This superimposed HF signal is referred to as 'HF bias' or simply as 'bias', and will be large enough to overcome the magnetic inertia of the domains and take the operating region into the linear portion of the BH curve.

Several theories have been offered to account for the way in which 'HF bias' linearises the recording process. Of these the most probable is that the whole composite signal is in fact recorded but that the very high frequency part of it decays rapidly, due to self cancellation, so that only the desired signal will be left on the tape, as shown in Fig. 7.3.

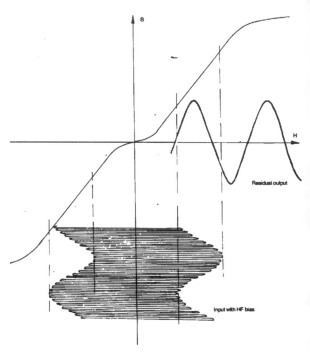

Figure 7.3 The linearising effect of superimposed HF bias on the recording process.

When the tape is passed over the replay head – which will often be the same head which was used for recording the signal in the first place – the fluctuating magnetic flux of the tape will induce a small voltage in the windings of the replay head, which will then be amplified electronically in the recorder. However, both the recorded signal, and the signal recovered from the tape at the replay head will have a non-uniform frequency response, which will demand some form of response equalisation in the replay amplifier.

Sources of Non-Uniformity in Frequency Response

If a sinusoidal AC signal of constant amplitude and frequency is applied to the recording head and the tape is passed over this at a constant speed, a sequence of magnetic poles will be laid down on the tape, as shown schematically in Fig. 7.4, and the distance between like poles, equivalent to one cycle of recorded signal, is known as the recorded wavelength. The length of this can be calculated from the applied frequency and the movement, in unit time of the tape, as 'λ', which is equal to tape velocity (cm/s) divided by frequency (cycles/s) which will give wavelength (λ) in cm.

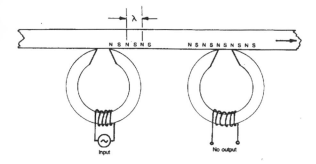

Figure 7.4 The effect of recorded wavelength on replay head output.

This has a dominant influence on the replay characteristics in that when the recorded wavelength is equal to the replay head gap, as is the case shown at the replay head in Fig. 7.4, there will be zero output. This situation is worsened by the fact that the effective replay head gap is, in practice, somewhat larger than the physical separation between the opposed pole pieces, due to the spread of the magnetic field at the gap, and to the distance between the head and the centre of the magnetic coating on the tape, where much of the HF signal may lie.

Additional sources of diminished HF response in the recovered signal are eddy-current and other magnetic losses in both the recording and replay heads, and 'self-demagnetisation' within the tape, which increases as the separation between adjacent N and S poles decreases.

If a constant amplitude sine-wave signal is recorded on the tape, the combined effect of recording head and tape losses will lead to a remanent flux on the tape which is of the form shown in Fig. 7.5. Here the HF turnover frequency (F_t) is principally dependent on tape speed.

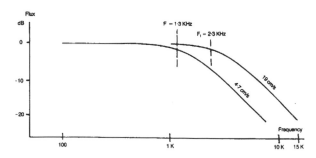

Figure 7.5 Effect of tape speed on HF replay output.

Ignoring the effect of tape and record/replay head losses, if a tape, on which a varying frequency signal has been recorded, is replayed at a constant linear tape speed, the signal output will increase at a linear rate with frequency. This is such that the output will double for each octave of frequency. The result is due to the physical laws of magnetic induction, in which the output voltage from any coil depends on the magnetic flux passing through it, according to the relationship $V = L.\mathrm{d}B/\mathrm{d}t$. Combining this effect with replay head losses leads to the kind of replay output voltage characteristics shown in Fig. 7.6.

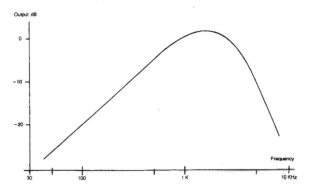

Figure 7.6 The effect of recorded wavelength and replay head gap width on replay signal voltage.

At very low frequencies, say below 50 Hz, the recording process becomes inefficient, especially at low tape speeds. The interaction between the tape and the profile of the pole faces of the record/replay heads leads to a characteristic undulation in the frequency response of the type shown in Fig. 7.7 for a good quality cassette recorder.

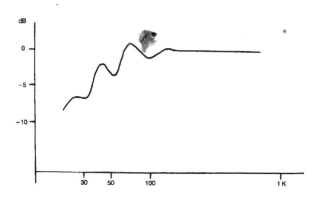

Figure 7.7 Non-uniformity in low-frequency response due to pole-piece contour effects.

Record/Replay Equalisation

In order to obtain a flat frequency response in the record/replay process, the electrical characteristics of the record and replay amplifiers are modified to compensate for the non-linearities of the recording process, so far as this is possible. This electronic frequency response adjustment is known as 'equalisation', and is the subject of much misunderstanding, even by professional users.

This misunderstanding arises because the equalisation technique is of a different nature to that which is used in FM broadcast receivers, or in the reproduction of LP gramophone records, where the replay de-emphasis at HF is identical in time-constant (turnover frequency) and in magnitude, but opposite in sense to the pre-emphasis employed in transmission or recording.

By contrast, in tape recording it is assumed that the total inadequacies of the recording process will lead to a remanent magnetic flux in the tape following a recording which has been made at a constant amplitude which has the frequency response characteristics shown in Fig. 7.8, for various tape speeds. This remanent flux will be as specified by the time-constants quoted for the various international standards shown in Table 7.3.

These mainly refer to the expected HF roll-off, but in some cases also require a small amount of bass pre-emphasis so that the replay de-emphasis – either electronically introduced, or inherent in the head response – may lessen 'hum' pick-up, and improve LF signal-to-noise (S/N) ratio.

The design of the replay amplifier must then be chosen so that the required flat frequency response output would be obtained on replaying a tape having the flux characteristics shown in Fig. 7.8.

This will usually lead to a replay characteristic of the kind shown in Fig. 7.9. Here a –6 dB/octave fall in output,

with increasing frequency, is needed to compensate for the increasing output during replay of a constant recorded signal – referred to above – and the levelling off, shown in curves a–d, is simply that which is needed to correct for the anticipated fall in magnetic flux density above the turn-over frequency, shown in Fig. 7.8 for various tape speeds.

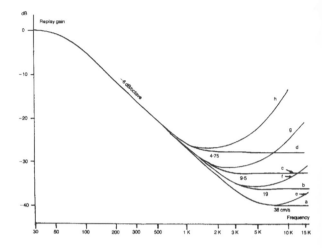

Figure 7.9   Required replay frequency response, for different tape speeds.

However, this does not allow for the various other head losses, so some additional replay HF boost, as shown in the curves e–h, of Fig. 7.9, is also used.

The recording amplifier is then designed in the light of the performance of the recording head used, so that the remanent flux on the tape will conform to the specifications shown in Table 7.3, and as illustrated in Fig. 7.8. This will generally also require some HF (and LF) pre-

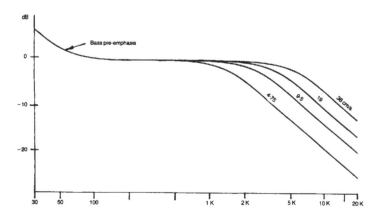

Figure 7.8   Assumed remanent magnetic flux on tape for various tape speeds, in cm/s.

emphasis, of the kind shown in Fig. 7.10. However, it should be remembered that, especially in the case of recording amplifier circuitry the component values chosen by the designer will be appropriate only to the type of recording head and tape transport mechanism – in so far as this may influence the tape/head contact – which is used in the design described.

Because the subjective noise level of the system is greatly influenced by the amount of replay HF boost which is employed, systems such as reel-to-reel recorders, operating at relatively high-tape speeds, will sound less 'hissy' than cassette systems operating at 1.875 in/s (4.76 cm/s), for which substantial replay HF boost is necessary. Similarly, recordings made using Chromium dioxide tapes, for which a 70 μs HF emphasis is employed, will sound less 'hissy' than with Ferric oxide tapes equalised at 120 μs, where the HF boost begins at a lower frequency.

Table 7.3 *Frequency correction standards*

Tape speed	Standard	Time constants (μs) HF	LF	−3 dB@ HF (kHz)	+3 dB@ LF (Hz)
15 in/s	NAB	50	3180	3.18	50
(38.1 cm/s)	IEC	35		4.55	
	DIN	35	—	4.55	—
7½ in/s	NAB	50	3180	3.18	50
(19.05 cm/s)	IEC	70	—	2.27	—
	DIN	70	—	2.27	—
3¾ in/s	NAB	90	3180	1.77	50
(9.53 cm/s)	IEC	90	3180	1.77	50
	DIN	90	3180	1.77	50
1⅞ in/s	DIN (Fe)	120	3180	1.33	50
(4.76 cm/s)	(CrO$_2$)	70	3180	2.27	50

Head Design

Record/replay heads

Three separate head types are employed in a conventional tape recorder for recording, replaying and erasing. In some cases, such as the less expensive reel-to-reel and cassette recorders, the record and replay functions will be combined in a single head, for which some design compromise will be sought between the different functions and design requirements.

In all cases, the basic structure is that of a ring electromagnet, with a flattened surface in which a gap has been cut, at the position of contact with the tape. Conventionally, the form of the record and replay heads is as shown in Fig. 7.11, with windings symmetrically placed on the two limbs, and with gaps at both the front (tape side) and the rear. In both cases the gaps will be filled with a thin shim of low magnetic permeability material, such as gold, copper or phosphor bronze, to maintain the accuracy of the gap and the parallelism of the faces.

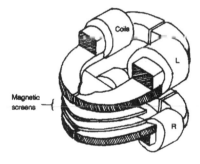

Figure 7.11 General arrangement of cassette recorder stereo record/replay head using Ferrite pole pieces.

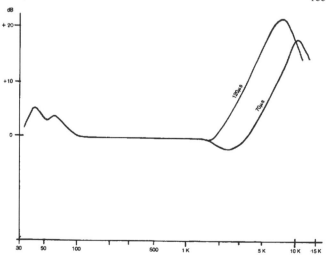

Figure 7.10 Probable recording frequency response in cassette recorder required to meet flux characteristics shown in Fig. 7.8.

The pole pieces on either side of the front gap are shaped, in conjunction with the gap filling material, to concentrate the magnetic field in the tape, as shown schematically in Fig. 7.12, for a replay head, and the material from which the heads are made is chosen to have as high a value of permeability as possible, for materials having adequate wear resistance.

The reason for the choice of high permeability core material is to obtain as high a flux density at the gap as possible when the head is used for recording, or to obtain as high an output signal level as practicable (and as high a resultant S/N ratio) when the head is used in the replay mode. High permeability core materials also help to confine the magnetic flux within the core, and thereby reduce crosstalk.

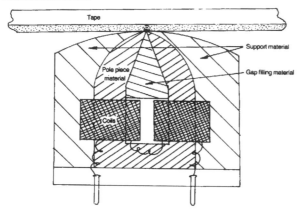

Figure 7.12　Relationship between pole-pieces and magnetic flux in tape.

With most available ferromagnetic materials, the permeability decreases with frequency, though some ferrites (sintered mixes of metallic oxides) may show the converse effect. The rear gap in the recording heads is used to minimise this defect. It also makes the recording efficiency less dependent on intimacy of the tape to head contact. In three-head machines, the rear gap in the replay head is often dispensed with, in the interests of the best replay sensitivity.

Practical record/replay heads will be required to provide a multiplicity of signal channels on the tape: two or four, in the case of domestic systems, or up to 24 in the case of professional equipment. So a group of identical heads will require to be stacked, one above the other and with the gaps accurately aligned vertically to obtain accurate time coincidence of the recorded or replayed signals.

It is, of course, highly desirable that there should be very little interaction, or crosstalk, between these adjacent heads, and that the recording magnetic field should be closely confined to the required tape track. This demands very careful head design, and the separation of adjacent heads in the stack with shims of low permeability material.

To preserve the smoothness of the head surface in contact with the tape, the non-magnetic material with which the head gap, or the space between head stacks, is filled is chosen to have a hardness and wear resistance which matches that of the head alloy. For example, in Permalloy or Super Permalloy heads, beryllium copper or phosphor bronze may be used, while in Sendust or ferrite heads, glass may be employed.

Table 7.4　*Magnetic materials for recording heads*

Material	Mumetal	Permalloy	Super Permalloy	Sendust (Hot pressed)	Ferrite	Hot Pressed Ferrite
Composition	75 Ni 2 Cr 5 Cu 18 Fe	79 Ni 4 Mo 17 Fe	79 Ni 5 Mo 16 Fe	85 Fe 10 Si 5 Al	Mn Ni Fe oxides Zn	Mn Ni Fe oxides Zn
Treatment	1100°C in hydrogen	1100°C	1300°C in hydrogen	800°C in hydrogen		500 kg/cm² in hydrogen
Permeability 1 KHz	50000	25000	200000*	50000*	1200	20000
Max flux density (gauss)	7200	16000	8700	> 5000*	4000	4000
Coercivity (oersteds)	0.03	0.05	0.004	0.03	0.5	0.015
Conductivity	High	High	High	High	Very low	Very low
Vickers hardness	118	132	200*	280*	400	700

(*Depends on manufacturing process.)

The characteristics of the various common head materials are listed in Table 7.4. Permalloy is a nickel, iron, molybdenum alloy, made by a manufacturing process which leads to a very high permeability and a very low coercivity. This term refers to the force with which the material resists demagnetisation: 'soft' magnetic materials have a very low coercivity, whereas 'hard' magnetic materials, i.e. those used for 'permanent' magnets, have a very high value of coercivity.

Super-Permalloy is a material of similar composition which has been heat treated at 1200–1300° C in hydrogen, to improve its magnetic properties and its resistance to wear. Ferrites are ceramic materials, sintered at a high temperature, composed of the oxides of iron, zinc, nickel and manganese, with suitable fillers and binders. As with the metallic alloys, heat treatment can improve the performance, and hot-pressed ferrite (typically 500 Kg/cm^2 at 1400° C) offers both superior hardness and better magnetic properties.

Sendust is a hot-pressed composite of otherwise incompatible metallic alloys produced in powder form. It has a permeability comparable to that of Super Permalloy with a hardness comparable to that of ferrite.

In the metallic alloys, as compared with the ferrites, the high conductivity of the material will lead to eddy-current losses (due to the core material behaving like a short-circuited turn of winding) unless the material is made in the form of thin laminations, with an insulating layer between these. Improved HF performance requires that these laminations are very thin, but this, in turn, leads to an increased manufacturing cost, especially since any working of the material may spoil its performance. This would necessitate re-annealing, and greater problems in attaining accurate vertical alignment of the pole piece faces in the record/replay heads. The only general rule is that the better materials will be more expensive to make, and more difficult to fabricate into heads.

Because it is only the trailing edge of the record head which generates the remanent flux on the tape, the gap in this head can be quite wide; up to 6–10 microns in typical examples. On the other hand, the replay head gap should be as small as possible, especially in cassette recorders, where high quality machines may employ gap widths of one micron (0.00004 in.) or less.

The advantage of the wide recording head gap is that it produces a field which will penetrate more fully into the layer of oxide on the surface of the tape, more fully utilising the magnetic characteristics of the tape material. The disadvantage of the narrow replay gap, needed for good HF response, is that it is more difficult for the alternating zones of magnetism on the recorded tape to induce changes in magnetic flux within the core of the replay head. Consequently the replay signal voltage is less, and the difficulty in getting a good S/N ratio will be greater.

Other things being equal, the output voltage from the replay head will be directly proportional to the width of the magnetic track and the tape speed. On both these counts, therefore, the cassette recorder offers an inferior replay signal output to the reel-to-reel system.

The erase head

This is required to generate a very high frequency alternating magnetic flux within the tape coating material. It must be sufficiently intense to take the magnetic material fully into saturation, and so blot out any previously existing signal on the tape within the track or tracks concerned.

The design of the head gap or gaps should allow the alternating flux decay slowly as the tape is drawn away from the gap, so that the residual flux on the tape will fall to zero following this process.

Because of the high frequencies and the high currents involved, eddy-current losses would be a major problem in any laminated metal construction. Erase heads are therefore made from ferrite materials.

A rear air gap is unnecessary in an erase head, since it operates at a constant frequency. Also, because saturating fields are used, small variations in the tape to head contact are less important.

Some modern cassette tape coating compositions have such a high coercivity and remanence (retention of impressed magnetism) that single gap heads may not fully erase the signal within a single pass. So dual gap heads have been employed, particularly on machines offering a 'metal tape' facility. This gives full erasure, (a typical target value of –70 dB being sought for the removal of a fully recorded signal), without excessive erase currents being required, which could lead to overheating of the erase head.

It is important that the erase head should have an erase field which is closely confined to the required recorded tracks, and that it should remain cool. Otherwise unwanted loss of signal or damage to the tape may occur when the tape recording is stopped during the recording process, by the use of the 'pause' control.

In some machines provision is made for switching off the erase head, so that multiple recordings may be overlaid on the same track. This is an imperfect answer to this requirement, however, since every subsequent recording pass will erase pre-existing signals to some extent.

This facility would not be practicable on inexpensive recorders, where the erase head coil is used as the oscillator coil in the erase and 'bias' voltage generator circuit.

Recording Track Dimensions

These conform to standards laid down by international agreements, or, in the case of the Philips compact cassette, within the original Patent specifications, and are as shown in Fig. 7.13.

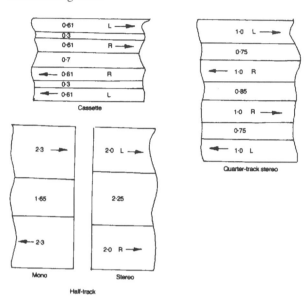

Figure 7.13   Tape track specifications. (All dimensions in mm.)

HF Bias

Basic bias requirements

As has been seen, the magnetic recording process would lead to a high level of signal distortion, were it not for the fact that a large, constant amplitude, HF bias waveform is combined with the signal at the recording head. The basic requirement for this is to generate a composite magnetic flux, within the tape, which will lie within the linear range of the tape's magnetic characteristics, as shown in Fig. 7.3.

However, the magnitude and nature of the bias signal influences almost every other aspect of the recording process. There is no value for the applied 'bias' current which will be the best for all of the affected recording characteristics. It is conventional, and correct, to refer to the bias signal as a current, since the effect on the tape is that of a magnetising force, defined in 'ampere turns'. Since the recording head will have a significant inductance – up to 0.5 H in some cases – which will offer a substantial impedance at HF, the applied voltage for a constant flux level will depend on the chosen bias frequency.

The way in which these characteristics are affected is shown in Fig. 7.14 and 7.15, for ferric and chrome tapes. These are examined separately below, together with some of the other factors which influence the final performance.

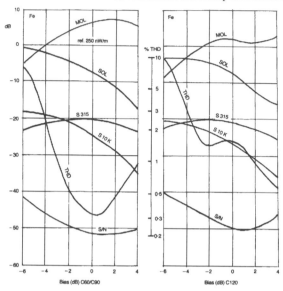

Figure 7.14   The influence of bias on recording characteristics. (C60/90 and C120 Ferric tapes.)

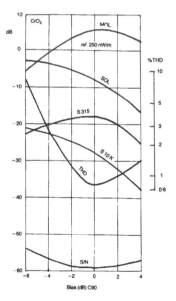

Figure 7.15   The influence of bias on recording characteristics. (C90 Chrome tape.)

HF bias frequency

The particular choice of bias frequency adopted by the manufacturer will be a compromise influenced by his own preferences, and his performance intentions for the equipment.

It is necessary that the frequency chosen shall be sufficiently higher than the maximum signal frequency which is to be recorded that any residual bias signal left on the tape will not be reproduced by the replay head or amplifier, where it could cause overload or other undesirable effects.

On the other hand, too high a chosen bias frequency will lead to difficulties in generating an adequate bias current flow through the record head. This may lead to overheating in the erase head.

Within these overall limits, there are other considerations which affect the choice of frequency. At the lower end of the usable frequency range, the overall efficiency of the recording process, and the LF S/N ratio is somewhat improved. The improvement is at the expense of some erasure of the higher audio frequencies, which increases the need for HF pre-emphasis to achieve a 'flat' frequency response. However, since it is conventional to use the same HF signal for the erase head, the effectiveness of erasure will be better for the same erase current.

At the higher end of the practicable bias frequency range, the recorded HF response will be better, but the modulation noise will be less good. The danger of saturation of the recording head pole-piece tips will be greater, since there is almost invariably a decrease in permeability with increasing frequency. This will require more care in the design of the recording head.

Typically, the chosen frequency for audio work will be between four and six times the highest frequency which it is desired to record: it will usually be in the range 60–120 kHz, with better machines using the higher values.

HF bias waveform

The most important features of the bias waveform are its peak amplitude and its symmetry. So although other waveforms, such as square waves, will operate in this mode quite satisfactorily, any inadvertent phase shift in the harmonic components would lead to a change in the peak amplitude, which would affect the performance.

Also, it is vitally important that the HF 'bias' signal should have a very low noise component. In the composite signal applied to the record head, the bias signal may be 10–20 times greater than the audio signal to be recorded. If, therefore, the bias waveform is not to degrade the overall S/N ratio of the system, the noise content of the bias waveform must be at least 40 times better than that of the incoming signal.

The symmetry of the waveform is required so that there is no apparent DC or unidirectional magnetic component of the resultant waveform, which could magnet-

ise the record head and impair the tape modulation noise figure. An additional factor is that a symmetrical waveform allows the greatest utilisation of the linear portion of the tape magnetisation curve.

For these reasons, and for convenience in generating the erase waveform, the bias oscillator will be designed to generate a very high purity sine wave. Then the subsequent handling and amplifying circuitry will be chosen to avoid any degradation of this.

Maximum output level (MOL)

When the tape is recorded with the appropriate bias signal, it will have an effective 'B-H' characteristic of the form shown in Fig. 7.16. Here it is quite linear at low signal levels, but will tend to flatten-off the peaks of the waveform at higher signal levels. This causes the third harmonic distortion, which worsens as the magnitude of the recorded signal increases, and effectively sets a maximum output level for the recorder/tape combination.

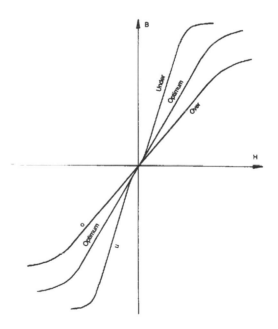

Figure 7.16 Relationship between remanent flux ('B'), and applied magnetic field at 315 Hz in a biased tape, showing linearisation and overload effects.

This is usually quoted as 'MOL (315 Hz)', and is defined as the output level, at 315 Hz. It is given some reference level, (usually a remanent magnetic flux on the tape of 250 nano Webers/m, for cassette tapes), at which the third harmonic distortion of the signal reaches 3 percent. The MOL generally increases with bias up to some maximum value, as shown in Fig. 7.14 and 7.15.

The optimum, or reference bias

Since it is apparent that there is no single best value for the bias current, the compromise chosen for cassette recorders is usually that which leads to the least difficulty in obtaining a flat frequency response. Lower values of bias give better HF performance but worse characteristics in almost all other respects, so the value chosen is that which leads to a fall in output, at 10 kHz in the case of cassettes of 12 dB in comparison with the output at 315 Hz.

In higher tape speed reel-to-reel recorders, the specified difference in output level at 315 Hz and 12.5 kHz may be only 2.5 dB. Alternatively, the makers recommended 'reference' bias may be that which gives the lowest value of THD at 1 kHz.

Sensitivity

This will be specified at some particular frequency, as 'S-315 Hz' or 'S-10 kHz' depending on the frequency chosen. Because of self-erasure effects, the HF components of the signal are attenuated more rapidly with increasing HF 'bias' values than the LF signals. On the other hand, HF signals can act, in part, as a bias waveform, so some recorders employ a monitoring circuit which adjusts the absolute value of the bias input, so that it is reduced in the presence of large HF components in the audio signal.

This can improve the HF response and HF output level for the same value of harmonic distortion in comparison with a fixed bias value system.

Total harmonic distortion (THD)

This is generally worsened at both low and high values of 'bias' current. At low values, this is because inadequate bias has been used to straighten out the kink at the centre of the 'B-H' curve, which leaves a residue of unpleasant, odd harmonic, crossover-type distortion. At high values, the problem is simply that the bias signal is carrying the composite recording waveform into regions of tape saturation, where the B-H curve flattens off.

Under this condition, the distortion component is largely third harmonic, which tends to make the sound quality rather shrill. As a matter of practical convenience, the reference bias will be chosen to lie somewhere near the low point on the THD/bias curve. The formulation and coating thickness adopted by the manufacturer can be chosen to assist in this, as can be seen by the comparison between C90 and C120 tapes in Fig. 7.14.

The third harmonic distortion for a correctly biased tape will increase rapidly as the MOL value is approached. The actual peak recording values for which the recorder VU or peak recording level meters are set is not, sadly, a matter on which there is any agreement between manufacturers, and will, in any case, depend on the tape and the way in which it has been biased.

Noise level

Depending on the tape and record head employed, the residual noise level on the tape will be influenced by the bias current and will tend to decrease with increasing bias current values. However, this is influenced to a greater extent by the bias frequency and waveform employed, and by the construction of the erase head.

The desired performance in a tape or cassette recorder is that the no-signal replay noise from a new or bulk erased tape should be identical to that which arises from a single pass through the recorder, set to record at zero signal level.

Aspects of recording and replay noise are discussed more fully below.

Saturation output level (SOL)

This is a similar specification to the MOL, but will generally apply to the maximum replay level from the tape at a high frequency, usually 10 kHz, and decreases with increasing bias at a similar rate to that of the HF sensitivity value.

The decrease in HF sensitivity with increasing bias appears to be a simple matter of partial erasure of short recorded wavelength signals by the bias signal. However, in the case of the SOL, the oscillation of the domains caused by the magnetic flux associated with the bias current appears physically to limit the ability of short wavelength magnetic elements to coexist without self-cancellation.

It is fortunate that the energy distribution on most programme material is high at LF and lower middle frequencies, but falls off rapidly above, say, 3 kHz, so HF overload is not normally a problem. It is also normal practice to take the output to the VU or peak signal metering circuit from the output of the recording amplifier, so the effect of pre-emphasis at HF will be taken into account.

Bias level setting

This is a difficult matter in the absence of appropriate test instruments, and will require adjustment for every significant change in tape type, especially in cassette machines, where a wide range of coating formulations is available.

Normally, the equipment manufacturer will provide a switch selectable choice of pre-set bias values, labelled in a cassette recorder, for example, as types '1' (all ferric oxide tapes, used with a 120 μs equalisation time constant), '2' (all chromium dioxide tapes, used with a 70 μs equalisation), '3' (dual coated 'ferro-chrome' tapes, for 70 μs equalisation) and '4' (all 'metal' tapes), where the machine is compatible with these.

Additionally, in the higher quality (three-head) machines, a test facility may be provided, such as a built-in dual frequency oscillator, having outputs at 330 Hz and 8 kHz, to allow the bias current to be set to provide an identical output at both frequencies on replay. In some recent machines this process has been automated.

The Tape Transport Mechanism

The constancy and precision of the tape speed across the record and replay heads is one of the major criteria for the quality of a tape recorder mechanism. Great care is taken in the drive to the tape 'capstans' to smooth out any vibration or flutter generated by the drive motor or motors. The better quality machines will employ a dual-capstan system, with separate speed or torque controlled motors, to drive the tape and maintain a constant tension across the record and replay heads, in addition to the drive and braking systems applied to the take-up and unwind spools.

This requirement can make some difficulties in the case of cassette recorder systems, in which the design of the cassette body allows only limited access to the tape. In this case a dual-capstan system requires the use of the erase head access port for the unwind side capstan, and forces the use of a narrowed erase head which can be fitted into an adjacent, unused, small slot in the cassette body.

The use of pressure pads to hold the tape against the record or replay heads is considered by many engineers to be an unsatisfactory practice, in that it increases the drag on the tape, and can accelerate head wear. In reel-to-reel recorders it is normally feasible to design the tape layout and drive system so that the tape maintains a constant and controlled position in relation to the heads, so the use of pressure pads can be avoided.

However, in the case of cassettes where the original patent specification did not envisage the quality of performance sought and attained in modern cassette decks, a pressure pad is incorporated as a permanent part of the cassette body. Some three-head cassette machines, with dual capstan drive, control the tape tension at the record and replay position so well that the inbuilt pressure pad is pushed away from the tape when the cassette is inserted.

An essential feature of the maintenance of any tape or cassette recorder is the routine cleaning of the capstan drive shafts, and pinch rollers, since these can attract deposits of tape coating material, which will interfere with the constancy of tape drive speed.

The cleanliness of the record/replay heads influences the closeness of contact between the tape and the head gap, which principally affects the replay characteristics, and reduces the replay HF response.

Transient Performance

The tape recording medium is unique as the maximum rate of change of the reproduced replay signal voltage is determined by the speed with which the magnetised tape is drawn past the replay head gap. This imposes a fixed 'slew-rate' limitation on all reproduced transient signal voltages, so that the maximum rate of change of voltage cannot exceed some fixed proportion of its final value.

In this respect it differs from a slew-rate-limited electronic amplifier, in which the effective limitation is signal amplitude dependent, and may not occur on small signal excursions even when it will arise on larger ones.

Slew-rate limitation in electronic amplifiers is, however, also associated with the paralysis of the amplifier during the rate-limited condition. This is a fault which does not happen in the same way with a replayed magnetic tape signal, within the usable region of the B-H curve. Nevertheless, it is a factor which impairs the reproduced sound quality, and leads to the acoustic superiority of high tape speed reel-to-reel machines over even the best of cassette recorders, and the superiority of replay heads with very narrow head gaps, for example in three-head machines, in comparison with the larger, compromise form, gaps of combined record/replay heads.

It is possibly also, a factor in the preference of some audiophiles for direct-cut discs, in which tape mastering is not employed.

A further source of distortion on transient waveforms arises due to the use of HF emphasis, in the recording process and in replay, to compensate for the loss of high frequencies due to tape or head characteristics.

If this is injudiciously applied, this HF boost can lead to 'ringing' on a transient, in the manner shown, for a square-wave in Fig. 7.17. This impairs the audible performance of the system, and can worsen the subjective noise figure of the recorder. It is therefore sensible to examine the square-wave performance of a tape recorder system following any adjustment to the HF boost circuitry, and adjust these for minimum overshoot.

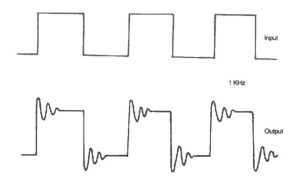

Input

1 KHz

Output

Figure 7.17 'Ringing' effects, on a 1 KHz square-wave, due to excessive HF boost applied to extend high-frequency response.

Tape Noise

Noise in magnetic recording systems has two distinct causes: the tape, and the electronic amplification and signal handling circuitry associated with it. Circuit noise should normally be the lesser problem, and will be discussed separately.

Tape noise arises because of the essentially granular, or particulate, nature of the coating material and the magnetic domains within it. There are other effects, though, which are due to the actual recording process, and these are known as 'modulation' or 'bias noise', and 'contact' noise – due to the surface characteristics of the tape.

The problem of noise on magnetic tape and that of the graininess of the image in the photographic process are very similar. In both cases this is a statistical problem, related to the distribution of the individual signal elements, which becomes more severe as the area sampled, per unit time, is decreased. (The strict comparison should be restricted to cine film, but the analogy also holds for still photography.)

Similarly, a small output signal from the replay head, (equivalent to a small area photographic negative), which demands high amplification, (equivalent to a high degree of photographic enlargement), will lead to a worse S/N ratio, (equivalent to worse graininess in the final image), and a worse HF response and transient definition, (image sharpness).

For the medium itself, in photographic emulsions high sensitivity is associated with worse graininess, and vice versa. Similarly, with magnetic tapes, low noise, fine grain tape coatings also show lower sensitivity in terms of output signal. Metal tapes offer the best solution here.

A further feature, well-known to photographers, is that prints from large negatives have a subtlety of tone and gradation which is lacking from apparently identical

small-negative enlargements. Similarly, comparable size enlargements from slow, fine-grained, negative material are preferable to those from higher speed, coarse-grained stock.

There is presumably an acoustic equivalence in the tape recording field.

Modulation or bias noise arises because of the random nature and distribution of the magnetic material throughout the thickness of the coating. During recording, the magnetic flux due to both the signal and bias waveforms will redistribute these domains, so that, in the absence of a signal, there will be some worsening of the noise figure. In the presence of a recorded signal, the magnitude of the noise will be modulated by the signal.

This can be thought of simply as a consequence of recording the signal on an inhomogeneous medium, and gives rise to the description as 'noise behind the signal', as when the signal disappears, this added noise also stops.

The way in which the signal to noise ratio of a recorded tape is dependent on the area sampled, per unit time, can be seen from a comparison between a fast tape speed dual-track reel-to-reel recording with that on a standard cassette.

For the reel-to-reel machine there will be a track width of 2.5 mm with a tape speed of 15 in./s, (381 mm/s) which will give a sampled area equivalent to 2.5 mm × 381 mm = 925 mm²/s. In the case of the cassette recorder the tape speed will be 1.875 in/s (47.625 mm/s), and the track width will be 0.61 mm, so that the sampled area will only be 29 mm²/s. This is approximately 32 times smaller, equivalent to a 15 dB difference in S/N ratio.

This leads to typical maximum S/N ratios of 67 dB for a two-track reel-to-reel machine, compared with a 52 dB value for the equivalent cassette recorder, using similar tape types. In both cases some noise reduction technique will be used, which will lead to a further improvement in these figures.

It is generally assumed that an S/N ratio of 70–75 dB for wideband noise is necessary for the noise level to be acceptable for high quality work, through some workers urge the search for target values of 90 dB or greater. However, bearing in mind that the sound background level in a very quiet domestic listening room is unlikely to be less than +30 dB (reference level, 0 dB = 0.0002 dynes/cm²) and that the threshold of pain is only +105–110 dB, such extreme S/N values may not be warranted.

In cassette recorders, it is unusual for S/N ratios better than 60 dB to be obtained, even with optimally adjusted noise reduction circuitry in use.

The last source of noise, contact noise, is caused by a lack of surface smoothness of the tape, or fluctuations in the coating thickness due to surface irregularities in the backing material. It arises because the tape coating completes the magnetic circuit across the gap in the heads during the recording process.

Any variations in the proximity of tape to head will, therefore, lead to fluctuations in the magnitude of the replayed signal. This kind of defect is worse at higher frequencies and with narrower recording track widths, and can lead to drop-outs (complete loss of signal) in severe cases.

This kind of broadband noise is influenced by tape material, tape storage conditions, tape tension and head design. It is an avoidable nuisance to some extent, but is invariably present.

Contact noise, though worse with narrow tapes and low tape speeds, is not susceptible to the kind of analysis shown above, in relation to the other noise components.

The growing use of digital recording systems, even in domestic cassette recorder form, is likely to alter substantially the situation for all forms of recorded noise, although it is still argued by some workers in this field that the claimed advantages of digitally encoded and decoded recording systems are offset by other types of problem.

There is no doubt that the future of all large-scale commercial tape recording will be tied to the digital system, if only because of the almost universal adoption by record manufacturers of the compact disc as the future style of gramophone record.

This is recorded in a digitally encoded form, so it is obviously sensible to carry out the basic mastering in this form also, since their other large-scale products, the compact cassette and the vinyl disc, can equally well be mastered from digital as from analogue tapes.

Electronic Circuit Design

The type of circuit layout used for almost all analogue (i.e. non-digital) magnetic tape recorders is as shown, in block diagram form, in Fig. 7.18.

The critical parts of the design, in respect of added circuit noise, are the replay amplifier, and the bias/erase oscillator. They will be considered separately.

It should also be noted that the mechanical design of both the erase and record heads can influence the noise level on the tape. This is determined by contact profile and gap design, consideration of which is beyond the scope of this chapter.

The replay amplifier

Ignoring any noise reduction decoding circuitry, the function of this is to amplify the electrical signal from the replay head, and to apply any appropriate frequency response equalisation necessary to achieve a substantially

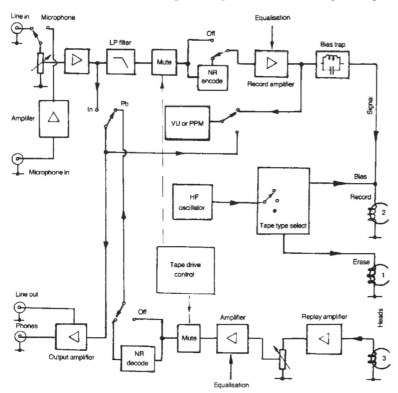

Figure 7.18   Schematic layout of the 'building blocks' of a magnetic tape recorder.

uniform frequency response. The design should also ensure the maximum practicable S/N ratio for the signal voltage present at the replay head.

It has been noted previously that the use of very narrow replay head gaps, desirable for good high frequency and transient response, will reduce the signal output, if only because there will be less tape material within the replay gap at any one time. This makes severe demands on the design of the replay amplifier system, especially in the case of cassette recorder systems, where the slow tape speed and narrow tape track width reduces the extent of magnetic induction.

In practice, the mid-frequency output of a typical high quality cassette recorder head will only be of the order of 1 mV, and, if the overall S/N ratio of the system is not to be significantly degraded by replay amplifier noise, the 'noise floor' of the replay amp. should be of the order of 20 dB better than this. Taking 1 mV as the reference level, this would imply a target noise level of –72 dB if the existing tape S/N ratio is of the order of –52 dB quoted above.

The actual RMS noise component required to satisfy this criterion would be 0.25 µV, which is just within the range of possible circuit performance, provided that care is taken with the design and the circuit components.

The principal sources of noise in an amplifier employing semiconductor devices are:

(1) 'Johnson' or 'thermal' noise, caused by the random motion of the electrons in the input circuit, and in the input impedance of the amplifying device employed (minimised by making the total input circuit impedance as low as possible, and by the optimum choice of input devices)
(2) 'shot' noise, which is proportional to the current flow through the input device, and to the circuit bandwidth
(3) 'excess', or '1/f', noise due to imperfections in the crystal lattice, and proportional to device current and root bandwidth, and inversely proportional to root frequency
(4) collector-base leakage current noise, which is influenced by both operating temperature and DC supply line voltage
(5) surface recombination noise in the base region.

Where these are approximately calculable, the equations shown below are appropriate.

$$\text{Johnson (thermal noise)} = \sqrt{4KTR\Delta f}$$

$$\text{Shot noise} = \sqrt{2qI_{DC} \times \Delta f}$$

$$\text{Modulation } (1/f) \text{ noise} = \frac{\sqrt{\Delta I \times \Delta f}}{f}$$

where 'Δf' is the bandwidth (Hz), $K = 1.38 \times 10^{-23}$, T is the temperature (Kelvin), q the electronic charge (1.59×10^{-19} Coulombs), f is the operating frequency and R the input circuit impedance.

In practical terms, a satisfactory result would be given by the use of silicon bipolar epitaxial-base junction transistor as the input device, which should be of PNP type. This would take advantage of the better surface recombination noise characteristics of the n-type base material, at an appropriately low collector to emitter voltage, say 3–4 V, with as low a collector current as is compatible with device performance and noise figure, and a base circuit impedance giving the best compromise between 'Johnson' noise and device noise figure requirements.

Other devices which can be used are junction field-effect transistors, and small-signal V-MOS and T-MOS insulated-gate type transistors. In some cases the circuit designer may adopt a slightly less favourable input configuration, in respect of S/N ratio, to improve on other performance characteristics such as slew-rate balance or transient response.

Generally, however, discrete component designs will be preferred and used in high quality machines, although low-noise integrated circuit devices are now attaining performance levels where the differences between IC and discrete component designs are becoming negligible.

Typical input amplifier circuit designs by Sony, Pioneer, Technics, and the author, are shown in Figs. 7.19–7.22.

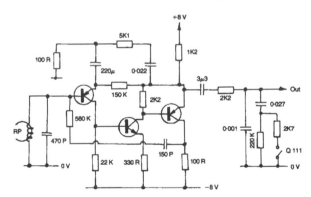

Figure 7.19   Cassette deck replay amplifier by Sony (TCK81).

In higher speed reel-to-reel recorders, where the signal output voltage from the replay head is higher, the need for the minimum replay amplifier noise is less acute, and low-noise IC op amplifiers, of the type designed for audio purposes will be adequate. Examples are the Texas Instruments TL071 series, the National Semiconductor LF351 series op amps. They offer the advantage of improved circuit symmetry and input headroom.

If the replay input amplifier provides a typical signal output level of 100 mV RMS, then all the other stages in the signal chain can be handled by IC op amps without significant degradation of the S/N ratio. Designer preferences or the desire to utilise existing well tried circuit structures may encourage the retention of discrete component layouts, even in modern equipment.

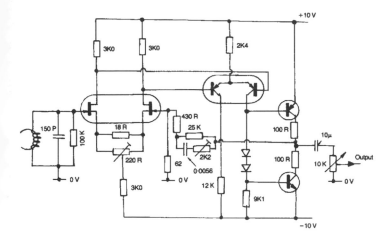

Figure 7.20 Very high quality cassette recorder replay amplifier due to Pioneer (CT-A9).

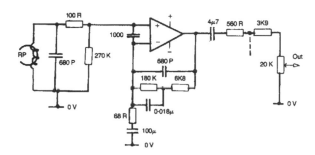

Figure 7.21 Combined record/replay amplifier used by Technics, shown in replay mode (RSX20).

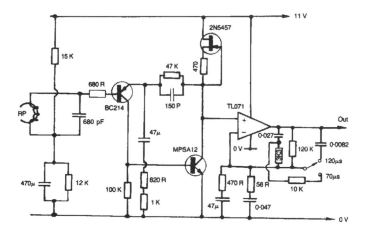

Figure 7.22 Complete cassete recorder replay system, used in portable machine.

However, in mass-produced equipment aimed at a large, low-cost, market sale, there is an increasing tendency to incorporate as much as possible of the record, replay, and noise reduction circuitry within one or two complex, purpose-built ICs, to lessen assembly costs.

Replay Equalisation

This stage will usually follow the input amplifier stage, with a signal level replay gain control interposed between this and the input stage, where it will not degrade the overall noise figure, it will lessen the possibility of voltage overload and 'clipping' in the succeeding circuitry.

Opinions differ among designers on the position of the equalisation network. Some prefer to treat the input amplifier as a flat frequency response stage, as this means the least difficulty in obtaining a low input thermal noise figure. Others utilise this stage for the total or initial frequency response shaping function.

In all cases, the requirement is that the equalised replay amplifier response shall provide a uniform output frequency response, from a recorded flux level having the characteristics shown in Fig. 7.8, ignoring for the moment any frequency response modifications which may have been imposed by the record-stage noise reduction processing elements, or the converse process of the subsequent decoding stage.

Normally, the frequency response equalisation function will be brought about by frequency sensitive impedance networks in the feedback path of an amplifier element, and a typical layout is shown in Fig. 7.23.

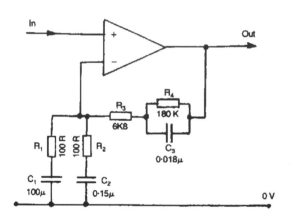

Figure 7.23   RC components network used in replay amplifier to achieve frequency response equalisation.

Considering the replay curve shown in Fig. 7.24, this can be divided into three zones, the flattening of the curve, in zone 'A', below 50 Hz, in accordance with the LF 3180 µs time-constant employed, following Table 7.3, in almost all recommended replay characteristics, the levelling of the curve, in zone 'B', beyond a frequency determined by the HF time constant of Table 7.3, depending on application, tape type, and tape speed.

Since the output from the tape for a constant remanent flux will be assumed to increase at a +6 dB/octave rate, this levelling-off of the replay curve is equivalent to a +6 dB/octave HF boost. This is required to correct for the presumed HF roll-off on record, shown in Fig. 7.8

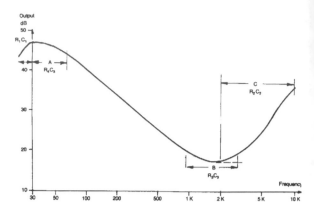

Figure 7.24   Cassette recorder replay response given by circuit layout shown in Fig. 7.23, indicating the regions affected by the RC component values.

Finally, there is the HF replay boost of zone 'C', necessary to compensate for poor replay head HF performance. On reel-to-reel recorders this effect may be ignored, and on the better designs of replay head in cassette recorders the necessary correction may be so small that it can be accomplished merely by adding a small capacitor across the replay head, as shown in the designs of Fig. 7.19–7.22. The parallel resonance of the head winding inductance with the added capacitor then gives an element of HF boost.

With worse replay head designs, some more substantial boost may be needed, as shown in 'C' in Fig. 7.23.

A further type of compensation is sometimes employed at the low-frequency end of the response curve and usually in the range 30–80 Hz, to compensate for any LF ripple in the replay response, as shown in Fig. 7.10. Generally all that is attempted for this kind of frequency response flattening is to interpose a tuned LF 'notch' filter, of the kind shown in Fig. 7.25 in the output of the record amp, to coincide with the first peak in the LF replay ripple curve.

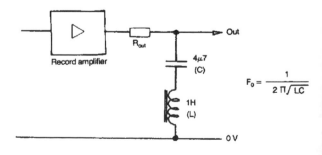

Figure 7.25   Notch filter in record amplifier, used to lessen LF ripple effects.

The Bias Oscillator

The first major requirement for this circuit is a substantial output voltage swing-since this circuit will also be used to power the 'erase' head, and with some tapes a large erase current is necessary to obtain a low residual signal level on the erased tape. Also needed are good symmetry of waveform, and a very low intrinsic noise figure, since any noise components on the bias waveform will be added to the recorded signal.

The invariable practical choice, as the best way of meeting these requirements, is that of a low distortion HF sine waveform.

Since symmetry and high output voltage swing are both most easily obtained from symmetrical 'push-pull' circuit designs, most of the better modern recorders employ this type of layout, of which two typical examples are shown in Fig. 7.26 and 7.27. In the first of these, which

is used in a simple low-cost design, the erase coil is used as the oscillator inductor, and the circuit is arranged to develop a large voltage swing from a relatively low DC supply line.

The second is a much more sophisticated and ambitious recorder design. Where the actual bias voltage level is automatically adjusted by micro-computer control within the machine to optimise the tape recording characterics. The more conventional approach of a multiple winding transformer is employed to generate the necessary positive feedback signal to the oscillator transistors.

In all cases great care will be taken in the choise of circuitry and by the use of high 'Q' tuned transformers to avoid degrading the purity of the waveform or the S/N ratio.

In good quality commercial machines there will generally be both a switch selectable bias level control, to suit the IEC designated tape types (1–4) and a fine adjustment. This latter control will usually be a pair of variable

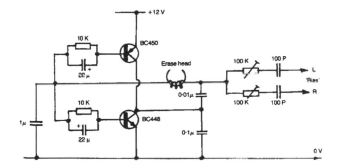

Figure 7.26 High-output, low-distortion, bias/erase oscillator used in cassette recorder.

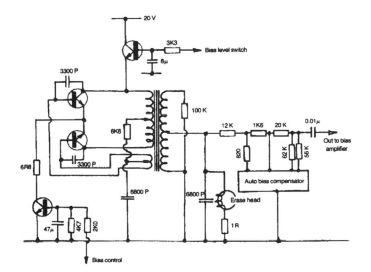

Figure 7.27 High-quality bias oscillator having automatic bias current adjustment facilities (Pioneer CT-A9).

resistors, in series with separate small capacitors, in the feed to each recording channel, since the high AC impedance of the capacitor tends to restrict the introduction of lower frequency noise components into the record head. It also serves to lessen the possibility of recording signal cross-talk from one channel to the other.

The Record Amplifier

Referring to the block diagram of Fig. 7.18, the layout of the recording amplifier chain will normally include some noise reduction encoding system (Dolby A or equivalent, and/or Dolby HX, in professional analogue-type tape recorders, or Dolby B or C in domestic machines). The operation of this would be embarassed by the presence of high-frequency audio components, such as the 19 kHz FM 'pilot tone' used in stereo broadcasts to regenerate the 38 kHz stereo sub-carrier.

For this reason, machines equipped with noise reduction facilities will almost invariably also include some low-pass filter system, either as some electronic 'active filter' or a simple LC system, of the forms shown in Fig. 7.28. Some form of microphone amplifier will also be provided as an alternative to the line inputs. The design of this will follow similar low-noise principles to those essential in the replay head amplifier, except that the mic. amplifier will normally be optimised for a high input circuit impedance. A typical example is shown in Fig. 7.29.

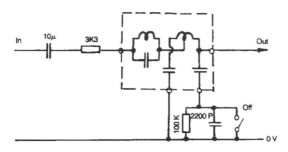

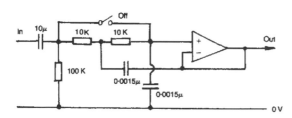

Figure 7.28   L-C and active low-pass filter systems used to protect noise reduction circuitry from spurious HF inputs.

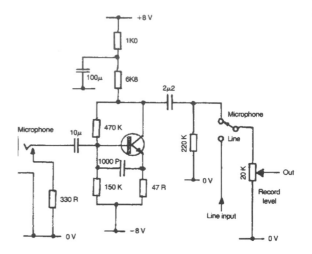

Figure 7.29   Single stage microphone input amplifier by Sony (TCK-81).

As mentioned above, it is common (and good) practice to include a 'muting' stage in the replay amp line, to restrict the audibility of the amplifier noise in the absence of any tape signal. A similar stage may be incorporated in the record line, but in this case the VU or peak recording meter will require to derive its signal feed from an earlier stage of the amplifier. Thus the input signal level can still be set with the recorder at 'pause'.

The recording amplifier will be designed to deliver the signal to the record head at an adequate amplitude, and with low added noise or distortion, and in a manner tolerant of the substantial HF bias voltage present at the record head.

Contemporary design trends tend to favour record heads with low winding impedances, so the record amplifier must be capable of coping with this.

As noted above, the design of the record amplifier must be such that it will not be affected by the presence of the HF bias signal at the record head.

There are approaches normally chosen. One is to employ an amplifier with a very high output impedance, the output signal characteristics of which will not be influenced by superimposed output voltage swings. Another is to interpose a suitable high value resistor in the signal line between the record amp and the head. This also has the benefit of linearising the record current *vs* frequency characteristics of the record system. If the record amplifier has a low output impedance, it will cause the intruding bias frequency signal component to be greatly attenuated in its reverse path.

However, the simplest, and by far the most common, choice is merely to include a parallel-tuned LC rejection circuit, tuned to the HF bias oscillator frequency, in the record amplifier output, as shown in the circuit design of Fig. 7.30.

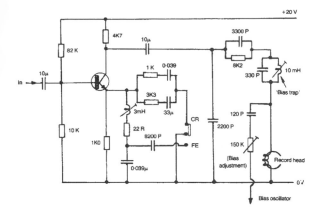

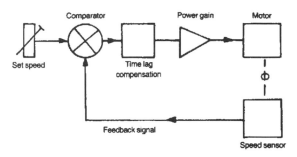

Figure 7.30   Record output stage, from Pioneer, showing bias trap and L-C equalisation components (CT4141-A).

Some very advanced recorder designs actually use a separate mixer stage, in which the signal and bias components can be added before they are fed to the record head. In this case the design of the output stage is chosen so that it can handle the total composite (signal + bias) voltage swing without cross-modulation or other distortion effects.

Recording Level Indication

This is now almost exclusively performed in domestic type equipment by instantaneously acting light-emitting diodes, adjusted to respond to the peak recording levels attained. There are several purpose-built integrated circuits designed especially for this function.

In earlier instruments, which moving coil display meters were used, great care was necessary to ensure that the meter response had adequate ballistic characteristics to respond to fast transient signal level changes. This invariably meant that good performance was expensive.

It is notable that professional recording engineers still prefer to use meter displays, the ballistic characteristics of which are normally precisely controlled, and to the ASA C16–5 specification. However, the indication given by these may be supplemented by LED-type peak level indication.

Tape Drive Control

Few things have a more conspicuous effect upon the final performance of a tape recorder system than variations in the tape speed. Electronic servo systems, of the type shown in schematic form in Fig. 7.31, are increasingly

used in better class domestic machines (they have been an obligatory feature of professional recorder systems for many years) to assure constancy of tape travel.

Figure 7.31   Schematic layout of speed control servo-mechanism.

In arrangements of this kind, the capstan drive motors will either be direct current types, the drive current of which is increased or decreased electronically depending on whether the speed sensor system detects a slowing or an increase in tape speed. Alternatively, the motors may be of 'brushless' AC type, fed from an electronically generated AC waveform, the frequency of which is increased or decreased depending on whether a positive or a negative tape speed correction is required.

With the growing use of microprocessor control and automated logic sequences to operate head position, signal channel and record/replay selection, and motor control, the use of electronically generated and regulated motor power supplies is a sensible approach. The only problem is that the increasing complexity of circuitry involved in the search for better performance and greater user convenience may make any repairs more difficult and costly to perform.

Professional Tape Recording Equipment

This is aimed at an entirely different type of use to that of the domestic equipment, and is consequently of different design, with a different set of priorities.

Domestic tape cassette recorders are used most commonly for transferring programme material from radio broadcasts gramophone or other tape recordings, or spontaneous (i.e., non-repeatable) 'live' events, to tape and the finished recording is made for the benefit of the recordist or the participants. No editing is desired or practicable. In contrast, in commercial/professional work it is likely that the final product will be closely scrutinised for imperfections, and will be highly edited.

Moreover, in broadcasting or news applications, it is essential that time synchronisation, for example with pictures, be practicable.

Also, since the recording process is just one step in a necessarily profitable commercial operation, professional machines must be rugged, resistant to misuse, and reliable. The cost of the equipment, except as dictated by competitive pressures between rival manufacturers, will be of a secondary consideration, and its physical bulk will be less important than its ease of use.

These differences in use and outlook are reflected in the different types of specification of professional recording systems, and these are listed below.

General Description

Mechanical

Professional tape recording machines will invariably be reel-to-reel recording as 'half-track', or occasionally as 'quarter-track' on ¼ in. tape, or as 8-track, 16-track, 24-track or 32-track, on ½ in., 1 in. or 2 in. wide tape. (Because of the strong influence of the USA in the recording field, metric dimensions are very seldom used in specifying either tape widths or linear tape speeds of professional equipment.)

Even on half-track machines, facilities are now being provided for time code marking as are more generally found on multi-track machines by way of an additional channel in the central 2.25 mm dead band between the two stereo tracks.

This allows identification of the time elapsed since the start of the recording as well as the time remaining on the particular tape in use. This information is available even if the tape is removed and replaced, part wound. (It is customary for professional recorders, copied now by some of the better domestic machines, to indicate tape usage by elapsed-time displays rather than as a simple number counter.) Also available is the real time – as in the conventional clock sense – and any reference markers needed for external synchronisation or tape editing.

The mechanisms used will be designed to give an extremely stable tape tension and linear tape speed, with very little wow or flutter, a switched choice of speeds, and provision for adjusting the actual speed, upwards or downwards, over a range – customarily chosen as seven semitones – with facility for accurate selection of the initial reference tape speed. It is also customary to allow for different sized tape spools, with easy and rapid replacement of these.

The high power motors customarily used to ensure both tape speed stability and rapid acceleration – from rest to operating speed in 100 ms is attainable on some machines – can involve significant heat dissipations. So

the bed-plate for the deck itself is frequently a substantial die-casting, of up to two inches in thickness, in which the recesses for the reels, capstans and heads are formed by surface machining. The use of ancillary cooling fans in a studio or sound recording environment is understandably avoided wherever possible.

To make the tape speed absolutely independent of mains power frequency changes, it is common practice to derive the motor speed reference from a crystal controlled source, frequently that used in any microprocessor system used for other control, editing, and search applications.

Since the product will often be a commercial gramophone record or cassette recording, from which all imperfections must be removed before production, editing facilities – either by physically splicing the tape, or by electronic processing – will feature largely in any 'pro' machine. Often the tape and spool drive system will permit automatic 'shuttle' action, in which the tape is moved to and fro across the replay head, to allow the precise location of any edit point.

An internal computer memory store may also be provided to log the points at which edits are to be made, where these are noted during the recording session. The number and position of these may also be shown on an internal display panel, and the tape taken rapidly to these points by some auto-search facility.

Electrical

The electronic circuitry employed in both professional and domestic tape recorder systems is similar, except that their tasks are different. Indeed many of the facilities which were originally introduced in professional machines have been taken over into the domestic scene, as the ambitions of the manufacturers and the expectations of the users have grown.

In general outline, therefore, the type of circuit layout described above will be appropriate. However, because of the optional choice of tape speeds, provision will be made for different, switch-selectable, equalisation characteristics and also, in some cases, for different LF compensation to offset the less effective LF recording action at higher tape speeds. Again, the bias settings will be both adjustable and pre-settable to suit the tape types used.

To assist in the optimisation of the bias settings, built-in reference oscillators, at 1 kHz, 10 kHz, and sometimes 100 Hz, may be provided, so that the flattest practicable frequency response may be obtained.

This process has been automated in some of the more exotic cassette recorders, such as the Pioneer CT-A9, in which the bias and signal level is automatically adjusted when the recording is initiated, using the first eleven seconds worth of tape, after which the recorder resets itself to the start of the tape to begin recording. Predictably, this style of thing does not appeal greatly to the professional recording engineer!

Whether or not noise reduction circuitry will be incorporated within the machine is a matter of choice. 'Pro' equipment will, however, offer easy access and interconnection for a variety of peripheral accessories, such as remote control facilities and limiter and compressor circuitry, bearing in mind the extreme dynamic range which can be encountered in live recording sessions.

Commonly, also, the high-frequency headroom expansion system, known as the Dolby HX Pro will be incorporated in high quality analogue machines. This monitors the total HF components of the recorded signal, including both programme material and HF bias, and adjusts the bias voltage output so that the total voltage swing remains constant. This can allow up to 6 dB more HF signal to be recorded without tape saturation.

In the interests of signal quality, especially at the high frequency end of the audio spectrum, the bias frequency employed will tend to be higher than in lower tape speed domestic machines. 240 kHz is a typical figure.

However, to avoid overheating in the erase head, caused by the eddy-current losses which would be associated with such a high erase frequency, a lower frequency signal is used, such as 80 kHz, of which the bias frequency employed is the third harmonic. This avoids possible beat frequency effects.

Very frequently, both these signals are synthesised by subdivision from the same crystal controlled master oscillator used to operate the internal microprocessor and to control the tape drive speed.

With regard to the transient response shortcomings of tape recording systems, referred earlier, there is a growing tendency in professional analogue machines to employ HF phase compensation, to improve the step-function response. The effect of this is illustrated in Fig. 7.32, and the type of circuitry used is shown in Fig. 7.33.

A major concern in professional work is the accurate control of both recording and output levels. Frequently the recording level meters will have switch selectable sensitivities for their OVU position, either as the basic standard (OVU = 1 mW into 600 ohms = 0.774 V RMS) or as 'dB V' (referred to 1 V RMS). It is also expected that the OVU setting will be related accurately to the magnetisation of the tape, with settings selectable at either 185, 250 or 320 nanoWebers/m.

Finally, and as a clear distinction from amateur equipment, professional units will all have provision for balanced line inputs from microphone and other programme sources, since this greatly lessens the susceptibility of such inputs to mains induced hum and interference pick-up.

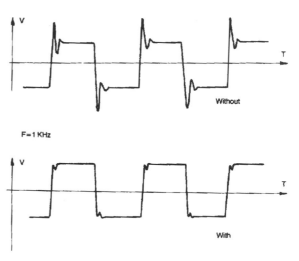

Figure 7.32 Improvement in HF transient response, and consequent stereo image quality, practicable by the use of HF phase compensation.

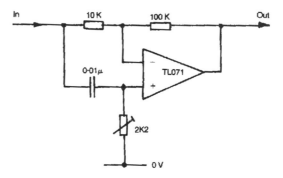

Figure 7.33 HF phase-compensation circuit.

Multi-Track Machines

Although single channel stereo machines have a place in the professional recording studio, the bulk of the work will be done on multi-track machines, either 8-, 16-, 24-, or 32-track. The reasons for this are various.

In the pop field, for example, it is customary for the individual instrumentalists or singers in a group to perform separately in individual sound-proofed rooms, listening to the total output, following an initial stereo mixing, on 'cans' (headphones). Each individual contributor to the whole is then recorded on a separate track, from which the final stereo master tape can be recorded by mixing down and blending, with such electronic sound effects as are desired.

This has several advantages, both from the point of view of the group, and for the recording company – who may offer a sample of the final product as a cassette, while retaining the master multi-track tape until all bills are paid. For the performers, the advantages are that the whole session is not spoilt by an off-colour performance by any individual within the group, since this can easily be re-recorded. Also, if the number is successful, another equally successful 'single' may perhaps be made, for example, by a re-recording with a different vocal and percussion section.

Moreover, quite a bit of electronic enhancement of the final sound can be carried out after the recording session, by frequency response adjustment and added reverberation, echo and phase, so that fullness and richness can be given, for instance, to an otherwise rather thin solo voice. Studios possessing engineers skilled in the arts of electronic 'fudging' soon become known and sought after by the aspiring pop groups, whose sound may be improved thereby.

On a more serious level, additional tracks can be invaluable for recording cueing and editing information. They are also useful in the case, say, of a recording of a live event, such as an outside broadcast, or a symphony orchestra, in gathering as many possible signal inputs, from distributed microphones, as circumstances allow. The balance between these can then be chosen, during and after the event, to suit the intended use. A mono radio broadcast needing a different selection and mix of signal than, for example, a subsequent LP record, having greater dynamic range.

This 'multi-miking' of large scale music is frowned upon by some recording engineers. Even when done well, it may falsify the tonal scale of solo instrumentalists, and when done badly may cause any soloist to have an acoustic shadow who follows his every note, but at a brief time interval later. It has been claimed that the 'Sound Field' system will do all that is necessary with just a simple tetrahedral arrangement of four cardioid microphones and a four-channel recorder.

A technical point which particularly relates to multi-track recorder systems is that the cross-talk between adjacent channels should be very low. This will demand care in the whole record/replay circuitry, not just the record/replay head. Secondly, the record head should be usable, with adequate quality, as a replay head.

This is necessary to allow a single instrumentalist to record his contribution as an addition to a tape containing other music, and for it to be recorded in synchronism with the music which he hears on the monitor headphones, without the record head to replay head tape travel time lag which would otherwise arise.

The need to be able to record high-frequency components in good phase synchronism, from channel to channel, to avoid any cancellation on subsequent mixing, makes high demands on the vertical alignment of the record and replay head gaps on a multi-channel machine. It also imposes constraints on the extent of wander or slewing of the tape path over the head.

Digital Recording Systems

The basic technique of digitally encoding an analogue signal entails repetitively sampling the input signal at sufficiently brief intervals, and then representing the instantaneous peak signal amplitude at each successive moment of sampling as a binary coded number sequence consisting of a train of '0s' and '1s'. This process is commonly termed 'pulse code modulation', or 'PCM'.

The ability of the encoding process to resolve fine detail in the analogue waveform ultimately depends on the number of 'bits' available to define the amplitude of the signal, which, in turn, determines the size of the individual step in the 'staircase' waveform, shown in Fig. 7.34, which results from the eventual decoding of the digital signal.

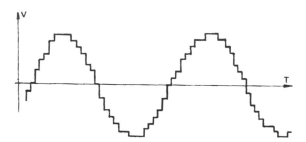

Figure 7.34 'Staircase' type waveform resulting from the digital encoding/decoding process. (1300 Hz sine-waveform illustrated as sampled at 44.1 kHz, and at 4-bit resolution).

Experience in the use of digitally encoded audio signals suggests that the listener will generally be unaware of the digital encoding/decoding process if '16-bit' (65 536 step) resolution is employed, and that resolution levels down to '13-bit' (8192 steps), can give an acceptable sound quality in circumstances where some low-level background noise is present, as, for example, in FM broadcast signals.

The HF bandwidth which is possible with a digitally encoded signal is determined by the sampling frequency, and since the decoding process is unable to distinguish between signals at frequencies which are equally spaced above and below half the sampling frequency, steep-cut low-pass filters must be used both before and after the digital encoding/decoding stages if spurious audio signals are to be avoided.

In the case of the compact disc, a sampling frequency of 44.1 kHz is used. This allows a HF turn-over frequency of 20 kHz, by the time 'anti-aliasing' filtering has been included. In the case of the '13-bit' pulse-code modulated distribution system used by the BBC for FM stereo broadcasting, where a sampling frequency of 32 kHz is used, the 'aliasing' frequency is 16 kHz, which limits the available HF bandwidth to 14.5 kHz.

It must be noted that the digital encoding (PCM) process suffers from the disadvantage of being very extravagant in its use of transmission bandwidth. For example, in the case of the 'compact disc', by the time error correction 'parity bits' and control signals have been added, the digital pulse train contains 4 321 800 bit/s. Even the more restricted BBC PCM system requires a minimum channel bandwidth of 448 kHz. Obviously, the use of a wider audio pass-band, or a greater resolution, will demand yet higher bit rates.

If it is practicable to provide the wide HF bandwidth required, the use of digitally encoded signals offers a number of compelling advantages. Of these probably the most important is the ability to make an unlimited number of copies, each identical to the 'first generation' master tape, without any copy to copy degradation, and without loss of high frequency and low-level detail from the master tape due to flux induced demagnetisation during the replay process.

In addition, the digital process allows a ruler flat replay response from, typically, 5 Hz–20 kHz, and offers signal to noise ratio, dynamic range, channel separation, frequency stability and harmonic and intermodulation distortion characteristics which are greatly superior to those offered by even the best of the traditional analogue reel-to-reel equipment.

Moreover, digital '0/1' type encoding shows promise of greater archival stability, and eliminates 'print-through' during tape storage, while the use of 'parity bit' (properly known as 'cross interleave Reed-Solomon code') error correction systems allow the correction or masking of the bulk of the faults introduced during the recording process, due, for example, to tape 'drop outs' or coating damage.

Specific bonuses inherent in the fact that the PCM signal is a data stream (not in the time domain until divided into time-related segments at intervals determined by some master frequency standard) are the virtual absence of 'wow', 'flutter' and tape speed induced errors in pitch, and also that the signals can be processed to remove any time or phase errors between tracks.

To illustrate the advantages of digitally encoded signal storage, a comparison between the performance of a good quality analogue recording, for example, a new (i.e., not previously played) 12 inch vinyl 'LP' disk, and a 16-bit digitally encoded equivalent is shown in Table 7.5.

Table 7.5 *Analogue vs. digital performance characteristics*

	12" LP	*16-bit PCM system (i.e., compact disc)*
Frequency response	30 Hz–20 kHz +/–2 dB	5 Hz–20 kHz +/–0.2 dB
Channel separation	25–30 dB	>100 dB
S/N ratio	60–65 dB	>100 dB
Dynamic range	Typically 55 dB @1 kHz	>95 dB
Harmonic distortion	0.3–1.5%	<0.002%
'Wow' and 'flutter'	0.05%	nil
Bandwidth requirement	20 kHz	4.3 MHz

There are, of course, some snags. Of these, the most immediate is the problem of 'editing' the digitally encoded signal on the tape. In the case of conventional analogue reel-to-reel tape machines, editing the tape to remove faults, or to replace unsatisfactory parts of the recording, is normally done by the recording engineer by inching the tape, manually, backwards and forwards across the general region of the necessary edit, while listening to the output signal, until the precise cut-in point is found. A physical tape replacement is then made by 'cut and splice' techniques, using razor blade and adhesive tape.

Obviously, editing a digital tape is, by its nature, a much more difficult process than in the case of an analogue record, and this task demands some electronic assistance. A range of computer-aided editing systems has therefore been evolved for this purpose, of which the preferred technique, at least for '16-bit' 'stereo' recordings, is simply to transfer the whole signal on to a high capacity (1 Gbyte or larger) computer 'hard disk', when the edit can be done within the computer using an appropriate programme.

The second difficulty is that the reproduced waveform is in the form of a 'staircase' (as shown in the simplified example shown in Fig. 7.34), and this is a characteristic which is inherent in the PCM process. The type of audible distortion which this introduces becomes worse as the signal level decreases, and is of a curious character, without a specific harmonic relationship to the fundamental waveform.

The PCM process also causes a 'granularity' in low level signals, most evident as a background noise, called 'quantisation noise', due to the random allocation of bit level in signals where the instantaneous value falls equally between two possible digital steps. This kind of noise is always present with the signal and disappears when the signal stops: it differs in this way from the noise in an analogue recording, which is present all the time.

In normal PCM practice, a random 'dither' voltage, equal to half the magnitude of the step, is added to the recorded signal, and this increases the precision in voltage level resolution which is possible with signals whose duration is long enough to allow many successive samples to be taken at the same voltage level. When the output signal is then averaged, by post-decoder filtering, much greater subtlety of resolution is possible, and this degree of resolution is increased as the sampling frequency is raised.

This increase in the possible subtlety of resolution has led to the growing popularity of 'over-sampling' techniques in the replay systems of 'CD' players (a process which is carried to its logical conclusion in the 256× over-sampling 'bitstream' technique, equivalent to a 11.29 MHz sampling rate), although part of the acoustic benefit given by increasing the sampling rate is simply due to the removal of the need for the very steep-cut, 20 kHz, anti-aliasing output filters essential at the original 44.1 kHz sampling frequency. Such so-called 'brick wall' filters impair transient performance, and can lead to a 'hard' or 'over bright' treble quality.

Some critical listeners find, or claim to find, that the acoustic characteristics of 'digital' sound are objectionable. They are certainly clearly detectable at low encoding resolution levels, and remain so, though less conspicuously, as the resolution level is increased.

For example, if the listener is allowed to listen to a signal encoded at 8-bit resolution, and the digital resolution is then progressively increased, by stages, to 16-bits, the 'digital' character of the sound still remains noticeable, once the ear has been alerted to the particular nature of the defects in reproduced signal. This fault is made much less evident by modern improvements in replay systems.

Professional equipment

So far as the commercial recording industry is concerned, the choice between analogue and digital tape recording has already been made, and recording systems based on digitally encoded signals have effectively superseded all their analogue equivalents.

The only exceptions to this trend are in small studios, where there may seem little justification for the relatively high cost of replacing existing analogue recording equipment, where this is of good quality and is still fully serviceable. This is particularly true for 'pop' music, which will, in any case, be 'mixed down' and extensively manipulated both before and after recording, by relatively low precision analogue circuitry.

Some contemporary professional digital tape recording systems can offer waveform sampling resolution as high as '20-bit' (1 048 576 possible signal steps), as compared with the '16-bit' encoding (65 536 steps) used in the compact disc system. The potential advantages, at least for archive purposes, offered by using a larger number of steps in the digital 'staircase' are that it reduces the so-called 'granularity' of the reproduced sound, and it also reduces the amount of 'quantisation noise' associated with the digital-analogue decoding process.

Unfortunately, in the present state of the art, part of the advantage of such high resolution recording will be lost to the listener since the bulk of such recordings will need to be capable of a transfer to compact disc, and, for this transfer, the resolution must be reduced once again to '16-bit', with the appropriate resolution enhancing sub-bit 'dither' added once more to the signal.

Some 'top of the range' professional fixed head multi-track reel-to-reel digital recorders can also provide on-tape cueing for location of edit points noted during recording, as well as tape identification, event timing and information on tape usage, shown on a built-in display system.

However, although multi-track fixed head recorders are used in some of the larger commercial studios, a substantial number of professional recording engineers still prefer the Sony 'U-matic' or '1630' processor, which is, effectively, a direct descendant of the original Sony 'PCM-1' and 'PCM-F1' audio transfer systems, which were designed to permit '16-bit' digitally encoded audio signals to be recorded and reproduced by way of a standard rotary head video cassette recorder.

It is unlikely that the situation in relation to digital recording will change significantly in the near future, except that the recording equipment will become somewhat less expensive and easier to use, with the possibility of semi-automatic editing based on 'real time' information recorded on the tape.

What does seem very likely is that variations of the 'near-instantaneous companding' (dynamic range compression and expansion) systems developed for domestic equipment will be employed to increase the economy in the use of tape, and enhance the apparent digital resolution of the less expensive professional machines.

Domestic equipment

In an ideal world, the only constraints on the design and performance of audio equipment intended for domestic use would be the need to avoid too high a degree of complexity – the cost of which might limit its saleability – or the limitations due simply to the physical characteristics of the medium.

However, in reality, although the detailed design of equipment offered by a mass-market manufacturer will be influenced by his technical competence, the basic function of the equipment will be the outcome of deliberations based on the sales forecasts made by the marketing divisions of the companies concerned, and of the policy agreements made behind the closed doors of the trade confederations involved.

A good example of the way in which these forces operate is given by the present state of 'domestic' (i.e., non-professional) cassette recording technology. At the time of writing (1992), it is apparent that the future for high-quality domestic tape recording systems lies, just as certainly as in the case of professional systems, in the use of some form of digitally encoded signal.

This is a situation which has been known, and for which the technical expertise has been available for at least 15 years, but progress has been virtually brought to a standstill by a combination of marketing uncertainties and objections from the record manufacturing companies.

As noted above, the major advantage offered by digital recording/replay systems is that, in principle, an unlimited number of sequential copies can be made, without loss of quality – a situation which is not feasible with the 'analogue' process.

The realisation that it is now possible, by direct digital transfer of signals from a compact disc, to make a multiplicity of blemish free copies, indistinguishable from the original, has frightened the recording companies, who fear, quite reasonably, that this could reduce their existing large profits.

As a result, their trade organisation, the International Federation of Phonograph Industries (IFPI) has pressed for some means of preventing this, at least so far as the domestic user is concerned, and has brought pressure to bear on their national governments to prevent importation of digital cassette recorders until the possibility of unlimited digital copying has been prevented.

This pressure effectively put a stop to the commercial development, ten years ago, of digital audio tape (DAT) machines, apart from such pioneering units as the Sony PCM-F1, which is designed for use with standard ¾ inch rotary-head video tape recorders, and which were classified as 'professional' units.

A technical solution to the record manufacturers objections has now been developed, in the form of the 'Serial Copy Management System' (SCMS). On recorders fitted with this device, a digitally coded signal is automatically added to the 'first generation' tape copy, when this is made from a 'protected' source, such as a compact disc, and this will prevent any further, 'second generation', digital copies being made on an 'SCMS' equipped recorder. However, even if the copy is made from an unprotected source, such as a direct microphone input, a digital copy of this material will still carry a 'copy code' signal to prevent more than one further generation of sequential copies from being made.

In the interim period, the debate about the relative merits of domestic 'S-DAT' recorders (digital tape recorders using stationary head systems) and 'R-DAT' systems (those machines based on a rotary head system, of the kind used in video cassette recorders) had faded away. In its original form, 'S-DAT' is used solely on multiple track professional systems, and 'R-DAT' apparatus is offered, by companies such as Sony, Aiwa and Casio, mainly as portable recorder systems, which come complete with copy-code protection circuitry.

The success of any mass-market recording medium is dependent on the availability of pre-recorded music or other entertainment material, and because of the uncertainties during the pre-'SCMS' period, virtually no pre-recorded digital tapes (DAT) have been available, and now that the political problems associated with 'DAT' seem to have been resolved, three further, mutually competitive, non-professional recording systems have entered the field.

These are the recordable compact disc (CDR), which can be replayed on any standard CD player: the Sony 'MiniDisc', which is smaller (3"), and requires a special player: and the Philips 'Digital Compact Cassette' (DCC), which is a digital cassette recorder which is compatible with existing compact cassettes for replay purposes.

This latter equipment was developed as a result of a commercial prediction, by the Philips marketing division, that vinyl discs, especially 12 inch 'LP's, would no longer be manufactured beyond 1995, and that compact cassettes, at present the major popular recorded music medium, would reach a sales peak in the early 1990s, and would then begin to decline in sales volume, as users increasingly demanded 'CD' style replay quality.

Compact discs would continue their slow growth to the mid-1990s but sales of these would become static, from then onwards, mainly due to market saturation. What was judged to be needed was some simple and inexpensive apparatus which would play existing compact cassettes, but which would also allow digital recording and replay at a quality standard comparable with that of the compact disc.

The Philips answer to this perceived need is the digital compact cassette player, a machine which would accept, and replay, any standard compact cassette, and would therefore be 'backwards compatible' both with existing cassette stocks, and user libraries. However, when used as a recorder, the unit will both record and replay a digitally encoded signal, at the existing 4.76 cm/s tape speed, using a fixed eight-track head.

Unfortunately digital recording systems are extravagant in their use of storage space, and the technical problems involved in trying to match the performance of a compact disc, and in then accommodating the digitally encoded signal on such a relatively low speed, narrow width tape, as that used in the compact cassette, are daunting. Even with the effective resolution reduced, on average, to the 4-bit level, a substantial reduction in bit rate is necessary from the 4.3218 Mbit/s of the compact disk to the 384 kbit/s possible with the digital compact cassette.

The technical solutions offered by Philips to this need for an eleven-fold bit-rate reduction are what is termed 'precision adaptive sub-band coding' (PASC), and 'adaptive bit allocation' (ABA). The 'PASC' system is based on the division of the incoming audio signal into 32 separate frequency 'sub-bands', and then comparing the magnitude of the signal in each of these channels with the total instantaneous peak signal level. The sub-band signals are then weighted according to their position in the frequency spectrum, and any which would not be audible are discarded.

The 'ABA' process takes into account the masking effect on any signal of the presence of louder signals at adjacent, especially somewhat lower, frequencies, and then discards any signals, even when they are above the theoretical audibility threshold, if they are judged to be likely to be masked by adjacent ones. The 'spare' bits produced by this 'gardening' exercise are then reallocated to increase the resolution available for other parts of the signal.

Reports on demonstrations of the digital compact cassette system which have been given to representatives of the technical press, suggest that Philips' target of sound quality comparable with that of the compact disc has been achieved, and the promised commercial backing from the record companies leads to the expectation that pre-recorded cassettes will be available, at a price intermediate between that of the CD and the existing lower quality compact cassette.

Blank tapes, using a chromium dioxide coating, will be available, initially, in the 'C60' and 'C90' recording lengths, and are similar in size to existing analogue cassettes, but have an improved mechanical construction, to allow a greater degree of precision in tape positioning.

In the analogue field, improved cassette record/replay heads capable of a genuine 20 Hz–20 kHz frequency response are now available, though tape characteristics, even using metal tape, prevent this bandwidth being attained at greater signal levels than 10 dB below the maximum output level (see pages 114–115, 118–119).

There is a greater availability, in higher quality cassette recorders, of the 'HX-PRO' bias control system, in which the HF signal components present in the incoming programme material are automatically taken into account in determining the overall bias level setting. When this is used in association with the Dolby 'C' noise reduction process S/N ratios better than 70 dB have been quoted.

A further improvement in performance on pre-recorded compact cassettes has been offered by Nakamichi by the use of their patented split-core replay head system, which is adjusted automatically in use to achieve the optimum azimuth alignment, for replaying tapes mastered on other recorders, and 'three-head' recording facilities are now standard in 'hi-fi' machines.

The extent to which manufacturers will continue to improve this medium is, however, critically dependent on the success, in the practical terms of performance, cost and reliability, of the digital compact cassette system. Whatever else, the future for domestic equipment is not likely to be one of technical stagnation.

Recommended Further Reading

Jorgensen, F. *The Complete Handbook of Magnetic Recording.* TAB Books (1990).

Mallinson, J.C. *The Foundations of Magnetic Recording.* Academic Press (1987). See also *General further reading* at the end of this book.

8 Noise Reduction Systems

Dave Fisher

One technology that, more than any other, has been able to keep analogue recording systems acceptable has been the increasingly complex noise-reduction systems being used. Dave Fisher here explains the basis of these systems and the most recent methods being used.

Introduction

Since the start of the audio industry, the recording medium has been the weak link in the audio chain. The frequency response, noise, and distortion performance of electronic systems has almost always been superior to contemporary recording systems. Even modern professional analogue recorders have a noise and distortion performance which is far worse than the electronics of the audio chain; it is for this reason that noise reduction systems have generally been devised specifically to reduce the noise of tape recorders, rather than of the audio chain as a whole. Nevertheless, there is no reason in principle why noise reduction systems should not be applied to any part of the transmission chain (i.e. the system which conveys programme material from one place to another, such as magnetic tape (including compact cassette), vinyl discs, radio broadcasts and, in broadcasting, land lines).

Noise reduction systems can generally be classified as either complementary or non-complementary. A complementary system (Fig. 8.1) modifies the audio signal in some way before it is recorded (or before it undergoes some other noisy process), and on replay performs the mirror image operation with a view to undoing the original modification. If the characteristics of the record and replay processors are chosen correctly, then the noise will be reduced because it has only passed through the replay processor, whereas the audio output signal will be the same as the input signal because it has passed through both the record and replay processors which are mirror images of one other. In practice, there will probably be some small change in the audio signal; it is this

which distinguishes one noise reduction system from another.

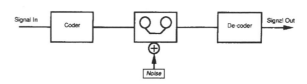

Figure 8.1 Complementary noise reduction system.

It is important to note that only the noise which is introduced between the record and replay processors can be reduced by a complementary system; the record processor cannot distinguish between noise inherent in the input signal and the wanted audio, thus, in a correctly aligned system, any noise fed into the record processor will emerge from the output at exactly the same level as it went in. In theory, it may seem that the noise introduced between the processors could be reduced to zero, but in practice this is neither possible nor desirable. It is generally best to process the audio as little as possible, since the less treatment it receives, the less audible will be the effects of any mis-tracking between the record and replay processors. On the other hand, if only a small amount of treatment is applied, the noise performance of the system as a whole will only be improved by a small amount; the aim of the designer of noise reduction systems must therefore be to strike a balance between improving the signal to noise ratio and degrading the audio quality because of audible artifacts in the output signal. In commercial noise reduction systems such audible effects are low or inaudible, and must be balanced against the improved noise performance; which you prefer, of course, is a matter of personal choice.

A non-complementary system (Fig. 8.2) attempts to reduce noise already superimposed on an audio signal by modifying both the signal and the noise to make the noise less objectionable. Only in exceptional circumstances can such a system reduce the noise without changing the quality of the audio. As non-complementary systems are less common and generally less satisfactory than complementary systems for high quality use, they will be discussed first.

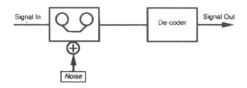

Figure 8.2   Non-complementary noise reduction system.

Non-Complementary Systems

There are two different approaches that can be adopted with non-complementary systems, either the frequency response or the level of the noisy audio signal can be modified to make the noise less noticeable or objectionable.

Equalisation

If the spectrum of the noise is significantly different to the spectrum of the programme material, then equalisation may reduce the noise without changing the programme material (see Fig. 8.3). For instance, a recording of male speech will contain very little energy above 10 kHz or so. If such a signal has wideband noise, then a 10 kHz low-pass filter can reduce the noise without significantly

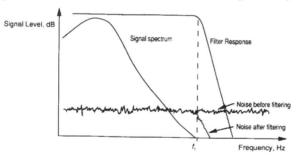

Figure 8.3   Using equalisation to reduce noise. If the spectrum of the noise and the signal do not overlap, a filter can be used to improve the signal to noise ratio. Provided that the turnover frequency of the filter is f₁ or higher, the wanted signal will not be affected.

changing the programme material. Similarly, a solo violin contains no significant energy below 196 Hz (the fundamental of its lowest string); if such a recording has traffic rumble or mains hum superimposed on it, then a high-pass filter set to the highest possible frequency below 196 Hz will produce an improvement. Unfortunately, the circumstances in which single-ended (i.e. non-complementary) equalisation produces an improvement

are limited, and, of course, the equalisation needs to be set for each circumstance. In extreme cases there may be a subjective improvement even if the spectra of programme and noise overlap; for instance high level hum superimposed on wideband programme may be more objectionable than a notch in the frequency response at the hum frequency. In general, however, the use of single-ended equalisation as a noise reduction system for high quality use is of only limited use (but see pre- and de-emphasis, later).

Dynamic noise limiters or filters

The dynamic noise filter/limiter is a development of simple equalisation. The human ear is not as sensitive to some frequencies as to others. Moreover, its sensitivity (and hence its frequency response) varies by a very large amount with the loudness of a sound. Equal loudness curves (see Fig. 8.4) are plots of sound pressure level against frequency for a variety of listening levels. Any point on any one curve will sound as loud as any other point on the same curve, regardless of frequency. Note

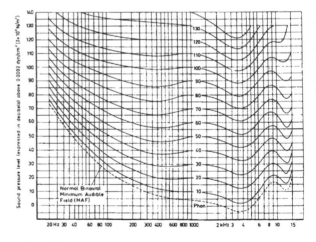

Figure 8.4   Equal loudness curves.

that the curves are much flatter at high level, and that at low levels the ear has very poor response to both low and high frequencies. Advantage can be taken of the ear's characteristics, by tailoring the frequency response of an adaptive filter to provide progressively more high frequency cut as the signal level falls (see Fig. 8.5). Because this mimics the ear's own response, the result is much more satisfactory than a fixed filter. It is not generally necessary to vary the low frequency response of a dynamic noise filter to achieve a subjective improvement in the noise performance, because off-tape noise has a spectrum that is predominantly high frequency. Although the frequency response after treatment is flat at high levels, and there is, therefore, no noise reduction, the high level programme may mask the noise, making noise reduction unnecessary.

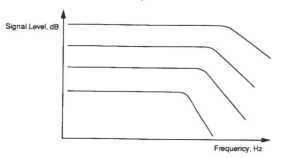

Figure 8.5 Dynamic noise filter. As the signal level falls it is subject to increasing amounts of HF cut.

Dynamic noise filters/limiters are now rarely used in domestic equipment, though in the early 1970s they were available built in to some cassette machines. They are available professionally, however, and are sometimes used to reduce the noise of systems, such as guitar amplifiers. Whilst they have the advantage that they can reduce the noise of a recording that has not been pre-processed, they have been superseded for normal recording use by complementary companding systems.

Noise gates

A noise gate can be thought of as a kind of level-dependent automatic switch. If there are gaps in the programme material, e.g. between tracks on a pop album or movements in a classical symphony, then it may be advantageous to reduce the replay gain, or, in the extreme, mute the output during these gaps. This can be achieved with a noise gate, such as that shown in Fig. 8.6. The input level is examined by the level sensor and if it is below some pre-set threshold the switch is turned off, and the gate is described as **closed**. When the signal level next exceeds the threshold the switch is turned on again (the gate is then **open**). This clearly has some disadvantages.

(1) It may be impossible to ensure that there is never any wanted signal below the threshold; if there is it will be cut when the gate closes.
(2) There is a danger that a steady signal at the threshold level will cause the gate to switch rapidly and repeatedly between open and closed. This would produce very objectionable changes in the noise level. To prevent this, hysteresis or delay must be built into the sensing circuit or it must have a slow response.
(3) The sensing circuit and switch cannot act instantaneously, so the signal which causes the gate to open

will always be clipped to some extent; this may be even more severe if the gate has a slow response.
(4) Although the noise during gaps in the programme material will be reduced, there will be no change in the noise when the signal is above the threshold level. Thus the effect of a noise gate may be to swap noise during programme gaps for the even more objectionable effect of very quiet gaps followed by noisy programme, drawing attention to the presence of the noise.

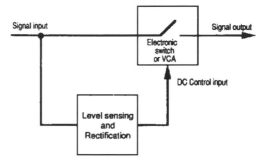

Figure 8.6 Principle of the noise gate. The electronic switch is 'on' when the input signal level is high, and 'off' when the signal level is low.

To alleviate some of these problems, noise gates are normally designed with a voltage controlled amplifier (VCA) as the active element instead of a switch. A VCA's gain is proportional to its DC control voltage, so it could be 0 dB for signals above the threshold, and lower (say –20 dB) for signals below the threshold. This has the advantage that the noise in the programme gaps can be attenuated by a controlled amount, so that it is reduced rather than removed; this is generally less objectionable because the contrast between the gaps and the programme is less pronounced. Furthermore, with a suitable DC control signal, the VCA can fade the signal down at the start of a gap and fade up into the programme at the end of a gap, producing a less objectionable result than a cut. Although noise gates are common in professional studios, where they are commonly used to reduce spill between microphones, they are uncommon in domestic equipment.

If the DC control signal is arranged to vary the gain of the VCA over a wide range, rather than just to switch its gain between two settings, then the gate becomes an expander. Though not common, expanders have been used domestically for some years. If used to expand the dynamic range of music from what is normally recorded to something closer to the original sound, they can produce very dramatic effects. Although single-ended expanders will not be considered here, the use of expanders is crucial to most complementary noise reduction systems, and will be described later.

Complementary Systems

The aim of a complementary noise reduction system is to modify or code the input signal so that it makes best use of the transmission path (in this case usually a tape machine), and then, after the signal has undergone the noisy process, to decode it so that it emerges from the output as unchanged as possible (see Fig. 8.1). As with non-complementary systems, both the spectrum and the dynamic range of the input signal may be modified to make the signal more suitable for the transmission path. The system must be carefully designed to produce good results with audio paths that are typically available in terms of frequency response, phase, distortion, gain and noise. For instance, a system which reduced the noise by 60 dB but which only worked satisfactorily with a tape recorder that had a frequency response which was flat within 0.1 dB to 30 kHz would be of no value, since such a recorder is unavailable. Furthermore, the noise which remains after the noise reduction process should not have a recognisable spectrum or other characteristic, since this will give the system a sort of *fingerprint*; the noise reduction should be perceptively similar for all types of noise likely to be encountered.

Finally, it should be remembered that one obvious way of reducing the noise of a tape machine is to increase the track width. Unfortunately the signal to noise ratio improves by only 3 dB for each doubling of the track width: to obtain a 10 dB improvement would require the track width to be increased by a factor of ten. Not only would this make tapes inconvenient in size, but their cost would be considerably greater, and the problems of engineering transport systems that could cope with the extra width and weight and still achieve, for instance, satisfactory azimuth would be very difficult to overcome; this should be borne in mind when assessing any artifacts which commercial systems may generate – it may be that the reduction in noise outweighs any audible change to the programme signal, especially as the change, if indeed there is any at all, will be small.

Emphasis

Emphasis is used in a wide range of audio systems; FM broadcasting (both UHF TV and Band II Radio), NICAM 728 (used for TV stereo sound in Britain and some parts of Europe), and digital audio (including CD) are just some of the systems that are, or can be, pre-emphasised.

Frequency modulated systems have a noise spectrum that is triangular, that is if the spectrum of the noise which is introduced between transmitter and receiver is

flat, then the noise on the decoded signal will rise in level with increasing frequency; therefore pre-emphasis is almost always used in conjunction with them. Analogue tape recording also has a noise spectrum that increases in level with increasing frequency, due to the equalisation needed to overcome the effects of thickness loss and head losses which both increase with frequency. Moreover, HF reproduced noise is more troublesome than LF, since at normal audio replay levels the noise volume will be at a level where the human ear is less sensitive to LF due to the shape of the equal loudness curves (e.g. around 30 to 40 phons). Luckily the spectrum of most programme material is predominantly LF, with little energy at HF. So, if the channel through which the audio signal is being conveyed has an overload characteristic that is flat with frequency, then HF boost can be added to the signal before recording or transmission, and corresponding HF cut on replay or reception will restore the original signal and reduce the noise introduced between the pre- and de-emphasis circuits. The overall effect is to tailor the spectrum of the signal to fit the capabilities of the channels (see Fig. 8.7).

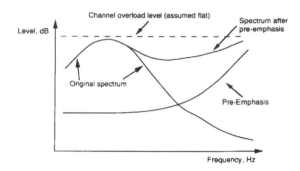

Figure 8.7   Use of pre-emphasis to make best use of the overload capabilities of an audio channel.

The major snag to this is in predicting the spectrum of the original programme material. If the signal level is measured with a meter before the pre-emphasis is applied there is a danger that the pre-emphasis will lead to overload of HF signals. If the signal level can be monitored after pre-emphasis has been applied this is not a problem. Most digital systems which use pre-emphasis therefore meter the signal after pre-emphasis has been applied; it is then up to the operator to ensure that the headroom of the system is not exceeded.

Emphasis curves are normally specified by their time constants. This is just a convenient short-hand way of specifying their turn-over frequency. It is important that emphasis curves are accurately applied if there is to be no overall change in the frequency response of the system, but to quote the gain at each frequency to a number of decimal places would be cumbersome. Luckily, the precise frequency response of a capacitor – resistor (CR) circuit, out of which pre-emphasis and de-emphasis circuits are constructed, is fully specified by its time constant, irrespective of the specific values of resistance or capacitance used in a particular design (see Fig. 8.8).

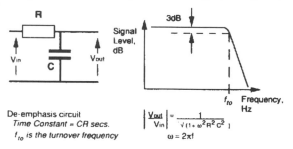

De-emphasis circuit
Time Constant = CR secs.
f_{to} is the turnover frequency

$$\left|\frac{V_{out}}{V_{in}}\right| = \frac{1}{\sqrt{(1+\omega^2 R^2 C^2)}}$$

$\omega = 2\pi f$

Figure 8.8 De-emphasis. Emphasis is the inverse.

Band II FM radio and the FM analogue sound transmitted with UHF television both have pre-emphasis of 50 µs, Fig. 8.9; the NICAM 728 digital sound transmitted with UHF television has the internationally agreed CCITT J 17 pre-emphasis, Fig. 8.10; many digital audio systems (e.g. CD) have optional pre-emphasis of 50/15 µs. Whether or not emphasis has been applied is identified by a control character incorporated into the digital datastream. Here, 50 µs sets the frequency of the start of the HF boost (3.183 kHz) and 15 µs sets the frequency at which the boost stops (10.61 kHz), producing a **shelf** characteristic of 10 dB boost, Fig. 8.11.

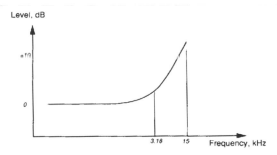

Figure 8.9 50 µs emphasis.

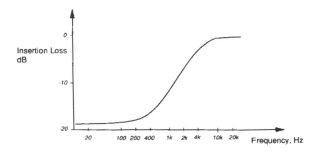

Figure 8.10 CCITT recommendation J 17.

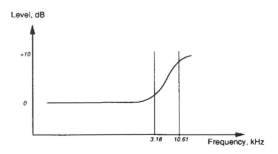

Figure 8.11 50/15 µs emphasis.

Companding Systems

Commercial systems are generally based on companding, that is the dynamic range of the signal is compressed before recording and then expanded after replay. If this is done correctly the original signal dynamics will be restored, but the noise level will be reduced (see Fig. 8.12).

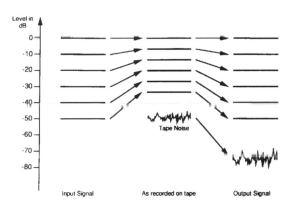

Figure 8.12 Principle of a companding system. The dynamic range is reduced by a compressor before recording; on replay an expander restores the original dynamic range, and reduces the noise introduced between the compressor and expander. This example shows a compression ratio of 1:1.5.

It is important that the expansion is a mirror image of the compression, and to that end it would be advantageous to send a control signal from the compressor to the expander, since this could ensure that there were no tracking errors. Unfortunately there is no reliable way of doing this in analogue tape systems; the whole bandwidth is needed for audio, and the out-of-band performance of tape machines is undefined and therefore unreliable. One system which did have a control link was the old monophonic *sound in syncs* digital audio system which was used by broadcasting organisations to convey television sound digitally to transmitters before the advent of stereo, though the digits were not radiated from the transmitters. The digital information was inserted into the television line synchronising pulses after being compressed by 20 dB. At the transmitter, the signal could be expanded precisely, with no mistracking, because a pilot tone, whose level was proportional to the amount of compression applied at any instant, could be conveyed reliably by the digital link.

Conventional companders

Figure 8.13 shows the characteristic of a high level compressor (and the complementary expander). Up to the threshold point the output level is equal to the input level. Above the threshold any change in the input level produces a smaller change in the output level depending on the compression ratio, which is defined as

$$\text{compression ratio} = \frac{\text{change in input level (in dB)}}{\text{corresponding change in output level (in dB)}}$$

It should be noted that a compressor is a type of automatic gain control, one way of thinking of it is as a fader controlled by the signal level, it does not clip the signal.

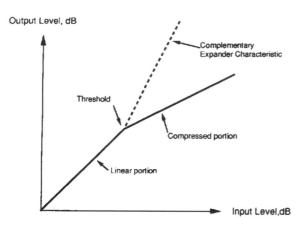

Figure 8.13   Compressor characteristic. Above the threshold, the output level increases more slowly than the input. In this example the compression ratio is 2:1; for every 2 dB change in the input, the output changes by only 1 dB.

Unfortunately, after compression, the level of high level signals is reduced, so if this programme were recorded as it was, the signal to noise ratio would be worsened at high level and be unchanged at low level. If the signal is amplified after the compressor, however, the peak level can be restored, whilst the lower level signals are amplified (Fig. 8.14). In practice the rotation point (the level at which the amplified output of the compressor is the same as the compressor input) is chosen so that the last few dBs below the peak level are reduced by the

compressor action, improving the distortion characteristics of the system; any worsening of the signal to noise ratio at these very high levels is masked by the high level signal. For use in a noise reduction system the threshold will probably be set at a very much lower level than shown in Fig. 8.14.

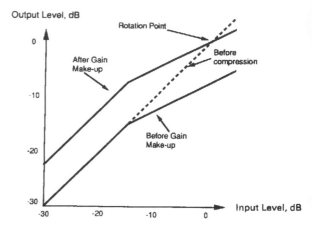

Figure 8.14   Compressor characteristic with gain make-up. The threshold has been chosen arbitrarily.

Bi-linear companders

Bi-linear companders were designed by Dolby and are exclusively used in their noise reduction systems. The principle is shown in Fig. 8.15. The compression law is generated by adding the output of a *differential network* to the input signal.

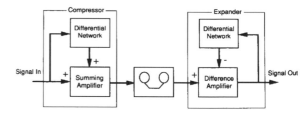

Figure 8.15   Bi-linear compressor and expander.

For input levels above about 40 dB below peak level, the output from the differential network is nearly constant, because it is effectively an audio limiter with a low threshold. When two signals of the same level are added together, the resultant increases by 6 dB; when two signals that are considerably different in level are added together the resultant is negligibly different from the higher level. Thus the level of low level signals can be increased whilst keeping the level of high level signals unchanged. If the gain of the sidechain is now adjusted, the low level signals can be increased by any chosen amount. If the contributions from the main and the differential network at low levels are in the ratio 1:2:16, then the low level signals will be increased by 10 dB. In the Dolby A system, for instance, the differential signal is combined with the main signal in this way, so that low levels (those that are more than about 40 dB below peak level) are increased in level by 10 dB, high level signals (those that are less than about 10 dB below peak level) are effectively unchanged, and those in between are changed proportionally. The maximum compression ratio occurs at about –30 dB and is approximately 2:1. The expander is made from the same circuit blocks, but by feeding the differential network from the output of the system and by subtracting its output from the input signal the complementary mirror image characteristic can be obtained (see Fig. 8.16).

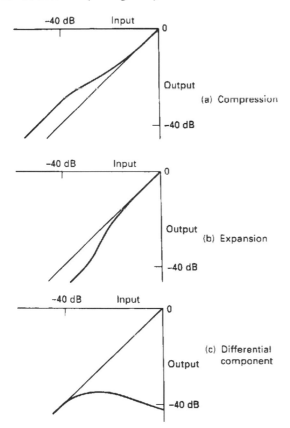

Figure 8.16 Bi-linear compander characteristics.

There are a number of advantages to this type of compander:

(1) It is easy to ensure that the compressor and expander characteristics are mirror images of each other,

because the same differential circuit is used in the compressor and the expander.

(2) High level signals, where distortion and tracking errors are most likely to be heard, have the minimum of treatment, since the differential component is then very low in level compared to signals in the direct path.

(3) The threshold can be set above the noise floor of the transmission channel so that the companding action is not controlled by noise.

(4) Only a small change of circuitry (a switch and a signal inverter) is necessary to convert a compressor into an expander.

A disadvantage is that, because the compression ratio changes over the audio dynamic range, a gain misalignment between compressor and expander will have different effects at different levels, and possibly cause dynamic errors in the reproduced signal.

Attack and decay times

The attack time of a compressor is the time which it takes to reduce its gain after a signal, which is initially just below the threshold, rapidly rises above the threshold, the release or recovery time is the time which the compressor takes to restore its gain to the steady-state value after a signal falls below the threshold. At first sight it may seem that the faster these times could be the better, but this is not necessarily so. When any signal is changed in level the theory of amplitude modulation shows that side frequencies are generated. If, for example, the amplitude of a signal at 1000 Hz were changed sinusoidally at a rate of 2 Hz, then side frequencies of 1002 and 998 Hz would be generated; in the case of companders, the change in level is not sinusoidal, but determined by the attack and decay characteristics. Thus there may be a group of frequencies (known as a sideband) on each side of the programme signal. Luckily the side frequencies are likely to be masked by the main signal (since they are very close in frequency) unless the attack and decay times are very fast. Furthermore, if the attack time is very fast, the gain of the compressor will be able to change during the first half cycle of a signal that exceeds the threshold, with consequential waveform distortion (see Fig. 8.17).

It is vital that the expansion is a mirror image of the compression; if it is not then tracking errors will occur. This means that the expander must be matched to the compressor in terms of gain, threshold, compression ratio, attack time and decay time both statically and dynamically; dynamic errors, e.g. of attack or decay time, will manifest themselves as objectionable breathing or programme modulated noise effects; static errors may cause only gain errors, but unless the threshold is set very low and noise reduction action occurs over a large level range, then the gain errors may be level dependent.

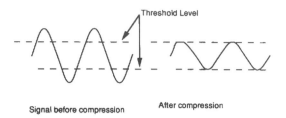

Figure 8.17   Very fast attack times cause waveform distortion. This diagram assumes a limiter with a very high compression ratio.

Masking

If a loud sound occurs at the same time as a quiet sound it is possible that the quiet sound may be inaudible. Figure 8.18 shows how the threshold of hearing, i.e. the level below which nothing can be heard, is shifted by the presence of a high level signal. The effect is not the same for all frequencies, though there is more effect above the masking signal frequency than below it; this means that high frequency signals are better masked by lower frequency signals than vice versa, and that the masking effect falls as the difference between the frequencies increases. Thus, a predominantly LF signal such as a bass drum will not significantly mask HF noise such as tape hiss, nor will a predominantly HF signal such as a triangle mask LF noise such as mains hum.

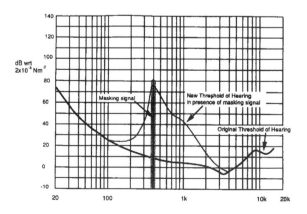

Figure 8.18   Masking. The effect of a narrow band of noise on the minimum audible field (or threshold of hearing).

In addition to this simultaneous, or spectral, masking, there is another effect known as non-simultaneous, or temporal, masking, in which sounds that occur a short time before or after a louder sound may be rendered inaudible. Temporal masking is dependent upon the duration, level and spectrum, of both the masked and masking signals, as well as their relative timings and durations. However, for forward masking, i.e. masked signals which occur after the signal which masks them, there is more effect if the signals are close together in time. The results obtained by different researchers in this field have not, unfortunately, been quite consistent. Nevertheless, it is known that masking may occur for 100 to 200 ms, that is quiet sounds which occur 100 ms after louder sounds may, for instance, be inaudible; the amount of masking increases as the duration of the initial masking signal increases up to durations of at least 20 ms and maybe as much as 200 ms; the amount of masking is also influenced by the relative frequencies and spectrum of the two signals. All of this has implications for the attack and decay times of companders, since it is unnecessary, and therefore probably undesirable, to attempt to reduce the noise if it is already inaudible having been masked by the programme signal. Certainly the response times of companders need be no better than that required by temporal masking.

A further problem which must be overcome is that of programme modulated noise. When the signal is at low level all companding systems will raise its level on record, with a consequent improvement in the signal to noise ratio on replay. When the signal is at peak level, however, the record processor cannot increase the level, so there will then be no noise reduction action. The result is that the noise level will be higher for high level programme than for low level programme. Consider the situation of a high level, low frequency signal, with low level, high frequency, noise. As the companding action changes the gain of the system under the influence of the low frequencies, the HF noise will also go up and down in level, but will probably not be masked by the LF signal. Unless consideration is given to this at the design stage of the system by ensuring that this noise is masked, there may be objectionable breathing or swishing sounds evident on the decoded signal. The effects of masking can be taken advantage of to ensure that the dynamic artifacts of compansion are not audible on the replayed signal in a number of ways. The frequency range over which the noise reduction action occurs can be limited, so that, for instance, only high frequency noise is reduced; the signal can be split into a number of different frequency bands, and each band companded separately, so the high level signals within the band have only to mask low level signals in the same band; or advantage can be taken of pre-emphasis and companding combined.

The Dolby A System

The Dolby A Noise Reduction System was developed in 1966; it rapidly became the most commonly used professional system, a position which it still holds, with the result that many records and cassettes carry signals that have been processed by Dolby A companders, though these signals have been decoded, or re-coded using some other noise reduction system, before being distributed on the domestic market.

The system uses band splitting to reduce the effects of programme modulated noise, by giving each band its own compander (see Fig. 8.19). The filter bands are:

(1) 80 Hz low pass, this band deals with hum and rumble;
(2) 80 Hz to 3 kHz bandpass, this band deals with print-through, crosstalk, and broadband noise;
(3) 3 kHz high pass, this band deals with hiss;
(4) 9 kHz high pass, this band also deals with hiss.

Band 2 is in fact generated by adding bands 1 and 3 together and subtracting the result from the input signal; this has the advantage that no part of the audio signal is omitted by small errors in the frequency of the filters. Bands 3 and 4 overlap above 9 kHz, with the result that there is increased noise reduction action there. Overall, the Dolby A system gives an improvement of 10 dB in the signal to noise ratio up to about 5 kHz, rising to 15 dB at 15 kHz and above (see Fig. 8.20).

The advantage of using relatively slow time constants in the control circuits (see the section on masking) has the side effect that the compressor output may contain signal overshoots. These would waste headroom if they were passed linearly by the recorder, and worse, if they were clipped by the recorder there would be increased distortion, and the signal presented to the replay processor would not be identical to that generated by the record processor, with the possibility of mis-tracking. The Dolby A system deals with this problem by incorporating a non-linear limiter (or clipper) after each of the linear limiters in the side chain (see Fig. 8.19). At first sight this may seem unsatisfactory, since it increases distortion. However, the clipped signal is in the side chain, it is therefore at least 40 dB below peak level and only occurs in the presence of a high level main chain signal, resulting in its being masked; the clipping is of very short duration (1 ms or less); and the clipper operates only rarely. The result is improved performance with no degradation of the signal path performance.

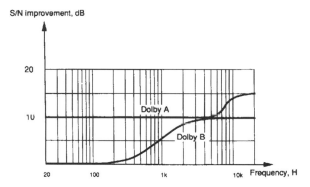

Figure 8.20 Maximum signal to noise ratio improvement of Dolby A and B.

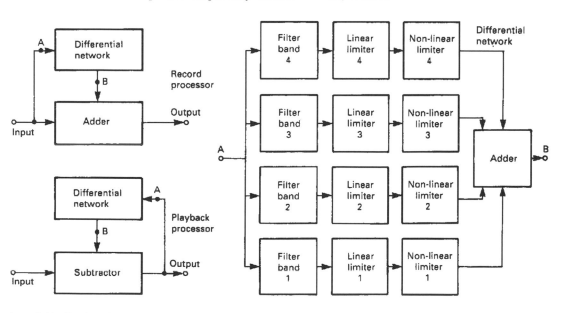

Figure 8.19 Dolby A band splitting.

To ensure that the decoder is presented with exactly the same signal level as the coder generated, every tape that is Dolby A encoded has Dolby tone recorded on it. Dolby tone is easily recognisable, because it is frequency modulated from 850 Hz to 930 Hz for 30 ms every 750 ms, but is constant in level, and therefore easy to set using a meter.

Telcom C4

Telcom C4 is a system for professional use, which was originally designed by Telefunken and is now made by ANT. It is a four-band system, each band with its own compander. The companders used are of the conventional type, i.e. they are not bi-linear, but they have a low compression ratio (1:1.5) and a very low threshold, with the result that they can compress programme material over a very wide dynamic range, see Fig. 8.21. Approximately 25 dB improvement in the signal to noise ratio is possible. Because the compression ratio is constant across the whole dynamic range of the input signal, any errors of gain between the recorder input and the replay machine output produce the same error in dB at all levels, i.e. for gain errors between the processors, only gain errors are produced in the decoder output. Thus, for a 2 dB error in the recorder, the decoded signal will be 3 dB too high or too low, but will be 3 dB at all levels, unlike a system that has a non-constant compression ratio, where there is a possibility of signal dynamic errors at intermediate levels. On the other hand, the decoded error is greater than the error of the tape machines at all levels, whereas at high and low (but not intermediate) levels, a system with non-constant compression ratio will have the same error as the recorder. This has implications for tape editing, where the level two tapes which are to be edited together are not identical.

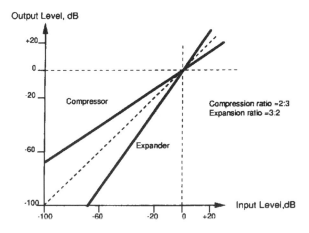

Figure 8.21 The Telcom C4 compander static characteristics.

The *rotation point*, shown as 0 dB in Fig. 8.21, can, of course be set to any suitable tape flux. Since the level of the signal above this point is reduced during coding (compression) and restored during replay (expansion) the peak recorded level can be reduced, compared to a system without compansion, with a consequent improvement of the off-tape distortion. Because tape distortion rises rapidly near peak level, a significant reduction in distortion is produced for only small changes in peak flux.

As Dolby A, Telcom C4 takes advantage of band splitting to aid noise masking. The bands used are 30 to 215 Hz, 215 to 1450 Hz, 1.45 to 4.8 kHz and 4.8 to 20 kHz, each band has its own compressor (on record) and expander (on replay). Although not so important for line-up as in Dolby A, the Telcom C4 system also has its own identify tone, which alternates between 550 Hz and 650 Hz, which aurally identifies the tape as being Telcom C4 encoded.

dbx

There are two versions of dbx, Type I intended for high bandwidth professional systems, and Type II for limited bandwidth systems, such as cartridge and cassette machines; there is also a special version of Type II for vinyl disc. Type I has detectors which operate over the range 22 Hz to 21 kHz, whereas Type II has detectors which respond over the range 30 Hz to 10 kHz. Although they use similar circuitry they are not, therefore, compatible with each other. Both systems use 2:1 compressors and 1:2 expanders with a very low threshold so that they operate over an input range of 100 dB and are able to produce a noise reduction of about 30 dB plus any increase of 10 dB in the headroom. The compansion is combined with pre-emphasis both of the main signal path and of the side chain in the companders. See Fig. 8.22.

The signal high-pass filter has a slope of 18 dB/octave and is 3 dB down at 17 Hz; it prevents low frequency out of band signals causing mistracking through being recorded unreliably. The record signal is then preemphasised. The RMS level detector bandpass filter has a slope of 18 dB/octave with 3 dB points at 16 Hz and 20 kHz, and is used to ensure that the compander VCA's do not respond to sub- or supersonic signals, and reduces mistracking due to poor frequency response of the tape machine (though note that the manufacturers specify tape machines with a response that is within ±1 dB from 20 Hz to 20 kHz for Type I). The RMS level detector preemphasis is used to avoid excessive high frequency levels which might cause saturation. The decoder is complementary to the coder (but note that to achieve this, the pre-emphasis in the side chain must be the same in the compressor and the expander).

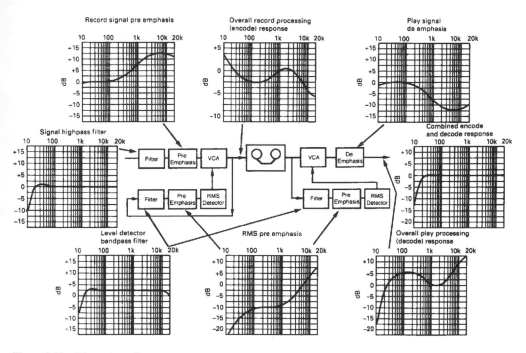

Figure 8.22 Dbx system diagram.

In order to reduce the tracking errors which could occur with average or peak level sensing in the companders, all dbx systems use RMS detection. RMS detectors respond to the total energy in the signal, irrespective of the phase of the individual harmonic components which make up the waveform. If these harmonics have been changed in phase by the recording process the waveshape will be changed, and it is likely that the peak level will also be changed, leading to possible mistracking. RMS detection is, therefore, least affected by phase shifts.

Because the compression ratio is constant across the whole dynamic range of the input signal, gain errors between the compressor and expander produce the same error in dB at all levels, i.e. for gain errors between the processors, only gain errors are produced in the decoder output. Thus, for a gain error of 2 dB in the recorder, there will be a change in the output level from the expander of 4 dB at all levels. The implications of this were described in the section on Telcom C4; those implications (e.g. for tape editing) are equally true with the dbx system, except that because the compression ratio is higher, the effects are proportionally greater.

The dbx system is available built into cassette machines, has been used to encode vinyl discs, and has also been used to encode part of the stereo TV system used in the USA. Figure 8.23 shows the response of the Type I and Type II processing (corrected for the compression ratio).

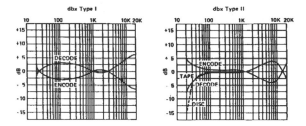

Figure 8.23 Dbx processing for Type I and Type II systems.

Dolby B

The Dolby B system was initially developed in 1967 as a domestic system for slow speed quarter inch tape, but was redesigned by the end of 1969 to match the needs of the compact cassette. The system is based on similar principles to Dolby A, e.g. bi-linear companders, but to reduce the complexity, it has only one band. This is not such a disadvantage as it may seem, because in domestic systems it should be possible to eliminate hum by good design and layout, and the listening levels commonly encountered will tend to make LF noise less significant than in professional environments. In order to make a single band acceptable in terms of aural performance, no attempt, therefore, is made to improve low frequency noise (see Fig. 8.20, there is no noise reduction below 500 Hz), and the turnover frequency of the filter, i.e. the frequency at which noise reduction action starts, is varied by the amplitude and spectrum of the programme signal, thus producing a sliding band compander (see Fig. 8.24). When the input signal level is below the threshold (about 40 dB below Dolby level (see below) at high frequencies), the rectifier and integrator produce no control signal, and the output of the secondary path is proportional to the signal level within its pass band; under these circumstances the operation of the circuit is essentially the same as the bi-linear compressors previously described. As the signal level within the pass band rises the control signal increases in level, raising the cut-off frequency of the filter, so that by peak level the signal in the secondary path is so low as to have no significant effect on the output level of the encoder. Thus, if a high level signal occurs within the pass band of the filter the cut-off frequency is shifted upwards, so that noise reduction action is continued above the loud signal, overcoming the problem of the noise level being modulated by lower frequency components of the signal. The overall characteristics are shown in Fig. 8.25.

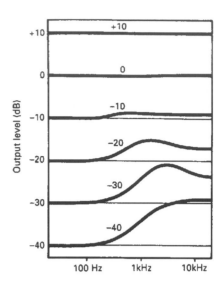

Figure 8.25   Dolby B characteristics.

In order to eliminate the need for adjustment to suit individual tapes (or transmissions in the case of radio) the levels at which Dolby B operates are fixed; in the case of cassette tape Dolby Level is a fluxivity of 200 nWb/m ANSI, and in the case of FM broadcasting a deviation of ±37.5 kHz.

Dolby C

The Dolby C noise reduction system, launched in 1983, is a domestic system based on experience gained from Dolby B. Like the A and B systems it uses bi-linear companders, with similar side chains, but increases the amount of noise reduction to 20 dB by using two separate stages of sliding band compansion, staggered in level (see Fig. 8.26). The level staggering ensures that the

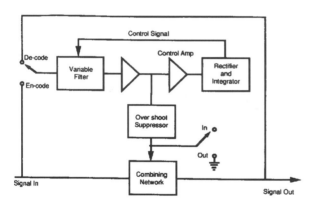

Figure 8.24   Dolby B system diagram.

Dolby B has become the standard noise reduction system for pre-recorded cassette tapes, and some American (though no British) FM radio stations transmit a signal which is Dolby B encoded. This means that many people listen to the signal without decoding it; whether or not this is aurally acceptable is a matter of opinion. The frequency response is, of course, not flat at low levels, there is HF boost at low levels, and a flat response at high levels. For FM this can be modified by reducing the pre-emphasis; since the receiver de-emphasis remains unchanged this redistributes the frequency response errors making low levels less boosted and high levels slightly cut at HF. However, many people find Dolby B coded signals aurally satisfactory.

Figure 8.26   Dolby C system diagram.

regions of dynamic action for the two stages do not overlap, but the high level stage begins where the low level stage finishes (see Fig. 8.27); thus the maximum compression ratio is once again 2:1. The high level stage operates at similar levels to the B type system, with the low level stage some 20 dB below this, which also simplifies the manufacture of circuits which are switchable between B and C. The range of frequencies which are treated has been increased for the C type system by setting the cut-off frequency of the filter in the sliding band companders to 375 Hz; this results in 3 dB improvement at 100 Hz, 8 dB at 200 Hz, 16 dB at 500 Hz, and nearly 20 dB at 1 kHz for sub-threshold signals (see Fig. 8.28). The response of the compressor (and hence the amount of noise reduction) at different levels is shown in Fig. 8.29.

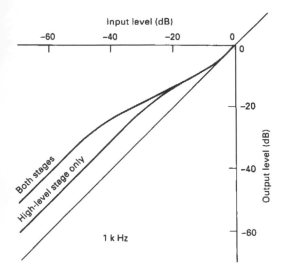

Figure 8.27 Dolby C characteristics.

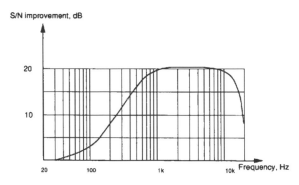

Figure 8.28 Maximum signal to noise ratio improvement of Dolby C.

As has been mentioned before, to achieve complementary characteristics in the compressor and expander is not enough to ensure that there is no mis-tracking; it is also necessary to ensure that the signal which the recorder delivers to the expander is the same in amplitude, frequency and phase as the compressor fed to the recorder. Unfortunately, there are many reasons why the high frequency response of a domestic cassette machine might be poor; incorrect bias, head wear, poor equalisation, dirty replay heads, tape for which the machine is not aligned are only some of them. These errors will produce mis-tracking in any companding system, but with sliding band systems there is the potential, though it is rare in

real programme material, for high frequency errors to modulate the level of lower, midband, frequencies. Dolby refers to this as the midband modulation effect. If the predominant level (after the pre-emphasis of the sliding band filter) fed to the compressor is an intermittent high frequency, then this will control the amount of compression applied to the signal as a whole. All frequencies that are being compressed (including midband frequencies) will be changed in level under the control of the HF. If the replay machine is unable to reproduce the high frequencies accurately, the midband frequencies will not be correctly decoded, because the control signal that caused them to change level in the compressor cannot be accurately recreated in the expander. To overcome this effect, Dolby C is equipped with spectral skewing (as follows).

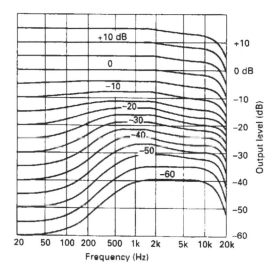

Figure 8.29 Dolby C compressor characteristics at different levels.

Luckily, it has been shown by many researchers (e.g. CCIR) that the subjective annoyance caused by noise of different frequencies is far from equal. CCIR Recommendation 468–4 is a weighting curve to make objective noise measurements match subjective assessments of noise more accurately (see Fig. 8.30). This shows that above about 7 kHz the objectionability of noise falls rapidly, with a consequent lessening of the need for noise reduction. Spectral skewing, which in practice is a sharp reduction of extreme HF in the compressor, with a complementary increase in HF in the expander, is applied at the input to the compressor and at the output of the expander (see Fig. 8.31). This characteristic is sufficient to ensure that the control signal generated in the compressor for the extreme HF (which may be unreliable on replay) is lower than that generated by the midband signals, which are therefore dominant in determining the amount of compression applied, and thus, because they are below the frequency where most cassette machines can be assumed to achieve a flat frequency response, the midband modulation effect is significantly reduced or eliminated. This leads to there being a maximum of about 8 dB of noise reduction at 20 kHz, an exchange of noise reduction at extreme HF (where it is relatively unnecessary) for an improvement in the system's ability to cope with errors outside its control.

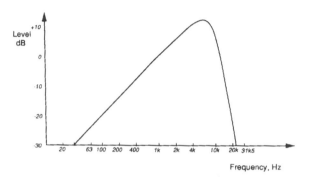

Figure 8.30   CCIR 468–4 noise weighting curve.

Noise Reduction Systems

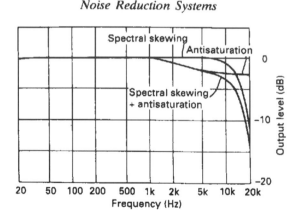

Figure 8.31   Spectral skewing and anti-saturation in Dolby C.

A further problem of magnetic recording, that of HF saturation, is also tackled by Dolby C. In effect, saturation is a reduction of the maximum HF level that can be recorded on tape as the level rises; at low levels the frequency response can therefore be flat, but as the level rises there is a progressive reduction in the HF response. When recording characteristics were set for magnetic tape the HF content of average programme material was probably lower than it is now, due to the increase in the use of electronic instruments which can intrinsically generate more HF than acoustic instruments. The spectral skewing circuits will have a beneficial effect as far as the reduction of saturation losses are concerned, but the

anti-saturation circuits must have an effect at lower frequencies (possibly down to 2 kHz) than can be implemented with spectral skewing. What is required is an adaptive frequency response, which attenuates the HF at high levels (where the loss of noise reduction, though small, will be masked by the signal level), but not at low levels (where saturation is not a problem, and where the loss of noise reduction might not be masked by the programme material). This can easily be achieved by putting the saturation reducing HF attenuation in the main chain of the low level compressor, after the feed to the differential network. In this position it will have maximum effect at high levels, where a negligible amount of the output signal is contributed by the side chain, and only a small effect at low levels, where most of the output signal is from the sidechain. Thus, a simple shelf network with time constants of 70 μs and 50 μs produces a 3 dB reduction in HF high level drive for a loss of noise reduction of only 0.8 dB. The overall system diagram is shown in Fig. 8.32. Dolby C also uses Dolby level (200 nWb/m ANSI) to eliminate the need for individual replay line up of each tape, by standardising the recorded levels.

Dolby SR

Dolby SR (spectral recording process) is a professional noise reduction system which incorporates features of the Dolby A and C systems. Spectral skewing and anti-saturation are used in a similar way to Dolby C, but as this is a professional system noise reduction in provided over the whole of the audio bandwidth, thus both spectral skewing and anti-saturation are applied at high and low frequencies (see Fig. 8.33). High level signals at both extremes of the frequency range are reduced in level so as to reduce or eliminate the effects of saturation and unreliable frequency response, whilst low level signals are amplified in a highly selective and complex way, so as to take best advantage of the constraints of magnetic tape. The system combines the advantages of fixed band companders as used in Dolby A and sliding band companders as used in Dolby B and C, by having both types of compander in each of its five stages, and by using what Dolby calls action substitution and modulation control to take best advantage of each compander. In common with all Dolby systems, bi-linear companders are implemented by generating a sidechain signal which is added to the main, high level, signal to produce the compressor characteristic, and which is subtracted from the main signal to produce the expander characteristic.

The SR system has three different levels which, as in the C type system, are staggered to distribute the dynamic action over a wider input range, resulting in

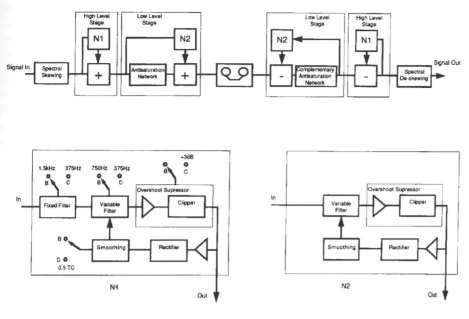

Figure 8.32 Dolby C system diagram. N1 and N2 are the noise reduction networks.

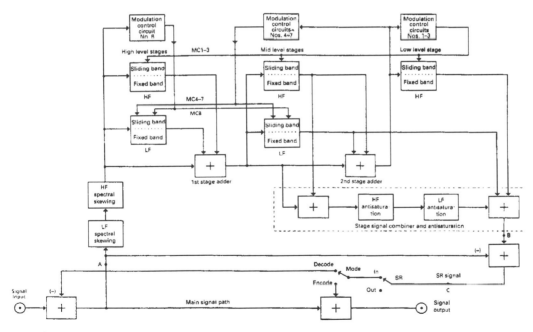

Figure 8.33 Dolby SR system diagram. This is a processor which is switchable between record and replay.

thresholds at about –30 dB, –48 dB, and –62 dB, produ-cing a maximum improvement in the signal to noise ratio of about 24 dB. The high and mid level stages have both HF and LF companders, with a crossover frequency of 800 Hz, although there is considerable overlap between the HF and LF stages, with the result that the LF stages can operate on signals up to 3 kHz, and the HF stages on signals down to 200 Hz. The low level stage acts only on HF signals, with an 800 Hz high pass characteristic.

To understand how the fixed band and sliding band compressors are combined (action substitution) consider the situation shown in Fig. 8.34. where the dominant frequency applied to an HF compressor is shown. A feature of fixed band compressors is that all frequencies within the band are treated in the same way, so for the signal shown there is, say, a loss of 2 dB of noise reduction action over the whole band, and thus the maximum noise reduction potential above the dominant frequency is not achieved. However, if a high frequency sliding band compressor were used, then because all frequencies are not treated similarly, there would be a loss of noise reduction effect below the dominant frequency, but an improvement above the dominant frequency compared to the fixed band compressor. The most advantageous combination is therefore to have the fixed band response below the dominant frequency, and the sliding band response above it for an HF stage, and to have the fixed band above and the sliding band below the dominant frequency for an LF stage. Thus, in each of the five stages, the fixed band compressor is used whenever it provides the best performance, and the sliding band is substituted whenever it provides an improvement.

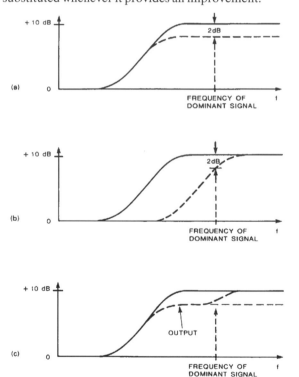

Figure 8.34 Combining fixed and sliding band characteristics by action substitution: (a) fixed band compressor characteristic; (b) sliding band compressor characteristic; (c) action substitution compressor characteristic

A further improvement of SR over A or C is in the use of modulation control. In the A, B and C type systems, the side chain signal is heavily limited when the input level is high (see Fig. 8.16). This is an advantage, since it

ensures that a negligible proportion of the output signal is derived from the limited signal, giving the system low distortion and overshoot. However, outside the frequency range of a particular band, the threshold can be allowed to rise, and the degree of limiting fall, once the change in gain necessary to produce the desired overall compression law has been achieved. Doing this ensures that large signals outside the band do not cause signal modulation within the band with a consequent loss of noise reduction. Figure 8.35 shows this effect for an HF fixed band compressor. In Fig. 8.35(a), each line shows the effect which a 100 Hz signal at different levels has on higher frequencies. Since the circuit has an 800 Hz high-pass configuration, a 100 Hz signal should ideally have no effect within its band. Without modulation control there is a considerable lessening of the noise reduction, but with modulation control (Fig. 8.35(b)) the capacity for noise reduction is much improved. The situation is similar for the sliding band compressors. Figure 8.36 shows how modulation control applied to them reduces unnecessary sliding. Modulation control is applied by generating, for out of band frequencies, further control signals which, when added to those generated by the compressors themselves, act in opposition and so bring about the improvements described above (see Fig. 8.33).

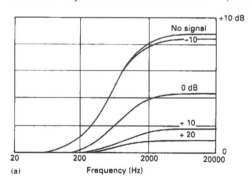

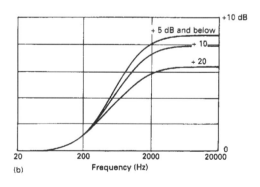

Figure 8.35 Modulation control in Dolby SR. The diagram shows the effect in the fixed band compressors. Frequency response curves with 100 Hz signal at the levels indicated, (a) with no modulation control, (b) with modulation control.

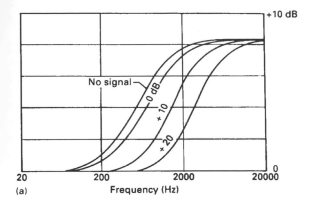

Frequency (Hz)

(a)

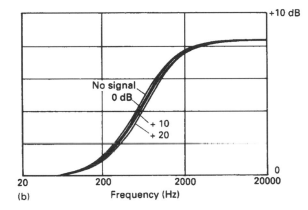

Frequency (Hz)

(b)

Figure 8.36 Modulation control in Dolby SR. The diagram shows the effect in the sliding band compressors. Frequency response curves with 100 Hz signal at the levels indicated, (a) with no modulation control, (b) with modulation control.

The overall performance of Dolby SR for low level signals is shown in Fig. 8.37. It should be noted that the shape of the decode response is very similar to that of the threshold of hearing, so that the smaller noise reduction achieved at LF is balanced by the ear's inability to hear noise at these frequencies, and that the maximum noise reduction is available at the very frequencies where it is needed most, that is where the ear is most sensitive. Finally, Fig. 8.38 shows the response of Dolby SR to an 800 Hz signal at different levels. Note how noise reduction is applied both above and below the signal, even at high level. So that the level presented to the decoder can be made the same as the level generated by the coder. Dolby SR uses a line-up signal of pink noise, interrupted with 20 ms gaps every 2 seconds at a standard level approximately 15 dB below reference level. On replay, the decoder can repeatedly switch automatically between the off-tape line-up noise and the reference noise, so that an easy aural comparison between the two

can be used as an aid to detecting tape machine alignment errors; the signal can also be used with a meter to set the levels accurately.

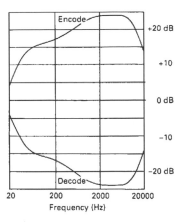

Frequency (Hz)

Figure 8.37 Low level response of Dolby SR. This shows the maximum noise reduction available at each frequency.

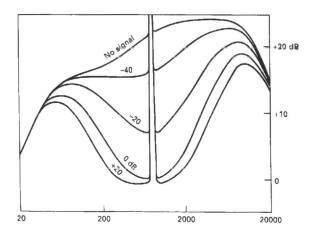

Figure 8.38 Low level response of Dolby SR in the presence of an 800 Hz signal at various levels.

Dolby S

The Dolby S system is a domestic version of Dolby SR, introduced in 1990. Nevertheless, it incorporates all the important features of SR, but reduces its complexity by having fewer companders (5, unlike SR's 10), and having only low and high level stages (operating roughly in the range –60 to –30 dB and –30 to 0 dB with respect to Dolby level). As a result it produces less noise reduction at low frequencies than SR, but still more than B or C. The system diagram is shown in Fig. 8.39, and the overall noise

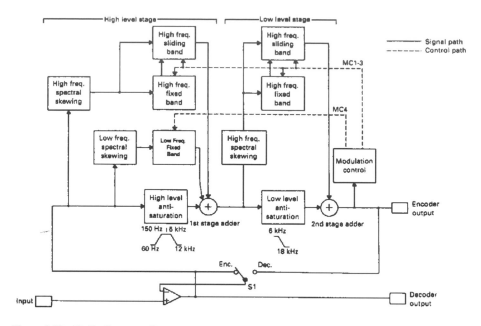

Figure 8.39   Dolby S system diagram.

Bibliography

Berkovitz, R. and Gundry, K. (1973) Dolby B-Type noise reduction system. *Audio*, September and October 1973.

Broch, J.T. (1971) *Acoustic Noise Measurements*, Brüel and Kjær, Hounslow.

Cossette, S. (1990) A new analogue recording process for consumer recording formats. *AES 89th Convention*, Los Angeles, Preprint No. 3004, Session Paper K-4.

Dolby, R.M. (1967) An audio noise reduction system. *JAES*, Oct 1967.

Dolby, R.M. (1983) A 20 dB audio noise reduction system for consumer applications. JAES, 31(3), March 1983.

Dolby, R.M. (1986) The spectral recording process. *AES 81st Convention*, Los Angeles, Preprint No. 2413 (C-6).

Fisher, D.M. (1987) *Television Broadcasting, Sound Recording Practice*, 3rd edition. Ed. J. Borwick, published for APRS by OUP, Oxford.

Gerber, W. (1983) *Companders in sound studio engineering*. West German Broadcast Company for the State of Hessen (HR). Reprint from Edition No 12/1983, Fernseh- und kinotechnik.

Jorgensen, F. (1980) *The Complete Handbook of Magnetic Recording*. First edition. TAB Books Inc, Blue Ridge Summit, PA 17214.

Moore, B.C.J. (1989) *An Introduction to the Psychology of Hearing*. Academic Press, London.

Sattler, M. (1981) *Compander Technology for Audio Engineers*, ANT, Nachrichtentechnikv GmbH, Wolfenbüttel.

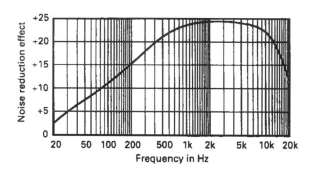

Figure 8.40   Maximum noise reduction in Dolby S.

reduction available in Fig. 8.40. It incorporates spectral skewing, anti-saturation, action substitution and modulation control, and uses, of course, Dolby's fixed and sliding band technique with bi-linear companding. See previous Dolby systems for explanations of these.

Dolby Level of 200 nWb/m ANSI is also used in Dolby S.

9 The Vinyl Disc

*(based on work by Alvin Gold
and Don Aldous)*

**Though the vinyl disc, formerly known as the LP, is no
longer the mainstream choice for issuing recorded music,
millions of homes contain players for these discs, and
some knowledge of the system is needed by anyone with
audio interests. This chapter shows how mechanical
sound reproduction developed, the equipment that was
developed to deal with replay, and how existing recordings can be kept in good order.**

Introduction

Looking back after many years of CD production, we can
see how crude so many of the stages of the record making
process were, in particular the reliance on the mechanically tortuous chiselling out of the record groove from a
lacquer blank. The whole process now looks very 'low-
tech', and there were in fact many areas of practice in disc
mastering and cutting that grew up 'willy-nilly' over the
years.

The system, however, defied its unpromising roots and
proved to be both highly successful and stood the test of
time very effectively, though it did not have as long a life
as its shellac predecessor. The LP record became for
many years universally accepted as a genuine world
standard with a widely distributed technology base.

The process of manufacturing a vinyl disc is relatively
economical and straightforward, even though it is error-
and blemish-prone and for most producers was always
inherently labour-intensive. It was, however, a very successful and widely used music storage medium and was
the predominant medium for distributing high-quality
sound until the development of digital methods.

Background

In the years before 1940, records were cut onto hard wax
slabs, in the region of 1–2 in. thick and about 13 in. in
diameter. The surface was polished to an optically flat
but extremely delicate finish which had to be protected
very carefully indeed from the effects of dust and
mechanical damage. When removed from the cutting
lathe, the wax slab was subjected to an electro-forming
process to produce a copper master. The problem here
was not so much the cutting as the electroplating, which
tended to result in a noisy cut which increasingly failed to
match improving standards elsewhere in the process.

The solution came with the development of the lacquer
in the late 1930s. From quite an early date, this took the
form of an aluminium disc about 14 in. diameter, covered
on both sides with a cellulose nitrate solution. The lacquer (or acetate) also contained a small amount of castor
oil along with some other additives, and was quite soft.

Changes in materials were matched over the years by
changes in the geometry of the cutting process. Back in
the cylinder era, the information used to be cut in the vertical plane, a method which was known as 'hill and dale'
recording. This was superseded by cutting in the lateral
plane, a process developed by Berliner. It was Blumlein
who formalised the lateral cutting system for stereo signals with his development of the 45°/45° cutting system,
which was patented in 1931.

Groove packing densities also increased as the rotational speed of the records dropped, from just under 100
grooves/in. in the 78 rpm era to around 250 grooves/in.
with the LP. Varigroove systems used groove pitch which
varied over the range 180–360 grooves/in. approximately.
Over the same time, frequency responses became wider
and more even, yet the later records are not cut so far in
towards the label, and the number of information channels doubled (to two) with the introduction of stereo.

Summary of major steps and processes

The manufacture of the LP record was latterly achieved
in several discrete steps. The process conventionally
began with a two-track master tape (sometimes known

as the production master), which was often mixed down from a multi-track original – more recently this tape would have been digitally recorded. The electrical signal recorded on the tape provided an analogue version of the original music to drive the amplifiers which in turn fed the cutter head which converted the electrical signal into its mechanical representation. The cutter head was part of a lathe with a turntable on which would be placed the 'acetate' or 'lacquer' onto which the musical waveform was cut. The lacquer was the parent disc from which all subsequent disc generations were derived.

The stages described above may well have been carried out by a facility attached to the recording studio where the tapes were produced or perhaps an outside production house, but the remaining steps were normally under the control of a specialised plating operation in the first instance, and finally a pressing plant for the duplication of the final product.

By convention, the lacquer is described as a positive, which means that like the final manufactured record it can be played by a stylus attached to a normal arm and cartridge. However, the lacquer would never be played in practice, except in the case of test cuts which would not be subsequently used to make the metalwork, since significant damage would inevitably be caused to the delicate surface.

The acetate itself then underwent a process known as electro-forming in which it was electroplated with silver. A much thicker layer of nickel was then grown onto the silver plated lacquer. The metal matrix was then separated from the lacquer, whose life ended at that point. The master was then used to grow a so-called 'mother' (or a number of them) from which the 'stampers' were produced, the processes here being similar to the production of the master itself.

As the name implies, stampers were the final steps before the production of discs for sale, and were used in pairs to make the records. The records were produced in presses by one of two common processes, extrusion or injection moulding, using a mixture based on a polyvinyl material, often in granular or pellet form.

The lathe

The lathe, with its associated amplifiers, was responsible for the task of transferring the electromagnetic waveform stored on tape to the lacquer, from which the stampers were ultimately derived.

A typical lathe (Fig. 9.1) consists of three essential assemblies. The first is the platter or turntable, driven by belt(s) or directly driven. They usually pack considerable rotational inertia (some weigh in at 75 lb), and use suc-

tion to hold the lacquer down. The turntable is also associated with a drive servo system of some kind which helps to reduce wow and flutter due to variations in stylus drag as the disc modulation levels vary. The second part of the lathe is the cutter head, which is fitted with a cutting stylus attached in turn to a carriage which allows the cutter to traverse the disc on a radius. The cutter head and its various attachments are drawn across the surface of the lacquer by a threaded bar, better known as a *lead screw*. The third part is of course the rest of the structure which, aside from its obvious constructional role, is also designed to keep the system stable and solid, and immune (or as immune as practical) to the effects of feedback and interference through the air and through the structure of the building.

Figure 9.1   A Neumann VMS70 cutting lathe, showing the cutting head.

The cutter head is functionally complementary to the phono cartridge. There are constructional similarities too, though it is probably more apt to think in terms of the loudspeaker – more correctly a pair of loudspeaker drive units – with cones replaced by a stylus attached to a cantilever, split to connect to two 'voice' coils. The coils are arranged to be mutually perpendicular and at 45° to the record surface, or to behave as though they were. All in all, the system has appreciable mass and must resist mechanical drag when cutting a swathe through the record, so the drive amplifiers can be pretty meaty too. A figure of 500 W for each channel (1000 W altogether) is not unusual, and some carefully designed current limiters are usually fitted to protect the very expensive cutter heads.

Cutting the acetate

One of the key starting points when cutting a lacquer is being scrupulous about cleanliness, the flatness of the lacquer and so on, because any residual roughness causes noise. The cutting action would normally add quite a lot of noise too, as the geometry of the cutting stylus, which has a sapphire cutting edge which works like a chisel, would normally tear into the surface of the lacquer.

One way of achieving a quieter cut is to preheat the stylus. This is done by passing an alternating current through a wire wound around the shaft near the stylus tip. The heat softens the acetate so that it cuts more cleanly, giving it something approaching the action of a knife passing through butter. Overdoing the heat, however, has its own problems, as the acetate material will stay hot and continue to deform after the tip has passed, in effect removing much of the fine detail, and limiting the resolution of the system, or it can even overheat, and stick to the cutter.

Swarf (the strand of material cut away from the groove during cutting) must be removed, and the commonest way to do this is with suction. Note that the swarf has a large surface area in relation to volume, and is made of various materials, some of which are highly volatile. The swarf is therefore highly flammable.

The groove walls have an almost 90° included angle, with the sloping walls inclined at 45° to the record surface. Dimensions of the grooves and the land that separates them inevitably vary, but the mean values run something like this: the groove bottom radius is 3.75 μm and depth is about 30 μm. The nominal width of a groove is some 60 μm, and the space (the 'land') between grooves averages just under 40 μm, leading to the groove packing density of 10 grooves/mm or 250/in. as quoted earlier.

Peak recorded velocities (as seen by the cutter or replay stylus) can reach as much as 50–60 cm/s at high frequencies. The largest amplitudes of course are at low frequencies, and are of the order of 0.005 cm. This can be readily seen just by examining a record visually. Accelerations of many hundreds of 'g' are possible at high frequencies, which highlights the requirement for a low moving mass at the replay stylus tip. By convention, the left-hand channel is recorded on the inner wall of the groove (nearest the centre hole) and the right on the outer. The groove shape can be checked with a microscope (Fig. 9.2), or by using an arrangement that looks at the way the groove scatters light from an incident light source.

Vertical cutting angle

One of the more important specifications is vertical cutting angle, which corresponds to vertical tracking angle (VTA) on playback. This is the angle that determines the line on the groove wall that acts as an axis for lateral changes in direction. If the playback stylus is spherical, then in principle it meets the groove at a single point on

each side and the tracking angle is more or less immaterial. In practice the radius is such that high frequencies are not read at all. This is not the case, however, with extended contact styli of various kinds and line contact styli in particular, where the plane of the line of contact should match that of the information imprinted on the groove–both to maximise high frequency output and to minimise noise.

Figure 9.2 Stereo binoculars used to check the cut groove – a TV monitor is also used for checking.

The standard cutting angle is now given as 20° to the vertical, which in practice allows for clean swarf removal, but there is an allowable tolerance of ±5°, or 10° in total, and when compounded with the other errors that can creep in during the cutting and replay, the high frequency integrity of the system is suspect. Amongst these other factors are variations in cutting angle at high recorded amplitudes due to the geometry of the cutter and stylus, and misalignment during playback, which can be severely affected by cartridge down-force even though the cartridge body may have been carefully lined up horizontally at the time of installation.

A signal which is equal and in phase on the tape recording is fed to the two coils so as to produce a wiggly lateral cut of constant depth (though a spherical replay stylus will 'see' a groove that narrows as the groove curves away from the net direction of movement, and will therefore rise and fall in the groove at twice the recorded frequency – the so-called pinch effect). Vertical modulation recorded in the 45/45 system is the out-of-phase or stereo difference information.

Cutting a 'hot' vertical signal, which corresponds to a large, low-frequency difference signal, can cause the cutter on one peak to dig through the acetate to the aluminium substrate underneath, and at the other extreme to lose contact with the groove surface altogether. Such cuts can also be difficult to track when it comes to playing them back. Strong vertical accelerations especially can cause havoc with arm/cartridge assemblies with low tracking forces, particularly where the tracking force is kept low.

Vertical limiting

The solution that was most widely adopted was to pass the signal to be recorded through a matrix to extract the vertical (difference) signal, and then limit it progressively more as its amplitude rises. The subjective effect is to reduce slightly the separation of very low frequency signals, a change which is generally thought to be unimportant (and essentially inaudible) due to the ear's inability to hear in stereo at very low frequencies. Conventional wisdom on this subject decrees that as sound propagation at these frequencies is so non-directional, and as the ear is not very good at determining incident angles from very low frequency information, the loss of stereo caused by reducing output in this region just is not very significant.

Nevertheless, observers have reported narrowed stereo separation from discs cut with this limiting in place, when a comparison has been made to an otherwise identical record without. This underlines as always the importance of taking due care when introducing untoward signal processing. The minimum of signal manipulation is invariably the best.

High velocities will mean possible non-linearities in the phono cartridge generator assembly. This could mean that the pre-amplifier will be producing output voltages that the partnering pre-amplifier may not be able to accommodate without clipping. Similarly, very high accelerations will inevitably mean mechanical damage to the groove walls from cartridge styli and cantilever assemblies that are too massy or crude to be guided accurately by the groove. In extreme cases, the stylus will be thrown out of contact with the record altogether, and this used to be an especially common condition in the days when arms and cartridges were more burdened with mass than they are now.

As should be obvious by now, disc cutting represents a range of delicate compromises that are complicated by additional subsidiary factors. For example, one way to limit lateral groove excursion whilst leaving the recorded level more or less intact is to apply a high pass filter to the cutter feed, operating below about 40 Hz. Conventional wisdom is that there is little but random noise below this frequency, and in any case cutting such frequencies simply encourages tracking problems. This idea is related to inhibiting the out-of-phase component on the record at these low frequencies, the advantages and disadvantages of which were discussed earlier.

If recorded velocity is to be held constant regardless of frequency – this would be logical as output voltage is proportional to voltage, and a flat output frequency response is obviously desirable – the amplitude variations with frequency would be very large. For a given cutting level, very low frequencies would use so much

land that record playing times would be very short or severe high-pass filtering would be required. Additionally, the amplitudes would be such that practical cartridges would have difficulties just tracking the groove unless cantilever compliance was so high that a whole new set of arm/cartridge problems would rear their head. Conversely, the amplitudes at high frequencies would be very small, to the point where they would be rivalled or swamped by random scratches and noise inducing blemishes on the record surface (some people say that is how it is already). Small scale granularity in the vinyl surface would also have a significant audible effect on the output.

Amplitude compensation

Some form of amplitude compensation is therefore required, and the characteristic chosen is known as the RIAA curve, the effect of which is to reduce groove excursions that would otherwise be excessive, and act as a noise reduction system. The mechanical resonance of the active part of the cutter head assembly falls around 1 kHz, and this figure has been chosen as the hinge point for amplitude correction applied to the recording signal. Below 1 kHz the output to disc is attenuated, and above it is boosted, though the curve shape is rather more complex than this very simple description implies. The time constants are 75 µs, 318 µs and 3180 µs. The characteristic approximates to constant amplitude to recorded groove below 1 K, and constant velocity above.

The amplitude correction, which is applied to the cutter along with resonance control using a feedback signal, works out to about 6 dB/octave. The feedback signal is derived from position sensing coils near the drive coils for the cutter stylus. Complementary processing is applied in the RIAA section of the preamplifier on playback to restore the flat (neutral) frequency/amplitude response. However, the correction is 'turned off' at very low frequencies to reduce the amount by which hum and rumble effects interfere with playback of legitimate LF musical information.

In pursuit of quality

Like other areas within the music recording industry, the disc cutting and pressing cycle was peculiarly responsive to the level of care and attention lavished on all stages of the product. Although the methods by which discs were manufactured were widely understood and practised, the quality of results can vary a great deal. It is not surprising then that a whole sub-industry grew up around audiophile recordings and pressings – especially in the

USA and Japan – where improved cutting and pressing standards have proved to be a major attraction.

Just prior to the CD revolution, the improvements that were achieved were fed back into the mainstream of the industry, and artistic attitudes towards the use of compression and band-limiting when mastering and cutting (to give just two examples) had shifted perceptibly in the last few years in the direction of less interventionalist ideals. Even on the classical side of the business, there was a perceptible trend towards live mixes and fewer microphones.

The changes at the cutting stage included such elements as reduced levels (sometimes the complete elimination) of out-of-band filtering, reductions in the amount of compression (often applied almost without thought, for example, as a built-in function of the lathe) and reduced gain riding. There are signs also that some studios are beginning to forgo noise reduction units, also to clean up the signal path, except where the noise reduction is clearly essential.

The influence of digital processing

Another more or less unrelated motivation behind purist engineering came from the direction of digital recording. At first, digital recording was only really established in the classical recording arena, but its effect there was little short of revolutionary. One of the most obvious and significant changes was, as described above, the move away from complex multi-mix multi-track recording setups (digital multi-track equipment was then prohibitively expensive for many studios) in favour of simple live balances, mixed straight down onto a two-track digital master – typically using a Sony PCM-1610 or 1630 and an ordinary (or mildly tweaked) video recorder.

With the very simple signal paths and relatively unprocessed, naturally balanced, tapes that have resulted from the move to digital, a number of influential engineers began to demand a similarly clean and unprocessed quality of sound at the cutting and pressing stage. There was also a requirement at that time that LP records should not sound too different from the same title on compact disc, and this factor too tends to support the idea of simplicity right down the line.

Disc cutting – problems and solutions

Cutting the lacquer (or acetate) always entailed certain critical compromises. The major one is between recorded level (i.e. velocity) and playing time. It is possible to accommodate around 30 minutes' worth of music on one side of a record without distress, especially if the mean record level is not too high. Classical orchestral music often has short-term peaks and prolonged quieter passages, where the groove pitch can be closed up for quite long periods of time, and it is with this kind of music that some of the longest recorded LP sides have been made (in excess of 35 minutes in some isolated cases).

There are dangers in this approach. If adjacent groove turns are too close together, damage to the groove walls can occur as a result of imprinting of information from the next turn of the spiral. The parallel here is sticking a hammer through a wall, and finding that this has made certain modifications to the shape of the wall as viewed from the other side. A less catastrophic but still undesirable audible side effect of having grooves that are too densely packed is pre-echo – a faint reproduction of the next groove during reproduction of the groove that precedes the loud passage of music. The time lead associated with this effect is the time it takes for the record to perform one revolution – just under two seconds for 33.3 rpm discs.

The problem can be minimised by ensuring a wider groove pitch (distance between adjacent grooves) especially just prior to a loud passage. A wider groove pitch is also associated with greater groove excursions each side of the mean position, in other words a higher level or 'hotter' cut – there are several conflicting factors here that have to be bought into balance. Too wide a pitch, however, will unnecessarily restrict playing time, and too high a level on disc will result in accelerations and velocities that many phono pick-ups will be simply unable to cope with. The effect of the varigroove system itself can produce problems. Merely to perform the task of placing the cutting stylus where required, varigroove produces low-frequency signals, and some varigroove systems have quite an abrupt effect, producing a range of frequencies which can appear on the final record.

Time delay

To allow optimum adjustments of groove pitch, it is necessary to monitor the sound being fed to the cutter head, sufficiently in advance of the cutter feed so that corrective action (varigroove) can be taken in good time. Traditionally this has been achieved by using two (sometimes three) playback heads on the tape machine employed during the cutting sessions, the one the tape passes first being used for monitoring, and the second for feeding the lathe where just two heads are used. Where a third is fitted, it looks at the signal after the one being cut so that coincidences of high amplitude on adjacent turns can be corrected properly.

The time delay is determined by the distance between the tape heads and by the linear speed of travel of the tape, and this can cause certain practical problems. A two second of lead time at 30 ips for example means five feet of tape loop to be wound around guides and rollers. All this additional complication inevitably has the undesirable side effect of degrading the mechanical stability of the system (the sharp angles the tape traverses could also cause slight high frequency erasure) with some formulations of high energy tape stock.

A later technique, therefore, was to feed the lathe via the output of a wideband digital delay unit. However, there are problems here too, unless the digital delay unit is totally acoustically transparent. The electronic component has not been invented yet which meets this criterion (unless the moon really is made of green cheese), and any digital process involves A/D and D/A processors with attendant filters and so on which are known to have side effects which are audible. Of course, if a digital tape is being used as the source, and provided the datastream stays in the digital domain until all the signal processing has been completed, then the objection does not apply.

Disc pressing

The artistic judgements that need to be made throughout the disc cutting process cease to apply with subsequent processing, which concerns the sequence of operations that end up with the final record. Up to the point where the sound was cut onto the acetate by the cutting lathe, there was at least a degree of control over the overall sonic balance, though nothing like as much as at the earlier recording and mixing sessions. From the lacquer on, it is simply a matter of doing the job as effectively possible to avoid any more degradation than strictly necessary.

The period around the late 1930s and early 1940s was a turning point in the development of the record in a number of ways. The lacquer coated disc was developed in this period. Where the old style wax blanks had to be slowly heated through before they could accept a clean cut, the acetate could be used without additional processing at normal temperatures (though pre-heating the cutting stylus does expedite the cut, as described earlier). Similarly, using wax it was hard to get clean metallisation using off-the-shelf methods of achieving suitable surface conductivity. Graphite and copper based powders, whose granular nature had a bad effect on surface noise levels, were once used, but the introduction of vacuum deposition made a considerable improvement. By contrast, lacquer caused few such problems.

An acetate could also be played without further processing using the new lighter weight arms and cartridges just then being introduced, and this was a key factor to the broadcast organisations whose interest was of key importance at the time. The record industry, which in time was to grow tremendously powerful, grew from the professional side rather than the other way around. It was later that tapes and tape recorders came along to take the place of the record in the broadcast companies affections, but it was the convenience of tape that they had been looking for in records all along.

Modern materials

It was at about the same time that polyvinyl compounds, much like the materials used to make records today, came into existence. Their role was also to lead to much quieter, less abrasive surfaces compared to shellac. The lacquer (Fig. 9.3) is prepared for metal deposition by cleaning with various agents, finishing off with deionised water. The surface is then treated with stannous chloride which keys the surface preparatory to being metallised with silver. The latter is produced when a solution of silver nitrate is reduced in a chemical reaction with glucose formaldehyde, for example. The process up to this point is usually done using chemical sprays which combine on the disc surface where the reaction is required. The discs are normally rotated at the same time, to help ensure an even thickness of coating.

Figure 9.3 A lacquer disc ready for metallising.

The final step in the metallisation of the now silvered acetate is to electroform a nickel coating to make the metal part robust enough in its own right to withstand the rigours of further processing. This is done in an electroplating bath, and when the metal parts have reached the required thickness, it is removed and the acetate stripped off and disposed of. The following stages – the mother and the stampers – are produced in analogous fashion.

The mother is, of course, a positive impression, and can be played as a sound check prior to further processing.

The vinyl used in the final pressing process consists of a vinyl copolymer, a stabiliser, lubricants and pigments (usually carbon black). The material is fed to the press pre-heated, usually in the form of small pellets or granules. The stampers are mounted to copper backing plates, and attached to the press. Heating is by steam; the vinyl is inserted and the heating/compression cycle which moulds the vinyl to the required shape starts when the press is closed. Once completed, the disc area is cooled with water, and the pressing is finally removed. The complete cycle time is typically around 30 s.

Injection moulding is often used in the production of 7 in. singles, which along with printed centre labels results in a process that can be performed very rapidly, is amenable to automation, and from which reject records can be totally reclaimed without the necessity of removing the centre label portion. This is a necessity where paper labels are required, as they invariably are on LPs.

Disc reproduction

The LP disc is relatively easy to handle; self-contained and easy to file; processing is relatively cheap; short items can be catered for, and the sections of the record are accessible for extracting short portions for programme or educational applications. These points always differentiated the LP disc from reel-to-reel tape, and made it the preferred medium for many purposes.

Disc playing equipment starts with the electromechanical portion known as the record deck, consisting of the platter ('turntable') which spins the disc, plus plinth, motor, pick-up arm and cartridge (Fig. 9.4). The function of a turntable is to rotate the record at the chosen speed and be as free as technically possible from short- and long-term speed variations. Any undesirable speed variations will affect the musical pitch. Quiet running and freedom from significant warp or up-and-down movements are other requirements.

Switching is used for the LP/EP standard speeds of 33⅓ and 45 rpm, sometimes with a third speed of 78 rpm and the platter often uses stroboscopic markings on the rim to check for even and correct speed. Where provision has been made for the 78 rpm speed, a suitable cartridge must be available, and fine speed adjustment is another necessity for playing old nominal speed 78s, which were often recorded at 76 or 80 rpm. With an auto-changer deck, complex programming may be included in the design to play a batch of records, but auto-changers were never accepted as part of a hi-fi assembly.

Figure 9.4 A typical older type of playing deck unit.

As well as having constant speed, the turntable should be free from mechanical vibration, which could be transmitted to the disc and detected as 'rumble' by the replay cartridge. This low-frequency effect is largely related to the turntable bearing, and the best decks are so constructed that external vibration or shocks are 'damped' by the deck and the turntable suspension.

Specifications often give 'rumble' figures as minus a number of decibels, for example, DIN B weighting. This is meaningless unless the reference level employed is stated. Usually the signals are of 1 kHz or 315 Hz with a recorded velocity of 10 cm/s. To measure rumble after the datum of 10 cm/s has been established at, say, 1 kHz, you must compare this level with a low-frequency signal from the pick-up (after low-pass filtering to remove groove/stylus noise) tracking an unrecorded (silent) groove. This ratio for a first-class belt-drive table can reach as high as 65 dB. Weightings to DIN, IEC or NAB standards are often used. DIN A weighting is more revealing of performance, but some manufacturers are loathe to quote this figure.

Careful design and machining of the drive surfaces on turntables can minimise wow (speed variations below about 20 Hz) and flutter (speed variations above 100 Hz). Where a 'weighted root-mean-square' (wrms) figure is stated in a specification, it will always be less than the equivalent 'peak weighted' reading. With a sinusoidal variation of speed, the wrms figure can be converted to peak-weighted by multiplying it by 1.4, and to convert peak to rms, multiplied by 0.70.

Drive systems

Various materials have been used for the turntable platter, including cast aluminium, or sheet aluminium for the lighter designs. Cast and sheet steel are often fitted, but the problem of magnetic attraction arises when a magnetic type of pick-up cartridge is used. Non-magnetic turntables are preferable, and manufacturers have also attacked the problem with mu-metal screens and novel forms of construction. Of course, with the earlier crystal and ceramic pick-up cartridges (which were never deemed to be of hi-fi standard), no magnetic pull problems arise.

For many years, some drive systems used a powerful capstan motor and a stepped pulley wheel driven by the capstan, which in turn drives the idler wheel coupled to the inside of the turntable rim. Models such as the popular Garrard 301 and 401 series were fitted with a heavy turntable, so that no significant speed variation problems occurred. Adjustable speed was provided by eddy-current brake.

An alternative to idler wheel drive was the belt drive, which is simple and reliable. The best designs have a low-speed motor, with a capstan coupled by the belt to a flange on the underside of the turntable. The motor can be a small AC type-mounted remote from the pick-up to minimise hum induction or a servo-controlled DC motor. Another design method – as used in the highly-regarded Thorens TD 125 Mk II – was based on an oscillator (Wien bridge type), whose output is power amplified to operate the drive motor. Speed change is achieved by frequency change of the oscillator.

Models which use mains power can change speed by using stepped pulleys or by a conical drive, whose vertical position on the idler wheel can be adjusted to provide both speed change and fine speed adjustment (as in the Goldring-Lenco turntables). Servo speed control can be used with electronic control systems, as fitted in models from Philips and Pye, in which a small tacho-generator is coupled to the drive motor. With this method, speed variations are sensed and suitable correction to the electronics introduced.

To avoid drift on direct-drive turntables, the feedback is a closed loop and this type of control system can be very effective. Yet another drive arrangement uses a tapered motor shaft, along which the idler wheel can be traversed to give continuously variable speed control (as in the Goldring-Lenco designs). Click stop positions were included for the three standard speeds.

One design engineer, Ivor Tiefenbrun (widely known for his Linn Products', particularly the Linn-Sondek turntable) looked at the problems of turntables in mechanical terms, rather than considering them electronically. At first glance, turntable design may seem simple: suspend the platter, sub-chassis arm and cartridge on a set of springs, leave the motor on the main plinth, and drive the platter with a belt.

As the Sondek designer has found, there are complications. Drive belt and signal wires from the pick-up arm are mechanical links to the outer atmosphere and so make up part of the 'springing'. These environmental vibrations mean that the spring modes must be right to avoid strong and unavoidable cartridge resonance (usually between 8 and 15 Hz).

The final turntable design must have reasonable stability and be self-centred in suspension, so that any movements do not turn into less stable and controllable modes. This means that the drive belt must not be significantly disturbed by vibrations, which could cause speed fluctuations and upset stability.

The bearing on the Linn table is unique and patented, as fitted on the LP12 unit. It is a single-point oil bath bearing, i.e. a spindle encased within a strong, oil-filled housing. The actual spindle of hardened tool steel, with a precisely centred and tempered tip sitting on a high speed thrust, plate machined, ground and lapped to a mirror finish. High-grade bearings of this standard are expensive, and the Linn model is fitted with a spindle that goes through 14 machining and hardening processes; the whole bearing involves 23 different machining operations.

The Linn-Sondek turntable incorporates several other subtle details, e.g. fitting a custom built motor with 24 poles, rather than the usual four, and running at 250 rpm, compared with the four-pole motor at 1500 rpm. In turn, this means that the larger pulley that is fitted can be machined with less percentage error. Even the mains power frequency has been stabilised by what is called the 'Valhalla' kit whose circuitry irons out any current fluctuations to improve the turntable's performance.

A turntable (or record deck, when complete with pick-up arm and cartridge) should be selected on its technical merits and suitability for purpose – always bearing in mind the buyer's budget. Some decks can be obtained without a pick-up arm and cartridge, which means that the items can be made compatible, so that a modest turntable is not fitted with an expensive moving-coil cartridge. Few decks with ready-fitted arms can offer as good overall results as those with carefully chosen – and more expensive – components.

Over the years there have been several designs of a high standard, such as, the famous Thorens TD.150 and TD.160 models, leading on to the later 320 series from this manufacturer. Nor should one forget, of course, the Ittok LVII arm and the cheaper Basik arms from Linn. These can be fitted with almost any cartridge and high-grade turntable. A lift/lower cueing lever is a useful adjunct.

It appears to be generally agreed that the belt drive systems are the best available compromise, often using a suspended sub-chassis. This approach was introduced many years ago by AR (Acoustic Research) and later popularised by Thorens. In the lower end bracket, some manufacturers employ a non-suspended sub-chassis but adopt fine tuning by fitting isolating feet, arm termination, etc. Other designers – such as Systemdek, Walker and Ariston – have managed to retain a sub-chassis even in modestly priced product.

Pick-up arms and cartridges

For any cartridge to perform at its best, it must be fitted with a suitable stylus (usually conical or elliptical) and fabricated from sapphire or diamond. Early 'crystal' types were simple, but for high-quality reproduction, only magnetic units will be considered.

Electromagnetic pick-ups all depend on the basic phenomenon that relative movement between a magnetic field and any metal produces an electrical voltage. Several types of magnetic pick-up cartridge have evolved, derived from this principle; early designs were 'moving-iron' having a fixed coil-magnet assembly and a soft-iron armature, which is moved by the stylus vibrations and so causes the magnetic field to fluctuate.

The most popular cartridges latterly were the moving magnet (fixed coil and tiny vibrating magnet) and the moving coil (fixed magnet and tiny vibrating coil); in both designs, stylus vibrations cause magnetic field fluctuations. The variable reluctance cartridge, also popular, used a fixed coil and pole-piece, magnetic circuit, and pivoted arm rocked by the stylus movement.

Characteristics

In examining the characteristics of cartridges, it must be remembered that the frequency response can be affected by the amplifier input, and so it is necessary to know and check the input capacitance and, to some degree, the input resistance of the amplifier to which the pick-up will be connected. Obtaining the specified performance and getting cartridge/amplifier matching is usually not difficult as some amplifiers have switchable loadings, or plug-in modules that can be fitted into the input sockets.

Moving magnet cartridges have a relatively high output and require a load normally of about 47k in parallel with a specified capacitance to provide optimum performance. Moving coil cartridges (Fig. 9.5) are usually low output, low impedance, with low stylus compliance. To increase the small output, a low-noise pre-amplifier or transformer is required before the disc input on the main amplifier. A suitable pre-amplifier is preferable on quality grounds, but it will increase the total cost of the generator system.

Figure 9.5 The Goldring Electro-II moving-coil cartridge. This features an unusually high output level that approaches the typical level of moving-magnet designs.

Each cartridge type has its advocates, and the differences appear to be due partly to the particular methods of mounting the stylus and cantilever. Each type is capable of giving high quality sound; the moving magnet has a user replaceable push-fit stylus on the end of its cantilever, but the moving coil designs has little or no cantilever and the stylus mount is coupled directly to the coil. The delicate operation of stylus changing usually means returning the cartridge to the manufacturer for replacement.

The cartridge/arm combination

To get the optimum results from cartridge/arm combination, care must be taken in setting up the system. Tracking error and offset angle can be adjusted by moving the cartridge in its headshell, so that the stylus/arm traverses across the disc in as near to a straight line as a pivoted arm will permit. A small alignment protractor will help in this procedure, and ensure that the stylus tip is perpendicular, as seen from the front view.

The playing weight (also termed stylus pressure, downforce or tracking weight) is important, and the deck/cartridge manufacturers' instructions will list a range of playing weights, and the maximum weight suggested is usually the optimum for reliable groove tracking (no skipping). A calibrated weight adjuster is fitted on most arms.

To counteract the tendency of the pick-up to be pulled against one wall of the record groove, causing distortion, an anti-skating force (or bias compensation) is applied. This sidethrust is countered by a spring, a magnet or the nylon thread/weight device (due to John Crabbe back in 1960). To the critical ear, incorrectly set bias compensation will result in vague stereo imaging in the reproduction and poor tracking. The importance of careful setting up of the arm and cartridge on the turntable deck has long been recognised by such audiophile manufacturers as Ivor Tiefenbrun of Linn Products.

For completeness, another way of tackling tracking error on pivoted arms is the radial or parallel arm deck, remembering that the grooves on a disc are cut by a chisel-shaped cutting stylus tracking across the disc surface on a kind of radial carriageway. The BBC has used such an arm for many years (for playing 78s), with the carriage having six ball races sliding on a polished rail. The Revox B790 employed this arrangement on its direct drive turntable, with a very accurate internal oscillator to feed a low-voltage AC motor. A Bang & Olufsen design fitted a motor-driven radial carriage, and Marantz

by a rotary wheel, with bias controlled by a similar wheel below that. Arm height is easily adjusted; it has an in-built fluid damper and an optional finger-lift. A well-produced instruction manual is provided.

Styli

Styli fitted on cylinder players before the turn of the century were ball-pointed, and made of sapphire or ruby. The latter jewel type was regarded as permanent, i.e. retaining its shape for many playings. In fact, diamond-tipped styli (needles) were also used for many years to

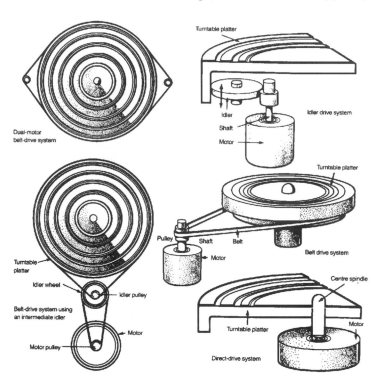

Figure 9.6(a)   Drive Systems.

(USA) produced their SLT (straight line tracking) arm. Perhaps the most highly regarded pick-up arms have been the SME types which, though very costly, are of a very high standard.

The tube is 'dead', with no headshell/arm joint and at the bearing end the arm/tube assembly is secured to the alloy steel cross-shaft – forming the bearing – by two widely spaced Allen-head bolts. This fixing method means that the large bearing shafts are less prone to damage, and cartridges can be changed with the arm still bolted to the turntable deck. Tracking weight is applied

reproduce cylinders and disc records (including the Edison vertical-cut discs).

As pressings were made in less abrasive material, it became possible to fabricate cheaper steel needles for replay. Different grades of steel were sold – softer types for 'soft tone' and harder grades for 'loud tone'. Nevertheless, by the end of one or two sides of 78s, most steel needles were worn and had to be replaced. So-called 'chromium' needles were introduced, and in the middle 1930s, semi-permanent sapphires increased the playing life some ten times of the steel brands. Other

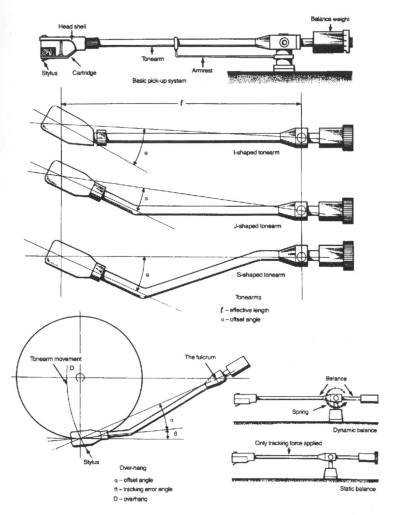

Figure 9.6(b) Pick-up/arm geometry.

attempts to reduce record wear included the thorn cactus spine, bamboo and fibre. The last type could be resharpened and one 1930 report claims that a fibre had been resharpened 16 times and played about 1091 sides (on 78s, of course).

Stylus dimensions are usually quoted in terms of the radius of the hemi-spherical tip, and the optimum size will have a tip that maintains two-point contact with the groove walls. If too big, it will ride on the groove edges, and if too small it will bottom in the groove, both undesirable positions. Groove walls usually have an included angle of 90° and the dimensions of 78s are tip radius 0.0025–0.0035 in., with bottom groove radius 0.001 in., and for microgroove LPs, tip radius 0.0005–0.001 in. with bottom groove radius of 0.00015 in.

To get optimum results from 78s, 0.0025–0.003 in. tips are recommended but earlier 78s may need a

Figure 9.7 The SME Series V precision pick-up arm, which embodies every known worthwhile feature of a pick-up arm.

0.0035–0.004 in. tip. Mono LPs had a recommended stylus radius between 0.005 in. and 0.001 in., with the 0.0005–0.0007 in. tip for playing stereo discs. A compromise figure of 0.0007 in. copes with worn mono discs. Oval or elliptical styli are now available with a minor axis radius of around 0.0003 in., and a major axis of about 0.0007 in., which must be aligned at right angles to the groove. Figure. 9.8 illustrates typical stylus tip shapes.

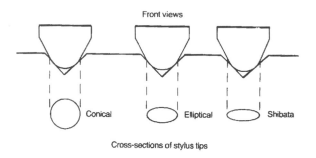

Figure 9.8　Three shapes for stylii, showing the groove contact.

Stylus wear depends on the tracking pressure, alignment of the pick-up arm, compliance of the stylus, side pressure on the pick-up/arm, motor vibration and irregular rotation of the turntable. Dust particles in the grooves are also a factor, which can be avoided by the latest record cleaning devices.

Old recordings

The upsurge of interest in recent times for old 78s or early LPs centres around nostalgia, coupled with unrepeatable performances from artists long dead. Modern technology has provided enthusiasts, professional and amateur, with equipment to clean up historic recordings capable of giving sonic enjoyment to 'golden-eared' audiophiles. All the electronic tricks of exponents in rejuvenating old records (78s and early LPs) to remove or limit clicks, 'bonks', and disturbing surface noise have not been revealed, but graphic equalisers, variable threshold Dolby B circuit, 'dippers', notch filters to cut or boost narrow band of frequencies, and pseudo stereo enhancement are employed in these recreations. Purists may cavil at these treatments, but if these digitally re-recorded masters issued by the BBC give more musical satisfaction, the technique must be widely acceptable.

Specifications

Pick-up and turntable terminology is a mixture of electrical and mechanical language, and a few definitions may help to clarify the descriptions.

For example, 20 Hz to 20 kHz is a *response* rather than a frequency range. This type of description, e.g. 20 Hz to 20 kHz ± 4 dB indicates the limits of deviation from the level response, namely 4 dB one way or the other. A graphical display is preferable, as it shows location and extent of deviation from linearity.

Compliance is the inverse of stiffness and is used to refer to the pivoting arrangement of a pick-up in which the stylus is held. For a given mass of moving elements, the higher the compliance, the lower the fundamental resonance of the system. The compliance unit (CU) equals 10^{-6} cm/dyne, i.e. a force of 1 dyne produces a displacement of 0.00001 cm. Compliance can also be quoted in the more modern units of micrometres per millinewton, (μm/mN) – the conversion is 1 cm/dyne = 10^7 μm/mN, so that 10^{-6} cm/dyne = 10 μm/mN. There is no universally agreed method of measuring dynamic compliance, so usually the static compliance figure is quoted.

Crosstalk is the unwanted transfer of sound or audio frequency energy from one channel to another, or it may refer to absolute separation between channels. Crosstalk influences the width and quality of the stereo image. It is expressed as a level of undesired signal in relation to wanted signal channel, and measured in decibels. If given as one figure, say, 20 dB, this applies to a 1 kHz sine wave, unless otherwise indicated.

Dynamic range is the ratio of the loudest to the softest passages occurring in music or speech, expressed in decibels (or phons in live performances), which the system can handle without distortion or masking by background noise.

Load impedance. For magnetic cartridges, the standard is 47 kΩ, plus a capacitance of, say, 200 pF. Moving-coil types are less sensitive to loads, and may need a pre-amplifier or booster transformer.

Output level. The output of a cartridge is referred to a stated recorded velocity, namely, the velocity of the groove modulations. This figure may quote, say, 1 mV per cm/s velocity, which means that for each centimetre per second velocity, there is a millivolt of signal output on each channel.

Resonance is a familiar term indicating the tendency of a mechanical or electrical device to vibrate in sympathy with a particular frequency. A tuning fork, when struck, resonates at a certain pitch and produces a recognisable note and the sound-board of a piano also resonates. In electrical systems, the combination of mass and compliance (or the electrical equivalents of inductance and capacitance) in a circuit results in maximum responses at a certain frequency. When interpreting specification details, low-frequency resonance data is very helpful for ensuring good arm mass/compliance compatibility.

Signal-to-noise ratio (S/N) is expressed in dB and refers to the power or voltage ratio of desired signal to unwanted background noise, that is, signal-to-noise ratio. Strictly speaking, all dB measurements should use power figures, calculating the dB figure from $10\log(P_1/P_2)$, but decibel ratios are also calculated from voltage ratios, using $20\log(V_1/V_2)$. The two are equivalent only when the voltage figures are measured across identical resistance values.

Tip mass, when quoted in specifications, refers to the effective moving mass due to the moving parts, e.g. the armature, cantilever, etc., which give the groove modulations work. It is difficult to measure.

Tracking force/weight. This is the vertical force that the combined weight of the cartridge and the long end of the pick-up arm exerts on the record grooves. Tracking weight is always stated in grams, and should be set at the figure specified by the cartridge maker. For a high grade stereo cartridge, the manufacturer's recommended range will be narrow, say, from 0.75 to 1.5 gm is typical. Some cartridges perform optimally with a very light tracking weight, 0.5–1.5 gm, while others will not track properly (stay in the groove) at less than 2–3 gm. In most cases the cartridge weighs considerably more than the recommended tracking force so the arm balances this out by means of a counterweight or spring. Remember that good tracking starts with a stable turntable/pick-up arm foundation.

Measurement methods

One of the most useful items of test equipment now available for checking the performance of cartridges is the manufacturer Ortofon's TC3000 cartridge test system. Simply and speedily this computer device checks the overall performance, including such parameters as tracking weight bias compensation vertical tracking angle input capacitance and bias compensation.

The virtue of this product is its repeatability for a series of cartridges, so it can form the basis of screening tests for all types of cartridge. It does not replace the familiar test discs, of which there are many produced by the major record companies for both amateur and professional applications. With such devices as the Neutrik pen-recorder, and suitable test discs, e.g., the JVC TRS 1007, II record, one can produce informative traces (showing the general trend of the response). These traces are usually taken with low capacitance loading at the pre-amplifier, which is 100 pF, including pick-up arm and leads. Any discrepancies in these readings arise from the fact that the computer-derived data relates to a carefully selected mean frequency response. The pen chart assumes that the inherent response is flat.

Some well-equipped reviewers use a Nicolet computing spectrum analyser but, with all these measurement methods, the listening ears of an informed audiophile are still of major importance in determining 'sound quality' of cartridge/arm combination, or, indeed, all the links in the reproducing chain.

Maintaining old recordings

The problem of the LP is that each playing alters the recording. The masters for many LPs of historic interest have been used to make CDs, so that the preservation of the recordings is ensured, but this is often of little interest to those who feel that the LP, for all its defects, represents the recording methods that they most admire. This presents the dilemma for such listeners that the more they enjoy their music, the shorter the time they have to enjoy it.

One solution is to make a copy of a treasured LP for normal use, reserving the playing of the treasured LP for special occasions only. This presents a problem, because it is unlikely that a recording on cassette, even using modern methods, will be acceptable, and digital audio tape (DAT) is a rare and expensive medium. Curiously enough, video recorders can be useful because a modern stereo video recorder will often be fitted with both stereo input and stereo output, allowing three hours or more of uninterrupted quality audio recording on an E180 tape. The quality of such digital recordings is excellent, and many enthusiasts have been using old Beta video recorders for this purpose for many years. Curiously enough, the record companies who held up the release of DAT until they could be assured it could not be used for copying CDs failed to note that video recorders were being used in the same way at a much lower cost.

Current developments in computing have produced optical drives which allow the recording of CD-format discs at comparatively low cost – currently around £400 for the drive and £5 per blank disc. Though these are intended for computer data, there is no difference between this and audio digital data, so that all that is required is the circuitry for converting audio signals to digital form. It would be curious if do-it-yourself CD recorders did not become available soon now that all the technology is in place, because the development of recordable CDs has been too rapid for the record industry to stop it.

Given that a precious recording has, to use computer terms, been backed up, how do we look after these LPs? There is, in fact, nothing new that can be recommended, and all the principles for looking after LPs are those which have been practised for years. These are as follows:

(1) Always store a record in its sleeve, on edge, and keep the sleeves clean. A record collection should be kept in a cool dry place.
(2) Use a dust brush on the turntable to take off any small amounts of dust while playing.
(3) Keep the turntable covered at all times, lifting the cover only to insert or remove a record.
(4) Replace the stylus at recommended intervals – you should note how often a stylus has been used.
(5) Use a recommended method of cleaning discs that have become dusty. Wet cleaning methods are often suspected of causing damage because the dust then acts as a form of grinding paste, and DIY methods should be avoided. Ultrasonic record cleaning baths used by specialists are probably the best method of removing troublesome dirt with least likelihood of damage to the record, but a cleaned record should not be played until it has been thoroughly dried, and drying must not use heat.

References

AES, *Disk Recording – An Anthology*, vols 1 and 2. Audio Engineering Society (1981).

Borwick, John, *The Gramophone Guide to Hi-Fi*, David & Charles, Newton Abbot (1982).

Gelatt, R. *The Fabulous Phonograph*, Cassell, London (1977).

Gordon Holt, J. '*Finial LT* Laser *Turntable*', *Stereophile* **V9** (5) Santa Fe, New Mexico, USA (1986).

Harris, Steve, '*78s* De-*Bonked*', *New Hi-Fi Sound*, (Sept. 1985).

Messenger, Paul, and Colloms, Martin, '*Cartridges* and *Turntables*', *Hi-Fi Choice* (43) Sportscene Publishers, London.

Roys, H.E. (ed.) *Disk Recording and Reproduction*, Dowden, Hutchinson and Ross (1978).

10 Valve Amplifiers

Morgan Jones

Why, when transistors are so much smaller and cheaper, does the audio fraternity persist in talking about the electronic anachronisms that we know as valves? Who is still using them and why? To paraphrase Mark Twain, reports of the valve's death have been greatly exaggerated. In this chapter, Morgan Jones reviews and reminds us of the technology that will not die. Note that the circuit drawing conventions used in this chapter are those of the valve age (no dots at joins, half-loops used to indicate crossing lines that do not connect).

Who uses valves and why?

Broadly speaking, there are three distinct audio groups using valves; each group takes advantage of a particular facet of the thermionic valve's performance. In the same way that a diamond is used by a jeweller for its sparkle, an LP stylus manufacturer for its hardness, and a semiconductor foundry for its thermal conductivity, the three users of valves have very little in common.

Musicians – meaning pop and rock musicians – use valves. This is well known, and it is equally well known that they do so because of the distortion that valves can produce. At this point, common knowledge halts, and we encounter a common engineering myth:

Musicians use valve amplifiers because they distort, ergo, *all* valve amplifiers distort, so anyone using valves is listening to distortion – and if I used it, my distortion test set would agree with me.

Musicians' amplifiers are not hi-fi amplifiers, they are *designed* to be bad – that's how musicians need them. An electric guitar doesn't have a body cavity to resonate and so give the sound character. The string vibration is picked up and leaves the guitar as an electrical signal that is quite close to a sine wave.

As far as a musician is concerned, the guitar is only part of the instrument, the whole instrument is the combination of the guitar and amplifier/loudspeaker. Since 2nd harmonic distortion needs to be >5 per cent of the fundamental before it is clearly heard, it follows that quite gross distortion will be needed to give the entire instrument character. For this reason, a musician's amplifier has an undersized output transformer to give bass punch, power supplies with poor regulation to give compression, and high gain stages near to the loudspeaker allow microphony and acoustic feedback.

In audio, there are two distinct types of valves: triodes (three electrodes) and pentodes (five electrodes). The distortion produced by the two types is quite different. A musician's amplifier exploits these differences, and has excessive gain combined with multiple volume controls to allow different stages to be deliberately overdriven. Overloading earlier (triode) stages produces even harmonics, whereas overloading the power stage (usually pure pentode) produces odd harmonics.

The second group of valve users is the recording industry. Recording engineers treat valves as another tool from the effects box, but are not as extreme as the musicians, and use them simply as 'fairy dust' (because a sprinkling of fairy dust might bring an otherwise lacklustre recording to life).

Recording engineers do not overdrive valves, quite the reverse. Because of the high power supply voltages, it is very difficult to overload valve circuitry, so recording engineers use valves in equipment where signal levels are liable to be uncontrolled, and clipping must be avoided at all costs. Common uses are microphone channels and compressor/limiters.

For many recordings, it is the sound of the lead vocalist that sells the recording, so special efforts are taken with the vocal chain, and large capsule capacitor microphones dominate. The output impedance of the capsule in a capacitor microphone is very high and needs an integral pre-amplifier to enable the microphone to drive a cable. Because it is normal for the singer to sing very close to the microphone, the microphone produces a healthy output, and noise is not an issue. By definition, microphone levels are uncontrolled, and valves are so popular that kits are even available to quickly retrofit valve pre-amplifiers to the current Neumann series of large capsule microphones.

Many engineers do not stop simply with a valve microphone, but use an entire chain of valve electronics from microphone channel amplifiers and compressor/limiters up to group level. Usually, outboard microphone channels are needed, but some innovative mixer manufacturers offer options with individual channels using ICs, discrete transistors or valves, according to taste.

A more unusual use of valves in the recording industry is in the analogue processing stages of an analogue-to-digital converter (ADC). Digital audio has been accused of sterility, so some enlightened manufacturers offer ADCs with valves to sweeten the sound – one has a control with the rather splendid legend 'angel zoom' to adjust the degree of sweetening!

The final group using valves is composed of audiophiles, or hi-fi enthusiasts. Their only common ground with the first two groups is that if they don't like the sound, then they refuse to listen to it. They listen to valves because they like the sound.

Subjectivism and objectivism

Some people argue that subjective testing of audio reproducing equipment is invalid. The only possible test can be an engineering measurement, and if this is perfect, then the equipment is perfect, and nothing remains to be said, thank you. Pressed further, diatribes about contemporary music not having an original performance surface, which is odd, considering that commercial recordings of classical music are commonly an edited mélange of the bits that the musicians got right. The final defence is that auditory memory is short term and so only double-blind A–B testing is valid, which seems rather a shame since it implies that we should sack all sound mixers and tonmeisters once they become sufficiently experienced to know what they are doing.

Both sound mixers and audiophiles become educated about the sound of their monitoring system, and although the sound mixer learns to compensate for their monitoring system, the audiophile becomes paranoid about achieving perfection. How can this be, when neither has an absolute reference?

Fixed pattern noise

To answer the previous question, it is useful to draw a parallel with video, as analogue video has all the problems of analogue audio, plus a few of its own. One of the more serious camera problems is producing a picture with an acceptable signal-to-noise (S/N) ratio. When cameras with tube (valve) sensors were used, it was generally agreed that a measured S/N of >46 dB was acceptable to the viewer, but when charge coupled device (CCD) semiconductor sensors were introduced, it was found that the 46 dB criterion no longer worked. The reason was that the CCDs produced a fixed pattern of noise on the picture which was not visible on a still frame, but when the camera panned, the scene changed and the noise remained stationary; the relative movement between noise and picture revealed the noise.

The eye/brain has great sensitivity to fixed pattern noise because it has survival value. If you stand on a grassy plain and sweep your gaze from side to side, you are not interested in the fact that the whole scene changes, but you are interested if a few blades of grass move, since it may indicate a lurking predator. Those moving blades of grass are a pattern that may change their relative position from one scene to the next, but the pattern itself 'moving blades of grass = predator' is fixed, and tracked by the brain. Another fixed (and dangerous) pattern might be, 'suspicious ripple on water + fin = shark'. For both video and audio, the brain has been found to be 15–20 dB more sensitive to fixed pattern than to random noise.

In the same way, we might consider an amplifier to have a fixed pattern of noise/defects, and if we pass enough music through this fixed pattern, then the pattern will be revealed, irrespective of whether the music contains its own patterns. One amplifier might measure almost perfectly, yet its fixed pattern be more noticeable, and therefore subjectively objectionable, than another whose technical performance is poorer.

What is a valve?

Thermionic emission

Before transistors became common, all electronics relied on the valve, there were even computers using valves!

All metals have free electrons within their crystal structure, so some of them must be at the surface of the metal, but they are bound there by the nuclear forces between them and the adjacent atoms. However, the atoms and electrons are constantly vibrating due to thermal energy, and if the metal is heated sufficiently, some electrons may gain sufficient kinetic energy to overcome the attractive forces of the atoms and escape.

As a consequence of these escaping electrons, an electron 'cloud', or space charge, forms above the surface of the heated metal **cathode**. Once this cloud reaches a certain size, it will prevent other electrons attempting to escape from the surface because like charges repel, and an equilibrium is therefore reached.

Some metals have stronger forces binding their electrons than others, so stripping an electron from their surface requires more energy, and the cathode has to be heated to a higher temperature:

Melting point of pure tungsten:	3410°C
Pure tungsten cathode (transmitter valves and lightbulbs):	2700°C
Thoriated tungsten cathode (small transmitter valves):	1700°C
Oxide coated cathode (receiving valves):	1000°C

As can be seen, the operating temperature is sufficiently high that the cathode could literally burn, so the structure of the valve is enclosed in glass and the air is evacuated.

We now introduce a positively charged plate, or **anode**, into the enclosure. Electrons will be attracted from the cloud, and will be accelerated through the vacuum to be captured by the charged anode, and thus a current flows. The cloud has now been depleted, and no longer repels electrons quite so strongly, so more electrons escape from the surface of the cathode to replenish it. Current flow is unidirectional because only the positively charged anode is able to attract electrons, and only the cathode can emit electrons. We now have a rectifier, but it requires rather more than 0.7 V to switch it on; typically 50 V is needed.

In order to control current flow, we interpose a **grid** or mesh of wires between cathode and anode, resulting in a structure with three electrodes known as a **triode**. The internal construction of a triode is suggested by its graphic symbol, Fig. 10.1.

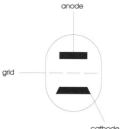

anode

grid

cathode

Figure 10. 1 Circuit symbol of triode.

If the grid is negatively charged, then it will repel electrons, and although there is a space charge above the cathode, no electrons reach the positively charged anode because they are unable to overcome the repulsion of the grid. The grid to cathode voltage V_{gk} therefore controls the number of electrons reaching the anode, or anode current I_a. The dependence of anode current on anode and grid voltage may be plotted graphically, as in Fig. 10.2.

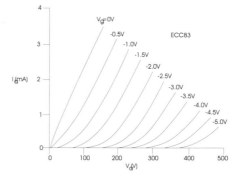

Figure 10.2 Anode characteristics of triode.

Comparison of different valve types

Because the triode response is poor at radio frequencies (RF), due to capacitive feedback from anode to grid, an additional (screen) grid can be added, resulting in the tetrode. The tetrode has a kink in its anode characteristics, which can be removed by the addition of yet another grid, resulting in the pentode. Originally developed to circumvent the pentode patent, the beam tetrode can swing its anode voltage (V_a) closer to 0 V, resulting in greater efficiency, but is otherwise very similar.

Manufacturers trying to jump on the valve bandwagon often claim that field effect transistors (FETs) are very similar to valves, and therefore valve sound can be simulated (reducing cost) using FETs. Agreed, the anode characteristics of the pentode are similar to those of the FET, but they are also similar to those of the bipolar transistor, see Fig.10.3.

The fact of the matter is that most audio valve circuitry

Table 10.1 *Summary of differences between main classes of audio valve.*

Type	Number of electrodes	Advantages	Disadvantages	Typical use
Triode	3	Good linearity (even harmonics)	Simple circuits poor at RF (cascode needed)	Most audio gain stages
Tetrode	4	Good at RF	Tetrode kink (non-linearity)	Transmitters
Pentode	5	Good at RF, high gain, efficiency	Noise, rather non-linear, (odd harmonics)	Output stages, RF
Beam tetrode	5	Similar to pentode, but greater efficiency	More difficult to make than pentode, expensive	Output stages, particularly RF

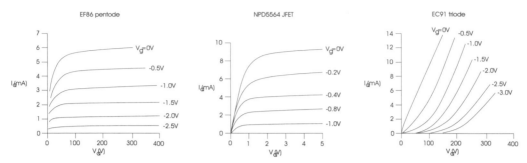

Figure 10.3   Comparison of output characteristics – pentode, JFET, and triode.

is implemented using triodes, whose characteristics are entirely unlike any semiconductor. Pentodes tend to be reserved for power stages where their greater efficiency outweighs their more objectionable (odd harmonic) distortion.

Valve models and AC parameters

A full analysis of valve circuit design is beyond the scope of this chapter, so an overview using an equivalent model will be used to demonstrate the differences between the three fundamental circuits.

Transistors are invariably analysed using the transconductance or Norton model, but because the output resistance of a triode is normally low compared to its load resistance, triodes are normally best analysed using the Thévenin model as illustrated in Fig. 10.4.

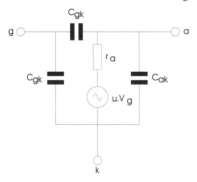

Figure 10.4   Triode equivalent model at audio frequencies.

Valve AC parameters

The equivalent model relies on AC parameters that are not constant and which must be evaluated at the operating point. These parameters are as follows.

The **amplification factor**, μ (no units) which is the ratio of the change in anode voltage (ΔV_a), to the change in grid voltage (ΔV_g), with anode current held constant:

$$\mu = \frac{\Delta V_a}{\Delta V_g}$$

The **mutual conductance, gm** (mA/V) which is the ratio of the change in anode current (ΔI_a), to the change in grid voltage (ΔV_g), with anode voltage held constant:

$$gm = \frac{\Delta I_a}{\Delta V_g}$$

The **anode resistance, r_a** (kΩ, Ω) is the ratio of the change in anode voltage (ΔV_a), to the change in anode current (ΔI_a), with grid voltage held constant:

$$r_a = \frac{\Delta V_a}{\Delta I_a}$$

The three parameters are linked together by the following equation, so a third parameter can be found provided that two are known:

$$\mu = gm.r_a$$

The three fundamental circuits

The common cathode triode amplifier

This inverting amplifier is very similar to the FET common source amplifier, with all components performing identical functions, but impedances and voltages are typically multiplied by a factor of ten. The cathode capacitor prevents reduction of gain by negative feedback, and the output coupling capacitor blocks DC, Fig. 10.5.

A potential divider is formed by R_L and r_a, so the voltage amplification A_v of the amplifier stage is:

$$A_v = \mu . \frac{R_L}{R_L + r_a}$$

If R_L could be increased to ∞, the potential divider would disappear, and gain would rise to the maximum value of μ. Looking into the input of the amplifier, C_{ag} suffers a multiplying effect (known as the Miller effect), and appears as a very much larger capacitance across the input of the stage, modifying the input capacitance to:

$$C_{in} = C_{gk} + C_{ag}(A_v + 1)$$

C_{in} may be reduced either by reducing the C_{ag} term (grounded grid and pentode), or the $(A_v + 1)$ term (cascode and cathode follower).

Looking back into the output of the amplifier, we see r_a and R_L in parallel, so the output resistance is:

$$r_{out} = \frac{r_a.R_L}{r_a + R_L}$$

Looking down through the anode, all resistances are multiplied by a factor of $(\mu + 1)$, so if the cathode resistor is left unbypassed (causing negative feedback), the new anode resistance becomes:

$$r_{a'} = r_a + R_k.(\mu + 1)$$

This new value of r_a can be used for calculating the new output resistance (much higher) and the new gain (lower).

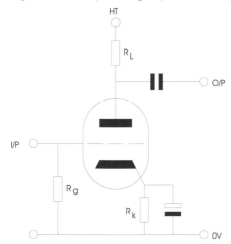

Figure 10.5   Common cathode amplifier.

The cathode follower

This non-inverting amplifier has moved R_L from the anode to the cathode circuit, resulting in 100 per cent negative feedback.

Gain is first calculated using the standard triode gain equation, and the result from this is substituted into the standard feedback equation, typically resulting in a gain of 0.9:

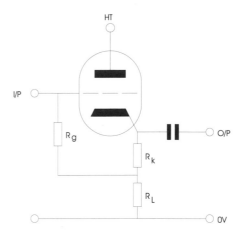

Figure 10.6   Self-biased cathode follower.

$$A_{v'} = \frac{A_v}{1 + \beta A_v}$$

In the same way that resistances were multiplied by a factor of $(\mu + 1)$ when looking down through the anode, they are now divided by $(\mu + 1)$ when looking up through the cathode, therefore:

$$r_k = \frac{R_a + r_a}{\mu + 1}$$

For the cathode follower, the anode circuit does not contain an external resistor $(R_a - 0)$, so $r_k \gg 1/gm$, and the output resistance is low.

Looking into the input, R_g passes to ground via R_L, whose voltage is very nearly that of the input signal, so little current flows through R_g, which is said to be bootstrapped, and the input resistance is very high (typically $10R_g$);

$$r_{in} = \frac{R_g}{1 - A_{v'} . \dfrac{R_L}{R_L + R_k}}$$

By the same token, bootstrapping decreases input capacitance, and input capacitance becomes:

$$C_{in} \approx C_{ag} + strays$$

The grounded grid amplifier

This is a non-inverting amplifier, Fig. 10.7, with the input applied to the cathode and the grid held at a constant voltage. Unless blocked by a capacitor and diverted, the valve's cathode current now flows through the source.

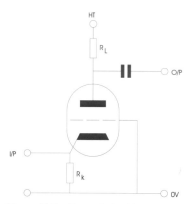

Figure 10.7 Grounded grid amplifier.

Because the input voltage is in series with (and adds to) the output voltage, the standard common cathode amplifier equation is modified:

$$A_v = \frac{(\mu + 1) . R_L}{R_L + r_a}$$

The advantage of the grounded grid amplifier is that the anode is now screened from the input by the (grounded) grid, and so this is a mostly a circuit for use at RF, although it has been used as a moving coil input stage. The above equation assumes that the source has zero impedance. A non-zero impedance source will allow cathode feedback, increasing r_a and reducing gain.

Practical circuit examples

All practical circuits use cascaded individual circuits or compound circuits made up from the preceding three fundamental circuits. We will begin by looking at the simplest possible circuit, and develop this as necessary.

A unity-gain pre-amplifier

Now that all sources (apart from phono) provide enough signal to drive a power amplifier, passive 'pre-amplifiers' consisting of nothing more than a volume control and a selector switch have become common. However, the output resistance of the volume control potentiometer is likely to be quite high, and this forms a low-pass filter in conjunction with the capacitance of the cable feeding the power amplifier. The obvious next step is to add a unity gain stage that can drive the cable capacitance without high frequency loss, and an example is illustrated in Fig. 10.8.

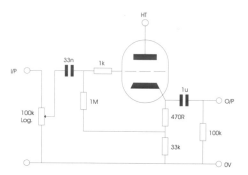

Figure 10.8 World Audio KLP1 pre-amplifier.

The active stage is a self-biased cathode follower. Consequently, the stage has approximately unity gain ($A_v \approx 0.95$) and a low output impedance ($r_{out} \approx 180\,\Omega$). Because of the bootstrapping effect of the self-bias circuit, the loading resistance seen by the input coupling capacitor is $\approx 20\,M\Omega$, and a low value can be used whilst still maintaining a good low-frequency cut-off (0.2 Hz). The very low input capacitance of the cathode follower barely loads the volume control potentiometer, and the stage has wide bandwidth.

A very similar circuit is used in capacitor microphones using valve pre-amplifiers, with the following modifications. The input coupling capacitor is removed so that the valve provides the necessary polarising voltage for the capsule. To ensure a high input resistance, the value of the grid-leak resistor must rise to 100 MΩ necessitating that the valve be selected for low grid current, which fortuitously also ensures low noise.

A line-level pre-amplifier with gain

Although a unity gain pre-amplifier can be good quality, some power amplifiers might need more signal to drive them to full level, and so a pre-amplifier with gain, Fig.10.9, would be required.

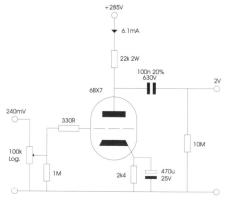

Figure 10.9 Line level pre-amplifier with gain.

The octal based 6BX7 is a low μ triode ($\mu \approx 8$), with a low r_a ($r_a \approx 1.5$ kΩ), consequently, it is capable of driving moderate (<500 pF) capacitances directly. Although the stage has a much higher input capacitance ($C_{in} \approx 30$ pF), the high frequency cut-off occurs at ≈ 200 kHz, which is still more than adequate.

If a lower output resistance is needed, this circuit can be followed by a direct coupled cathode follower.

A balanced line-level pre-amplifier with gain

Two common cathode amplifiers can be combined to form a differential pair, Fig. 10.10, to allow balanced working (as is common in studios). Balanced working is able to reject noise (such as hum, etc.) induced into cables, and the system should therefore offer a lower noise floor.

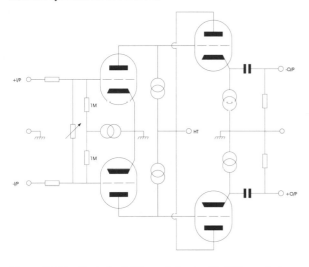

Figure 10.10 Optimised differential pair with DC coupled cathode followers.

In order to maximise this rejection of induced noise, it is usual to fit a constant current sink into the cathode circuit of the differential pair. Not only does the constant current sink improve rejection of induced noise, but it also greatly improves the linearity of the amplifier. The differential pair can be further linearised by replacing the anode load resistors with constant current sources, because the r_a term in the gain equation then becomes insignificant compared to R_L, and it is largely the variation in r_a that causes distortion.

Now that the differential pair has been made so linear, it may be thought necessary to improve the linearity of the cathode followers by replacing their load resistors with constant current sinks. The constant current sources

and sinks could all be made with pentodes (noisy), or FETs could be used, as in the Audio Research LS15.

The Earmax headphone amplifier

In 1994, a delectable little German designed headphone amplifier called Earmax was released to the world – anyone who saw it fell in love with its diminutive dimensions, and it redefined hi-fi jewellery. Unfortunately, at £325, it also redefined bank balances, so when the Earmax was later reviewed, the review information was used to reverse engineer the circuit.

The photograph showed three valves, each of which contains two triodes in one envelope. The input valve was stated to be an ECC81, and the ECC88 was said to be acceptable for the output valves. The amplifier is stereo, so each channel can only use three triodes. The amplifier was stated to be able to drive a low impedance load (200 Ω – 2 kΩ), so the output stage must be a cathode follower. Low output impedance from a cathode follower implies high *gm*, which confirms the suitability of the ECC88 (≈ 10 mA/V at high(ish) anode currents). But even a *gm* of 10 mA/V still gives an output impedance of 100 Ω in normal cathode follower mode, and we need much lower than that to be able to drive 200 Ω; a reasonable design target would be 10 Ω, or less.

There are two ways of achieving a 10 Ω output impedance. We could either use a conventional cathode follower, and apply 20 dB of overall negative feedback, or we could use a White cathode follower. Maintaining stability with 20 dB of negative feedback might be a problem, so it is more likely that a White cathode follower was used. There are two versions of the White cathode follower, one of which has a very low output impedance and can only be operated in Class A, whilst the other can be operated in Class AB, but requires a phase splitter. Both forms use two triodes, leaving only one valve for the input stage. It is possible to make a phase splitter with only one valve, but this would give the entire amplifier a voltage gain of ≈ 1, which would be insufficient to drive 2 kΩ headphones from the typical line level available from a Walkman Pro.

From these arguments, it is possible to say with a fair degree of certainty that the circuit topology is as shown in Fig. 10.11.

Earmax is alleged to be able to deliver 100 mW into an unspecified load, but loads of 200 Ω – 2 kΩ are deemed to be acceptable. 100 mW into 200 Ω would require a far higher current than the ECC88 can deliver, so the assumed target was 100 mW into 2kΩ.

Using $P = I^2R$, 100 mW into 2 kΩ requires 7 mA$_{RMS}$, and implies a minimum of 10 mA quiescent current in the output stage (Class A). Perusal of the anode characteristics of the ECC88 reveals that at this operating current,

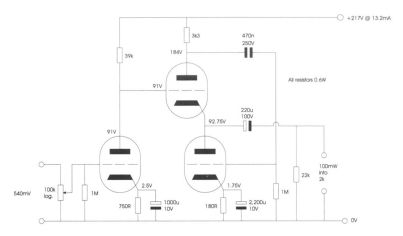

Figure 10.11   Earmax derivative headphone amplifier.

gm = 10 mA/V, and r_a = 3.3 kΩ. The equation for the output impedance of the version of the White cathode follower used is:

$$r_{out} = \frac{1}{gm^2 \, R'}$$

Where R^1 is the parallel combination of r_a (anode resistance), and R_L (anode load).

Although a high value of R_L reduces r_{out}, it wastes power and increases heat dissipation. A good compromise is $R_L = r_a$ = 3.3 kΩ, giving a theoretical output resistance of 6 Ω, so the choice seems justified.

In order to drive 100 mW into 2 kΩ, we must swing quite a large voltage. Using $V^2 = PR$, we need 14 V_{RMS}, or 20 V_{pk}. Now, this might seem to be a trivial voltage swing for a valve, but in combination with the 10mA standing current, it is easy to choose an operating point that exceeds maximum allowable anode dissipation. Although $P_{a(max.)}$ = 1.5W for an individual triode within the ECC88 envelope, total anode dissipation is only 2W. V_a = 91V, I_a = 10 mA gives an anode dissipation of 0.9 W, and would just allow a 20 V_{pk} swing before running into grid current.

Once the operating point has been deduced, the required HT voltage and biasing is easy to calculate, and an input stage is needed.

Although an ECC81 was used for the original circuit, it forces a rather low anode current (≈ 1 mA), and rather high gain, which must have resulted in increased noise and sensitivity to valve variation. As an alternative, the input stage could use another ECC88. The advantage is that the ECC88 can be operated very linearly at the anode voltage required to directly bias the output stage, whilst drawing a reasonable anode current and hence minimising noise. Because the gain of the ECC88 is only half that of the ECC81, input sensitivity is reduced, and noise becomes still less of a problem. The amplifier now

has a total voltage gain of ≈ 28, and needs ≈ 540 mV_{RMS} for its full output of 14 V_{RMS} into a 2 k Ω load.

Turning to power supply requirements, an HT of 217 V @ 26 mA is required, whilst the heaters could be wired in series, requiring 18.9 V at 0.3 A, which could then easily be supplied by a regulated DC supply to ensure an absence of heater induced hum. With the calculated sensitivity, there should be no problem with hum, even with AC on the heaters, and simple two stage RC smoothing of the HT supply will be quite adequate.

The Croft Series III

The Earmax was a low powered amplifier capable of driving headphones, but it is perfectly possible to develop the design to drive loudspeakers. By no stretch of the imagination can valves be said to be ideal for driving 8 Ω loads directly. Nevertheless, some manufacturers believe that Output TransformerLess (OTL) amplifiers are worthwhile, and continue to produce them.

Valves are naturally high impedance devices, and are normally transformer-matched to their load to improve efficiency and reduce output resistance. If we omit the output transformer and direct couple the output valves to the loudspeaker, we can immediately expect a drop in efficiency and a high output resistance. The rise in output resistance can be partly controlled by choosing valves with a low r_a, and using pairs of identical valves in parallel (to further reduce r_a), but plenty of global negative feedback will still be needed to reduce the output resistance to a useable value. Since loudspeakers draw currents measured in amps rather than milliamps, we will need a rugged valve capable of passing a large cathode current without damage. Power supply valves such as the 6080/6AS7G are common, but television line scan output

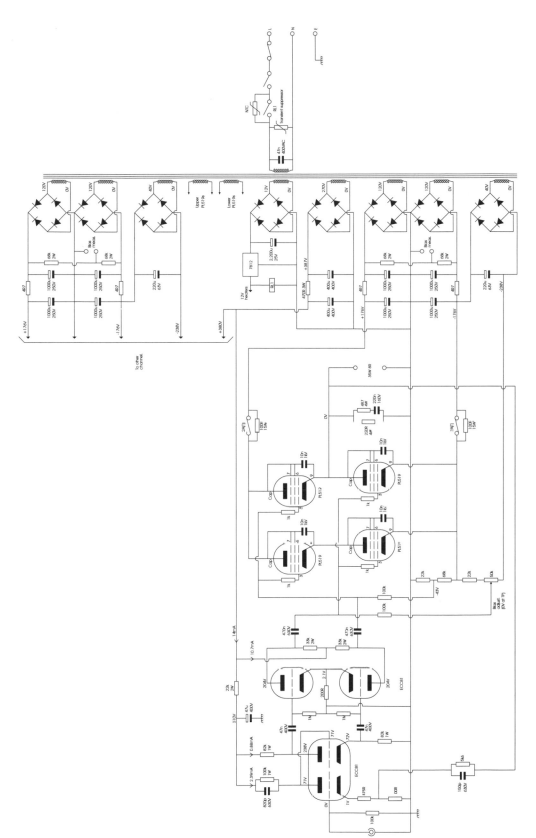

Figure 10.12 Croft Series III power amplifier.

valves such as the PL519 and PL504 (which are very rugged) are especially popular.

To summarise, any practical OTL amplifier will use multiple pairs of rugged output valves having questionable linearity, and despite the high feedback, will still tend to have a comparatively high output resistance. Efficiency will be appallingly low.

The Croft Series III, Fig. 10.12, is fundamentally a pair of paralleled push–pull White cathode followers driven by a derivative of the classic Williamson driver circuit, and is claimed to develop ≈ 30 W into 8 Ω.

As can be seen, the power supply has multiple secondaries, necessitating a very large custom-wound mains transformer, and has an interesting method of reducing switch-on surge. Initially RL1 is unpowered, and the amplifier draws current through the cold, and therefore high resistance thermistor, so all power supplies are at a reduced voltage. As the resistor warms, its resistance falls, and power supply voltages rise until RL1 operates; full mains voltage is then applied, causing all supplies to rise to their operating voltage. Note the use of a stabilised heater supply for the driver valves to eliminate heater induced hum.

Although it might seem alarming to couple a loudspeaker directly to an HT supply via a valve, the arrangement seems to be perfectly satisfactory, and the author has not yet heard of a Croft damaging a loudspeaker, although he has heard many tales of woe caused by faulty transistor amplifiers!

The Croft is disgracefully inefficient, despite the fact that the output stage is biased into Class AB, and quickly warms a room on a hot summer's day. However, it has a beguiling sound to its midrange, and would be ideal as a midrange amplifier in an active crossover loudspeaker system.

The Mullard 5–20, and derivatives

The Mullard 5–20, Fig. 10.13, was a five-valve 20 W amplifier designed to sell the EL34 power pentode developed by Mullard/Philips. The EL34 and the Mullard 5–20 were an immediate success, with many commercial amplifiers being sold that were either licensed versions of the design, or variations on the theme. More recently, the World Audio K5881, and particularly the Maplin Millennium 4–20, owe a great deal to this seminal design.

The driver circuitry is radically different to the Williamson/Croft and uses an EF86 pentode as an input stage, which is responsible for the high sensitivity, but poor noise performance, of this amplifier. A slightly unusual feature is that the g_2 decoupling capacitor is connected between g_2 and cathode, rather than g_2 and ground. In most circuits, the cathode is at (AC) ground,

and so there is no reason why the g_2 decoupling capacitor should not go to ground. In this circuit, there is appreciable negative feedback to the cathode, and so the g_2 capacitor is connected to the cathode in order to hold g_2–k (AC) volts at zero, otherwise there would be positive feedback to g_2.

The cathode coupled phase splitter is combined with the driver circuit using an ECC83 giving an overall gain of 27, and has 10 dB of overload capability. When the output stage gain begins to fall, due to cathode feedback, or the input capacitance of the EL34 loading the driver, the global feedback loop will try to correct this by supplying greater drive to the output stage. The 10 dB margin will quickly be eroded, and although Mullard quoted distortion of 0.4 per cent for the driver circuitry, distortion will quickly rise. The driver circuitry was designed to produce an amplifier of high sensitivity even after 30 dB of feedback had been applied, and this forces linearity and noise to be compromised.

The output stage uses a pair of EL34 in 'ultra-linear' configuration, with 43 per cent taps for minimum distortion. This is a very common topology, as it gives some of the advantages of triode operation (low output resistance, even harmonic distortion) whilst having the advantages of the pentode (easy load for the driver stage, high efficiency).

The Beast

It should be emphasised that the Beast is a design under development, and is offered as an example of design evolution rather than a proven example of practice.

The design was provoked by a request for a pure Class A, triode, amplifier that could drive home-made electrostatic loudspeakers directly. Electrostatic loudspeakers need very high signal voltages, so transmitter valves operating from 1 kV HT supplies become an advantage rather than a disadvantage. Electrostatic loudspeakers are inherently push–pull, so the search for a push–pull pair of suitable output valves began, culminating in the choice of a pair of 845 triodes. However, the fundamental design looked promising for driving conventional electromagnetic loudspeakers, so an additional design, Fig.10.14, was developed in parallel.

The 845 is a directly heated triode designed in the early 1930s (the RCA anode characteristics are dated November 1933), yet the valve displays outstanding linearity. The downside is that the 845 is extraordinarily difficult to drive.

Driving circuitry for the Beast

The 845 is a low μ triode ($\mu = 3$), which means that although it has reasonably low input capacitance, each

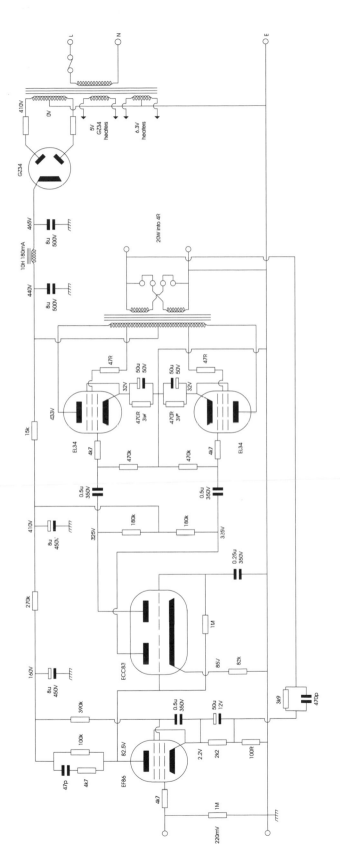

Figure 10.13 Mullard 5–20 power amplifier.

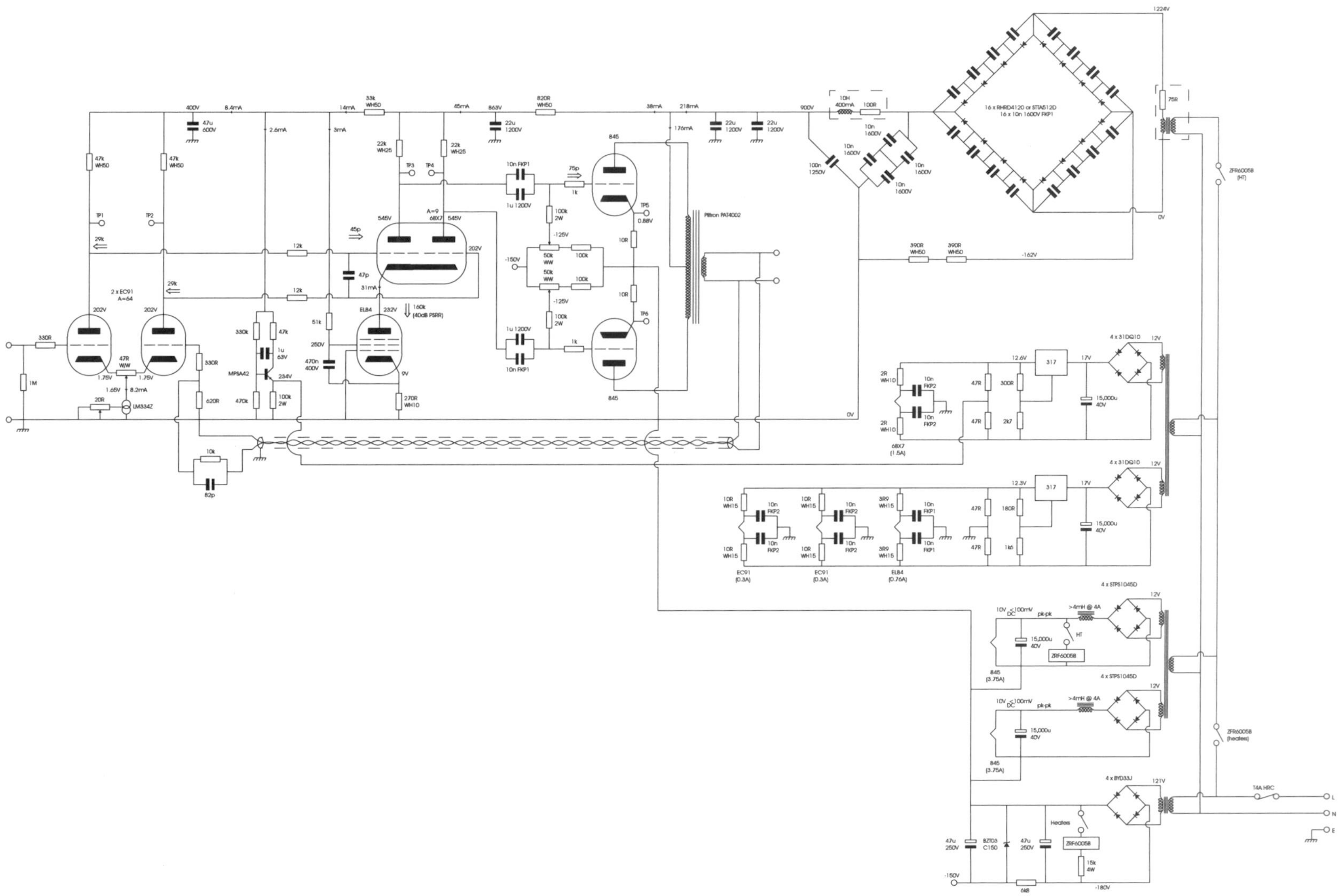

Figure 10.14 The Beast, a 3 W pure Class A triode amplifier.

valve requires ≈ 300 V_{pk-pk} to drive it to full power. Developing 300 V_{pk-pk} is not trivial, and it is no wonder that most designers resort to using power valves as drivers – with the 300B being a popular valve. The author took one look at the price of a 300B, and decided that an alternative had to be found – using a driver valve that cost as much as the *pair* of output valves seemed bizarre.

For a while, the E182CC double triode looked promising, but a careful search through the 'Electronic Universal Vade Mecum' uncovered the 6BX7. The 6BX7 is also a double triode, but built to rather more heroic proportions than the E182CC; an AC Cobra compared to a Triumph Spitfire, if you like. As an added bonus, because the 6BX7 has a low μ, it is very linear and capable of swinging many volts on its anode. A differential pair was the obvious choice for the driver stage, and it seemed desirable to direct couple this to the input stage, resulting in the cathodes floating >200 V above ground, so a constant current sink for the tail seemed both possible and desirable.

Various possibilities for the constant current sink were investigated (hybrid transistor/triode cascodes etc.) but the simplest, and best, seemed to be a power pentode, and even though the author had a number of N78 pentodes desperately seeking some use, an EL84 was best.

The input stage is also the phase splitter, and to ensure best balance should use high μ valves. The traditional choice would be the dual triode ECC83 (μ = 100), but the little known EC91 is a high μ (μ = 100) single triode with a *much* lower r_a than the ECC83. Belatedly, it was also realised that single triodes were an advantage rather than a disadvantage, since it means that input valves can easily be swapped until a match is found, whereas finding a pair of closely matched triodes in one envelope would be quite unusual.

The audio stage circuitry is fairly conventional, with no great surprises, although passive components often need to be of high voltage or power ratings – if there was ever a time to compare the cost of a valve amplifier with a transistor amplifier of similar power, then this is it! Moving to the power supply, it was thought that if the amplifier could not have infinite rejection of power supply noise, then the power supply ought to be designed for minimum noise.

The Beast's HT power supply

Providing an HT of ≈ 1 kV with a conventional capacitor input supply implies pulse switching of the diodes, between the mains transformer output and the smoothing capacitor, when the transformer reaches its peak voltage. It is hard to imagine any better way of causing noise than deliberately producing large pulses of current in the power supply.

Accordingly, the HT supply uses a choke input supply, which ensures that a constant (non-switching) current is drawn from the mains transformer rather than pulses. In practice, the rectifier diodes must switch off when the voltage across each of them falls to 0.7 V, meaning that the transformer secondary current is briefly interrupted at a 100 Hz rate. In order to reduce the disturbance caused by this interruption, the rectifier diodes are fast switching, soft recovery types, and are bypassed by capacitors.

As a further measure, the choke (and its stray capacitances) have been tuned to match the leakage inductance of the chosen mains transformer to ensure optimum filtering. In essence, the choke is a parallel resonant circuit (usually resonant between 3–15 kHz), whilst the mains transformer has leakage inductance. If a small capacitance is placed before the choke input filter, this forms a series resonant circuit whose resonance can be tuned to null the parallel resonance of the choke. The result is that the entire filter circuit no longer has a peak of output noise within the audio band, and the tuning capacitor also prevents the choke from producing large voltage spikes as the rectifier diodes switch off.

It should be noted that the tuning capacitor has to withstand a maximum voltage of:

$$V_{max} = V_{HT} + \sqrt{2}.V_{RMS(transformer)}$$

This is why the tuning capacitor shown in the diagram is made up of 1600 V capacitors in series to give a total voltage rating of 3200 V.

Another neglected aspect of power supply design is the low frequency resonance. Although a choke input supply ensures a low output impedance, it is not constant with frequency, and rises to a peak when the reactance of the output capacitor is equal to the reactance of the series choke. At best, power amplifiers represent a constant current load on the power supply, implying that they are a high impedance load and therefore do not damp the resonant peak.

The assertion that an amplifier is a constant current load might seem odd until it is considered that the amplifier strives to maintain a constant voltage into the (hopefully resistive) load, despite the fact that the power supply may be changing in voltage. If the amplifier maintains a constant voltage into a resistance, then that resistance draws a constant current, which is supplied by the power supply.

The low frequency resonance (in this case, a potential 9 dB peak at 8 Hz) can be damped either by series resistance, or by shunt resistance across either the inductor or capacitor. Shunt resistance across the inductor will greatly increase noise, whilst shunt resistance across the capacitor would waste a great deal of power, so series resistance seems the best compromise. Fortuitously, part of the necessary resistance can be made up of the unavoidable transformer and choke series resistance. The power supply was tuned to be critically damped so as to impart as little interference as possible to the amplifier.

Interestingly, some designers (author included) are now opting for shunt regulators, particularly in combination with choke input supplies, since a shunt regulator ensures that the choke input supply is always loaded with more than its critical minimum current. Viewed from the power supply filter, a shunt regulator (without additional series resistance) looks like a constant (almost zero) resistance load, and completely damps the low frequency resonance of a choke input LC filter. It seems likely that the reported sonic superiority of shunt regulators is not so much due to their internal operation as to their loading on the power supply filter.

In conclusion, a power supply that has series inductance (whether intentional or due to mains transformer leakage inductance) will have a resonant peak that is not damped by the load of the amplifier. Worse, if the amplifier is Class AB, the minimal damping provided by the amplifier will change with programme level, and the resonant power supply will cause programme level dependent coloration that will not be easily detected by steady-state tests.

The Beast's 845 LT supplies

Although choke input supplies are normally associated with HT supplies in valve equipment, they are actually even better in low voltage supplies. This is because there is a relationship between minimum current that must be drawn and input supply voltage that is associated with choke input power supplies:

$$I_{min(mA)} = \frac{V_{RMS}}{L\,(H)}$$

This relationship means that a power supply operating with a low input voltage needs only a low value choke. Low value chokes have very much smaller stray capacitances, so their self-resonant frequency is higher, making them a more nearly-perfect component.

It makes a great deal of sense to use a choke input filter for a low voltage supply, not just because the resonant frequency of a practical choke will be higher, but also because it reduces the peak current demanded from the mains transformer.

If we demand a large peak current from a mains transformer, then we stand a very good chance of saturating that transformer, and generating noise due to leakage flux. A choke input supply draws a constant current from the transformer equal to the DC current, whereas a capacitor input supply typically demands pulse currents of $>10I_{DC}$ – *ten* times the DC current!

If large pulses are demanded, they must flow somewhere. In a capacitor input supply, they flow through the reservoir capacitor, so this capacitor has to be able to withstand large pulses of current, known as ripple current. Ripple current causes heating within the capacitor because the capacitor has internal resistance, and current through resistance dissipates heat ($P = I^2R$). Internal heat within the capacitor helps the (liquid) electrolyte to evaporate, and so pressure builds up within the capacitor, which is either released by a safety vent (thus allowing electrolyte to escape and reducing the remaining electrolyte), or by the capacitor exploding destructively. Either way, electrolytic capacitors should be kept as cool as possible to ensure longevity.

Summarising, it is expensive to construct an electrolytic capacitor that can cope with high ripple currents, and it would be far better to reduce the current, particularly if $4A_{DC}$ (for an 845 heater) is required, which might conceivably cause a ripple current of 40A.

The Beast's driver LT supplies

Because each of the stages of the Beast is a differential pair and does not have the cathode bypassed down to ground, r_k becomes important. It is an unfortunate, but inevitable, consequence of increased immunity to HT noise that r_k rises, and the stage becomes more susceptible to noise capacitively induced from the heater. Admittedly, heater to cathode capacitance is quite small, but once r_k begins to rise, then C_{hk} becomes significant, especially if we consider that individual mains transformer windings are *not* screened from one another. Typical transformer adjacent winding capacitance is 1 nF, so if we were especially unfortunate, we might have 1 nF between the 1200 V HT winding and the 6.3 V winding destined for the input stage. There are three ways of avoiding the problem:

(1) Wind a custom transformer with electrostatic screens between each winding (expensive).
(2) Use a separate heater transformer for each valve (expensive, and bulky).
(3) The noise signal is common mode (on both wires), so simply regulating heater supplies does not help. We need common mode RF filtering on the supplies, and this can be achieved by matched series resistors in each wire combined with capacitance down to the chassis as close as possible to the valve base. The series resistance also has the advantage of reducing heater current surge at switch on.

Other circuits and sources of information

In one short chapter it is scarcely possible to cover more than a fraction of this wide field, so particular attention was focused on areas that are not well covered elsewhere. Inevitably, this has meant that detailed circuit theory, phono stages, and single ended amplifiers have been omitted, but this shortfall can be remedied by consulting the following sources.

Books

Valley, and Wallman, *Vacuum Tube Amplifiers*, McGraw-Hill (1948).
Out of print. Excellent for individual stage analysis.
Terman, F. E., *Electronic and Radio Engineering*, 4th edition, McGraw-Hill (1955).
Out of print. Sparse on audio, but good all-round theory.
Langford-Smith, F., *Radio Designer's Handbook*, 4th edition, Iliffe (1957).
Now available on CD-ROM. A classic on audio, but equations tend to be stated rather than derived.
Jones, M., *Valve Amplifiers*, Newnes (1995).

A practical introduction to valve circuit theory and practise.

Periodicals

Journal of the Audio Engineering Society, AES (UK section). P.O. Box 645 Slough.
An invaluable reference, particularly for loudspeaker design.
Electronics World (formerly *Wireless World*) Reed Business Information.
The grandfather of pure electronics periodicals.
Glass Audio. Audio Amateur Publications Inc., Peterborough, NH, USA.
An excellent source of information on current thinking on valves, articles range from fundamental theory to the best ways of finishing a chassis.
Sound Practices. Sound Practices, Austin, TX, USA.
A delightfully 'off the wall' publication; single-ended designs are popular.
Hi-Fi World (supplement). Audio Publishing, London.
The only UK publication to regularly feature valve constructional features and kits.
Hi-Fi News. Link House Magazines, Croydon.
Good technical reviews, but now sparse on constructional features.

11 Tuners and Radio Receivers

John Linsley Hood

Not all sound is recorded, and not all recorded sound is replayed directly. The broadcasting of sound is a more modern development than its recording, but tuners for high-quality sound reception have come a long way since the first use of FM. John Linsley Hood takes up the story.

Background

The nineteenth century was a time of great technical interest and experiment in both Europe and the USA. The newly evolved disciplines of scientific analysis were being applied to the discoveries which surrounded the experimenters and which, in turn, led to further discoveries.

One such happening occurred in 1864, when James Clerk Maxwell, an Edinburgh mathematician, sought to express Michael Faraday's earlier law of magnetic induction in proper mathematical form. It became apparent from Maxwell's equations that an alternating electromagnetic field would give rise to the radiation of electromagnetic energy.

This possibility was put to the test in 1888 by Heinrich Hertz, a German physicist. He established that this did indeed happen, and radio transmission became a fact. By 1901, Marconi in Morse code, using primitive spark oscillator equipment, had transmitted a radio signal across the Atlantic. By 1922 the first commercial public broadcasts had begun for news and entertainment.

By this time, de Forrest's introduction of control grid into Fleming's thermionic diode had made the design of high power radio transmitters a sensible engineering proposition. It had also made possible the design of sensitive and robust receivers. However, the problems of the system remain the same, and the improvements in contemporary equipment are merely the result of better solutions to these, in terms of components or circuit design.

Basic Requirements for Radio Reception

These are

- **selectivity** – to be able to select a preferred signal from a jumble of competing programmes;
- **sensitivity** – to be sure of being able to receive it reliably;
- **stability** – to be able to retain the chosen signal during the required reception period;
- **predictability** – to be able to identify and locate the required reception channel;
- **clarity** – which requires freedom from unwanted interference and noise, whether this originates within the receiver circuit or from external sources;
- **linearity** – which implies an absence of any distortion of the signal during the transmission/reception process.

These requirements for receiver performance will be discussed later under the heading of receiver design. However, the quality of the signal heard by the listener depends very largely on the nature of the signal present at the receiver. This depends, in the first place, on the transmitter and the transmission techniques employed.

In normal public service broadcasting – where it is required that the signal shall be received, more or less uniformly, throughout the entire service area – the transmitter aerial is designed so that it has a uniform, 360°, dispersal pattern. Also the horizontal shape of the transmission 'lobe' (the conventional pictorial representation of relative signal strength, as a function of the angle) is as shown in Fig. 11.1.

The influence of ground attenuation, and the curvature of the earth's surface, mean that in this type of transmission pattern the signal strength, gets progressively weaker as the height above ground level of the receiving aerial gets less, except in the immediate neighbourhood of the transmitter. There are a few exceptions to this rule, as will be shown later, but it is generally true, and implies that the higher the receiver aerial can be placed, in general the better.

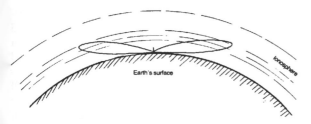

Figure 11.1 Typical transmitter aerial lobe pattern.

The Influence of the Ionosphere

The biggest modifying influence on the way the signal reaches the receiver is the presence of a reflecting – or, more strictly, refracting – ionised band of gases in the outer regions of the earth's atmosphere. This is called the ionosphere and is due to the incidence of a massive bombardment of energetic particles on the other layers of the atmosphere, together with ultra-violet and outer electromagnetic radiation, mainly from the sun.

This has the general pattern shown in Fig. 11.2, if plotted as a measure of electron density against height from the surface. Because it is dependent on radiation from the sun, its strength and height will depend on whether the earth is exposed to the sun's radiation (daytime) or protected by its shadow (night).

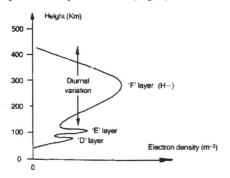

Figure 11.2 The electron density in the ionosphere.

As the predominant gases in the earth's atmosphere are oxygen and nitrogen, with hydrogen in the upper reaches, and as these gases tend to separate somewhat according to their relative densities, there are three effective layers in the ionosphere. These are the 'D' (lowest) layer, which contains ionised oxygen/ozone; the 'E' layer, richer in ionised nitrogen and nitrogen compounds; and the 'F' layer (highest), which largely consists of ionised hydrogen.

Since the density of the gases in the lower layers is greater, there is a much greater probability that the ions will recombine and disappear, in the absence of any sustaining radiation. This occurs as the result of normal

collisions of the particles within the gas, so both the 'D' and the 'E' layers tend to fade away as night falls, leaving only the more rarified 'F' layer. Because of the lower gas pressure, molecular collisions will occur more rarely in the 'F' layer, but here the region of maximum electron density tends to vary in mean height above ground level.

Critical frequency

The way in which radio waves are refracted by the ionosphere, shown schematically in Fig. 11.3, is strongly dependent on their frequency, with a 'critical frequency' ('F_c') dependent on electron density, per cubic metre, according to the equation

$$F_c = 9 \sqrt{N_{max}}$$

where N_{max} is the maximum density of electrons/cubic metre within the layer. Also, the penetration of the ionosphere by radio waves increases as the frequency is increased. So certain frequency bands will tend to be refracted back towards the earth's surface at different heights, giving different transmitter to receiver distances for optimum reception, as shown in Fig. 11.4, while some will not be refracted enough, and will continue on into outer space.

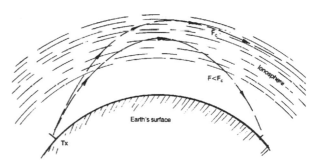

Figure 11.3 The refraction of radio waves by the ionosphere.

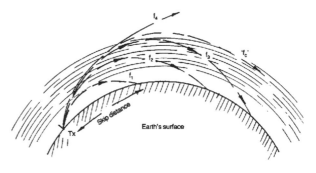

Figure 11.4 The influence of frequency on the optimum transmitter to receiver distance – the 'skip distance'.

The dependence of radio transmission on ionosphere conditions, which, in turn depends on time of day, time of year, geographical latitude, and 'sun spot' activity, has led to the term 'MUF' or maximum usable frequency, for such transmissions.

Also, because of the way in which different parts of the radio frequency spectrum are affected differently by the possibility of ionospheric refraction, the frequency spectrum is classified as shown in Table 11.1. In this VLF and LF signals are strongly refracted by the 'D' layer, when present, MF signals by the 'E' and 'F' layers, and HF signals only by the 'F' layer, or not at all.

Table 11.1   *Classification of radio frequency spectrum*

VLF	3–30	kHz
LF	30–300	kHz
MF	300–3000	kHz
HF	3–30	MHz
VHF	30–300	MHz
UHF	300–3000	MHz
SHF	3–30	GHz

Additionally, the associated wavelengths of the transmissions (from 100 000–1000 m in the case of the VLF and LF signals, are so long that the earth's surface appears to be smooth, and there is a substantial reflected 'ground wave' which combines with the direct and reflected radiation to form what is known as a 'space wave'. This space wave is capable of propagation over very long distances, especially during daylight hours when the 'D' and 'E' layers are strong.

VHF/SHF effects

For the VHF to SHF regions, different effects come into play, with very heavy attenuation of the transmitted signals, beyond direct line-of-sight paths, due to the intrusions of various things which will absorb the signal, such as trees, houses, and undulations in the terrain. However, temperature inversion layers in the earth's atmosphere, and horizontal striations in atmospheric humidity, also provide mechanisms, especially at the higher end of the frequency spectrum, where the line of sight paths may be extended somewhat to follow the curvature of the earth.

Only certain parts of the available radio frequency (RF) spectrum are allocated for existing or future commercial broadcast use. These have been subclassified as shown in Table 11.2, with the terms 'Band 1' to 'Band 6' being employed to refer to those regions used for TV and FM radio use.

The internationally agreed FM channel allocations ranged, originally, from Channel 1 at 87.2–87.4 MHz, to Channel 60, at 104.9–105.1 MHz, based on 300 kHz centre-channel frequency spacings. However, this allocation did not take into account the enormous future proliferation of local transmitting stations, and the centre-channel spacings are now located at 100 kHz intervals.

Depending on transmitted power, transmitters will usually only be operated at the same frequency where they are located at sites which are remote from each other. Although this range of operating frequencies somewhat overlaps the lower end of 'Band 2', all of the UK operating frequencies stay within this band.

Table 11.2   *Radio broadcast band allocations*

Wavelength	Allocation	Band
LW	150–285 kHz	
MW	525–1605 kHz	
SW	5.95–6.2 MHz	49 M
	7.1–7.3 MHz	40 M
	9.5–9.775 MHz	30 M
	11.7–11.975 MHz	25 M
	15.1–15.45 MHz	19 M
	17.7–17.9 MHz	16 M
	21.45–21.75 MHz	13 M
	25.5–26.1 MHz	11 M

Note National broadcasting authorities may often overspill these frequency limits.

Band	Wavelength allocation
I	41–68 MHz
II	87.5–108 MHz
III	174–223 MHz
IV	470–585 MHz
V	610–960 MHz
VI	11.7–12.5 GHz

Why VHF Transmissions?

In the early days of broadcasting, when the only reliable long to medium distance transmissions were thought to be those in the LF-MF regions of the radio spectrum, (and, indeed, the HF bands were handed over to amateur use because they were thought to be of only limited usefulness), broadcast transmitters were few and far between. Also the broadcasting authorities did not aspire to a universal coverage of their national audiences. Under these circumstances, individual transmitters could broadcast a high quality, full frequency range signal, without problems due to adjacent channel interference.

However, with the growth of the aspirations of the broadcasting authorities, and the expectations of the listening public, there has been an enormous proliferation of radio transmitters. There are now some 440 of these in the UK alone, on LW, MW and Band 2 VHF allocations, and this ignores the additional 2400 odd separate national and local TV transmissions.

If all the radio broadcast services were to be accommodated within the UK's wavelength allocations on the LW and MW bands, the congestion would be intolerable. Large numbers would have to share the same frequencies, with chaotic mutual interference under any reception conditions in which any one became receivable in an adjacent reception area.

Frequency choice

The decision was therefore forced that the choice of frequencies for all major broadcast services, with the exception of pre-existing stations, must be such that the area covered was largely line-of-sight, and unaffected by whether it was day or night.

It is true that there are rare occasions when, even on Band 2 VHF, there are unexpected long-distance intrusions of transmissions. This is due to the occasional formation of an intense ionisation band in the lower regions of the ionosphere, known as 'sporadic E'. Its occurrence is unpredictable and the reasons for its occurrences are unknown, although it has been attributed to excess 'sun spot' activity, or to local thermal ionisation, as a result of the shearing action of high velocity upper atmosphere winds.

In the case of the LW and MW broadcasts, international agreements aimed at reducing the extent of adjacent channel interference have continually restricted the bandwidth allocations available to the broadcasting authorities. For MW at present, and for LW broadcasts as from 1st February, 1988 – ironically the centenary of Hertz's original experiment – the channel separation is 9 kHz and the consequent maximum transmitted audio bandwidth is 4.5 kHz.

In practice, not all broadcasting authorities conform to this restraint, and even those that do interpret it as 'flat from 30 Hz to 4.5 kHz and –50 dB at 5.5 kHz' or more leniently, as 'flat to 5 kHz, with a more gentle roll-off to 9 kHz'. However, by comparison with the earlier accepted standards for MW AM broadcasting, of 30 Hz–12 kHz, ±1 dB, and with less than 1 per cent THD at 1 kHz, 80 per cent modulation, the current AM standards are poor.

One also may suspect that the relatively low standards of transmission quality practicable with LF/MF AM broadcasting encourages some broadcasting authorities

to relax their standards in this field, and engage in other, quality degrading, practices aimed at lessening their electricity bills.

AM or FM?

Having seen that there is little scope for high quality AM transmissions on the existing LW and MW broadcast bands, and that VHF line-of-sight transmissions are the only ones offering adequate bandwidth and freedom from adjacent channel interference, the question remains as to what style of modulation should be adopted to combine the programme signal with the RF carrier.

Modulation systems

Two basic choices exist, that of modulating the amplitude of the RF carrier, (AM), or of modulating its frequency, (FM), as shown in Fig. 11.5. The technical advantages of FM are substantial, and these were confirmed in a practical field trial in the early 1950s carried out by the BBC, in which the majority of the experimental listening panel expressed a clear preference for the FM system.

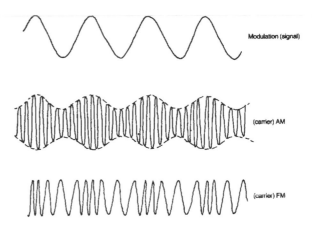

Figure 11.5   Carrier modulation systems.

The relative qualities of the two systems can be summarised as follows. AM is:

- the simplest type of receiver to construct
- not usually subject to low distortion in the recovered signal
- prone to impulse-type (e.g. car ignition) interference and to 'atmospherics'
- possibly subject to 'fading'
- prone to adjacent channel or co-channel interference

- affected by tuning and by tuned circuit characteristics in its frequency response.

FM:

- requires more complex and sophisticated receiver circuitry
- can give very low signal distortion
- ensures all received signals will be at the same apparent strength
- is immune to fading
- is immune to adjacent channel and co-channel interference, provided that the intruding signals are less strong.
- has, potentially, a flat frequency response, unaffected by tuning or tuned circuit characteristics
- will reject AM and impulse-type interference
- makes more efficient use of available transmitter power, and gives, in consequence, a larger usable reception area.

On the debit side, the transmission of an FM signal requires a much greater bandwidth than the equivalent AM one, in a ratio of about 6:1. The bandwidth requirements may be calculated approximately from the formula

$$Bn = 2M + 2DK$$

where B_n is the necessary bandwidth, M is the maximum modulation frequency in Hz, D is the peak deviation in Hz, and K is an operational constant (= 1 for a mono signal).

However, lack of space within the band is not a significant problem at VHF, owing to the relatively restricted geographical area covered by any transmitter. So Band 2 VHF/FM has become a worldwide choice for high quality transmissions, where the limitations are imposed more by the distribution method employed to feed programme signals to the transmitter, and by the requirements of the stereo encoding/decoding process than by the transmitter or receiver.

FM Broadcast Standards

It is internationally agreed that the FM frequency deviation will be 75 kHz, for 100 per cent modulation. The stereo signal will be encoded, where present, by the Zenith-GE 38 kHz sub-carrier system, using a 19 kHz ±2 Hz pilot tone, whose phase stability is better than 3° with reference to the 38 kHz sub-carrier. Any residual 38 kHz sub-carrier signal present in the composite stereo output shall be less than 1 per cent.

Local agreements, in Europe, specify a 50 μs transmission pre-emphasis. In the USA and Japan, the agreed pre-emphasis time constant is 75 μs. This gives a slightly better receiver S/N ratio, but a somewhat greater proneness to overload, with necessary clipping, at high audio frequencies.

Stereo Encoding and Decoding

One of the major attractions of the FM broadcasting system is that it allows the transmission of a high quality stereo signal, without significant degradation of audio quality – although there is some worsening of S/N ratio. For this purpose the permitted transmitter bandwidth is 240 kHz, which allows an audio bandwidth, on a stereo signal, of 30 Hz–15 kHz, at 90 per cent modulation levels. Lower modulation levels would allow a more extended high-frequency audio bandwidth, up to the 'zero transmission above 18.5 kHz' limit imposed by the stereo encoding system.

It is not known that any FM broadcasting systems significantly exceed the 30 Hz–15 kHz audio bandwidth levels.

Because the 19 kHz stereo pilot tone is limited to 10 per cent peak modulation, it does not cause these agreed bandwidth requirements to be exceeded.

The Zenith-GE 'Pilot Tone' Stereophonic System

This operates in a manner which produces a high-quality 'mono' signal in a receiver not adapted to decode the stereo information, by the transmission of a composite signal of the kind represented in Fig. 11.6. In this the combined left-hand channel and right-hand channel (L + R) – mono – signal is transmitted normally in the 30 Hz–15 kHz part of the spectrum, with a maximum modulation of 90 per cent of the permitted 75 kHz deviation.

An additional signal, composed of the difference signal between these channels, (L – R), is then added as a modulation on a suppressed 38 kHz sub-carrier. So that the total modulation energy will be the same, after demodulation, the modulation depth of the combined (L – R) signal is held to 45 per cent of the permitted maximum excursion. This gives a peak deviation for the transmitted carrier which is the same as that for the 'mono' channel.

Decoding

This stereo signal can be decoded in two separate ways, as shown in Fig. 11.7 and 11.8. In essence, both of these operate by the addition of the two (L + R) and (L – R) signals to give LH channel information only, and the subtraction of these to give the RH channel information only.

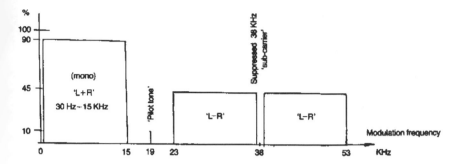

Figure 11.6 The Zenith-GE 'pilot tone' stereophonic system.

In the circuit of Fig. 11.7, this process is carried out by recovering the separate signals, and then using a matrix circuit to add or subtract them. In the circuit of Fig. 11.8, an equivalent process is done by sequentially sampling the composite signal, using the regenerated 38 kHz sub-carrier to operate a switching mechanism.

Advocates of the matrix addition method of Fig. 11.7,

have claimed that this allows a better decoder signal-to-noise (S/N) ratio than that of the sampling system. This is only true if the input bandwidth of the sampling system is sufficiently large to allow noise signals centred on the harmonics of the switching frequency also to be commutated down into the audio spectrum.

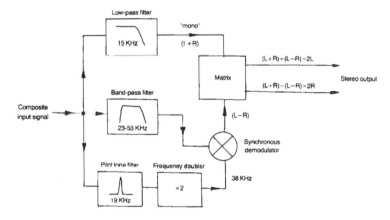

Figure 11.7 Matrix addition type of stereo decoder.

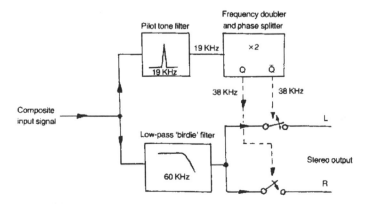

Figure 11.8 Synchronous switching stereo decoder.

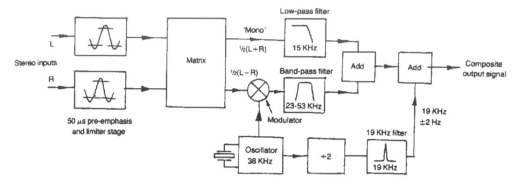

Figure 11.9 Zenith-GE stereophonic encoding system.

Provided that adequate input filtration is employed in both cases there is no operational advantage to either method. Because the system shown in Fig. 11.8 is more easily incorporated within an integrated circuit, it is very much the preferred method in contemporary receivers.

In both cases it is essential that the relative phase of the regenerated 38 kHz sub-carrier is accurately related to the composite incoming signal. Errors in this respect will degrade the 35–40 dB (maximum) channel separation expected with this system.

Because the line bandwidth or sampling frequency of the studio to transmitter programme link may not be compatible with the stereo encoded signal, this is normally encoded on site, at the transmitter, from the received LH and RH channel information. This is done by the method shown in Fig. 11.9.

In this, the incoming LH and RH channel signals (pre-emphasis and peak signal level limitation will generally have been carried out before this stage) are combined together in a suitable matrix – the simple double transformer of Fig. 11.10 would serve – and fed to addition networks in the manner shown.

In the case of the (L – R) signal, however, it is first converted into a modulated signal based on a 38 kHz sub-carrier, derived from a stable crystal-controlled oscillator, using a balanced modulator to ensure suppression of the residual 38 kHz sub-carrier frequency. It is then filtered to remove any audio frequency components before addition to the (L + R) channel.

Finally, the 38 kHz sub-carrier signal is divided, filtered and phase corrected to give a small amplitude, 19 kHz sine-wave pilot tone which can be added to the composite signal before it is broadcast.

The reason for the greater background 'hiss' associated with the stereo than with the mono signal is that wide-band noise increases as the square-root of the audio bandwidth. In the case of a mono signal this bandwidth is 30 Hz–15 kHz. In the case of a Zenith-GE encoded stereo signal it will be at least 30 Hz–53 kHz,

even if adequate filtering has been used to remove spurious noise components based on the 38 kHz sub-carrier harmonics.

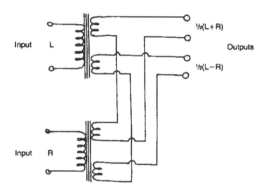

Figure 11.10 Simple matrixing method.

The BBC Pulse Code Modulation (PCM) Programme Distribution System

In earlier times, with relatively few, high-power, AM broadcast transmitters, it was practicable to use high-quality telephone lines as the means of routing the programme material from the studio to the transmitter. This might even, in some cases, be in the same building.

However, with the growth of a network of FM transmitters serving local areas it became necessary to devise a method which would allow consistently high audio quality programme material to be sent over long distances without degradation. This problem became more acute with the spread of stereo broadcasting, where any time delay in one channel with respect to the other would cause a shift in the stereo image location.

The method adopted by the BBC, and designed to take advantage of the existing 6.5 MHz bandwidth TV signal transmission network, was to convert the signal into digital form, in a manner which was closely analogous to that adopted by Philips for its compact disc system. However, in the case of the BBC a rather lower performance standard was adopted. The compact disc uses a 44.1 kHz sampling rate, a 16-bit (65536 step) sampling resolution, and an audio bandwidth of 20 Hz–20 kHz, (±0.2 dB), while the BBC system uses a 32 kHz sampling rate, a 13-bit (8192 step) resolution and a 50 Hz–14.5 kHz bandwidth, (±0.2 dB).

The BBC system offers a CCIR weighted S/N ratio of 57 dB, and a non-linear distortion figure of 0.1 per cent ref. full modulation at 1 kHz.

As in all digital systems, the prominence of the 'staircase type' step discontinuity becomes greater at low signal levels and higher audio frequencies, giving rise to a background noise associated with the signal, known as 'quantisation noise'.

Devotees of hi-fi tend to refer to these quantisation effects as 'granularity'. Nevertheless, in spite of the relatively low standards, in the hi-fi context, adopted for the BBC PCM transmission links, there has been relatively little criticism of the sound quality of the broadcasts.

The encoding and decoding systems used are shown in Fig. 11.11 and 11.12.

any audio components above half the sampling rate (say at 17 kHz) would be resolved identically to those at an equal frequency separation below this rate, (e.g. at 15 kHz) and this could cause severe problems both due to spurious signals, and to intermodulation effects between these signals. This problem is known as 'aliasing', and the filters are known as 'anti-aliasing' filters.

Because the quality of digitally encoded waveforms deteriorates as the signal amplitude gets smaller, it is important to use the largest practicable signal levels (in which the staircase-type granularity will be as small a proportion of the signal as possible). It is also important to ensure that the analogue-to-digital encoder is not overloaded. For this reason delay-line type limiters are used, to delay the signal long enough for its amplitude to be assessed and appropriate amplitude limitation to be applied.

In order to avoid hard peak clipping which is audibly displeasing, the limiters operate by progressive reduction of the stage gain, and have an output limit of +2 dB above the nominal peak programme level. Carrying out the limiting at this stage avoids the need for a further limiter at the transmitter. Pre-emphasis is also added before the limiter stage. The effect of this pre-emphasis on the frequency response is shown in Fig. 11.13.

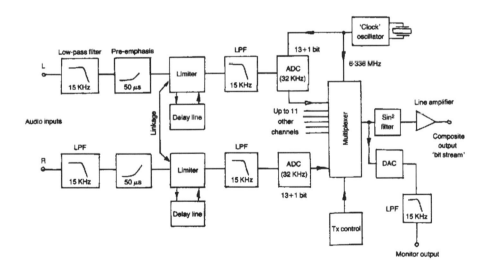

Figure 11.11 The BBC 13-channel pulse code modulation (PCM) encoder system.

Encoding

The operation of the PCM encoder, shown schematically in Fig. 11.11, may be explained by consideration of the operation of a single input channel. In this, the incoming signal is filtered, using a filter with a very steep attenuation rate, to remove all signal components above 15 kHz.

This is necessary since with a 32 kHz sampling rate,

An interesting feature of the limiter stages is that those for each stereo line pair are linked, so that if the peak level is exceeded on either, both channels are limited. This avoids any disturbance in stereo image location which might arise through a sudden amplitude difference between channels.

The AF bandwidth limited signal from the peak limiter is then converted into digital form by a clocked 'double ramp' analogue-to-digital converter, and fed,

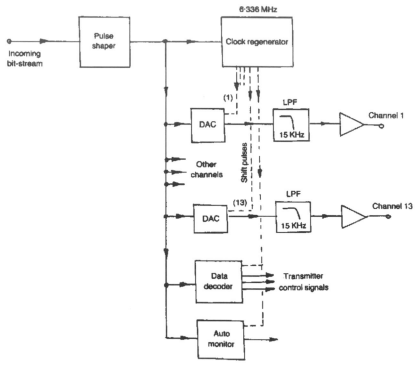

Figure 11.12   The BBC PCM decoding system.

along with the bit streams from up to 12 other channels, to a time-domain multiplexing circuit. The output from this is fed through a filter, to limit the associated RF bandwidth, to an output line amplifier.

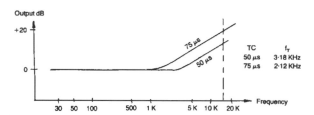

Figure 11.13   HF pre-emphasis characteristics.

The output pulses are of sine-square form, with a half amplitude duration of 158 ns, and 158 ns apart. Since the main harmonic component of these pulses is the second, they have a negligible energy distribution above 6.336 MHz, which allows the complete composite digital signal to be handled by a channel designed to carry a 625-line TV programme signal.

The output of the multiplexer is automatically sampled, sequentially, at 256 ms per programme channel, in order to give an automatic warning of any system fault. It can also be monitored as an audio output signal reference.

To increase the immunity of the digitally encoded signal to noise or transmission line disturbances, a parity bit is added to each preceding 13-bit word. This contains data which provide an automatic check on the five most significant digits in the preceding bit group. If an error is detected, the faulty word is rejected and the preceding 13-bit word is substituted. If the error persists, the signal is muted until a satisfactory bit sequence is restored.

Effectively, therefore, each signal channel comprises 14 bits of information. With a sampling rate of 32 kHz, the 6.336 Mbits/s bandwidth allows a group of 198 bits in each sample period. This is built from 13 multiplexed 14-bit channels (= 182 bits) and 16 'spare' bits. Of these, 11 are used to control the multiplexing matrix, and four are used to carry transmitter remote control instructions.

Decoding

The decoding system is shown in Fig. 11.12. In this the incoming bit stream is cleaned up, and restored to a sequence with definite '0' and '1' states. The data stream is then used to regenerate the original 6336 kHz clock signal, and a counter system is used to provide a sequence of shift pulses fed, in turn, to 13 separate digital-to-analogue (D-A) converters, which recreate the original input channels, in analogue form.

A separate data decoder is used to extract the transmitter control information, and, as before, an automatic sampling monitor system is used to check the correct operation of each programme channel.

The availability of spare channels can be employed to carry control information on a digitally controlled contrast compression/expansion (Compander) system (such as the BBC NICAM 3 arrangement). This could be used either to improve the performance, in respect of the degradation of small signals, using the same number of sample steps, or to obtain a similar performance with a less good digital resolution (and less transmission bandwidth per channel).

The standards employed for TV sound links, using a similar type of PCM studio-transmitter link, are:

- AF bandwidth, 50 Hz–13.5 kHz (± 0.7 dB ref. 1 kHz)
- S/N ratio, 53 dB CCIR weighted
- non-linear distortion and quantisation defects, 0.25 per cent (ref. 1 kHz and max. modulation).

Supplementary Broadcast Signals

It has been suggested that data signals could be added to the FM or TV programme signal, and this possibility is being explored by some European broadcasting authorities. The purpose of this would be to provide signal channel identification, and perhaps also to allow automatic channel tuning. The system proposed for this would use an additional low level subcarrier, for example, at 57 kHz. In certain regions of Germany additional transmissions based on this sub carrier frequency are used to carry the VWF (road/traffic conditions) information broadcasts.

In the USA, 'Storecast' or 'Subsidiary Communication Authorisation' (SCA) signals may be added to FM broadcasts, to provide a low-quality 'background music' programme for subscribers to this scheme. This operates on a 67 kHz sub-carrier, with a max. modulation level of 10 per cent.

Alternative Transmission Methods

Apart from the commercial (entertainment and news) broadcasts on the LW, MW and specified short wave bands shown in Table 11.2, where the transmission techniques are exclusively double-sideband amplitude modulated type, and in Band 2 (VHF) where the broadcasts are exclusively frequency modulated, there are certain police, taxi and other public utility broadcasts on Band 2 which are amplitude-modulated.

It is planned that these other Band 2 broadcasts will eventually be moved to another part of the frequency spectrum, so that the whole of Band 2 can be available for FM radio transmissions.

However, there are also specific bands of frequencies, mainly in the HF/VHF parts of the spectrum, which have been allocated specifically for amateur use. These are shown in Table 11.3.

Table 11.3 *Amateur frequency band allocations*

Band	MHz
80 M	3.5–3.725
40 M	7.0–7.15
20 M	14.0–14.35
15 M	21.0–21.45
10 M	28.0–29.7
6 M	50.0–54.0
2 M	144.0–148.0

In these amateur bands, other types of transmitted signal modulation may be employed. These are narrow-bandwidth FM (NBFM), phase modulation (PM), and single-sideband suppressed carrier AM (SSB).

The principal characteristics of these are that NBFM is restricted to a total deviation of ± 5 kHz, and a typical maximum modulation frequency of 3 kHz, limited by a steep-cut AF filter. This leads to a low modulation index (the ratio between the carrier deviation and the modulating frequency), which leads to poor audio quality and difficulties in reception (demodulation) without significant distortion. The typical (minimum) bandwidth requirement for NBFM is 13 kHz.

Phase modulation (PM), shares many of the characteristics of NBFM, except that in PM, the phase deviation of the signal is dependent both upon the amplitude and the frequency of the modulating signal, whereas in FM the deviation is purely dependent on programme signal amplitude.

Both of these modulation techniques require fairly complex and well designed receivers, if good reception and S/N ratio is to be obtained.

SSB

The final system, that of suppressed carrier single-sideband transmission (SSB), is very popular among the amateur radio transmitting fraternity, and will be found on all of the amateur bands.

This relies on the fact that the transmitted frequency spectrum of an AM carrier contains two identical, mirror-image, groups of sidebands below (lower sideband or

LSB) and above (upper sideband or USB) the carrier, as shown in Fig. 11.14. The carrier itself conveys no signal and serves merely as the reference frequency for the sum and difference components which together reconstitute the original modulation.

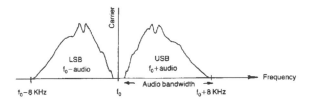

Figure 11.14   Typical AM bandwidth spectrum for double-sideband transmission.

If the receiver is designed so that a stable carrier frequency can be reintroduced, both the carrier and one of the sidebands can be removed entirely, without loss of signal intelligibility. This allows a very much larger proportion of the transmitter power to be used for conveying signal information, with a consequent improvement in range and intelligibility. It also allows more signals to be accommodated within the restricted bandwidth allocation, and gives better results with highly selective radio receivers than would otherwise have been possible.

The method employed is shown schematically in Fig. 11.15.

There is a convention among amateurs that SSB transmissions up to 7 MHz shall employ LSB modulation, and those above this frequency shall use USB.

This technique is not likely to be of interest to those concerned with good audio quality, but its adoption for MW radio broadcast reception is proposed from time to time, as a partial solution to the poor audio bandwidth possible with double sideband (DSB) transmission at contemporary 9 kHz carrier frequency separations.

The reason for this is that the DSB transmission system used at present only allows a 4.5 kHz AF bandwidth, whereas SSB would allow a full 9 kHz audio pass-band. On the other hand, even a small frequency drift of the reinserted carrier frequency can transpose all the programme frequency components upwards or downwards by a step frequency interval, and this can make music sound quite awful; so any practical method would have to rely on the extraction and amplification of the residue of the original carrier, or on some equally reliable carrier frequency regeneration technique.

Radio Receiver Design

The basic requirements for a satisfactory radio receiver were listed above. Although the relative importance of these qualities will vary somewhat from one type of receiver to another – say, as between an AM or an FM receiver – the preferred qualities of radio receivers are, in general, sufficiently similar for them to be considered as a whole, with some detail amendments to their specifications where system differences warrant this.

Selectivity

An ideal selectivity pattern for a radio receiver would be one which had a constant high degree of sensitivity at the desired operating frequency, and at a sufficient bandwidth on either side of this to accommodate transmitter sidebands, but with zero sensitivity at all frequencies other than this.

A number of techniques have been used to try to approach this ideal performance characteristic. Of these, the most common is that based on the inductor-capacitor (LC) parallel tuned circuit illustrated in Fig. 11.16 (a).

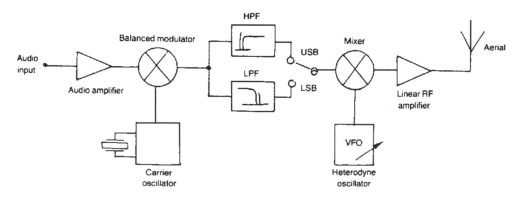

Figure 11. 15   Single sideband (USB or LSB) transmitter system.

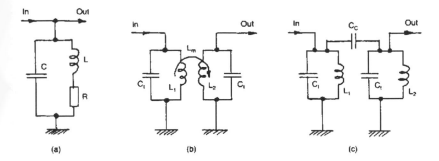

(a)   (b)   (c)

Fig 11.16   Single (a) and band-pass (b) (c) tuned circuits.

Tuned circuit characteristics

For a given circulating RF current, induced into it from some external source, the AC potential appearing across an L–C tuned circuit reaches a peak value at a frequency (F_0) given by the equation

$$F_0 = 1/2\pi\sqrt{LC}$$

Customarily the group of terms $2\pi F$ are lumped together and represented by the symbol ω, so that the peak output, or resonant frequency would be represented by

$$\omega_0 = \frac{1}{\sqrt{LC}}$$

Inevitably there will be electrical energy losses within the tuned circuit, which will degrade its performance, and these are usually grouped together as a notional resistance, r, appearing in series with the coil.

The performance of such tuned circuits, at resonance, is quantified by reference to the circuit magnification factor or quality factor, referred to as Q. For any given L–C tuned circuit this can be calculated from

$$Q = \frac{\omega_0 L}{r} \quad \text{or} \quad \frac{1}{\omega_0 Cr}$$

Since, at resonance, (ω_0), $\omega_0 = 1/\sqrt{LC}$, the further equation

$$Q = \frac{1}{r}\sqrt{\frac{L}{C}}$$

can be derived. This shows that the Q improves as the equivalent loss resistance decreases, and as the ratio of L to C increases.

Typical tuned circuit Q values will lie between 50 and 200, unless the Q value has been deliberately degraded, in the interests of a wider RF signal pass-band, usually by the addition of a further resistor, R, in parallel with the tuned circuit.

The type of selectivity offered by such a single tuned circuit is shown, for various Q values, in Fig. 11.17(a). The actual discrimination against frequencies other than the desired one is clearly not very good, and can be calculated from the formula

$$\delta F = F_0/2Q$$

where δF is the 'half-power' bandwidth.

One of the snags with single tuned circuits, which is exaggerated if a number of such tuned circuits are arranged in cascade to improve the selectivity, is that the drop in output voltage from such a system, as the frequencies differ from that of resonance, causes a very rapid attenuation of higher audio frequencies in any AM type receiver, in which such a tuning arrangement was employed, as shown in curve 'a' of Fig. 11.18.

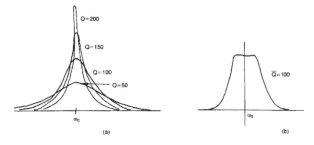

Figure 11. 17   Response curves of single (a) and band-pass (b) tuned circuits.

Clearly, this loss of higher audio frequencies is quite unacceptable, and the solution adopted for AM receivers is to use pairs of tuned circuits, coupled together by mutual inductance, L_m, or, in the case of separated circuits, by a coupling capacitor, C_c.

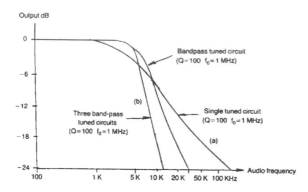

Figure 11.18 Effect on Am receiver AF response of tuned circuit selectivity characteristics.

This leads to the type of flat-topped frequency response curve shown in Fig. 11.17(b), when critical coupling is employed, when the coupling factor

$$k = 1/\sqrt{Q_1 Q_2}$$

For a mutually inductive coupled tuned circuit, this approximates to

$$L_m = \bar{L}/Q$$

and for a capacitively coupled band pass tuned circuit

$$C_c = \bar{C}_t/Q$$

where L_m is the required mutual inductance, $\bar{L}$ is the mean inductance of the coils, $\bar{C}_t$ is the mean value of the tuning capacitors, and C_c is the required coupling capacitance.

In the case of a critically-coupled bandpass tuned circuit, the selectivity is greater than that given by a single tuned circuit, in spite of the flat-topped frequency response. Taking the case of $F_o = 1$ MHz, and $Q = 100$, for both single-and double-tuned circuits the attenuation will be –6 dB at 10 kHz off tune. At 20 kHz off tune, the attenuation will be –12 dB and –18 dB respectively, and at 30 kHz off tune it will be –14 dB and –24 dB, for these two cases.

Because of the flat-topped frequency response possible with bandpass-coupled circuits, it is feasible to put these in cascade in a receiver design without incurring audio HF loss penalties at any point up to the beginning of the attenuation 'skirt' of the tuned circuit. For example, three such groups of 1 MHz bandpass-tuned circuits, with a circuit Q of 100, would give the following selectivity characteristics.

- 5 kHz = 0 dB
- 10 kHz = –18 dB
- 20 kHz = –54 dB
- 30 kHz = –72 dB

as illustrated in curve 'b' in Fig. 11.18. Further, since the bandwidth is proportional to frequency, and inversely proportional to Q, the selectivity characteristics for any other frequency or Q value can be derived from these data by extrapolation.

Obviously this type of selectivity curve falls short of the ideal, especially in respect of the allowable audio bandwidth, but selectivity characteristics of this kind have been the mainstay of the bulk of AM radio receivers over the past sixty or more years. Indeed, because this represents an ideal case for an optimally designed and carefully aligned receiver, many commercial systems would be less good even than this.

For FM receivers, where a pass bandwidth of 220–250 kHz is required, only very low Q tuned circuits are usable, even at the 10.7 MHz frequency at which most of the RF amplification is obtained. So an alternative technique is employed, to provide at least part of the required adjacent channel selectivity. This is the surface acoustic wave filter.

The surface acoustic wave (SAW) filter

This device, illustrated in schematic form in Fig. 11.19, utilises the fact that it is possible to generate a mechanical wave pattern on the surface of a piece of piezo-electric material (one in which the mechanical dimensions will change under the influence of an applied electrostatic field), by applying an AC voltage between two electrically conductive strips laid on the surface of the material.

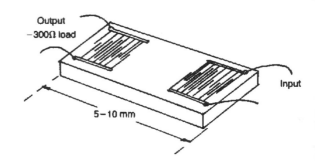

Figure 11.19 Construction of surface acoustic wave (SAW) filter.

Since the converse effect also applies – that a voltage would be induced between these two conducting strips if a surface wave in a piezo-electric material were to pass under them – this provides a means for fabricating an electro-mechanical filter, whose characteristics can be tailored by the number, separation, and relative lengths of the conducting strips.

These SAW filters are often referred to as IDTs (inter-digital transducers), from the interlocking finger pattern of the metallising, or simply as ceramic filters, since some of the earlier low-cost devices of this type were made from piezo-electric ceramics of the lead zirconate-titanate (PZT) type. Nowadays they are fabricated from thin, highly polished strips of lithium niobate, (LiNbO), bismuth germanium oxide, (Bi GeO), or quartz, on which a very precise pattern of metallising has been applied by the same photo-lithographic techniques employed in the manufacture of integrated circuits.

Two basic types of SAW filter are used, of which the most common is the transversal type. Here a surface wave is launched along the surface from a transmitter group of digits to a receiver group. This kind is normally used for bandpass applications. The other form is the resonant type, in which the surface electrode pattern is employed to generate a standing-wave effect.

Because the typical propagation velocity of such surface waves is of the order of 3000 m/s, practical physical dimensions for the SAW slices and conductor patterns allow a useful frequency range without the need for excessive physical dimensions or impracticably precise electrode patterns. Moreover, since the wave only penetrates a wavelength or so beneath the surface, the rear of the slice can be cemented onto a rigid substrate to improve the ruggedness of the element.

In the transversal or bandpass type of filter, the practicable range of operating frequencies is, at present, from a few MHz to 1–2 GHz, with minimum bandwidths of around 100 kHz. The resonant type of SAW device can operate from a few hundred KHz to a similar maximum frequency, and offers a useful alternative to the conventional quartz crystal (bulk resonator) for the higher part of the frequency range, where bulk resonator systems need to rely on oscillator overtones.

The type of performance characteristic offered by a bandpass SAW device, operating at a centre frequency of 10.7 MHz, is shown in Fig. 11.20. An important quality of wide pass-band SAW filters is that the phase and attenuation characteristics can be separately optimised by manipulating the geometry of the pattern of the conducting elements deposited on the surface.

This is of great value in obtaining high audio quality from FM systems, as will be seen below, and differs, in this respect, from tuned circuits or other types of filter where the relative phase-angle of the transmitted signal is directly related to the rate of change of transmission as a function of frequency, in proximity to the frequency at which the relative phase-angle is being measured.

However, there are snags. Good quality phase linear SAW filters are expensive, and there is a relatively high insertion loss, typically in the range of –15 to –25 dB, which requires additional compensatory amplification. On the other hand, they are physically small and do not suffer, as do coils, from unwanted mutual induction effects. The characteristic impedance of such filters is, typically, 300–400 ohms, and changes in the source and load impedances can have big effects on the transmission curves.

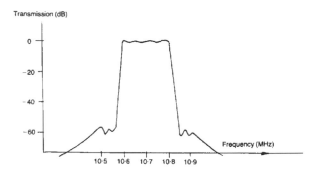

Figure 11.20 Transmission characteristics of typical 10.7 MHz SAW filter.

The superhet system

It will have been appreciated from the above that any multiple tuned circuit, or SAW filter, system chosen to give good selectivity will be optimised only for one frequency. To tune a group of critically coupled band-pass tuned circuits simultaneously to cover a required receiver frequency range would be very difficult. To tune one or more SAW filters simultaneously would simply not be possible at all, except, perhaps, over an exceedingly limited frequency range.

A practical solution to this problem was offered in 1918 by Major Armstrong of the US Army, in the form of the supersonic heterodyne or superhet receiver system, shown in Fig. 11.21.

In this, the incoming radio signal is combined with an adjustable frequency local oscillator signal in some element having a non-linear input/output transfer characteristic, (ideally one having a square-law slope). This stage is known as the frequency changer or mixer or, sometimes, and inappropriately, as the first detector.

This mixture of the two (input and LO) signals gives rise to additional sum and difference frequency outputs, and if the local oscillator frequency is chosen correctly, one or other of these can be made to coincide with the fixed intermediate frequency (usually known as the IF), at which the bulk of the amplification will occur.

The advantages of this arrangement, in allowing the designer to tailor his selectivity characteristics without regard to the input frequency, are enormous, and virtually all commercial radio receivers employ one or other of the possible permutations of this system.

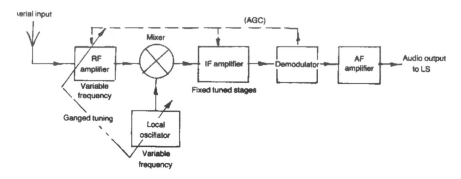

Figure 11.21   The superhet system.

Problem

The snag is that the non-linear mixer stage produces both sum and difference frequency outputs, so that, for any given local oscillator frequency there will be two incoming signal frequencies at which reception would be possible.

These are known as signal and image frequencies, and it is essential that the selectivity of the tuned circuits preceding the mixer stage is adequate to reject these spurious second channel or image frequency signals. This can be difficult if the IF frequency is too low in relation to the incoming signal frequency, since the image frequency will occur at twice the IF frequency removed from the signal, which may not be a very large proportion of the input signal frequency, bearing in mind the selectivity limitations of conventional RF tuned circuits.

In communications receivers and similar high-quality professional systems, the problem of image breakthrough is solved by the use of the double superhet system shown in Fig. 11.22. In this the incoming signal frequency is changed twice, firstly to a value which is sufficiently high to allow complete elimination of any spurious image frequency signals, and then, at a later stage, to a lower frequency at which bandwidth limitation and filtering can be done more easily.

A difficulty inherent in the superhet is that, because the mixer stage is, by definition, non-linear in its charac-

teristics, it can cause intermodulation products between incoming signals to be generated, which will be amplified by the IF stages if they are at IF frequency, or will simply be combined with the incoming signal if they are large enough to drive the input of the mixer into overload.

This type of problem is lessened if the mixer device has a true square law characteristic. Junction FETs have the best performance in this respect.

A further problem is that the mixer stages tend to introduce a larger amount of noise into the signal chain than the other, more linear, gain stages, and this may well limit the ultimate sensitivity of the system. The noise figure of the mixer stage is partly a function of the kind of device or circuit configuration employed, and partly due to noise introduced by the relatively large amplitude local oscillator signal.

It is undoubtedly true to say that the quality and care in design of the mixer stage from a major determining factor in receiver performance, in respect of S/N ratio and freedom from spurious signals.

Other possibilities

Because of the problems of the superhet system, in respect of mixer noise, and image channel interference,

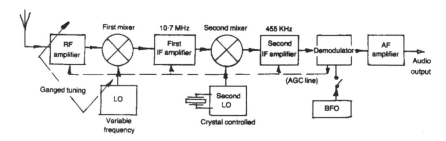

Figure 11.22   The double-superhet system, as used in a communication receiver.

direct conversion systems have been proposed, in which the signal is demodulated by an electronic switch operated by a control waveform which is synchronous in frequency and phase with the input signal. These are known as homodyne or synchrodyne receivers, depending on whether the control waveform is derived from the carrier of the incoming signal or from a separate synchronous oscillator.

Since both of these systems result in an audio signal in which adjacent channel transmissions are reproduced at audio frequencies dependent on the difference of the incoming signal from the control frequency, they offer a means for the control of selectivity, with a truly flat-topped frequency response, by means of post-demodulator AF filtering. On the debit side, neither of these offer the sensitivity or the ease of operation of the conventional superhet, and are not used commercially to any significant extent.

Sensitivity

Many factors influence this characteristic. The most important of these are

- the S/N ratio, as depending on the design, small signals can get buried in mixer or other circuit noise;
- inadequate detectability, due to insufficient gain preceding the demodulator stage;
- intermodulation effects in which, due to poor RF selectivity or unsatisfactory mixer characteristics, the wanted signal is swamped by more powerful signals on adjacent channels.

The ultimate limitation on sensitivity will be imposed by aerial noise, due either to man-made interference, (RFI) or to the thermal radio emission background of free space. The only help in this case is a better aerial design.

As mentioned above, the frequency changer stage in a superhet is a weak link in the receiver chain, as it introduces a disproportionately high amount of noise, and is prone to intermodulation effects if the signals present exceed its optimum input signal levels.

The problem of mixer noise is a major one with equipment using thermionic valves, but semiconductor devices offer substantial improvements in this respect. For professional equipment, diode ring balanced modulator layouts, using hot carrier or Schottky diodes, of the kind shown in Fig. 11.23, are the preferred choice, since they combine excellent noise characteristics with the best possible overload margin. However, this is a complex system.

In domestic equipment, at frequencies up to about 100 MHz, junction FETs are the preferred devices, though they have inconveniently large inter-electrode capaci-

tances for RF use. In the frequency range 100–500 MHz, dual-gate MOSFETs are preferable because their form of construction allows very good input – output screening, though their noise figure and other characteristics are somewhat less good than those of junction FETs.

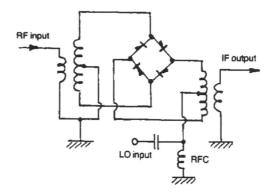

Figure 11.23 Diode ring double balanced mixer system.

Beyond 500 MHz, bipolar junction transistors are the only practical choice, though their use demands careful circuit design.

Integrated circuit balanced modulator systems have attracted some interest for high-quality designs, and have even been proposed as an alternative to ring diode modulators, though they have relatively poor overload characteristics. The various practical mixer systems are examined later, under 'Circuit design'.

In general, the best performance in a receiver, in respect of S/N ratio, sensitivity, and overload characteristics, requires a careful balance between the gain and selectivity of the various amplifying and mixing stages.

Stability

In a superhet system, in which there are a series of selective fixed-frequency amplifier stages, the major problems of frequency stability centre around the performance of the local (heterodyne) oscillator, which is combined with the incoming signal to give IF frequency. In relatively narrow bandwidth AM receivers, especially those operating in the short wave (3–30 MHz) region, a simple valve or transistor oscillator is unlikely to be adequately stable, unless very good quality components, and carefully balanced thermal compensation is employed.

Various techniques have been used to overcome this difficulty. For fixed frequency reception, oscillators based on individual quartz crystal or SAW resonators – which give an extremely stable output frequency, combined with high purity and low associated noise – are an excellent solution, though expensive if many reception frequencies are required. Alternatively, various ways of taking advantage of the excellent stability of the quartz crystal oscillator, while still allowing frequency variability, have been proposed, such as the phase-locked loop (PLL) frequency synthesizer, or the Barlow-Wadley loop systems.

Quartz Crystal Control

This operates by exciting a mechanical oscillation in a precisely dimensioned slab of mono-crystalline silica, either naturally occurring, as quartz, or, more commonly synthetically grown from an aqueous solution under conditions of very high temperature and pressure.

Since quartz exhibits piezo-electric characteristics, an applied alternating voltage at the correct frequency will cause the quartz slab to 'ring'in a manner analogous to that of a slab or rod of metal struck by a hammer. However, in the case of the crystal oscillator, electronic means can be used to sustain the oscillation.

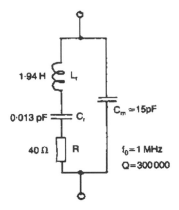

Figure 11.24 Equivalent electrical circuit of a quartz crystal resonator.

mounting), and series loss resistance R_1, for an 'X' cut crystal, are shown in the equivalent circuit of Fig. 11.24, for a crystal having a resonant frequency of 1 MHz, and an effective Q, as a series resonant circuit, of 300 000.

As in other materials, the physical dimensions of quartz crystals will change with temperature, but since its expansion is anisotropic (greater in some dimensions than others) it is possible to choose a particular section, or 'cut', through the crystal to minimise the effects of temperature on resonant frequency. Such crystals are known as zero temperature coefficient (ZTC) or AT cut types.

Zero temperature coefficient crystals (in practice this will imply temperature coefficients of less than ±2 parts per million per degree centigrade over the temperature range 5–45° C) do not offer quite such high Q values as those optimised for this quality, so external temperature stabilisation circuits (known as crystal ovens) are sometimes used in critical applications. Both of these approaches may also be used, simultaneously, where very high stability is required.

Although, in principle, the quartz crystal is a fixed frequency device, it is possible to vary the resonant frequency by a small amount by altering the value of an externally connected parallel capacitor. This would be quite insufficient as a receiver tuning means, so other techniques have been evolved.

The Barlow-Wadley Loop

This circuit arrangement, also known as a drift cancelling oscillator, has been used for many years in relatively inexpensive amateur communications receivers, such as the Yaesu Musen FRG-7, and the layout adopted for such a 500 kHz–30 MHz receiver is shown in Fig. 11.25.

In this, the incoming signal, after appropriate RF

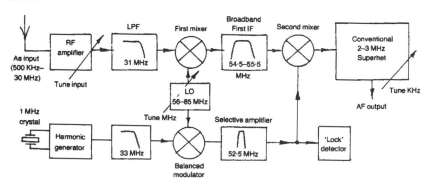

Figure 11.25 The Barlow-Wadley loop.

The monotonic frequency characteristics of the quartz crystal resonator derive from its very high effective Q value. Typical apparent values of L_r, C_r and C_m (The resonant inductance and capacitance, and that due to the

amplification and pre-selection, is passed though a steep-cut, low-pass filter, which removes all signals above 31 MHz, to the first mixer stage. This has a conventional L–C type tuned oscillator whose operating frequency gives a first IF output in the range 54.5–55.5 MHz, which is fed to a second mixer stage.

The L–C oscillator output is also fed to a double-balanced modulator where it is combined with the output from 1 MHz quartz-crystal controlled harmonic generator, and this composite output is taken through a selective amplifier having a centre frequency of 52.5 MHz, and a signal output detector system.

Certain combinations of the local oscillator frequency and the harmonics of 1 MHz will be within the required frequency range, and will therefore pass through this amplifier. When such a condition exists, the output voltage operates an indicator to show that this signal is present. This output signal is then fed to the second mixer stage to generate a second IF frequency in the range 2–3 MHz, from which the desired signal is selected by a conventional superhet receiver from the 1 MHz slab of signals presented to it.

The frequency drift in the first, high-frequency, L–C local oscillator is thereby cancelled, since it will appear, simultaneously, and in the same sense, at both the first and second mixers.

Frequency Synthesizer Techniques

These are based on developments of the phase-locked loop (PLL) shown in Fig. 11.26. In this arrangement an input AC signal is compared in phase with the output from a voltage controlled oscillator (VCO). The output from the phase comparator will be the sum and difference frequencies of the input and VCO signals.

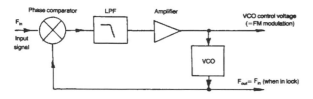

Figure 11.26   The phase-locked loop (PLL).

Where the difference frequency is low enough to pass the low-pass 'loop filter', the resultant control voltage applied to the VCO will tend to pull it into frequency synchronism, and phase quadrature, with the incoming signal – as long as the loop gain is high enough. In this condition, the loop is said to be 'locked'.

This circuit can be used to generate an AC signal in frequency synchronism with, but much larger in amplitude than, the incoming reference signal. It can also generate an oscillator control voltage which will accurately follow incoming variations in input signal frequency when the loop is in lock, and this provides an excellent method of extracting the modulation from an FM signal.

A further development of the basic PLL circuit is shown in Fig. 11.27. In this, a frequency divider is interposed between the VCO and the phase comparator, so that when the loop is in lock, the VCO output frequency will be a multiple of the incoming frequency. For example, if the divider has a factor n, then the VCO will have an output frequency equal to $n\,(F_{in})$.

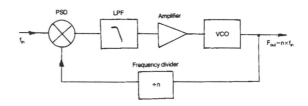

Figure 11.27   Phase-locked frequency multiplier.

In the PLL frequency synthesizer shown in Fig. 11.28, this process is taken one stage further, with a crystal controlled oscillator as the reference source, feeding the phase detector through a further frequency divider. If the two dividers have ratios of m and n, then when the loop is in lock, the output frequency will be $(n/m) \times F_{ref}$.

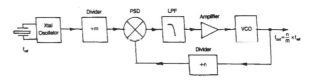

Figure 11.28   Phase-locked frequency synthesizer.

Provided that m and n are sufficiently large, the VCO output can be held to the chosen frequency, with crystal-controlled stability, and with any degree of precision required. Now that such frequency synthesizer circuitry is available in single IC form, this type of frequency control is beginning to appear in high quality FM tuners, as well as in communications receivers.

A minor operating snag with this type of system is that, because of the presence within the synthesizer IC of very many, relatively large amplitude, switching waveforms, covering a wide spectrum of frequencies, such receivers tend to be plagued by a multitude of minor tuning whistles, from which conventional single tuned oscillator systems are free. Very thorough screening of the synthesizer chip is necessary to keep these spurious signals down to an unobtrusive level.

Generally, in FM tuners, the relatively wide reception bandwidth makes oscillator frequency drift a less acute problem than in the case of AM receivers operating at comparable signal frequencies, although, since the distortion of the received signals will in most cases deteriorate if the set is not correctly tuned, or if it drifts off tune during use, there is still an incentive, in high-quality tuners, to employ quartz crystal stabilised oscillator systems.

A more serious problem, even in wide bandwidth FM tuners – where these do not employ PLL frequency control – is that the varicap diodes (semiconductor junction diodes in which the effective capacitance is an inverse function of the applied reverse voltage) often used to tune the RF and oscillator circuits, are quite strongly temperature dependant in their characteristics.

Varicap-tuned receivers must therefore employ thermally compensated DC voltage sources for their tuning controls if drift due to this cause is to be avoided.

AGC Effects

A particular problem which can occur in any system in which automatic gain control (AGC) is used, is that the operation of the AGC circuit may cause a sympathetic drift in tuned frequency.

This arises because the AGC system operates by extracting a DC voltage which is directly related to the signal strength, at some point in the receiver where this voltage will be of adequate size. This voltage is then used to control the gain of preceding RF or IF amplifier stages so that the output of the amplifier remains substantially constant. This is usually done by applying the control voltage to one or other of the electrodes of the amplifying device so that its gain characteristics are modified.

Unfortunately, the application of a gain control voltage to any active device usually results in changes to the input, output, or internal feedback capacitances of the device, which can affect the resonant frequency of any tuned circuits attached to it. This effect can be minimised by care in the circuit design.

A different type of problem concerns the time-constants employed in the system. For effective response to rapid changes in the signal strength of the incoming signal, the integrating time constant of the system should be as short as practicable. However, if there is too little constraint on the speed of response of the AGC system, it may interpret a low-frequency modulation of the carrier, as for example in an organ pedal note, as a fluctuation in the received signal strength, and respond accordingly by increasing or decreasing the receiver gain to smooth this fluctuation out.

In general, a compromise is attempted between the speed of the AGC response, and the lowest anticipated modulation frequency which the receiver is expected to reproduce, usually set at 30 Hz in good quality receivers. In transistor portable radios, where the small LS units seldom reproduce tones below some 200–250 Hz, a much more rapid speed of response is usable without audible tonal penalties.

Sadly, some broadcasting authorities take advantage of the rapid AGC response typically found in transistor portables to employ a measure of companding (tonal range compression on transmission followed by expansion on reception to restore the original dynamic range).

As practised by the BBC on its Radio 1, Radio 2, Radio 4 and some local radio transmissions, this consists of a reduction in both carrier strength and modulation on all dynamic peaks. This reduces the amount of electricity consumed by the transmitter, whose average power output decreases on sound level peaks.

If, then, the AGC system in the radio restores the received carrier level to a constant value, the original modulation range will be recovered. This will only work well if the AGC 'attack' and 'decay' time constants used in the receiver are correctly related to those employed at the transmitter – and this is purely a matter of chance. The result, therefore, is an additional and unexpected source of degradation of the broadcast quality of these signals.

Automatic Frequency Control (AFC)

Because of the shape of the output voltage versus input frequency relationship, at the output of the demodulator of an FM receiver, shown in idealised form in Fig. 11.29, the possibility exists that this voltage, when averaged so that it is just a DC signal, with no carrier modulation, can be used to control the operating frequency of the oscillator, or other tuned circuits. So if the tuning of the receiver drifts away from the ideal mid-point (F_t in Fig. 11.29) an automatic correction can be applied to restore it to the desired value.

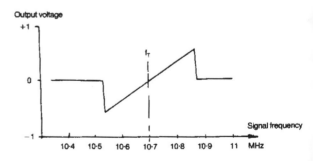

Figure 11.29   Voltage/frequency relationships in an ideal FM demodulator.

This technique is widely used, especially in the case of receivers where the tuning is performed by the application of a voltage to one or more varicap diodes, and which, in consequence, lends itself well to the superimposition of an additional AFC voltage. It does, however, have snags.

The first of these is that there is a close similarity in the action of an AFC voltage in an FM tuner to the action of an automatic gain control in an AM one. In both cases the control system sees the change – in carrier frequency in the case of an FM tuner – which is produced by the programme modulation, as being an error of the type which the control system is designed to correct. Too effective an AFC system can therefore lead to a worsening of bass response in the tuner, unless very heavy damping of the response is incorporated.

In the case of FM tuners, it is likely that the changes to be corrected will mainly be slow, and due only to thermal effects, provided that the receiver was correctly tuned in the first place, whereas in an AM radio the changes in received signal strength can often be quite rapid.

The second problem is that the AFC system, attempting to sit the tuning point on the zero DC level, may not lead to the best results if the demodulator circuit is misaligned, and gives the kind of curve shown in Fig. 11.30. Here the user might place the tuning at the point F^1 by ear, where the AFC signal will always restore it to F_t.

user could know the point of tune and return without difficulty to the same place. However, with contemporary IC technology, the cost of integrated circuit frequency counter systems has become so low, relative to the other costs of the design, that almost all modern tuners now employ some kind of frequency meter display.

This operates in the manner shown in Fig. 11.31. In this, a quartz crystal controlled oscillator, operating perhaps at 32.768 kHz, (the standard frequency for 'quartz' watches, for which cheap crystals and simple frequency dividers are readily available – 32768 is 2^{15}) is used to generate a precise time interval, say one second.

Meanwhile, the signal from the oscillator of the receiver, which will invariably employ a superhet layout, will be clocked over this period by an electronic counter, in which the IF frequency (that frequency by which the local oscillator differs from the incoming RF signal) will be added to, or subtracted from, the oscillator frequency – depending on whether the oscillator frequency is higher or lower than that of the signal – so that the counter effectively registers the incoming signal frequency.

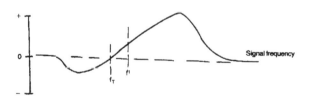

Figure 11.30 Effect of demodulator misalignment on AFC operation.

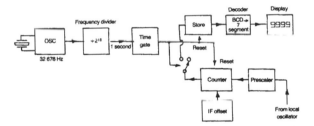

Figure 11.31 Frequency meter system.

This kind of demodulator misalignment is all too common in practice, and is one of the reasons why very linear demodulators (in which the choice of the correct point is less critical) are a worthwhile pursuit. Some more recent systems employ demodulator circuits which are comparatively insensitive to detuning. It should be remembered, also, that the output from the demodulator is likely to be used to operate the tuning meter, and encourage the user to tune to F_t.

In the case of this kind of tuning meter, the user can carry out a quick check on the alignment of the tuned circuits by noting the maximum meter deflection on either side of the incoming signal frequency. These readings should be symmetrical.

Predictability

The possession of a good, clear, tuning scale has always been a desirable virtue in any radio receiver, so that the

The electronic switch circuitry operated by the time interval generator is then used alternately to reset the frequency counter to zero, and to transfer its final count to a digital frequency store. A liquid crystal (LCD) or light-emitting diode (LED) display can then be used to register the signal frequency in numerical form.

As a matter of convenience, to avoid display flicker the counter output will be held in a store or latch circuit during the counting period, and the contents of the store only updated each time the count is completed. Since frequency dither is only likely to affect the last number in the display, in circumstances where there is an uncertainty of ±1 digit, some systems actually count to a larger number than is displayed, or up-date the last – least significant – digit less frequently than the others.

Additional information can also be shown on such an LED/LCD display, such as the band setting of the receiver, whether it is on AM or FM, whether AFC or stereo decoding has been selected, the local clock time. In the light of the continual search for marketable improvements, it seems probable that such displays will soon also include a precise time signal derived from one or other of the time code transmitters.

Clarity

One of the major advantages of FM radio is its ability to reject external radio frequency noise – such as that due to thunderstorms, which is very prominent on LF radio, or motor car ignition noise, which is a nuisance on all HF and VHF transmissions, which includes Band 2.

This ability to reject such noise depends on the fact that most impulse noise is primarily amplitude modulated, so that, while it will extend over a very wide frequency range, its instantaneous frequency distribution is constant. The design of the FM receiver is chosen, deliberately, so that the demodulator system is insensitive to AM inputs, and this AM rejection quality is enhanced by the use of amplitude limiting stages prior to the demodulator.

A valuable 'figure of merit' in an FM receiver is its AM rejection ratio, and a well-designed receiver should offer at least 60 dB (1000:1).

Because FM receivers are invariably designed so that there is an excess of gain in the RF and IF stages, so that every signal received which is above the detection threshold will be presented to the demodulator as an amplitude limited HF square wave, it is very seldom that such receivers – other than very exotic and high-cost designs – will employ AGC, and only then in the pre-mixer RF stage(s).

A further benefit in well-designed FM receivers is that intruding AM radio signals on the same band will be ignored, as will less strong FM broadcasts on the same channel frequency. This latter quality is referred to as the capture ratio, and is expressed as the difference in the (voltage) signal strength, in dB, between a more powerful and a less powerful FM transmission, on the same frequency allocation, which is necessary for the weaker transmission to be ignored.

The capture ratio of an FM receiver – a desirable feature which does not exist in AM reception – depends on the linearity, both in phase and amplitude, of the RF and IF stages prior to limiting, and also, to a great extent, on the type of demodulator employed. The best current designs offer 1 dB discrimination. Less good designs may only offer 3–4 dB. Two decibels is regarded as an acceptable figure for high-quality systems, and would imply that the designers had exercised suitable care.

It should be noted in this context that designs in which very high IF gain is employed to improve receiver sensitivity may, by causing overload in these stages, and consequent cross-modulation if they do not 'clip' cleanly, lead to a degradation in capture ratio.

Needless to say, those IF stages which are designed to limit the amplitude of the incoming signal should be designed so that they do not generate inter-modulation products. However, these invariably take the form of highly developed IC designs, based on circuit layouts of the form shown in Fig. 11.32, using a sequence of symmetrically driven, emitter-coupled, 'long-tailed pairs', chosen to operate under conditions where clean and balanced signal clipping will occur.

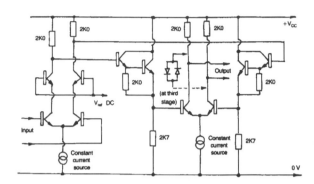

Figure 11.32   Cascode input stage, and the first (of two) symmetrical gain stages in a modern FM IF gain block IC (RCA CA3189E).

FM tuner designers very seldom feel inspired to attempt to better their performance by the use of alternative layouts.

In AM receivers, the best that can be done to limit impulse type interference is to use some form of impulse amplitude limiter, which operates either on the detection of the peak amplitude of the incoming signal or on the rate-of-change of that amplitude, such qualities will be higher on impulse noise than on normal programme content. An excellent AM noise limiter circuit from Philips, which incorporates an electronic delay line to allow the noise limiter to operate before the noise pulse arrives, is shown in Fig. 11.33.

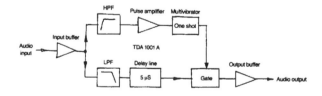

Figure 11.33   Philips' impulse noise limiting circuit.

The rejection of intruding signals in AM depends entirely on the receiver selectivity, its freedom from intermodulation defects, and – in respect of its ability to avoid internally generated hum, noise, and mush – on the quality of the circuit design and components used, and the wisdom of the choice of distribution of the gain and selectivity within the RF and IF stages.

Linearity, FM systems

In view of the continuing search for improved audio amplifier quality in both quantitative and subjective terms, it is not surprising that there has been a compar-able effort to obtain a high performance in FM receiver systems.

This endeavour is maintained by competitive rivalry and commercial pressures, and has resulted in many cases in the development of FM tuners with a quality which exceeds that of the incoming signal presented to them by the broadcasting authorities. Broadcasters' activities are not the subject of competitive pressures, and their standards are determined by the reluctance of governments to spend money on luxuries, and by the cynicism of their engineers.

These constraints have resulted in a continuing erosion of the quality of AM radio broadcasts, though there still remain some honourable exceptions. This has, sadly, often led to the AM sections of FM tuners being designed as low-cost functional appendages having a very poor audio performance, even when the quality of the FM section is beyond reproach.

A number of factors influence the quality of the demodulated signal, beginning with the stability and phase linearity of the RF, IF and AF gain stages. In the case of AM radios, incipient instability in the RF or IF stages will lead to substantial amplitude distortion effects, with consequent proneness to intermodulation defects. (Any non-linearity in an amplifier will lead to a muddling of the signals presented to it.) In the case of those FM radios based on a phase-sensitive demodulator, any RF or IF instability will lead to substantial phase distortions as the signal passes through the frequency of incipient oscillation.

Much care is therefore needed in the design of such stages to ensure their stable operation. If junction FETs are employed rather than dual-gate MOSFETs, some technique, such as that shown in Fig. 11.34, must be used to neutralise the residual drain-gate capacitance. The circuit of Fig. 11.34 utilises a small inductance in the source lead, which could be simply a lengthy track on the printed circuit board, to ensure that the unwanted feedback signal due to the gate-drain capacitance (C'), is cancelled by a signal, effectively in phase opposition, due to the drain-source capacitance (C'').

However, assuming competent RF/IF stage design, the dominant factor in recovered signal quality is that of the demodulator design. Representative demodulator systems, for FM and for AM, are shown below.

Slope Detection

This circuit, shown in Fig. 11.35(a), is the earliest and crudest method of detecting or demodulating an FM sig-

nal. The receiver is simply tuned to one side or the other of the tuned circuit resonant frequency (F_o), as illustrated in Fig. 11.35(b). Variations in the incoming frequency will then produce changes in the output voltage of the receiver, which can be treated as a simple AM signal. This offers no AM rejection ability, and is very nonlinear in its audio response, due to the shape of the resonance curve.

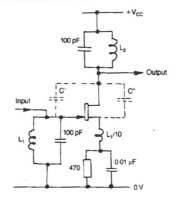

Figure 11.34 Feedback neutralisation system for junction FET RF amplifier.

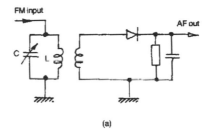

(a)

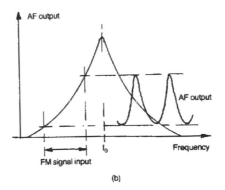

(b)

Figure 11.35 FM slope detector.

The Round-Travis Detector

This arrangement, shown in Fig. 11.36(a), employs a pair of tuned circuits with associated diode rectifiers ('detectors') which are tuned, respectively, above and below the incoming signal frequency, giving a balanced slope-detector characteristic, as seen in Fig. 11.36(b). This is more linear than the simple slope-detector circuit, but still gives no worthwhile AM rejection.

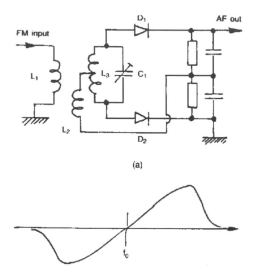

(a)

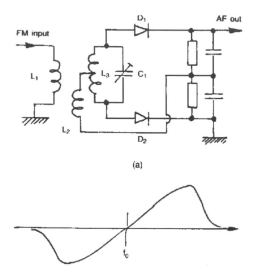

(b)

Figure 11.36   The Round-Travis FM detector.

The Foster-Seeley or Phase Detector

This circuit, of which one form is shown in Fig. 11.37(a), was evolved to provide an improved measure of AM rejection, by making its output dependent, at least in part, on the changes in phase induced in a tuned circuit by variations in the frequency of the incoming signal.

In this arrangement the tuned circuit, L_3C_1, provides a balanced drive to a matched pair of diode rectifiers (D_1, D_2), arranged in opposition so that any normal AM effects will cancel out. A subsidiary coil, L_2, is then arranged to feed the centre tap of L_3, so that the induced signal in L_2, which will vary in phase with frequency, will either reinforce or lessen the voltages induced in each half of L_3, by effectively disturbing the position of the electrical centre tap.

This gives the sort of response curve shown in Fig. 11.37(b), which has better linearity than its predecessors.

The Ratio Detector

This circuit, of which one form is shown in Fig. 11.38(a) is similar to the Foster-Seeley arrangement, except that the diode rectifiers are connected so that they produce an output voltage which is opposed, and balanced across the load. The output response, shown in Fig. 11.38(b), is very

similar to that of the Foster-Seeley circuit, but it has a greatly improved AM rejection.

(a)

Figure 11.37   The Foster-Seeley or phase detector.

The ratio detector was for many years the basic demodulator circuit for FM receivers, and offers a very good internal noise figure. This is superior to that of the contemporary IC-based phase-coincidence system, which has entirely superseded it, because of the much greater demodulation linearity offered by this latter system.

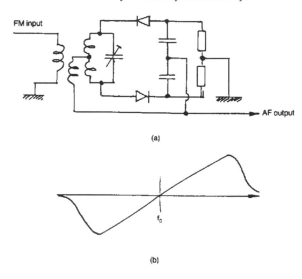

(a)

(b)

Figure 11.38   The ratio detector.

The Phase Coincidence Demodulator (PCD)

This method employs a circuit layout of the general form shown in Fig. 11.39, in which a group of identical bipolar transistors is interconnected so that the current from Q_1, used a simple constant-current source, will be routed either through Q_2 or Q_3, depending on the relative potential of the signal.

From Q_2/Q_3, the current flow will be directed either through Q_4/Q_7 or through Q_5/Q_6 and recombined at the load resistors R_1 and R_2.

It will be seen, from inspection, that if the transistors are well matched, the potential drop across R_1 and R_2 will be that due to half the output current of Q_1, regardless of the relative potentials applied to Q_2 or Q_3 or to Q_4/Q_7 or Q_5/Q_6, so long as these potentials are not simultaneously applied.

If synchronous HF signals are applied to all four input ports (a–d), the output across R_1 or R_2, (output ports e and f) will only be identical if inputs c and d are at phase quadrature to those at a and b.

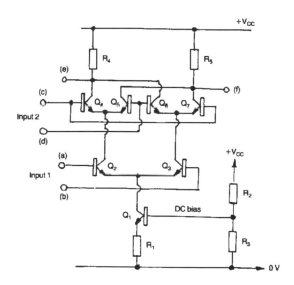

Figure 11.39 Gate-coincidence transistor array

Considering the circuit layout of Fig. 11.40, if ports b and d are taken to some appropriate DC reference potential, and an amplitude limited FM signal is applied to point a with respect of b, and some identical frequency signal, at phase quadrature at F_o, is applied to c with respect to d, then there will be an output voltage at point e with respect to point f, if the input frequency is varied in respect of F_o.

The linearity of this *V/F* relationship is critically dependent on the short-term frequency/phase stability of the potential developed across the quadrature circuit (L_1C_1), and this depends on the Q value of the tuned circuit. The snag here is that too high a value of Q will limit the usable FM bandwidth. A normal improvement which is employed is the addition of a further tuned circuit (L_2C_2) to give a band-pass coupling characteristic,

and further elaborations of this type are also used to improve the linearity, and lessen the harmonic distortion introduced by this type of demodulator.

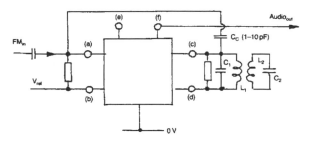

Figure 11.40 Phase coincidence demodulator circuit.

The performance of this type of demodulator is improved if both the signal drive (to ports a and b) and the feed to the quadrature circuit (ports c and d) are symmetrical, and this is done in some high quality systems.

The Phase-Locked Loop (PLL) Demodulator

This employs a system of the type shown above in Fig. 11.26. If the voltage controlled oscillator (VCO) used in the loop has a linear input voltage versus output frequency characteristic, then when this is in frequency lock with the incoming signal the control voltage applied to the VCO will be an accurate replica of the frequency excursions of the input signal, within the limits imposed by the low-pass loop filter.

This arrangement has a great advantage over all of the preceding demodulator systems in that it is sensitive only to the instantaneous input frequency, and not to the input signal phase. This allows a very low demodulator THD, unaffected by the phase-linearity of preceding RF and IF tuned circuits or SAW filters, and greatly reduces the cost, for a given performance standard, of the FM receiver system.

Such a circuit also has a very high figure for AM rejection and capture ratio, even when off-tune, provided that the VCO is still in lock.

Photographs of the output signal distortion, and the demodulator output voltage versus input frequency curves, taken from a frequency modulated oscillator display, are shown in Figs. 11.41–11.43, for actual commercial receivers employing ratio detector and phase coincidence demodulator systems, together with the comparable performance of an early experimental PLL receiver due to the author.

All of these units had been in use for some time, and showed the effects of the misalignment which could be expected to occur with the passage of time. When new, both the conventional FM tuners would have probably given a better performance than at the time of the subsequent test. However, the superiority of the PLL system is evident.

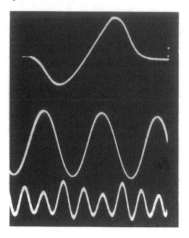

Figure 11.41   Practical demodulator frequency/voltage transfer characteristics, recovered signal, and overall receiver distortion waveform, (THD = 1.5 per cent , mainly third harmonic), for ratio detector.

Commercial manufacturers of domestic style equipment have been slow to exploit the qualities of the PLL, perhaps deterred by the unpleasant audio signal generated when the tuner is on the edge of lock. It is not, though, a difficult matter to incorporate a muting circuit which cuts off the signal when the receiver is off tune, and an experimental system of this kind has been in use for many years.

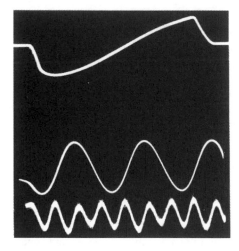

Figure 11.42   Practical demodulator frequency/voltage transfer characteristics, recovered signal, and overall receiver distortion waveform, (THD = 0.6 per cent , mainly third harmonic), for phase coincidence FM demodulator.

Pulse Counting Systems

In the early years of FM transmissions, when there was great interest in the exploitation of the very high quality signals then available, pulse counting systems were commonly employed as the basis for high fidelity amateur designs. Typical circuit arrangements employed were of the form shown in Fig. 11.44. After suitable RF amplification, the incoming signal would be mixed with a crystal controlled local oscillator signal, in a conventional superhet layout, to give a relatively low frequency IF in, say, the range 30–180 kHz, which could be amplified by a conventional broad bandwidth HF amplifier.

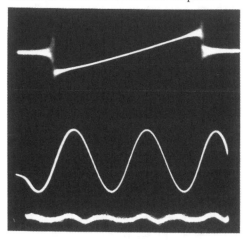

Figure 11.43   Practical demodulator frequency/voltage transfer characteristics, recovered signal, and overall receiver distortion waveform, (THD = 0.15 per cent , mainly second harmonic), for phase-locked loop coincidence FM demodulator.

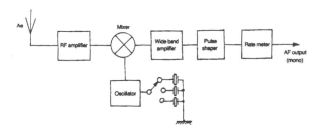

Figure 11.44   Pulse counting FM tuner.

The output signal would then be analysed by a linear rate meter circuit, to give a DC output level which was dependent on the instantaneous input signal frequency, yielding a low distortion recovered AF output.

Such systems shared the quality of the PLL demodulator that the output signal linearity was largely unaffected by the frequency/phase linearity of the preceding RF/IF circuitry. Unfortunately, they did not lend themselves well to the demodulation of stereo signals, and the method fell into disuse.

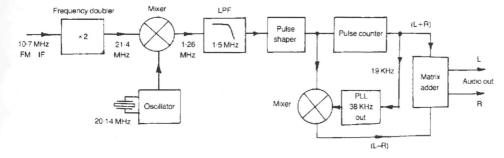

Figure 11.45 The Pioneer pulse counting FM tuner system.

However, this technique has been resurrected by Pioneer, in its F-90/F-99 receivers, in a manner which exploits the capabilities of modern digital ICs.

The circuit layout employed by Pioneer is shown in schematic form in Fig. 11.45. In this the incoming 10.7 MHz signal is frequency doubled, to double the modulation width, and mixed down to 1.26 MHz with a stable crystal controlled oscillator. After filtering, the signal is cleaned up and converted into a series of short duration pulses, having constant width and amplitude, of which the average value is the desired composite (L+R) audio signal.

A conventional PLL arrangement of the type shown in Fig. 11.27, is then used to reconstruct the 38 kHz sub- carrier signal, from which the LH and RH stereo outputs can be derived by adding the (L + R) + (L – R) and (L + R) + (R – L) components.

Practical FM receiver performance data

The claimed performance from the Pioneer F-90 receiver, using this pulse counting system, is that the THD is better than 0.01 per cent (mono) and 0.02 per cent (stereo), with a capture ratio of better than 1 dB, and a stereo channel separation greater than 60 dB. These figures are substantially better than any obtainable with more conventional demodulator systems.

In general, the expected THD performance for a good modern FM receiver is 0.1 per cent (mono) and 0.2–0.3 per cent (stereo), with capture ratios in the range 1 dB (excellent) to 2.5 dB (adequate), and stereo separations in the range 30–50 dB. These figures largely depend on the type of demodulator employed, and the quality of the RF and IF circuit components and alignment.

Typical AF bandwidths will be in the range 20–40 Hz to 14.5 KHz (–3 dB points). The LF frequency response depends on the demodulator type, with PLL and pulse counting systems allowing better LF extension. The upper frequency limit is set by the requirements of the

Zenith-GE encoding system, rather than by the receiver in use.

The ultimate signal to noise ratio of a good receiver could well be of the order of 70 dB for a mono signal, but normal reception conditions will reduce this figure somewhat, to 60 dB or so.

Input (aerial) signal strengths greatly influence the final noise figure, which worsens at low signal levels, so that input receiver sensitivities of the order of 2.5 µV and 25 µV might be expected for a 50 dB final S/N figure, on mono and stereo signals respectively, from a first quality receiver. Values of 25 µV/100 µV might be expected from less good designs.

A high (aerial) input sensitivity and a good intrinsic receiver S/N ratio is of value if the receiver is to be used with poor or badly sited aerial systems, even though the difference between receivers of widely different performance in this respect may not be noticeable with a better aerial installation.

Linearity, AM systems

The performance of an AM radio is invariably less good than that possible from a comparable quality FM receiver. The reasons for this are partly to do with the greater difficulty in obtaining a low demodulator distortion level and a wide AF bandwidth, with a sensibly flat frequency response, coupled with good signal to noise and selectivity figures – a difficulty which is inherent in the AM system – and partly to do with the quality of the radio signal, which is often poor, both as received and as transmitted.

There are, however, differences between the various types of AM demodulator, and some of these have significant performance advantages over those normally used in practice. These are examined below.

The Diode 'Envelope' Demodulator
This is a direct descendant of the old crystal detector of the 'crystal and cats-whisker' era, and is shown in

Fig. 11.46. In this the peak voltage level occurring across the tuned circuit L_2C_1 is passed through the rectifier diode, D_1, and stored in the output filter circuit C_3R_3. Since the rectifying diode will require some forward voltage to make it conduct (about 0.15 V in the case of germanium, and 0.55 V in the case of silicon types) it is good practice to apply a forward bias voltage, through $R_1R_2C_2$, to bring the diode to the threshold of forward conduction.

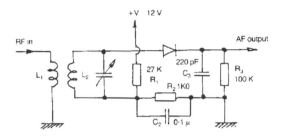

Figure 11.46   Simple forward-biased diode AM detector.

Because it is essential that the RF component is removed from the output signal, there is a minimum practical value for C_3. However, since this holds the peak audio modulation voltage until it can decay through R_3, a measure of waveform distortion is inevitable, especially at higher end of the audio range and at lower RF input signal frequencies.

Typical performance figures for such demodulators are 1–2 per cent at 1 kHz.

'Grid-Leak' Demodulation

This was a common system used in the days of amateur radio receivers, and shown in its thermionic valve form in Fig. 11.47(a). A more modern version of this arrangement, using a junction FET, is shown in Fig. 11.47(b). Both these circuits operate by allowing the RF signal appearing across the tuned circuit L_2C_1 to develop a reverse bias across R_1, which reduces the anode or drain currents.

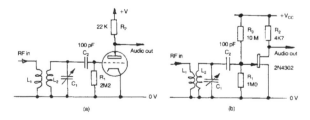

Figure 11.47   Valve and FET grid-leak AM detectors.

The THD performance of such circuits is similar to that of the forward-biased diode demodulator, but they have a lower damping effect on the preceding tuned circuits.

Anode-Bend or Infinite Impedance Demodulator Systems

These are shown in their junction FET versions, in Fig. 11.48(a) and 11.48(b). They are similar in their action and only differ in the position of the output load. They operate by taking advantage of the inevitable curvature of the I_d/V_g curve for a valve or FET to provide a distorted version of the incoming radio signal, as shown in Fig. 11.49, which, when averaged by some integrating circuit, gives an audio signal equivalent to the modulation.

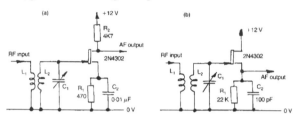

Figure 11.48   FET based anode-bend and infinite impedance detectors.

Since the circuit distorts the incoming RF waveform, it is unavoidable that there will be a related distortion in the recovered AF signal too. Typical THD figures for such a demodulator system will lie in the range 0.5–1.5 per cent, depending on signal level. Most IC AM radio systems will employ either diode envelope detectors or variations of the anode bend arrangement based on semiconductor devices.

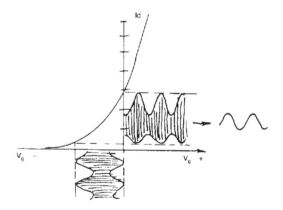

Figure 11.49   Method of operation of anode-bend or infinite impedance detectors.

The performance of all these systems can be improved by the addition of an unmodulated carrier, either at the same frequency, or at one substantially different from it, to improve the carrier to modulation ratio. This is not a technique which is used for equipment destined for the domestic market, in spite of quality improvement possible.

Synchronous Demodulation.

This method, illustrated schematically in Fig. 11.50(a), operates by synchronously inverting the phase of alternate halves of the RF signal, so that the two halves of the modulated carrier can be added together to give a unidirectional output. Then, when the RF component is removed by filtering, only the audio modulation waveform remains, as shown in Fig. 11.50(b).

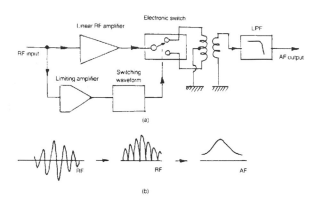

Figure 11.50   Circuit layout and method of operation of Homodyne type synchronous demodulator.

This technique is widely used in professional communication systems, and is capable of THD values well below 0.1 per cent.

Circuit Design

Although there is a continuing interest in circuit and performance improvements made possible by new circuit and device technology, there is also a tendency towards a degree of uniformity in circuit design, as certain approaches establish themselves as the best, or the most cost-effective.

This is particularly true in the RF and IF stages of modern FM tuners, and an illustration of the type of design approach employed is given in Fig. 11.51. The only feature in this not covered by the previous discussion is the general use of dual (back-to-back) Varicap diodes. These are preferred to single diodes as the tuning method in high-quality systems because they avoid the distortion of the RF waveform brought about by the signal voltage modulating the diode capacitance.

Variable air-spaced ganged capacitors would be better still, but these are bulky and expensive, and do not lend themselves to frequency synthesizer layouts.

New Developments

In such a fast changing market it is difficult to single out recent or projected design features for special mention, but certain trends are apparent. These mainly concern the use of microprocessor technology to memorise and store user selections of channel frequencies, as in the Quad FM4, (Fig. 11.52) and the use of 'sliding stereo separation'

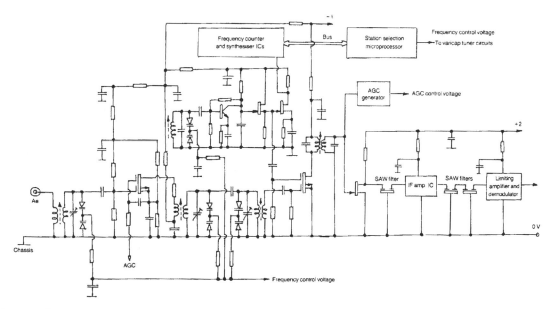

Figure 11.51   Typical circuit layout of RF and IF stages of good quality contemporary FM tuner.

systems to lessen the L–R channel separation, with its associated 'stereo hiss', when the incoming signal strength falls below the optimum value.

Figure 11.52 The Quad FM tuner.

Some synthesizer tuners offer normal spin-wheel tuning, as in the Harmon-Kardon TU915, by coupling the tuning knob shaft to the synthesizer IC by the use of an optical shaft encoder.

Low-noise gallium arsenide dual-gate Mosfets have made an appearance in the Hitachi FT5500 receiver, and these devices are likely to be more widely adopted in such systems.

Clearly, commercial pressures will encourage manufacturers to develop more complex large-scale integration (LSI) integrated circuits, so that more of the receiver circuitry may be held on a single chip. This undoubtedly saves manufacturing costs, but the results are not always of benefit to the user, as evidenced by the current performance of single IC AM radio sections.

An increasing number of the better FM tuners are now offering a choice of IF bandwidths, to permit user optimisation of selectivity, sensitivity or stereo separation. Variable receiver bandwidth would be a valuable feature on the MF bands, and may be offered if there is any serious attempt to improve the quality of this type of receiver.

Appendix 11.1   BBC Transmitted MF and VHF Signal Parameters

(Data by courtesy of the BBC).

Audio bandwidths

MF:
- 40–5800 Hz ±3 dB, with very sharp cut off. (> 24 dB/octave beyond 6 kHz)
- 50–5000 Hz ±1 dB.

VHF:
- 30–1500 Hz ±0.5 dB, with very sharp cut off beyond this frequency.

Distortion

MF:
- < 3% THD at 75% modulation.
- < 4% THD at 100% modulation.

VHF:
- < 0.5% THD.

Stereo crosstalk

VHF: > 46 dB. (0.5%).

Modulation depth

MF: Up to 100% over the range 100–5000 Hz.
VHF: Peak deviation level corresponds to ±60.75 kHz deviation.
(The total deviation, including pilot tone, is ±75 kHz.)

S/N ratio

MF: > 54 dB below 100% modulation.
VHF: up to 64 dB, using CCIR/468 weighting, with reference to peak programme modulation level.

Appendix 11.2.   The 57 KHZ Sub-Carrier Radio Data System (RDS)

(Data by courtesy of the BBC).

It is proposed to begin the introduction of this system in the Autumn of 1987, and, when in use on receivers adapted to receive this signal, will allow the reception of station, programme and other data.

This data will include programme identification, to allow the receiver to automatically locate and select the best available signal carrying the chosen programme, and to display in alphanumeric form a suitable legend, and to permit display of clock time and date.

Anticipated future developments of this system include a facility for automatic selection of programme type (speech/light music/serious music/news), and for the visual display of information, such as traffic conditions, phone-in numbers, news flashes, programme titles or contents, and data print out via a computer link.

12 Pre-amps and Inputs

John Linsley Hood

The production of an audio signal from disc, tape or tuner is the start of a process of amplification and signal shaping that will occupy the following three sections. In this Chapter, John Linsley Hood explains the problems and the solutions as applied to the Pre-amp.

Requirements

Most high-quality audio systems are required to operate from a variety of signal inputs, including radio tuners, cassette or reel-to-reel tape recorders, compact disc players and more traditional record player systems. It is unlikely at the present time that there will be much agreement between the suppliers of these ancillary units on the standards of output impedance or signal voltage which their equipment should offer.

Except where a manufacturer has assembled a group of such units, for which the interconnections are custom designed and there is in-house agreement on signal and impedance levels – and, sadly, such ready-made groupings of units seldom offer the highest overall sound quality available at any given time – both the designer and the user of the power amplifier are confronted with the need to ensure that his system is capable of working satisfactorily from all of these likely inputs.

For this reason, it is conventional practice to interpose a versatile pre-amplifier unit between the power amplifier and the external signal sources, to perform the input signal switching and signal level adjustment functions.

This pre-amplifier either forms an integral part of the main power amplifier unit or, as is more common with the higher quality units, is a free-standing, separately powered, unit.

Signal Voltage and Impedance Levels

Many different conventions exist for the output impedances and signal levels given by ancillary units. For tuners and cassette recorders, the output is either be that of the German DIN (Deutsches Industrie Normal) standard, in which the unit is designed as a current source which will give an output voltage of 1 mV for each 1000 ohms of load impedance, such that a unit with a 100 K input impedance would see an input signal voltage of 100 mV, or the line output standard, designed to drive a load of 600 ohms or greater, at a mean signal level of 0.775 V RMS, often referred to in tape recorder terminology as OVU.

Generally, but not invariably, units having DIN type interconnections, of the styles shown in Fig. 12.1, will conform to the DIN signal and impedance level convention, while those having 'phono' plug/socket outputs, of the form shown in Fig. 12.2 will not. In this case, the permissible minimum load impedance will be within the range 600 ohms to 10000 ohms, and the mean output signal level will commonly be within the range 0.25–1 V RMS.

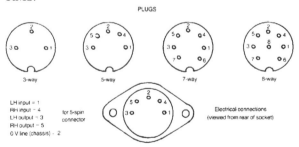

Figure 12.1 Common DIN connector configurations.

An exception to this exists in respect of compact disc players, where the output level is most commonly 2 V RMS.

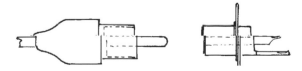

Figure 12.2 The phono connector.

Gramophone Pick-up Inputs

Three broad categories of pick-up cartridge exist: the ceramic, the moving magnet or variable reluctance, and the moving coil. Each of these has different output characteristics and load requirements.

Ceramic piezo-electric cartridges

These units operate by causing the movement of the stylus due to the groove modulation to flex a resiliently mounted strip of piezo-electric ceramic, which then causes an electrical voltage to be developed across metallic contacts bonded to the surface of the strip. They are commonly found only on low-cost units, and have a relatively high output signal level, in the range 100–200 mV at 1 kHz.

Generally the electromechanical characteristics of these cartridges are tailored so that they give a fairly flat frequency response, though with some unavoidable loss of HF response beyond 2 kHz, when fed into a pre-amplifier input load of 47 000 ohms.

Neither the HF response nor the tracking characteristics of ceramic cartridges are particularly good, though circuitry has been designed with the specific aim of optimising the performance obtainable from these units, (see Linsley Hood, J., *Wireless World*, July 1969). However, in recent years, the continuing development of pick-up cartridges has resulted in a substantial fall in the price of the less exotic moving magnet or variable reluctance types, so that it no longer makes economic sense to use ceramic cartridges, except where their low cost and robust nature are of importance.

Moving magnet and variable reluctance cartridges

These are substantially identical in their performance characteristics, and are designed to operate into a 47 K load impedance, in parallel with some 200–500 pF of anticipated lead capacitance. Since it is probable that the actual capacitance of the connecting leads will only be of the order of 50–100 pF, some additional input capacitance, connected across the phono input socket, is customary. This also will help reduce the probability of unwanted radio signal breakthrough.

PU cartridges of this type will give an output voltage which increases with frequency in the manner shown in Fig. 12.3(a), following the velocity characteristics to which LP records are produced, in conformity with the RIAA recording standards. The pre-amplifier will then be required to have a gain/frequency characteristic of the form shown in Fig. 12.3(b), with the de-emphasis time constants of 3180, 318 and 75 microseconds, as indicated on the drawing.

The output levels produced by such pick-up cartridges will be of the order of 0.8–2 mV/cm/s of groove modulation velocity, giving typical mean outputs in the range of 3–10 mV at 1 kHz.

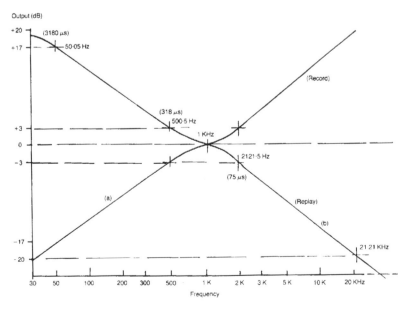

Figure 12.3   The RIAA record/replay characteristics used for 33/45 rpm vinyl discs.

Moving coil pick-up cartridges

These low-impedance, low-output PU cartridges have been manufactured and used without particular comment for very many years. They have come into considerable prominence in the past decade, because of their superior transient characteristics and dynamic range, as the choice of those audiophiles who seek the ultimate in sound quality, even though their tracking characteristics are often less good than is normal for MM and variable reluctance types.

Typical signal output levels from these cartridges will be in the range 0.02–0.2 mV/cm/s, into a 50–75 ohm load impedance. Normally a very low-noise head amplifier circuit will be required to increase this signal voltage to a level acceptable at the input of the RIAA equalisation circuitry, though some of the high output types will be capable of operating directly into the high-level RIAA input. Such cartridges will generally be designed to operate with a 47 K load impedance.

Input Circuitry

Most of the inputs to the pre-amplifier will merely require appropriate amplification and impedance transformation to match the signal and impedance levels of the source to those required at the input of the power amplifier. However, the necessary equalisation of the input frequency response from a moving magnet, moving coil or variable reluctance pick-up cartridge, when replaying an RIAA pre-emphasised vinyl disc, requires special frequency shaping networks.

Various circuit layouts have been employed in the pre-amplifier to generate the required 'RIAA' replay curve for velocity sensitive pick-up transducers, and these are shown in Fig. 12.4. Of these circuits, the two simplest are the 'passive' equalisation networks shown in (a) and (b), though for accuracy in frequency response they require that the source impedance is very low, and that the load impedance is very high in relation to R_1.

The required component values for these networks have been derived by Livy (Livy, W.H., *Wireless World*, Jan. 1957, p. 29.) in terms of *RC* time constants, and set out in a more easily applicable form by Baxandall (Baxandall, P.J., *Radio, TV and Audio Reference Book*, S. W. Amos [ed.], Newnes-Butterworth Ltd., Ch. 14), in his analysis of the various possible equalisation circuit arrangements.

From the equations quoted, the component values required for use in the circuits of Fig. 12.4(a) and (c), would be:

$$R_1/R_2 = 6.818 \quad C_1.R_1 = 2187\,\mu s \text{ and } C_2.R_2 = 109\,\mu s$$

For the circuit layouts shown in Fig. 12.4(b) and (d), the component values can be derived from the relationships:

$$R_1/R_2 = 12.38 \quad C_1.R_1 = 2937\,\mu s \text{ and } C_2.R_2 = 81.1\,\mu s$$

The circuit arrangements shown in Figs 12.4(c) and (d), use 'shunt' type negative feedback (i.e., that type in which the negative feedback signal is applied to the amplifier in parallel with the input signal) connected around an internal gain block.

These layouts do not suffer from the same limitations in respect of source or load as the simple passive equalisation systems of (a) and (b). However, they do have the practical snag that the value of R_{in} will be determined by the required p.u. input load resistor (usually 47k. for a typical moving magnet or variable reluctance type of PU cartridge), and this sets an input 'resistor noise' threshold which is higher than desirable, as well as requiring inconveniently high values for R_1 and R_2.

For these reasons, the circuit arrangements shown in Figs. 12.4(e) and (f), are much more commonly found in commercial audio circuitry. In these layouts, the frequency response shaping components are contained within a 'series' type feedback network (i.e., one in which the negative feedback signal is connected to the amplifier in series with the input signal), which means that the input circuit impedance seen by the amplifier is essentially that of the pick-up coil alone, and allows a lower mid-range 'thermal noise' background level.

The snag, in this case, is that at very high frequencies, where the impedance of the frequency-shaping feedback network is small in relation to R_{fb}, the circuit gain approaches unity, whereas both the RIAA specification and the accurate reproduction of transient waveforms require that the gain should asymptote to zero at higher audio frequencies.

This error in the shape of the upper half of the response curve can be remedied by the addition of a further *CR* network, C_3/R_3, on the output of the equalisation circuit, as shown in Figs. 12.4(e) and (f). This amendment is sometimes found in the circuit designs used by the more perfectionist of the audio amplifier manufacturers.

Other approaches to the problem of combining low input noise levels with accurate replay equalisation are to divide the equalisation circuit into two parts, in which the first part, which can be based on a low noise series feedback layout, is only required to shape the 20 Hz–1 kHz section of the response curve. This can then be followed by either a simple passive *RC* roll-off network, as shown in Fig. 12.4(g), or by some other circuit arrangement having a similar effect – such as that based on the use of shunt feedback connected around an inverting

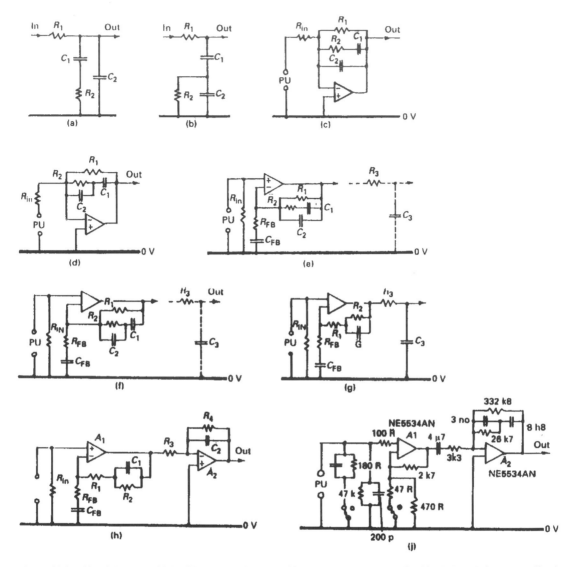

Figure 12.4   Circuit layouts which will generate the type of frequency response required for R.I.A.A. input equalization.

amplifier stage, as shown in Fig. 12.4(h) – to generate that part of the response curve lying between 1 kHz and 20 kHz.

A further arrangement, which has attracted the interest of some Japanese circuit designers – as used, for example, in the Rotel RC-870BX preamp., of which the RIAA equalising circuit is shown in a simplified form in Fig. 12.4(j) – simply employs one of the recently developed very low noise IC op. amps as a flat frequency response input buffer stage. This is used to amplify the input signal to a level at which circuit noise introduced by succeeding stages will only be a minor problem, and also to convert the PU input impedance level to a value at which a straightforward shunt feedback equalising circuit can be used, with resistor values chosen to minimise

any thermal noise background, rather than dictated by the PU load requirements.

The use of 'application specific' audio ICs, to reduce the cost and component count of RIAA stages and other circuit functions, has become much less popular among the designers of higher quality audio equipment because of the tendency of the semiconductor manufacturers to discontinue the supply of such specialised ICs when the economic basis of their sales becomes unsatisfactory, or to replace these devices by other, notionally equivalent, ICs which are not necessarily either pin or circuit function compatible.

There is now, however, a degree of unanimity among the suppliers of ICs as to the pin layout and operating conditions of the single and dual op. amp. designs, comnonly packaged in 8-pin dual-in-line forms. These are typified by the Texas Instruments TL071 and TL072 ICs, or their more recent equivalents, such as the TL051 and TL052 devices – so there is a growing tendency for circuit designers to base their circuits on the use of ICs of this type, and it is assumed that devices of this kind would be used in the circuits shown in Fig. 12.4.

An incidental advantage of the choice of this style of IC is that commercial rivalry between semiconductor manufacturers leads to continuous improvements in the specification of these devices. Since these nearly always offer plug-in physical and electrical interchangeability, the performance of existing equipment can easily be upgraded, either on the production line or by the service department, by the replacement of existing op. amp. ICs with those of a more recent vintage, which is an advantage to both manufacturer and user.

Moving Coil PU Head Amplifier Design

The design of pre-amplifier input circuitry which will accept the very low signal levels associated with moving coil pick-ups presents special problems in attaining an adequately high signal-to-noise ratio, in respect of the microvolt level input signals, and in minimising the intrusion of mains hum or unwanted RF signals.

The problem of circuit noise is lessened somewhat in respect of such RIAA equalised amplifier stages in that, because of the shape of the frequency response curve, the effective bandwidth of the amplifier is only about 800 Hz. The thermal noise due to the amplifier input impedance, which is defined by the equation below, is proportional to the squared measurement bandwidth, other things being equal, so the noise due to such a stage is less than would have been the case for a flat frequency response system, nevertheless, the attainment of an adequate S/N ratio, which should be at least 60 dB, demands that the input circuit impedance should not exceed some 50 ohms.

$$\overline{V} = \sqrt{4KT\,\delta\,FR}$$

where δF is the bandwidth, T is the absolute temperature, (room temperature being approx. 300°K), R is resistance in ohms and K is Boltzmann's constant (1.38×10^{-23}).

The moving coil pick-up cartridges themselves will normally have winding resistances which are only of the order of 5–25 ohms, except in the case of the high output units where the problem is less acute anyway, so the problem relates almost exclusively to the circuit impedance of the MC input circuitry and the semiconductor devices used in it.

Circuit Arrangements

Five different approaches are in common use for moving coil PU input amplification.

Step-up transformer

This was the earliest method to be explored, and was advocated by Ortofon, which was one of the pioneering companies in the manufacture of MC PU designs. The advantage of this system is that it is substantially noiseless, in the sense that the only source of wide-band noise will be the circuit impedance of the transformer windings, and that the output voltage can be high enough to minimise the thermal noise contribution from succeeding stages.

The principal disadvantages with transformer step-up systems, when these are operated at very low signal levels, are their proneness to mains 'hum' pick up, even when well shrouded, and their somewhat less good handling of 'transients', because of the effects of stray capacitances and leakage inductance. Care in their design is also needed to overcome the magnetic non-linearities associated with the core, which will be particularly significant at low signal levels.

Systems using paralleled input transistors

The need for a very low input circuit impedance to minimise thermal noise effects has been met in a number of commercial designs by simply connecting a number of small signal transistors in parallel to reduce their effective base-emitter circuit resistance. Designs of this type came from Ortofon, Linn/Naim, and Braithwaite, and are shown in Fig. 12.5–12.7.

If such small signal transistors are used without selection and matching – a time-consuming and expensive process for any commercial manufacturer – some means must be adopted to minimise the effects of the variation in base-emitter turn-on voltage which will exist between nominally identical devices, due to variations in doping level in the silicon crystal slice, or to other differences in manufacture.

In the Ortofon circuit this is achieved by individual collector-base bias current networks, for which the penalty is the loss of some usable signal in the collector circuit. In the Linn/Naim and Braithwaite designs, this evening out of transistor characteristics in circuits having common base connections is achieved by the use of individual emitter resistors to swamp such differences in device characteristics. In this case, the penalty is that

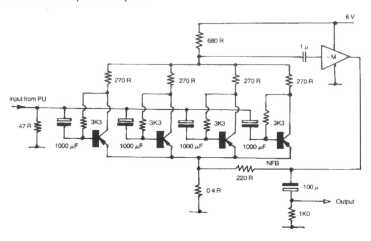

Figure 12.5 Ortofon MCA-76 head amplifier.

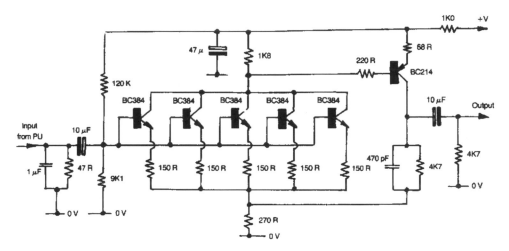

Figure 12.6 The Naim NAC 20 moving coil head amplifier.

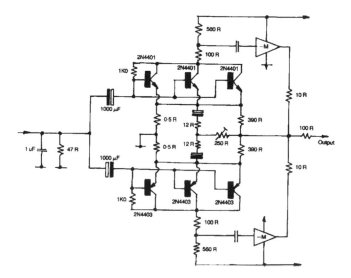

Figure 12.7 Braithwaite RAI4 head amplifier. (Output stage shown in a simplified form.)

such resistors add to the base-emitter circuit impedance, when the task of the design is to reduce this.

Monolithic super-matched input devices

An alternative method of reducing the input circuit impedance, without the need for separate bias systems or emitter circuit swamping resistors, is to employ a monolithic (integrated circuit type) device in which a multiplicity of transistors have been simultaneously formed on the same silicon chip. Since these can be assumed to have virtually identical characteristics they can be paralleled, at the time of manufacture, to give a very low impedance, low noise, matched pair.

An example of this approach is the National Semiconductors LM194/394 super-match pair, for which a suitable circuit is shown in Fig. 12.8. This input device probably offers the best input noise performance currently available, but is relatively expensive.

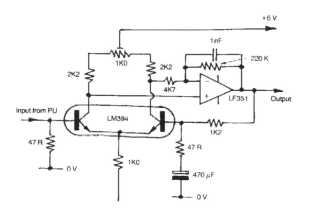

Figure 12.8 Head amplifier using LM394 multiple transistor array.

Small power transistors as input devices

The base-emitter impedance of a transistor depends largely on the size of the junction area on the silicon chip. This will be larger in power transistors than in small signal transistors, which mainly employ relatively small chip sizes. Unfortunately, the current gain of power transistors tends to decrease at low collector current levels, and this would make them unsuitable for this application.

However, the use of the plastic encapsulated medium power (3–4A Ic max.) styles, in T0126, T0127 and T0220 packages, at collector currents in the range 1–3 mA,

achieves a satisfactory compromise between input circuit impedance and transistor performance, and allows the design of very linear low-noise circuitry. Two examples of MC head amplifier designs of this type, by the author, are shown in Figs. 12.9 and 12.10.

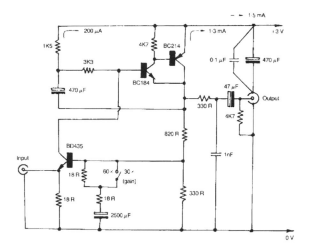

Figure 12.9 Cascade input moving coil head amplifier.

The penalty in this case is that, because such transistor types are not specified for low noise operation, some preliminary selection of the devices is desirable, although, in the writer's experience, the bulk of the devices of the types shown will be found to be satisfactory in this respect.

In the circuit shown in Fig. 12.9, the input device is used in the common base (cascade) configuration, so that the input current generated by the pick-up cartridge is transferred directly to the higher impedance point at the collector of this transistor, so that the stage gain, prior to the application of negative feedback to the input transistor base, is simply the impedance transformation due to the input device.

In the circuit of Fig. 12.10, the input transistors are used in a more conventional common-emitter mode, but the two input devices, though in a push-pull configuration, are effectively connected in parallel so far as the input impedance and noise figure are concerned. The very high degree of symmetry of this circuit assists in minimising both harmonic and transient distortions.

Both of these circuits are designed to operate from 3 V DC 'pen cell' battery supplies to avoid the introduction of mains hum due to the power supply circuitry or to earth loop effects. In mains-powered head amps. great care is always necessary to avoid supply line signal or noise intrusions, in view of the very low signal levels at both the inputs and the outputs of the amplifier stage.

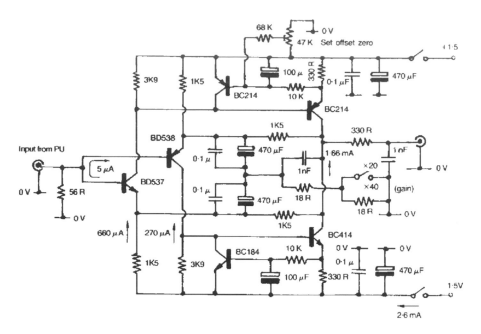

Figure 12.10   Very low-noise, low-distortion, symmetrical MC head amplifier.

It is also particularly advisable to design such amplifiers with single point '0 V' line and supply line connections, and these should be coupled by a suitable combination of good quality decoupling capacitors.

Very low noise IC op amps

The development, some years ago, of very low noise IC operational amplifiers, such as the Precision Monolithics OP-27 and OP-37 devices, has led to the proliferation of very high quality, low-noise, low-distortion ICs aimed specifically at the audio market, such as the Signetics NE-5532/5534, the NS LM833, the PMI SSM2134/2139, and the TI TL051/052 devices.

With ICs of this type, it is a simple matter to design a conventional RIAA input stage in which the provision of a high sensitivity, low noise, moving coil PU input is accomplished by simply reducing the value of the input load resistor and increasing the gain of the RIAA stage in comparison with that needed for higher output PU types. An example of a typical Japanese design of this type is shown in Fig. 12.11.

Other approaches

A very ingenious, fully symmetrical circuit arrangement which allows the use of normal circuit layouts and components in ultra-low noise (e.g., moving coil p.u. and similar signal level) inputs, has been introduced by 'Quad'

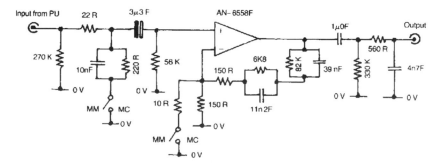

Figure 12.11   Moving coil/moving magnet RIAA input stage in Technics SU-V10 amplifier.

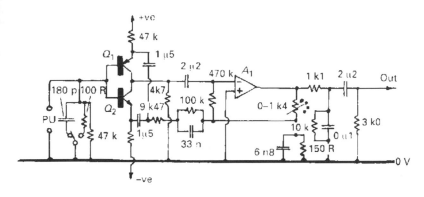

Figure 12.12 The 'Quad' ultra-low noise input circuit layout.

(Quad Electroacoustics Ltd.) and is employed in all their current series of preamps. This exploits the fact that, at low input signal levels, bipolar junction transistors will operate quite satisfactorily with their base and collector junctions at the same DC potential, and permits the type of input circuit shown in Fig. 12.12.

In the particular circuit shown, that used in the 'Quad 44' disc input, a two-stage equalisation layout is employed, using the type of structure illustrated in Fig. 12.4(g), with the gain of the second stage amplifier (a TL071 IC op. amp.) switchable to suit the type of input signal level available.

Input Connections

For all low-level input signals care must be taken to ensure that the connections are of low contact resistance. This is obviously an important matter in the case of low-impedance circuits such as those associated with MC pick-up inputs, but is also important in higher impedance circuitry since the resistance characteristics of poor contacts are likely to be non-linear, and to introduce both noise and distortion.

In the better class modern units the input connectors will invariably be of the 'phono' type, and both the plugs and the connecting sockets will be gold plated to reduce the problem of poor connections due to contamination or tarnishing of the metallic contacts.

The use of separate connectors for L and R channels also lessens the problem of inter-channel breakthrough, due to capacitive coupling or leakage across the socket surface, a problem which can arise in the five- and seven-pin DIN connectors if they are carelessly fitted, and particularly when both inputs and outputs are taken to that same DIN connector.

Input Switching

The comments made about input connections are equally true for the necessary switching of the input signal sources. Separate, but mechanically interlinked, switches of the push-on, push-off, type are to be preferred to the ubiquitous rotary wafer switch, in that it is much easier, with separate switching elements, to obtain the required degree of isolation between inputs and channels than would be the case when the wiring is crowded around the switch wafer.

However, even with separate push switches, the problem remains that the input connections will invariably be made to the rear of the amplifier/preamplifier unit, whereas the switching function will be operated from the front panel, so that the internal connecting leads must traverse the whole width of the unit.

Other switching systems, based on relays, or bipolar or field effect transistors, have been introduced to lessen the unwanted signal intrusions which may arise on a lengthy connecting lead. The operation of a relay, which will behave simply as a remote switch when its coil is energised by a suitable DC supply, is straightforward, though for optimum performance it should either be hermetically sealed or have noble metal contacts to resist corrosion.

Transistor switching

Typical bipolar and FET input switching arrangements are shown in Figs. 12.13 and 12.14. In the case of the bipolar transistor switch circuit of Fig. 12.13, the non-linearity of the junction device when conducting precludes its use in the signal line, the circuit is therefore arranged so that the transistor is non-conducting when the signal is passing through the controlled signal channel, but acts as a

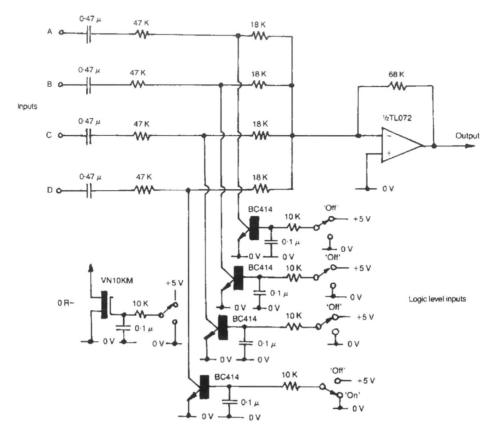

Figure 12.13   Bipolar transistor operated shunt switching. (Also suitable for small-power MOSFET devices.)

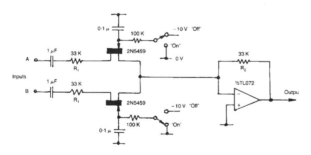

Figure 12.14   Junction FET input switching circuit.

short-circuit to shunt the signal path to the O V line when it is caused to conduct.

In the case of the FET switch, if R_1 and R_2 are high enough, the non-linearity of the conducting resistance of the FET channel will be swamped, and the harmonic and other distortions introduced by this device will be negligible. (Typically less than 0.02 per cent at 1 V RMS and 1 kHz.)

The CMOS bilateral switches of the CD4066 type are somewhat non-linear, and have a relatively high level of breakthrough. For these reasons they are generally thought to be unsuitable for high quality audio equipment, where such remote switching is employed to minimise cross-talk and hum pick-up.

However, such switching devices could well offer advantages in lower quality equipment where the cost savings is being able to locate the switching element on the printed circuit board, at the point where it was required, might offset the device cost.

Diode switching

Diode switching of the form shown in Fig. 12.15, while very commonly employed in RF circuitry, is unsuitable for audio use because of the large shifts in DC level between the 'on' and 'off' conditions, and this would produce intolerable 'bangs' on operation.

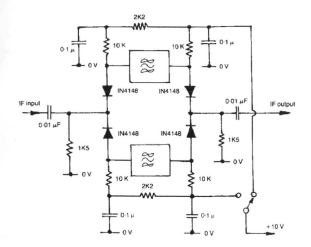

Figure 12.15 Typical diode switching circuit, as used in RF applications.

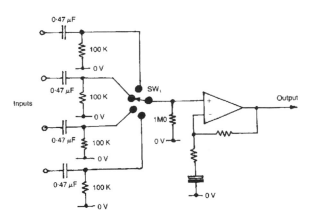

Figure 12.16 Use of DC blocking capacitors to minimise input switching noises.

For all switching, quietness of operation is an essential requirement, and this demands that care shall be taken to ensure that all of the switched inputs are at the same DC potential, preferably that of the 0 V line. For this reason, it is customary to introduce DC blocking capacitors on all input lines, as shown in Fig. 12.16, and the time constants of the input RC networks should be chosen so that there is no unwanted loss of low frequency signals due to this cause.

13 Voltage Amplifiers and Controls

John Linsley Hood

Voltage gain is the fundamental feature of an amplifier for audio, and nowadays we tend to take this very much for granted. Continuing the path of the audio signal, John Linsley Hood deals here with voltage amplification, including the controls that alter the signal gain selectively.

Preamplifier Stages

The popular concept of hi-fi attributes the major role in final sound quality to the audio power amplifier and the output devices or output configuration which it uses. Yet in reality the pre-amplifier system, used with the power amplifier, has at least as large an influence on the final sound quality as the power amplifier, and the design of the voltage gain stages within the pre- and power amplifiers is just as important as that of the power output stages. Moreover, it is developments in the design of such voltage amplifier stages which have allowed the continuing improvement in amplifier performance.

The developments in solid-state linear circuit technology which have occurred over the past 25 years seem to have been inspired in about equal measure by the needs of linear integrated circuits, and by the demands of high-quality audio systems, and engineers working in both of these fields have watched each other's progress and borrowed from each other's designs.

In general, the requirements for voltage gain stages in both audio amplifiers and integrated-circuit operational amplifiers are very similar. These are that they should be linear, which implies that they are free from waveform distortion over the required output signal range, have as high a stage gain as is practicable, have a wide AC bandwidth and a low noise level, and are capable of an adequate output voltage swing.

The performance improvements which have been made over this period have been due in part to the availability of new or improved types of semiconductor device, and in part to a growing understanding of the techniques for the circuit optimisation of device

performance. It is the interrelation of these aspects of circuit design which is considered below.

Linearity

Bipolar transistors

In the case of a normal bipolar (NPN or PNP) silicon junction transistor, for which the chip cross-section and circuit symbol is shown in Fig. 13.1, the major problem in obtaining good linearity lies in the nature of the base voltage/collector current transfer characteristic, shown in the case of a typical 'NPN' device (a 'PNP' device would have a very similar characteristic, but with negative voltages and currents) in Fig. 13.2.

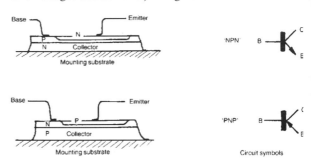

Figure 13.1 Typical chip cross-section of NPN and PNP silicon planar epitaxial transistors.

In this, it can be seen that the input/output transfer characteristic is strongly curved in the region 'X – Y' and an input signal applied to the base of such a device, which is forward biased to operate within this region, would suffer from the very prominent (second harmonic) waveform distortion shown in Fig. 13.3.

The way this type of non-linearity is influenced by the signal output level is shown in Fig. 13.4. It is normally found that the distortion increases as the output signal increases, and conversely.

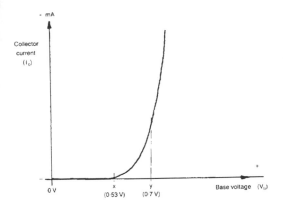

Figure 13.2 Typical transfer characteristic of silicon transistor.

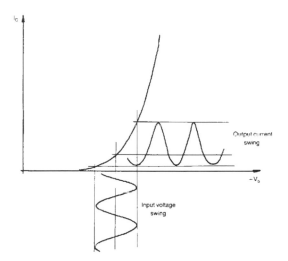

Figure 13.3 Transistor amplifier waveform distortion due to transfer characteristics.

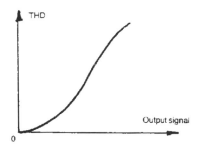

Figure 13.4 Relationship between signal distortion and output signal voltage in bipolar transistor amplifier.

There are two major improvements in the performance of such a bipolar amplifier stage which can be envisaged from these characteristics. Firstly, since the non-linearity is due to the curvature of the input charac-

teristics of the device – the output characteristics, shown in Fig. 13.5, are linear – the smaller the input signal which is applied to such a stage, the lower the non-linearity, so that a higher stage gain will lead to reduced signal distortion at the same output level. Secondly, the distortion due to such a stage is very largely second harmonic in nature.

This implies that a 'push-pull' arrangement, such as the so-called 'long-tailed pair' circuit shown in Fig. 13.6, which tends to cancel second harmonic distortion components, will greatly improve the distortion characteristics of such a stage.

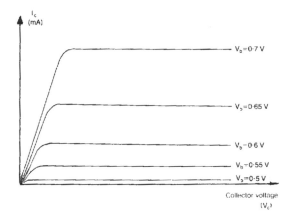

Figure 13.5 Output current/voltage characteristics of typical silicon bipolar transistor.

Also, since the output voltage swing for a given input signal (the stage gain) will increase as the collector load (R_2 in Fig. 13.6) increases, the higher the effective impedance of this, the lower the distortion which will be introduced by the stage, for any given output voltage signal.

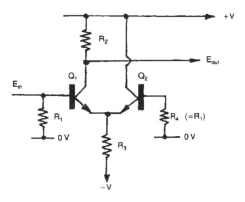

Figure 13.6 Transistor voltage amplifier using long-tailed pair circuit layout.

If a high value resistor is used as the collector load for Q_1 in Fig. 13.6, either a very high supply line voltage must be applied, which may exceed the voltage ratings of the devices or the collector current will be very small, which will reduce the gain of the device, and therefore tend to diminish the benefit arising from the use of a higher value load resistor.

Various circuit techniques have been evolved to circumvent this problem, by producing high dynamic impedance loads, which nevertheless permit the amplifying device to operate at an optimum value of collector current. These techniques will be discussed below.

An unavoidable problem associated with the use of high values of collector load impedance as a means of attaining high stage gains in such amplifier stages is that the effect of the 'stray' capacitances, shown as C_s in Fig. 13.7, is to cause the stage gain to decrease at high frequencies as the impedance of the stray capacitance decreases and progressively begins to shunt the load. This effect is shown in Fig. 13.8, in which the 'transition' frequency, f_o, (the –3 dB gain point) is that frequency at which the shunt impedance of the stray capacitance is equal to that of the load resistor, or its effective equivalent, if the circuit design is such that an 'active load' is used in its place.

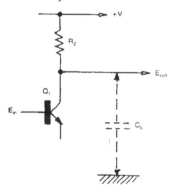

Figure 13.7   Circuit effect of stray capacitance.

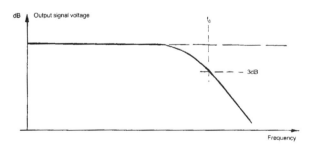

Figure 13.8   Influence of circuit stray capacitances on stage gain.

Field effect devices

Other devices which may be used as amplifying components are field effect transistors and MOS devices. Both of these components are very much more linear in their transfer characteristics but have a very much lower mutual conductance (G_m).

This is a measure of the rate of change of output current as a function of an applied change in input voltage. For all bipolar devices, this is strongly dependent on collector current, and is, for a small signal silicon transistor, typically of the order of 45 mA/V, per mA collector current. Power transistors, operating at relatively high collector currents, for which a similar relationship applies, may therefore offer mutual conductances in the range of amperes/volt.

Since the output impedance of an emitter follower is approximately $1/G_m$, power output transistors used in this configuration can offer very low values of output impedance, even without externally applied negative feedback.

All field effect devices have very much lower values for G_m, which will lie, for small-signal components, in the range 2–10 mA/V, not significantly affected by drain currents. This means that amplifier stages employing field effect transistors, though much more linear, offer much lower stage gains, other things being equal.

The transfer characteristics of junction (bipolar) FETs, and enhancement and depletion mode MOSFETS are shown in Fig. 13.9 (a), (b), and (c).

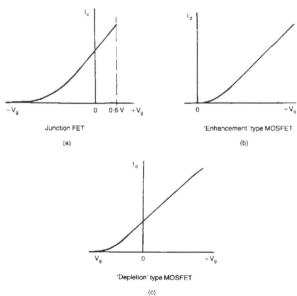

Figure 13.9   Gate voltage versus drain current characteristics of field effect devices.

Mosfets

MOSFETs, in which the gate electrode is isolated from the source/drain channel, have very similar transfer characteristics to that of junction FETs. They have an advantage that, since the gate is isolated from the drain/source channel by a layer of insulation, usually silicon oxide or nitride, there is no maximum forward gate voltage which can be applied – within the voltage breakdown limits of the insulating layer. In a junction FET the gate, which is simply a reverse biassed PN diode junction, will conduct if a forward voltage somewhat in excess of 0.6 V is applied.

The chip constructions and circuit symbols employed for small signal lateral MOSFETs and junction FETs (known simply as FETs) are shown in Figs. 13.10 and 13.11.

It is often found that the chip construction employed for junction FETs is symmetrical, so that the source and drain are interchangeable in use. For such devices the circuit symbol shown in Fig. 13.11 (c) should properly be used.

A practical problem with lateral devices, in which the current flow through the device is parallel to the surface of the chip, is that the path length from source to drain, and hence the device impedance and current carrying capacity, is limited by the practical problems of defining and etching separate regions which are in close proximity, during the manufacture of the device.

ones such as those employed in power output stages. It has led to the development of MOSFETs in which the current flow is substantially in a direction which is vertical to the surface, and in which the separation between layers is determined by diffusion processes rather than by photo-lithographic means.

Devices of this kind, known as V-MOS and T-MOS constructions, are shown in Fig. 13.12 (a) and (b).

Although these were originally introduced for power output stages, the electrical characteristics of such components are so good that these have been introduced, in smaller power versions, specifically for use in small signal linear amplifier stages. Their major advantages over bipolar devices, having equivalent chip sizes and dissipation ratings, are their high input impedance, their greater linearity, and their freedom from 'hole storage' effects if driven into saturation.

These qualities are increasingly attracting the attention of circuit designers working in the audio field, where there is a trend towards the design of amplifiers having a very high intrinsic linearity, rather than relying on the use of negative feedback to linearise an otherwise worse design.

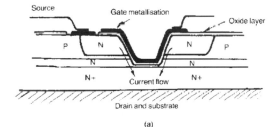

(a)

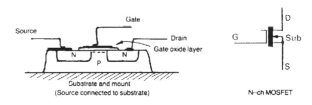

Figure 13.10 Chip cross-section and circuit symbol for lateral MOSFET (small signal type).

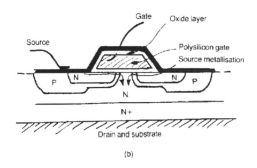

(b)

Figure 13.12 Power MOSFET constructions using (a) V and (b) T configurations. (Practical devices will employ many such cells in parallel.)

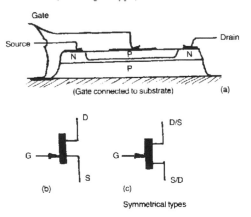

Figure 13.11 Chip cross-section and circuit symbols for (bipolar) junction FET.

V-Mos and T-Mos

This problem is not of very great importance for small signal devices, but it is a major concern in high current

Breakdown

A specific problem which arises in small signal MOSFET devices is that, because the gate-source capacitance is very small, it is possible to induce breakdown of the insulating layer, which destroys the device, as a result of transferred static electrical charges arising from mishandling.

Though widely publicised and the source of much apprehension, this problem is actually very rarely encountered in use, since small signal MOSFETs usually incorporate protective zener diodes to prevent this eventuality, and power MOSFETs, where such diodes may not be used because they may lead to inadvertent 'thyristor' action, have such a high gate-source capacitance that this problem does not normally arise.

In fact, when such power MOSFETs do fail, it is usually found to be due to circuit design defects, which have either allowed excessive operating potentials to be applied to the device, or have permitted inadvertent VHF oscillation, which has led to thermal failure.

Noise Levels

Improved manufacturing techniques have lessened the differences between the various types of semiconductor device, in respect of intrinsic noise level. For most practical purposes it can now be assumed that the characteristics of the device will be defined by the thermal noise figure of the circuit impedances. This relationship is shown in the graph of Fig. 13.13.

For very low noise systems, operating at circuit impedance levels which have been deliberately chosen to be as low as practicable – such as in moving coil pick-up head amplifiers – bipolar junction transistors are still the preferred device. These will either be chosen to have a large base junction area, or will be employed as a parallel-connected array; as, for example, in the LM194/394 'super-match pair' ICs, where a multiplicity of parallel connected transistors are fabricated on a single chip, giving an effective input (noise) impedance as low as 40 ohms.

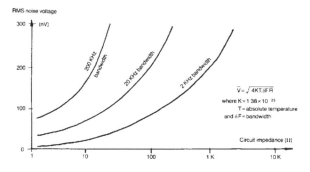

Figure 13.13   Thermal noise output as a function of circuit impedance.

However, recent designs of monolithic dual J-FETs, using a similar type of multiple parallel-connection system, such as the Hitachi 2SK389, can offer equivalent

thermal noise resistance values as low as 33 ohms, and a superior overall noise figure at input resistance values in excess of 100 ohms.

At impedance levels beyond about 1 kilohm there is little practical difference between any devices of recent design. Earlier MOSFET types were not so satisfactory, due to excess noise effects arising from carrier trapping mechanisms in impurities at the channel/gate interface.

Output Voltage Characteristics

Since it is desirable that output overload and signal clipping do not occur in audio systems, particularly in stages preceding the gain controls, much emphasis has been placed on the so-called 'headroom' of signal handling stages, especially in hi-fi publications where the reviewers are able to distance themselves from the practical problems of circuit design.

While it is obviously desirable that inadvertent overload shall not occur in stages preceding signal level controls, high levels of feasible output voltage swing demand the use of high voltage supply rails, and this, in turn, demands the use of active components which can support such working voltage levels.

Not only are such devices more costly, but they will usually have poorer performance characteristics than similar devices of lower voltage ratings. Also, the requirement for the use of high voltage operation may preclude the use of components having valuable characteristics, but which are restricted to lower voltage operation.

Practical audio circuit designs will therefore regard headroom simply as one of a group of desirable parameters in a working system, whose design will be based on careful consideration of the maximum input signal levels likely to be found in practice.

Nevertheless, improved transistor or IC types, and new developments in circuit architecture, are welcomed as they occur, and have eased the task of the audio design engineer, for whom the advent of new programme sources, in particular the compact disc, and now digital audio tape systems, has greatly extended the likely dynamic range of the output signal.

Signal characteristics

The practical implications of this can be seen from a consideration of the signal characteristics of existing programme sources. Of these, in the past, the standard vinyl ('black') disc has been the major determining factor. In this, practical considerations of groove tracking have limited the recorded needle tip velocity to about 40 cm/s, and typical high-quality pick-up cartridges capable of tracking this recorded velocity will have a voltage output of some 3 mV at a standard 5 cm/s recording level.

If the pre-amplifier specification calls for maximum output to be obtainable at a 5 cm/s input, then the design should be chosen so that there is a 'headroom factor' of at least 8×, in such stages preceding the gain controls.

In general, neither FM broadcasts, where the dynamic range of the transmitted signal is limited by the economics of transmitter power, nor cassette recorders, where the dynamic range is constrained by the limited tape overload characteristics, have offered such a high practicable dynamic range.

It is undeniable that the analogue tape recorder, when used at 15 in/s., twin-track, will exceed the LP record in dynamic range. After all, such recorders were originally used for mastering the discs. But such programme sources are rarely found except among 'live recording' enthusiasts. However, the compact disc, which is becoming increasingly common among purely domestic hi-fi systems, presents a new challenge, since the practicable dynamic range of this system exceeds 80 dB (10 000:1), and the likely range from mean (average listening level) to peak may well be as high as 35 dB (56:1) in comparison with the 18 dB (8:1) range likely with the vinyl disc.

Fortunately, since the output of the compact disc player is at a high level, typically 2 V RMS, and requires no signal or frequency response conditioning prior to use, the gain control can be sited directly at the input of the preamp. Nevertheless, this still leaves the possibility that signal peaks may occur during use which are some 56× greater than the mean programme level, with the consequence of the following amplifier stages being driven hard into overload.

This has refocused attention on the design of solid state voltage amplifier stages having a high possible output voltage swing, and upon power amplifiers which either have very high peak output power ratings, or more graceful overload responses.

Voltage Amplifier Design Techniques

The sources of non-linearity in bipolar junction transistors have already been referred to, in respect of the influence of collector load impedance, and push-pull symmetry in reducing harmonic distortion. An additional factor with bipolar junction devices is the external impedance in the base circuit, since the principal non-linearity in a bipolar device is that due to its input voltage/output current characteristics. If the device is driven from a high impedance source, its linearity will be sub-

stantially greater, since it is operating under conditions of current drive.

This leads to the good relative performance of the simple, two-stage, bipolar transistor circuit of Fig. 13.14, in that the input transistor, Q_1, is only required to deliver a very small voltage drive signal to the base of Q_2, so the signal distortion due to Q_1 will be low. Q_2, however, which is required to develop a much larger output voltage swing, with a much greater potential signal non-linearity, is driven from a relatively high source impedance, composed of the output impedance of Q_1, which is very high indeed, in parallel with the base-emitter resistor, R_4. R_1, R_2, and R_3/C_2 are employed to stabilise the DC working conditions of the circuit.

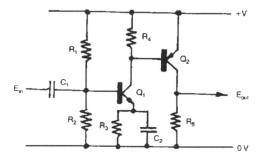

Figure 13.14 Two-stage transistor voltage amplifier.

Normally, this circuit is elaborated somewhat to include both DC and AC negative feedback from the collector of Q_2 to the emitter of Q_1, as shown in the practical amplifier circuit of Fig. 13.15.

This is capable of delivering a 14 V p-p output swing, at a gain of 100, and a bandwidth of 15 Hz – 250 kHz, at –3 dB points; largely determined by the value of C_2 and the output capacitances, with a THD figure of better that 0.01 per cent at 1 kHz.

The practical drawbacks of this circuit relate to the relatively low value necessary for R_3 – with the consequent large value necessary for C_2 if a good LF response is desired, and the DC offset between point 'X' and the output, due to the base-emitter junction potential of Q_1, and the DC voltage drop along R_5, which makes this circuit relatively unsuitable in DC amplifier applications.

An improved version of this simple two-stage amplifier circuit is shown in Fig. 13.16, in which the single input transistor has been replaced by a 'long-tailed pair' configuration of the type shown in Fig. 13.16. In this, if the two-input transistors are reasonably well matched in current gain, and if the value of R_3 is chosen to give an equal collector current flow through both Q_1 and Q_2, the DC offset between input and output will be negligible, and this will allow the circuit to be operated between symmetrical (+ and –) supply rails, over a frequency range extending from DC to 250 kHz or more.

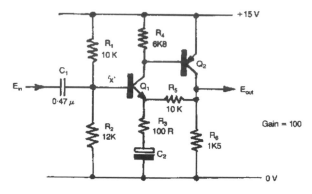

Figure 13.15   Practical two-stage feedback amplifier.

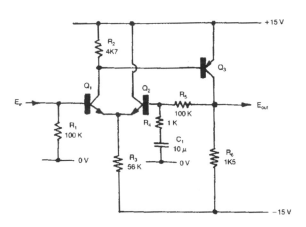

Figure 13.16   Improved two-stage feedback amplifier.

Because of the improved rejection of odd harmonic distortion inherent in the input 'push-pull' layout, the THD due to this circuit, particularly at less than maximum output voltage swing, can be extremely low, and this probably forms the basis of the bulk of linear amplifier designs. However, further technical improvements are possible, and these are discussed below.

Constant-Current Sources and 'Current Mirrors'

As mentioned above, the use of high-value collector load resistors in the interests of high stage gain and low inherent distortion carries with it the penalty that the performance of the amplifying device may be impaired by the low collector current levels which result from this approach.

Advantage can, however, be taken of the very high output impedance of a junction transistor, which is inherent in the type of collector current/supply voltage

characteristics illustrated in Fig. 13.5, where even at currents in the 1–10 mA region, dynamic impedances of the order of 100 kilohms may be expected.

A typical circuit layout which utilises this characteristic is shown in Fig. 13.17, in which R_1 and R_2 form a potential divider to define the base potential of Q_1, and R_3 defines the total emitter or collector currents for this effective base potential.

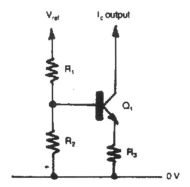

Figure 13.17   Transistor constant current source.

This configuration can be employed with transistors of either PNP or NPN types, which allows the circuit designer considerable freedom in their application.

An improved, two-transistor, constant current source is shown in Fig. 13.18. In this R_1 is used to bias Q_2 into conduction, and Q_1 is employed to sense the voltage developed across R_2, which is proportional to emitter current, and to withdraw the forward bias from Q_2 when that current level is reached at which the potential developed across R_2 is just sufficient to cause Q_1 to conduct.

The performance of this circuit is greatly superior to that of Fig. 13.17, in that the output impedance is about $10\times$ greater, and the circuit is insensitive to the potential, +Vref., applied to R_1, so long as it is adequate to force both Q_2 and Q_1 into conduction.

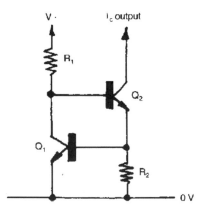

Figure 13.18   Two-transistor constant current source.

An even simpler circuit configuration makes use of the inherent very high output impedance of a junction FET under constant gate bias conditions. This employs the circuit layout shown in Fig. 13.19, which allows a true 'two-terminal' constant current source, independent of supply lines or reference potentials, and which can be used at either end of the load chain.

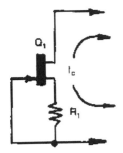

Figure 13.19 Two-terminal constant current source.

The current output from this type of circuit is controlled by the value chosen for R_1, and this type of constant current source may be constructed using almost any available junction FET, provided that the voltage drop across the FET drain-gate junction does not exceed the breakdown voltage of the device. This type of constant current source is also available as small, plastic-encapsulated, two-lead devices, at a relatively low cost, and with a range of specified output currents.

All of these constant current circuit layouts share the common small disadvantage that they will not perform very well at low voltages across the current source element. In the case of Figs. 13.17 and 13.18, the lowest practicable operating potential will be about 1 V. The circuit of Fig. 13.19 may require, perhaps, 2–3 V, and this factor must be considered in circuit performance calculations.

The 'boot-strapped' load resistor arrangement shown in Fig. 13.20, and commonly used in earlier designs of audio amplifier to improve the linearity of the last class 'A' amplifier stage (Q_1), effectively multiplies the resistance value of R_2 by the gain which Q_2 would be deemed to have if operated as a common-emitter amplifier with a collector load of R_3 in parallel with R_1.

This arrangement is the best configuration practicable in terms of available RMS output voltage swing, as compared with conventional constant current sources, but has fallen into disuse because it leads to slightly less good THD figures than are possible with other circuit arrangements.

All these circuit arrangements suffer from a further disadvantage, from the point of view of the integrated circuit designer: they employ resistors as part of the circuit design, and resistors, though possible to fabricate in IC structures, occupy a disproportionately large area of

the chip surface. Also, they are difficult to fabricate to precise resistance values without resorting to subsequent laser trimming, which is expensive and time-consuming.

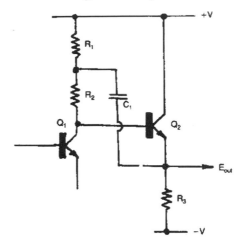

Figure 13.20 Load impedance increase by boot-strap circuit.

Because of this, there is a marked preference on the part of IC design engineers for the use of circuit layouts known as 'current mirrors', of which a simple form is shown in Fig. 13.21.

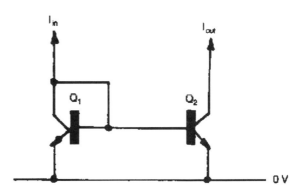

Figure 13.21 Simple form of current mirror.

IC solutions

These are not true constant current sources, in that they are only capable of generating an output current (I_{out}) which is a close equivalence of the input or drive current, (I_{in}). However, the output impedance is very high, and if the drive current is held to a constant value, the output current will also remain constant.

A frequently found elaboration of this circuit, which offers improvements in respect of output impedance and the closeness of equivalence of the drive and output currents, is shown in Fig. 13.22. Like the junction FET based constant current source, these current mirror devices are available as discrete, plastic-encapsulated, three-lead components, having various drive current/output current ratios, for incorporation into discrete component circuit designs.

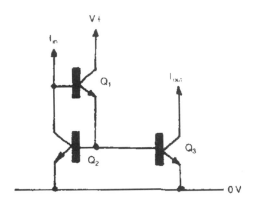

Figure 13.22   Improved form of current mirror.

The simple amplifier circuit of Fig. 13.16 can be elaborated, as shown in Fig. 13.23, to employ these additional circuit refinements, which would have the effect of increasing the open-loop gain, i.e. that before negative feedback is applied, by 10–100× and improving the har-

monic and other distortions, and the effective bandwidth by perhaps 3–10×. From the point of view of the IC designer, there is also the advantage of a potential reduction in the total resistor count.

These techniques for improving the performance of semiconductor amplifier stages find particular application in the case of circuit layouts employing junction FETs and MOSFETs, where the lower effective mutual conductance values for the devices would normally result in relatively poor stage gain figures.

This has allowed the design of IC operational amplifiers, such as the RCA CA3140 series, or the Texas Instruments TL071 series, which employ, respectively, MOSFET and junction FET input devices. The circuit layout employed in the TL071 is shown, by way of example, in Fig. 13.24.

Both of these op. amp. designs offer input impedances in the million megohm range – in comparison with the input impedance figures of 5–10 kilohm which were typical of early bipolar ICs – and the fact that the input impedance is so high allows the use of such ICs in circuit configurations for which earlier op. amp. ICs were entirely inappropriate.

Although the RCA design employs MOSFET input devices which offer, in principle, an input impedance which is perhaps 1000 times better than this figure, the presence of on-chip Zener diodes, to protect the device against damage through misuse or static electric charges,

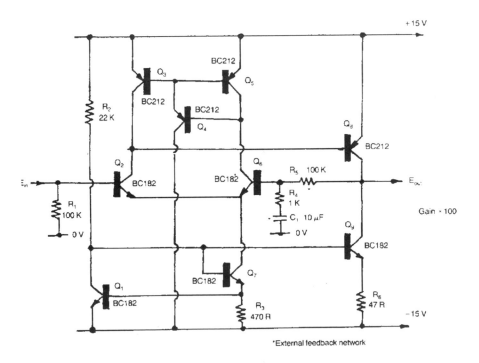

Figure 13.23   Use of circuit elaboration to improve two-stage amplifier of Fig. 13.16.

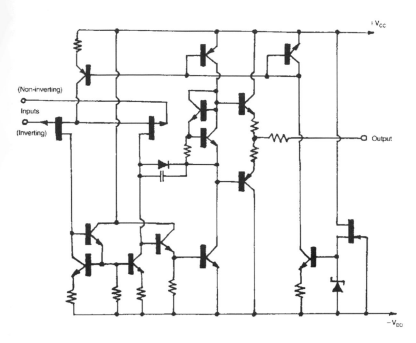

Figure 13.24   Circuit layout of Texas Instruments TL071 op. amp.

reduces the input impedance to roughly the same level as that of the junction FET device.

It is a matter for some regret that the design of the CA3140 series devices is now so elderly that the internal MOSFET devices do not offer the low level of internal noise of which more modern MOSFET types are capable. This tends rather to rule out the use of this MOSFET op. amp. for high quality audio use, though the TL071 and its equivalents such as the LF351 have demonstrated impeccable audio behaviour.

Performance Standards

It has always been accepted in the past, and is still held as axiomatic among a very large section of the engineering community, that performance characteristics can be measured, and that improved levels of measured performance will correlate precisely, within the ability of the ear to detect such small differences, with improvements which the listener will hear in reproduced sound quality.

Within a strictly engineering context, it is difficult to do anything other than accept the concept that measured improvements in performance are the only things which should concern the designer.

However, the frequently repeated claim by journalists and reviewers working for periodicals in the hi-fi field – who, admittedly, are unlikely to be unbiased witnesses –

that measured improvements in performance do not always go hand-in-hand with the impressions that the listener may form, tends to undermine the confidence of the circuit designer that the instrumentally determined performance parameters are all that matter.

It is clear that it is essential for engineering progress that circuit design improvements must be sought which lead to measurable performance improvements. Yet there is now also the more difficult criterion that those things which appear to be better, in respect of measured parameters, must also be seen, or heard, to be better.

Use of ICs

This point is particularly relevant to the question of whether, in very high quality audio equipment, it is acceptable to use IC operational amplifiers, such as the TL071, or some of the even more exotic later developments such as the NE5534 or the OP27, as the basic gain blocks, around which the passive circuitry can be arranged, or whether, as some designers believe, it is preferable to construct such gain blocks entirely from discrete components.

Some years ago, there was a valid technical justification for this reluctance to use op. amp. ICs in high quality audio circuitry, since the method of construction of such ICs was as shown, schematically, in Fig. 13.25, in which all the structural components were formed on the surface

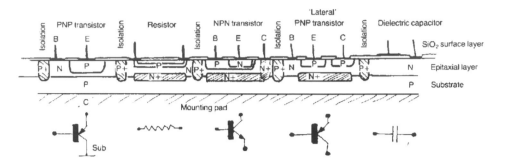

Figure 13.25   Method of fabrication of components in silicon integrated circuit.

of a heavily 'P' doped silicon substrate, and relied for their isolation, from one another or from the common substrate, on the reverse biassed diodes formed between these elements.

This led to a relatively high residual background noise level, in comparison with discrete component circuitry, due to the effects of the multiplicity of reverse diode leakage currents associated with every component on the chip. Additionally, there were quality constraints in respect of the components formed on the chip surface – more severe for some component types than for others – which also impaired the circuit performance.

A particular instance of this problem arose in the case of PNP transistors used in normal ICs, where the circuit layout did not allow these to be formed with the substrate acting as the collector junction. In this case, it was necessary to employ the type of construction known as a 'lateral PNP', in which all the junctions are diffused in, from the exposed chip surface, side by side.

In this type of device the width of the 'N' type base region, which must be very small for optimum results, depends mainly on the precision with which the various diffusion masking layers can be applied. The results are seldom very satisfactory. Such a lateral PNP device has a very poor current gain and HF performance.

In recent IC designs, considerable ingenuity has been shown in the choice of circuit layout to avoid the need to employ such unsatisfactory components in areas where their shortcomings would affect the end result. Substantial improvements, both in the purity of the base materials and in diffusion technology, have allowed the inherent noise background to be reduced to a level where it is no longer of practical concern.

Modern standards

The standard of performance which is now obtainable in audio applications, from some of the recent IC op. amps.

– especially at relatively low closed-loop gain levels – is frequently of the same order as that of the best discrete component designs, but with considerable advantages in other respects, such as cost, reliability and small size.

This has led to their increasing acceptance as practical gain blocks, even in very high quality audio equipment.

When blanket criticism is made of the use of ICs in audio circuitry, it should be remembered that the 741 which was one of the earliest of these ICs to offer a satisfactory performance – though it is outclassed by more recent types – has been adopted with enthusiasm, as a universal gain block, for the signal handling chains in many recording and broadcasting studios.

This implies that the bulk of the programme signals employed by the critics to judge whether or not a discrete component circuit is better than that using an IC, will already have passed through a sizeable handful of 741-based circuit blocks, and if such ICs introduce audible defects, then their reference source is already suspect.

It is difficult to stipulate the level of performance which will be adequate in a high-quality audio installation. This arises partly because there is little agreement between engineers and circuit designers on the one hand, and the hi-fi fraternity on the other, about the characteristics which should be sought, and partly because of the wide differences which exist between listeners in their expectations for sound quality or their sensitivity to distortions. These differences combine to make it a difficult and speculative task to attempt either to quantify or to specify the technical components of audio quality, or to establish an acceptable minimum quality level.

Because of this uncertainty, the designer of equipment, in which price is not a major consideration, will normally seek to attain standards substantially in excess of those which he supposes to be necessary, simply in order not to fall short. This means that the reason for the small residual differences in the sound quality, as between high quality units, is the existence of malfunctions of types which are not currently known or measured.

Audibility of Distortion Components

Harmonic and intermodulation distortions

Because of the small dissipations which are normally involved, almost all discrete component voltage amplifier circuitry will operate in class 'A' (that condition in which the bias applied to the amplifying device is such as to make it operate in the middle of the linear region of its input/output transfer characteristic), and the residual harmonic components are likely to be mainly either second or third order, which are audibly much more tolerable than higher order distortion components.

Experiments in the late 1940s suggested that the level of audibility for second and third harmonics was of the order of 0.6 per cent and 0.25 per cent respectively, and this led to the setting of a target value, within the audio spectrum, of 0.1 per cent THD, as desirable for high quality audio equipment.

However, recent work aimed at discovering the ability of an average listener to detect the presence of low order (i.e. second or third) harmonic distortions has drawn the uncomfortable conclusion that listeners, taken from a cross section of the public, may rate a signal to which 0.5 per cent second harmonic distortion has been added as 'more musical' than, and therefore preferable to, the original undistorted input. This discovery tends to cast doubt on the value of some subjective testing of equipment.

What is not in dispute is that the inter-modulation distortion, (IMD), which is associated with any non-linearity in the transfer characteristics, leads to a muddling of the sound picture, so that if the listener is asked, not which sound he prefers, but which sound seems to him to be the clearer, he will generally choose that with the lower harmonic content.

The way in which IMD arises is shown in Fig. 13.26, where a composite signal containing both high-frequency and low-frequency components, fed through a non-linear system, causes each signal to be modulated by the other. This is conspicuous in the drawing in respect of the HF component, but is also true for the LF one.

This can be shown mathematically to be due to the generation of sum and difference products, in addition to the original signal components, and provides a simple method, shown schematically in Fig. 13.27, for the detection of this type of defect. A more formal IMD measurement system is shown in Fig. 13.28.

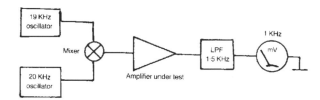

Figure 13.27 Simple HF two-tone inter-modulation distortion test.

With present circuit technology and device types, it is customary to design for total harmonic and IM distortions to be below 0.01 per cent over the range 30 Hz–20 kHz, and at all signal levels below the onset of clipping. Linear

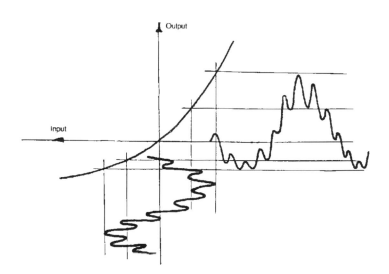

Figure 13.26 Inter-modulation distortions produced by the effect of a non-linear input/output transfer characteristic on a complex tone.

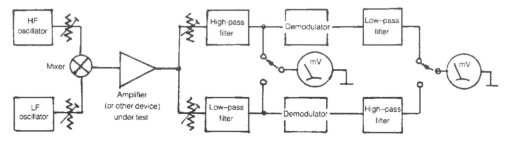

Figure 13.28   Two-tone inter-modulation distortion test rig.

IC op. amps., such as the TL071 and the LF351, will also meet this spec. over the frequency range 30 Hz–10kHz.

Transient defects

A more insidious group of signal distortions may occur when brief signals of a transient nature, or sudden step type changes in base level, are superimposed on the more continuous components of the programme signal. These defects can take the form of slew-rate distortions, usually associated with loss of signal during the period of the slew-rate saturation of the amplifier – often referred to as transient inter-modulation distortion or TID.

This defect is illustrated in Fig. 13.29, and arises particularly in amplifier systems employing substantial amounts of negative feedback, when there is some slew-rate limiting component within the amplifier, as shown in Fig. 13.30.

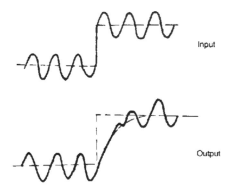

Figure 13.29   Effect of amplifier slew-rate saturation or transient inter-modulation distortion.

A further problem is that due to 'overshoot', or 'ringing', on a transient input, as illustrated in Fig. 13.31. This arises particularly in feedback amplifiers if there is an inadequate stability margin in the feedback loop, particu-

larly under reactive load conditions, but will also occur in low-pass filter systems if too high an attenuation rate is employed.

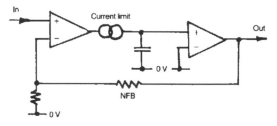

Figure 13.30   Typical amplifier layout causing slew-rate saturation.

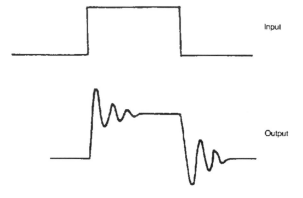

Figure 13.31   Transient 'ringing'.

The ear is very sensitive to slew-rate induced distortion, which is perceived as a 'tizziness' in the reproduced sound. Transient overshoot is normally noted as a somewhat over-bright quality. The avoidance of both these problems demands care in the circuit design, particularly when a constant current source is used, as shown in Fig. 13.32.

In this circuit, the constant current source, CC_1, will impose an absolute limit on the possible rate of change of potential across the capacitance C_1, (which could well be simply the circuit stray capacitance), when the output voltage is caused to move in a positive-going direction. This problem is compounded if an additional current limit mechanism, CC_2, is included in the circuitry to protect the amplifier transistor (Q_1) from output current overload.

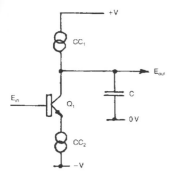

Figure 13.32 Circuit design aspects which may cause slew rate limiting.

Since output load and other inadvertent capacitances are unavoidable, it is essential to ensure that all such current limited stages operate at a current level which allows potential slewing to occur at rates which are at least 10× greater than the fastest signal components. Alternatively, means may be taken, by way of a simple input integrating circuit, (R_1C_1), as shown in Fig. 13.33, to ensure that the maximum rate of change of the input signal voltage is within the ability of the amplifier to handle it.

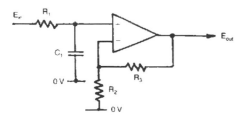

Figure 13.33 Input HF limiting circuit to lessen SLR.

Spurious signals

In addition to harmonic, IM, and transient defects in the signal channel, which will show up on normal instrumental testing, there is a whole range of spurious signals which may not arise in such tests. The most common of these is that of the intrusion of noise and alien signals, either from the supply line, or by direct radio pick-up.

This latter case is a random and capricious problem which can only be solved by steps appropriate to the circuit design in question. However, supply line intrusions, whether due to unwanted signals from the power supply, or from the other channel in a stereo system, may

be greatly reduced by the use of circuit designs offering a high immunity to voltage fluctuations on the DC supply.

Other steps, such as the use of electronically stabilised DC supplies, or the use of separate power supplies in a stereo amplifier, are helpful, but the required high level of supply line signal rejection should be sought as a design feature before other palliatives are applied. Modern IC op. amps. offer a typical supply voltage rejection ratio of 90 dB (30000:1). Good discrete component designs should offer at least 80 dB (10000:1).

This figure tends to degrade at higher frequencies, and this has led to the growing use of supply line bypass capacitors having a low effective series resistance (ESR). This feature is either a result of the capacitor design, or is achieved in the circuit by the designer's adoption of groups of parallel connected capacitors chosen so that the AC impedance remains low over a wide range of frequencies.

A particular problem in respect of spurious signals, which occurs in audio power amplifiers, is due to the loudspeaker acting as a voltage generator, when stimulated by pressure waves within the cabinet, and injecting unwanted audio components directly into the amplifier's negative feedback loop. This specific problem is unlikely to arise in small signal circuitry, but the designer must consider what effect output/line load characteristics may have – particularly in respect of reduced stability margin in a feedback amplifier.

In all amplifier systems there is a likelihood of microphonic effects due to the vibration of the components. This is likely to be of increasing importance at the input of 'low level', high sensitivity, pre-amplifier stages, and can lead to colouration of the signal when the equipment is in use, which is overlooked in the laboratory in a quiet environment.

Mains-borne interference

Mains-borne interference, as evidenced by noise pulses on switching electrical loads, is most commonly due to radio pick-up problems, and is soluble by the techniques (attention to signal and earth line paths, avoidance of excessive HF bandwidth at the input stages) which are applicable to these.

General Design Considerations

During the past three decades, a range of circuit design techniques has been evolved to allow the construction of highly linear gain stages based on bipolar transistors whose input characteristics are, in themselves, very nonlinear. These techniques have also allowed substantial improvements in possible stage gain, and have led to greatly improved performance from linear, but low gain, field-effect devices.

These techniques are used in both discrete component designs and in their monolithic integrated circuit equivalents, although, in general, the circuit designs employed in linear ICs are considerably more complex than those used in discrete component layouts.

This is partly dictated by economic considerations, partly by the requirements of reliability, and partly because of the nature of IC design.

The first two of these factors arise because both the manufacturing costs and the probability of failure in a discrete component design are directly proportional to the number of components used, so the fewer the better, whereas in an IC, both the reliability and the expense of manufacture are only minimally affected by the number of circuit elements employed.

In the manufacture of ICs, as has been indicated above, some of the components which must be employed are much worse than their discrete design equivalents. This has led the IC designer to employ fairly elaborate circuit structures, either to avoid the need to use a poor quality component in a critical position, or to compensate for its shortcomings.

Nevertheless, the ingenuity of the designers, and the competitive pressures of the market place, have resulted in systems having a very high performance, usually limited only by their inability to accept differential supply line potentials in excess of 36 V, unless non-standard diffusion processes are employed.

For circuitry requiring higher output or input voltage swings than allowed by small signal ICs, the discrete component circuit layout is, at the moment, unchallenged. However, as every designer knows, it is a difficult matter to translate a design which is satisfactory at a low working voltage design into an equally good higher voltage system. This is because:

- increased applied potentials produce higher thermal dissipations in the components, for the same operating currents;
- device performance tends to deteriorate at higher inter-electrode potentials and higher output voltage excursions;
- available high/voltage transistors tend to be more restricted in variety and less good in performance than lower voltage types.

Controls

These fall into a variety of categories:
- gain controls needed to adjust the signal level between source and power amplifier stages

- tone controls used to modify the tonal characteristics of the signal chain
- filters employed to remove unwanted parts of the incoming signal, and those adjustments used to alter the quality of the audio presentation, such as stereo channel balance or channel separation controls.

Gain controls

These are the simplest in basic form, and are often just a resistive potentiometer voltage divider of the type shown in Fig. 13.34. Although simple, this component can generate a variety of problems. Of these, the first is that due to the value chosen for R_1. Unless this is infinitely high, it will attenuate the maximum signal voltage, (E_{max}), obtainable from the source, in the ratio

$$E_{max} = E_{in} \times R_1 /(R_1 + Z_{source})$$

where Z_{source} is the output impedance of the driving circuit. This factor favours the use of a high value for R_1, to avoid loss of input signal.

However, the following amplifier stage may have specific input impedance requirements, and is unlikely to operate satisfactorily unless the output impedance of the gain control circuit is fairly low. This will vary according to the setting of the control, between zero and a value, at the maximum gain setting of the control, due to the parallel impedances of the source and gain control.

$$Z_{out} = R_1 /(R_1 + Z_{source})$$

The output impedance at intermediate positions of the control varies as the effective source impedance and the impedance to the 0 V line is altered. However, in general, these factors would encourage the use of a low value for R_1.

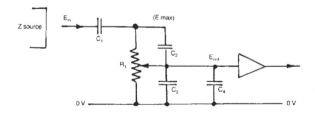

Figure 13.34   Standard gain control circuit.

An additional and common problem arises because the perceived volume level associated with a given sound pressure (power) level has a logarithmic characteristic. This means that the gain control potentiometer, R_1, must have a resistance value which has a logarithmic, rather than linear, relationship with the angular rotation of the potentiometer shaft.

Potentiometer Law

Since the most common types of control potentiometer employ a resistive composition material to form the potentiometer track, it is a difficult matter to ensure that the grading of conductivity within this material will follow an accurate logarithmic law.

On a single channel this error in the relationship between signal loudness and spindle rotation may be relatively unimportant. In a stereo system, having two ganged gain control spindles, intended to control the loudness of the two channels simultaneously, errors in following the required resistance law, existing between the two potentiometer sections, will cause a shift in the apparent location of the stereo image as the gain control is adjusted, and this can be very annoying.

In high-quality equipment, this problem is sometimes avoided by replacing R_1 by a precision resistor chain $(R_a–R_z)$, as shown in Fig. 13.35, in which the junctions between these resistors are connected to tapping points on a high-quality multi-position switch.

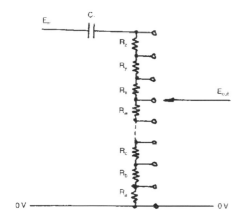

Figure 13.35 Improved gain control using multi-pole switch.

By this means, if a large enough number of switch tap positions is available, and this implies at least a 20-way switch to give a gentle gradation of sound level, a very close approximation to the required logarithmic law can be obtained, and two such channel controls could be ganged without unwanted errors in differential output level.

Circuit Capacitances

A further practical problem, illustrated in Fig. 13.34, is associated with circuit capacitances. Firstly, it is essential to ensure that there is no standing DC potential across R_1 in normal operation, otherwise this will cause an unwanted noise in the operation of the control. This

imposes the need for a protective input capacitor, C_1, and this will cause a loss of low frequency signal components, with a –3 dB LF turn-over point at the frequency at which the impedance of C_1 is equal to the sum of the source and gain control impedances. C_1 should therefore be of adequate value.

Additionally, there are the effects of the stray capacitances, C_2 and C_3, associated with the potentiometer construction, and the amplifier input and wiring capacitances, C_4. The effect of these is to modify the frequency response of the system, at the HF end, as a result of signal currents passing through these capacitances. The choice of a low value for R_1 is desirable to minimise this problem.

The use of the gain control to operate an on/off switch, which is fairly common in low-cost equipment, can lead to additional problems, especially with high resistance value gain control potentiometers, in respect of AC mains 'hum' pick up. It also leads to a more rapid rate of wear of the gain control, in that it is rotated at least twice whenever the equipment is used.

Tone controls

These exist in the various forms shown in Figs. 13.36–13.40, respectively described as standard (bass and treble lift or cut), slope control, Clapham Junction, parametric and graphic equaliser types. The effect these will have on the frequency response of the equipment is shown in the drawings, and their purpose is to help remedy shortcomings in the source programme material, the receiver or transducer, or in the loudspeaker and listening room combination.

To the hi-fi purist, all such modifications to the input signal tend to be regarded with distaste, and are therefore omitted from some hi-fi equipment. However, they can be useful, and make valuable additions to the audio equipment, if used with care.

Standard Tone Control Systems

These are either of the passive type, of which a typical circuit layout is shown in Fig. 13.41, or are constructed as part of the negative feedback loop around a gain block, using the general design due to Baxandall. A typical circuit layout for this kind of design is shown in Fig. 13.42.

It is claimed that the passive layout has an advantage in quality over the active (feedback network) type of control, in that the passive network merely contains resistors and capacitors, and is therefore free from any possibility of introduced distortion, whereas the 'active' network requires an internal gain block, which is not automatically above suspicion.

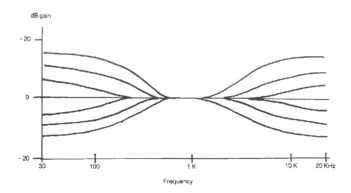

Figure 13.36   Bass and treble lift/cut tone control.

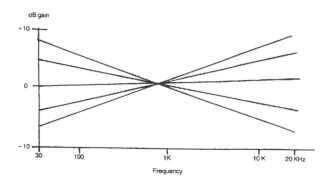

Figure 13.37   Slope control.

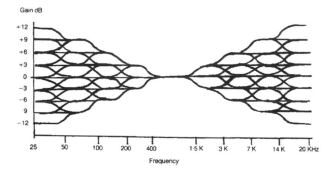

Figure 13.38   Clapham Junction type of tone control.

In reality, however, any passive network must introduce an attenuation, in its flat response form, which is equal to the degree of boost sought at the maximum 'lift' position, and some external gain block must therefore be added to compensate for this gain loss.

This added gain block is just as prone to introduce distortion as that in an active network, with the added disadvantage that it must provide a gain equal to that of the flat-response network attenuation, whereas the active system gain block will typically have a gain of unity in the flat response mode, with a consequently lower distortion level.

As a final point, it should be remembered that any treble lift circuit will cause an increase in harmonic distortion, simply because it increases the gain at the frequencies associated with harmonics, in comparison with that at the frequency of the fundamental.

The verdict of the amplifier designers appears to be substantially in favour of the Baxandall system, in that this is the layout most commonly employed.

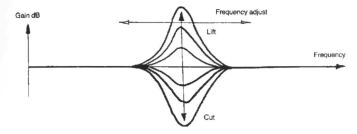

Figure 13.39 Parametric equaliser control.

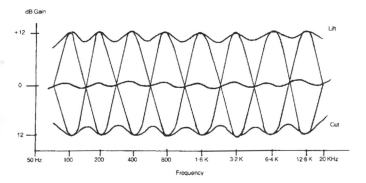

Figure 13.40 Graphic equaliser response characteristics.

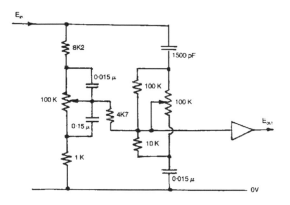

Figure 13.41 Circuit layout of passive tone control.

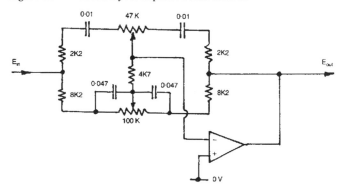

Figure 13.42 Negative feedback type tone control circuit.

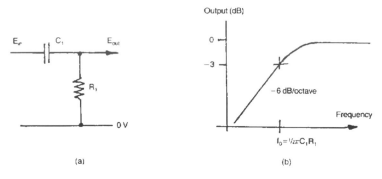

(a)   (b)

Figure 13.43   Layout and frequency response of simple bass-cut circuit (high-pass).

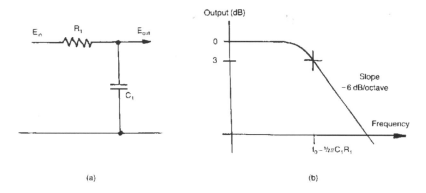

(a)   (b)

Figure 13.44   Layout and frequency response of simple treble-cut circuit (low-pass).

Both of these tone control systems – indeed this is true of all such circuitry – rely for their operation on the fact that the AC impedance of a capacitor will depend on the applied frequency, as defined by the equation:

$$Z_c = 1/(2\pi f_c),$$

or more accurately,

$$Z_c = 1/(2j\pi f_c),$$

where j is the square root of -1.

Commonly, in circuit calculations, the $2\pi f$ group of terms are lumped together and represented by the Greek symbol ω.

The purpose of the j term, which appears as a 'quadrature' element in the algebraic manipulations, is to permit the circuit calculations to take account of the 90 phase shift introduced by the capacitative element. (The same is also true of inductors within such a circuit, except that the phase shift will be in the opposite sense.) This is important in most circuits of this type.

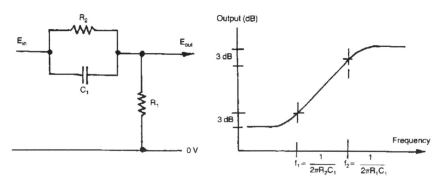

Figure 13.45   Modified bass-cut (high-pass) RC circuit.

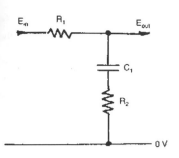

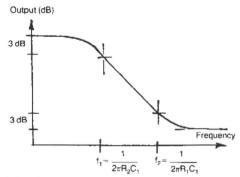

Figure 13.46 Modified treble-cut (low-pass) RC circuit.

The effect of the change in impedance of the capacitor on the output signal voltage from a simple RC network, of the kind shown in Figs. 13.43(a) and 13.44(a), is shown in Figs. 13.43(b) and 13.44(b). If a further resistor, R_2, is added to the networks the result is modified in the manner shown in Figs. 13.45 and 13.46. This type of structure, elaborated by the use of variable resistors to control the amount of lift or fall of output as a function of frequency, is the basis of the passive tone control circuitry of Fig. 13.41.

If such networks are connected across an inverting gain block, as shown in Figs. 13.47(a) and 13.48(a), the resultant frequency response will be shown in Figs. 13.47(b) and 13.48(b), since the gain of such a negative feedback configuration will be

$$\text{Gain} = Z_a / Z_b$$

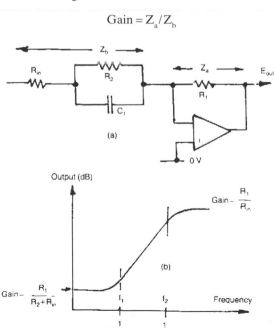

Figure 13.47 Active RC treble-lift or bass-cut circuit.

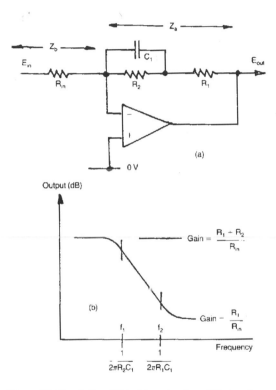

Figure 13.48 Active RC treble-cut or bass-lift circuit.

assuming that the open-loop gain of the gain block is sufficiently high. This is the design basis of the Baxandall type of tone control, and a flat frequency response results when the impedance of the input and output limbs of such a feedback arrangement remains in equality as the frequency is varied.

Slope Controls
This is the type of tone control employed by Quad in its type 44 pre-amplifier, and operates by altering the relative balance of the LF and HF components of the audio signal, with reference to some specified mid-point

frequency, as is shown in Fig. 13.37. A typical circuit for this type of design is shown in Fig. 13.49.

The philosophical justification for this approach is that it is unusual for any commercially produced programme material to be significantly in error in its overall frequency characteristics, but the tonal preferences of the recording or broadcasting balance engineer may differ from those of the listener.

In such a case, he might consider that the signal, as presented, was somewhat over heavy, in respect of its bass, or alternatively, perhaps, that it was somewhat light or thin in tone, and an adjustment of the skew of the frequency response could correct this difference in tonal preference without significantly altering the signal in other respects.

The Clapham Junction Type

This type of tone control, whose possible response curves are shown in Fig. 13.38, was introduced by the author to provide a more versatile type of tonal adjustment than that offered by the conventional standard systems, for remedying specific peaks or troughs in the frequency response, without the penalties associated with the graphic equaliser type of control, described below.

In the Clapham Junction type system, so named because of the similarity of the possible frequency response curves to that of railway lines, a group of push switches is arranged to allow one or more of a multiplicity of RC networks to be introduced into the feedback loop of a negative feedback type tone control system, as shown in Fig. 13.50, to allow individual ±3 dB frequency adjustments to be made, over a range of possible frequencies.

By this means, it is possible by combining elements of frequency lift or cut to choose from a variety of possible frequency response curves, without losing the ability to attain a linear frequency response.

Parametric Controls

This type of tone control, whose frequency response is shown in Fig. 13.39, has elements of similarity to both the standard bass/treble lift/cut systems, and the graphic equaliser arrangement, in that while there is a choice of lift or cut in the frequency response, the actual frequency at which this occurs may be adjusted, up or down, in order to attain an optimal system frequency response.

A typical circuit layout is shown in Fig. 13.51.

The Graphic Equaliser System

The aim of this type of arrangement is to compensate fully for the inevitable peaks and troughs in the frequency response of the audio system – including those due to deficiencies in the loudspeakers or the listening room acoustics – by permitting the individual adjustment of the channel gain, within any one of a group of eight single-octave segments of the frequency band, typically covering the range from 80 Hz–20 kHz, though ten octave equalisers covering the whole audio range from 20 Hz–20 kHz have been offered.

Because the ideal solution to this requirement – that of employing a group of parallel connected amplifiers, each of which is filtered so that it covers a single octave band of the frequency spectrum, whose individual gains could be separately adjusted – would be excessively expensive to implement, conventional practice is to make use of a series of LC tuned circuits, connected within a feedback control system, as shown in Fig. 13.52.

This gives the type of frequency response curve shown in Fig. 13.40. As can be seen, there is no position of lift or cut, or combination of control settings, which will permit a flat frequency response, because of the interaction, within the circuitry, between the adjacent octave segments of the pass-band.

While such types of tone control are undoubtedly useful, and can make significant improvements in the

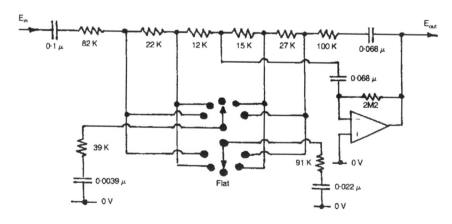

Figure 13. 49   The Quad tilt control.

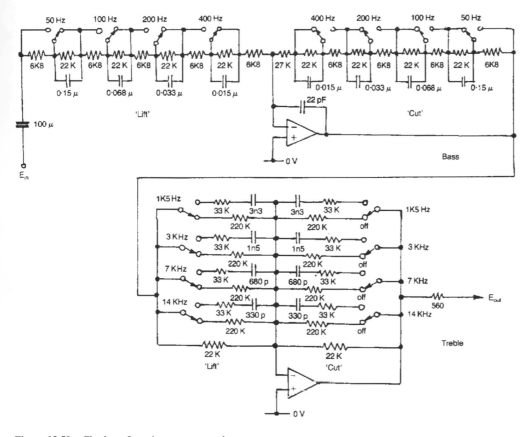

Figure 13.50 Clapham Junction tone control.

performance of otherwise unsatisfactory hi-fi systems, the inability to attain a flat frequency response when this is desired, even at the mid-position of the octave-band controls, has given such arrangements a very poor status in the eyes of the hi-fi fraternity. This unfavourable opinion has been reinforced by the less than optimal performance offered by inexpensive, add-on, units whose engineering standards have reflected their low purchase price.

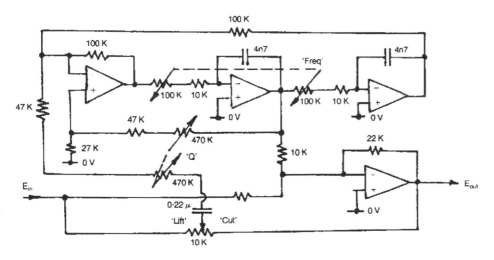

Figure 13.51 Parametric equaliser circuit.

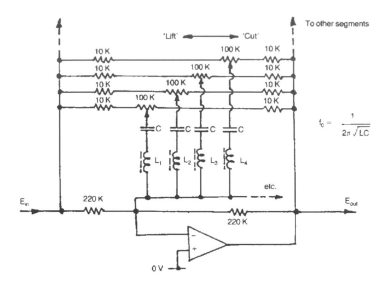

Figure 13.52   Circuit layout for graphic equaliser (four sections only shown).

Channel balance controls

These are provided in any stereo system, to equalise the gain in the left- and right-hand channels, to obtain a desired balance in the sound image. (In a quadraphonic system, four such channel gain controls will ideally be provided.) In general, there are only two options available for this purpose: those balance controls which allow one or other of the two channels to be reduced to zero output level, and those systems, usually based on differential adjustment of the amount of negative feedback across controlled stages, in which the relative adjustment of the gain, in one channel with reference to the other, may only be about 10 dB.

This is adequate for all balance correction purposes, but does not allow the complete extinction of either channel.

The first type of balance control is merely a gain control, of the type shown in Fig. 13.34. A negative feedback type of control is shown in Fig. 13.53.

Channel separation controls

While the closest reproduction, within the environment of the listener, of the sound stage of the original performance will be given by a certain specific degree of separation between the signals within the 'L' and 'R' channels, it is found that shortcomings in the design of the reproducing and amplifying equipment tend universally to lessen the degree of channel separation, rather than the reverse.

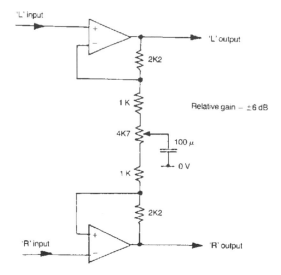

Figure 13.53   Negative feedback type channel balance control.

Some degree of enhancement of channel separation is therefore often of great value, and electronic circuits for this purpose are available, such as that, due to the author, shown in Fig. 13.54.

There are also occasions when a deliberate reduction in the channel separation is of advantage, as, for example in lessening 'rumble' effects due to the vertical motion of a poorly engineered record turntable, or in lessening the hiss component of a stereo FM broadcast. While this is also provided by the circuit of Fig. 13.54, a much less elaborate arrangement, as shown in Fig. 13.55, will suffice for this purpose.

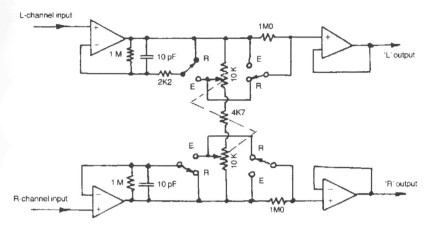

Figure 13.54 Circuit for producing enhanced or reduced stereo channel separation.

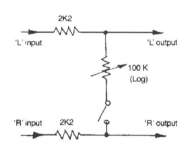

Figure 13.55 Simple stereo channel blend control.

A further, and interesting, approach is that offered by Blumlein, who found that an increase or reduction in the channel separation of a stereo signal was given by adjusting the relative magnitudes of the 'L + R' and 'L–R' signals in a stereo matrix, before these were added or subtracted to give the '2L' and '2R' components.

An electronic circuit for this purpose is shown in Fig. 13.56.

Filters

While various kinds of filter circuit play a very large part in the studio equipment employed to generate the programme material, both as radio broadcasts and as recordings on disc or tape, the only types of filter normally offered to the user are those designed to attenuate very low frequencies, below, say, 50 Hz and generally described as 'rumble' filters, or those operating in the region above a few kHz, and generally described as 'scratch' or 'whistle' filters.

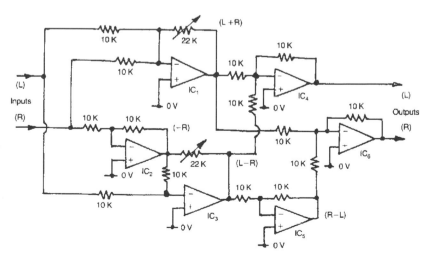

Figure 13.56 Channel separation or blending by using matrix addition or subtraction.

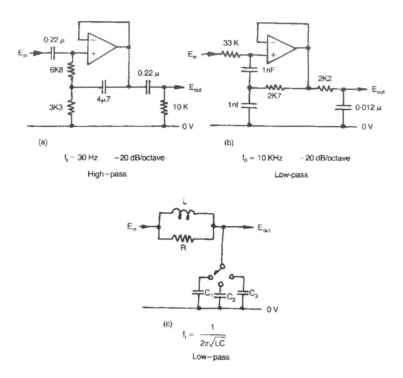

Figure 13.57   Steep-cut filter circuits.

Three such filter circuits are shown in Figs. 13.57(a), (b), and (c). Of these the first two are fixed frequency active filter configurations employing a bootstrap type circuit, for use respectively in high-pass (rumble) and low-pass (hiss) applications, and the third is an inductor-capacitor passive circuit layout, which allows adjustment of the HF turn-over frequency by variation of the capacitor value.

Such frequency adjustments are, of course, also possible with active filter systems, but require the simultaneous switching of a larger number of components. For

such filters to be effective in their intended application, the slope of the response curve, as defined as the change in the rate of attenuation as a function of frequency, is normally chosen to be high – at least 20 dB/octave – as shown in Fig. 13.58, and, in the case of the filters operating in the treble region, a choice of operating frequencies is often required, as is also, occasionally, the possibility of altering the attenuation rate.

This is of importance, since rates of attenuation in excess of 6 dB/octave lead to some degree of colouration of the reproduced sound, and the greater the attenuation

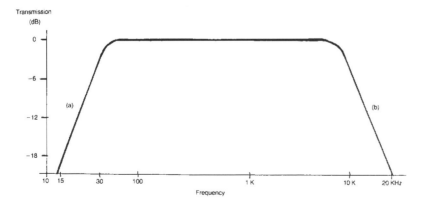

Figure 13.58   Characteristics of circuits of Figs. 13.57 (a) and (b).

rate, the more noticeable this colouration becomes. This problem becomes less important as the turn-over frequency approaches the limits of the range of human hearing, but very steep rates of attenuation produce distortions in transient waveforms whose major frequency components are much lower than notional cut-off frequency.

It is, perhaps, significant in this context that recent improvements in compact disc players have all been concerned with an increase in the sampling rate, from 44.1 kHz to 88.2 kHz or 176.4 kHz, to allow more gentle filter attenuation rates beyond the 20 kHz audio pass band, than that provided by the original 21 kHz 'brick wall' filter. The opinion of the audiophiles seems to be unanimous that such CD players, in which the recorded signal is two- or four-times 'oversampled', which allows much more gentle 'anti-aliasing' filter slopes, have a much preferable HF response and also have a more natural, and less prominent, high-frequency characteristic, than that associated with some earlier designs.

14 Power Output Stages

John Linsley Hood

The ultimate requirement of an amplifier is that it should feed power to loudspeakers. This is the most difficult of the purely electronic portions of the sound reproduction system, and in this Chapter, John Linsley Hood explains the difficulties of the process and the solutions that are employed in the amplifiers of today.

In principle, the function of an audio power amplifier is a simple one: to convert a low-power input voltage signal from a pre-amplifier or other signal source into a higher power signal capable of driving a loudspeaker or other output load, and to perform this amplification with the least possible alteration of the input waveform or to the gain/frequency or phase/frequency relationships of the input signal.

Valve-operated Amplifier Designs

In the earlier days of audio amplifiers, this function was performed by a two or three-valve circuit design employing a single output triode or pentode, transformer coupled to the load, and offering typical harmonic distortion levels, at full power – which would be, perhaps, in the range 2–8 W–of the order of 5–10 per cent THD, at 1 khz, and amplifier output stages of this kind formed the mainstay of radio receivers and radiograms in the period preceding and immediately following the 1939–45 World War.

With the improvement in the quality of gramophone recordings and radio broadcasts following the end of the war, and the development of the new 'beam tetrode' as a replacement for the output pentode valve, the simple 'single-ended' triode or pentode output stage was replaced by push-pull output stages. These used beam tetrodes either triode-connected, as in the celebrated Williamson design, or with a more complex output transformer design, as in the so-called Ultra-linear layout, or in the Quad design, shown in Fig. 14.1(a), (b) and (c).

The beam tetrode valve construction, largely the work of the Marconi-Osram valve company of London,

offered a considerable improvement in the distortion characteristics of the output pentode, while retaining the greater efficiency of that valve in comparison with the output triode. This advantage in efficiency was largely retained when the second grid was connected to the anode so that it operated as a triode.

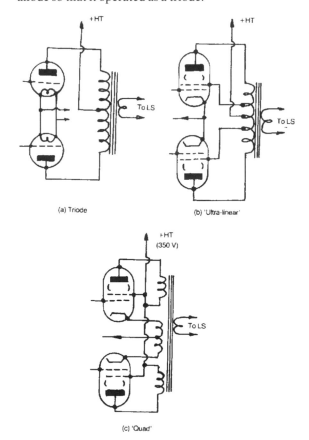

Figure 14.1 Push-pull valve amplifier output stages.

This electrode interconnection was adopted in the Williamson design, shown in Fig. 14.2, in which a substantial amount of overall negative feedback was

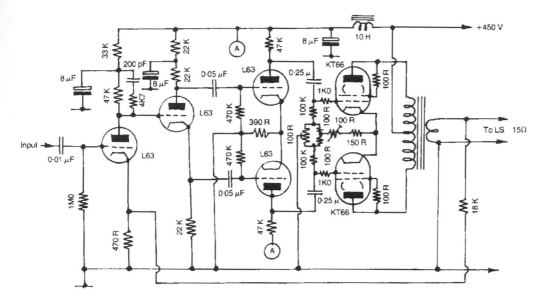

Figure 14.2 The Williamson amplifier.

employed, to give a harmonic distortion figure of better than 0.1 per cent, at the rated 15 W output power.

In general, the principle quality determining factor of such an audio amplifier was the output transformer, which coupled the relatively high output impedance of the output valves to the low impedance of the loudspeaker load, and good performance demanded a carefully designed and made component for this position. Nevertheless, such valve amplifier designs did give an excellent audio performance, and even attract a nostalgic following to this day.

Early Transistor Power Amplifier Designs.

With the advent of semiconductors with significant power handling capacity, and the growing confidence of designers in their use, transistor audio amplifier designs began to emerge, largely based on the 'quasi complementary' output transistor configuration designed by H C Lin, which is shown in Fig. 14.3. This allowed the construction of a push/pull power output stage in which only PNP power transistors were employed, these being the only kind then available. It was intended basically for use with the earlier diffused junction Germanium transistors, with which it worked adequately well.

However, Germanium power transistors at that time had both a limited power capability and a questionable reliability, due to their intolerance of junction temperatures much in excess of 100°C. The availability of silicon power transistors, which did not suffer from this problem

to such a marked degree, prompted the injudicious appearance of a crop of solid-state audio amplifiers from manufacturers whose valve operated designs had enjoyed an erstwhile high reputation in this field.

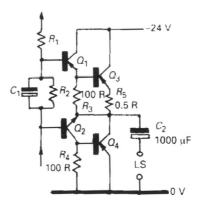

Figure 14.3 Quasi-complementary push-pull transistor output stage from HC Lin.

Listener Fatigue and Crossover Distortion

Fairly rapidly after these new semiconductor-based audio amplifiers were introduced, murmurings of discontent began to be heard from their users, who claimed, quite justifiably, that these amplifiers did not give the same warmth and fullness of tone as the previous valve designs, and the term 'listener fatigue' was coined to describe the user's reaction.

The problem, in this case, was a simple one. In all previous audio amplifier experience, it had been found to be sufficient to measure the distortion and other performance characteristics at full output power, since it could reasonably be assumed that the performance of the amplifier would improve as the power level was reduced. This was not true for these early transistor audio amplifiers, in which the performance at full power was probably the best it would ever give.

This circumstance arose because, although the power transistors were capable of increased thermal dissipation, it was still inadequate to allow the output devices to be operated in the 'Class A' mode, always employed in conventional valve operated power output stages, in which the quiescent operating current, under zero output power conditions, was the same as that at maximum output.

Crossover problems

It was normal practice, therefore, with transistor designs, to bias the output devices into 'Class B' or 'Class AB', in which the quiescent current was either zero, or substantially less than that at full output power, in order to increase the operating efficiency, and lessen the thermal dissipation of the amplifier.

This led to the type of defect known as 'crossover distortion', which is an inherent problem in any class 'B' or 'AB' output stage, and is conspicuous in silicon junction transistors, because of the abrupt turn-on of output current with increasing base voltage, of the form shown in Fig. 14.4.

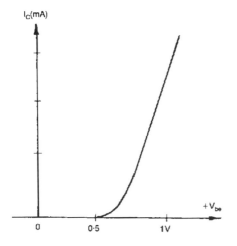

Figure 14.4   Turn-on characteristics of silicon junction transistor.

In a push-pull output stage employing such devices, the transfer characteristics would be as shown in Fig. 14.5(b), rather than the ideal straight line transfer of Fig. 14.5(a). This problem is worsened anyway, in the quasi-complementary Lin design, because the slopes of the upper and lower transfer characteristics are different, as shown in Fig. 14.5(c).

The resulting kink in the transfer characteristic shown in Fig. 14.5(d) lessens or removes the gain of the amplifier for low-level signals which cause only a limited excursion on either side of the zero voltage axis, and leads to a 'thin' sound from the amplifier.

It also leads to a distortion characteristic which is substantially worse at low power levels – at which much listening actually takes place – than at full power output. Moreover, the type of distortion products generated by crossover distortion comprise the dissonant seventh, ninth, eleventh and thirteenth harmonics, which are much more objectionable, audibly, than the second and third harmonics associated with the previous valve amplifier designs.

Additionally, the absence of the output transformer from such transistor amplifier designs allowed the designers to employ much greater amounts of negative feedback (NFB) from the output to the input of the circuit, to attempt to remedy the non-linear transfer characteristics of the system. This increased level of NFB impaired both the load stability and the transient response of the amplifier, which led to an 'edgy' sound. High levels of NFB also led to a more precise and rigid overload clipping level. Such a proneness to 'hard' clipping also impaired the sound quality of the unit.

This failure on the part of the amplifier manufacturers to test their designs adequately before offering them for sale in what must be seen in retrospect as a headlong rush to offer new technology is to be deplored for several reasons.

Of these, the first is that it saddled many unsuspecting members of the public with equipment which was unsatisfactory, or indeed unpleasant, in use. The second is that it tended to undermine the confidence of the lay user in the value of specifications – simply because the right tests were not made, and the correct parameters were left unspecified. Finally, it allowed a 'foot in the door' for those whose training was not in the field of engineering, and who believed that technology was too important to be left to the technologists.

The growth of this latter belief, with its emphasis on subjective assessments made by self-appointed pundits, has exerted a confusing influence on the whole field of audio engineering, it has led to irrationally based choices in consumer purchases and preferences, so that many of the developments in audio engineering over the past two decades, which have been substantial, have occurred in

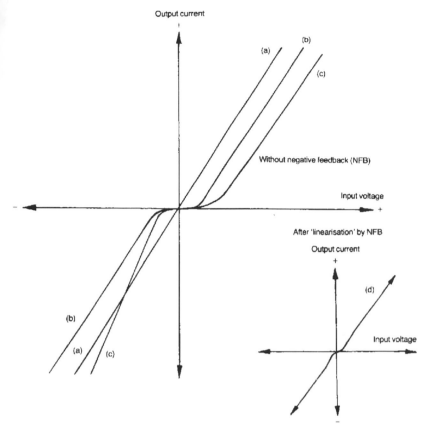

Figure 14.5 Transistor 'Class B' push-pull output stage characteristics. (Line 'a' is a straight line for comparison.)

spite of, and frequently in the face of, well-publicised and hostile opposition from these critics.

Improved Transistor Output Stage Design

The two major developments which occurred in transistor output circuit technology, which allowed the design of transistor power amplifiers which had a low inherent level of crossover distortion, were the introduction in the mid 1960s of fully complementary (NPN and PNP) output power transistors by Motorola Semiconductors Inc, and the invention by IM Shaw (*Wireless World*, June 1969), subsequently refined by PJ Baxandall (*Wireless World*, September 1969) of a simple circuit modification to the basic quasi-complementary output push-pull pair to increase its symmetry.

These layouts are shown in Figs. 14.6(a) and (b). The modified version shown in Fig. 14.6(c) is that used by the author in his *Hi-Fi News* 75 W power amplifier design, for which the complete circuit is given in Fig. 14.7.

In many ways the Shaw/Baxandall quasi-complementary output transistor circuit is preferable to the use of fully complementary output transistors, since, to quote J Vereker of Naim Audio, 'NPN and PNP power transistors are only really as equivalent as a man and a woman of the same weight and height' – the problem being that the different distribution of the N- and P-doped layers leads to significant differences in the HF performance of the devices. Thus, although the circuit may have a good symmetry at low frequencies, it becomes progressively less symmetrical with increasing operating frequency.

With modern transistor types, having a higher current gain transition frequency (that frequency at which the current gain decreases to unity), the HF symmetry of fully complementary output stages is improved, but it is still less good than desired, so the relative frequency/phase characteristics of each half of the driver circuit may need to be adjusted to obtain optimum performance.

Power Mosfet Output Devices

These transistor types, whose operation is based on the mobility of electrostatically induced charge layers

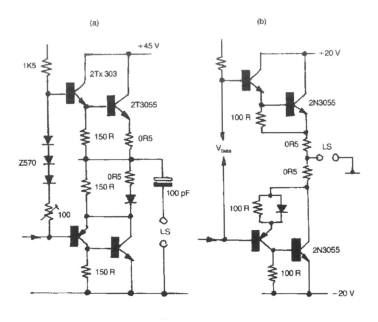

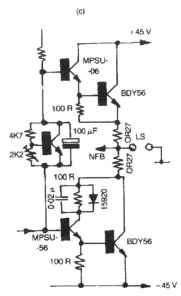

Figure 14.6   Improved push-pull transistor output stages.

(electrons or 'holes') through a very narrow layer of undoped or carrier-depleted silicon, have a greatly superior performance to that of bipolar junction transistors, both in respect of maximum operating frequency and linearity, and allow considerable improvements in power amplifier performance for any given degree of circuit complexity.

Two forms of construction are currently employed for power MOSFETs, the vertical or 'V' MOSFET, which can employ either a 'V' or a 'U' groove formation, of the types shown in Fig. 14.8(a) and (b); though, of these two,

the latter is preferred because of the lower electrostatic stress levels associated with its flat-bottomed groove formation; or the 'D' MOSFET shown in Fig. 14.8(c).

These devices are typically of the 'enhancement' type, in which no current flows at zero gate/source voltage, but which begin to conduct, progressively, as the forward gate voltage is increased. Once the conduction region has been reached, the relationship between gate voltage and drain current is very linear, as shown in Fig. 14.9.

By comparison with 'bipolar' junction power transistors, of conventional types, the MOSFET, which can be

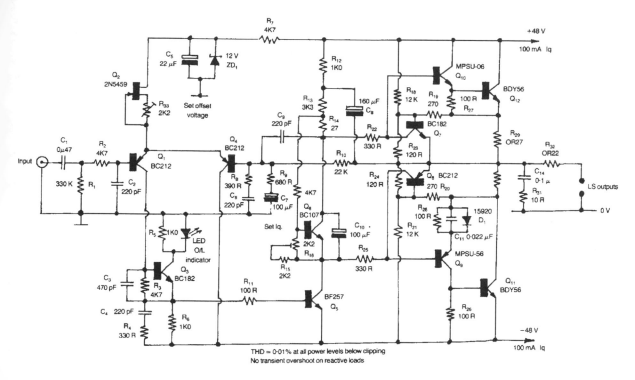

Figure 14.7   JLH *Hi Fi News* 75 W audio amplifier.

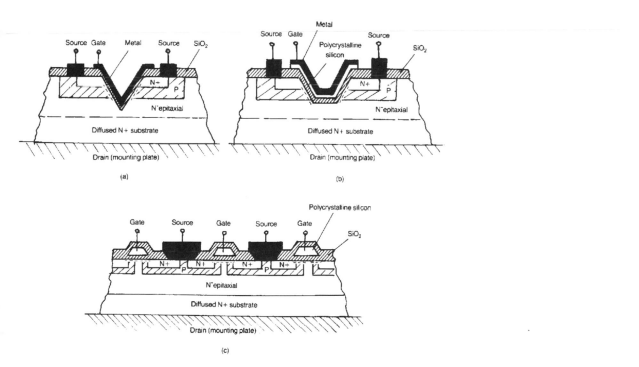

Figure 14.8   Power MOSFET structure (a) V-MOSFET (b) U-MOSFET (c) D-MOSFET.

made in both N channel and P channel types to allow symmetrical circuit layout – though there is a more limited choice of P channel devices – does not suffer from stored charge effects which limit the 'turn-off' speed of bipolar transistors.

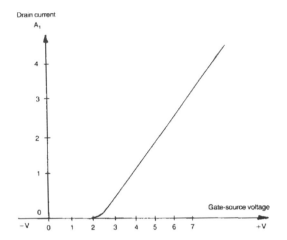

Figure 14.9   Power MOSFET operating characteristics.

The greater speed and lesser internal phase shifts of power amplifier mean that power MOSFETs allow greater freedom in the design of overall NFB layouts. This in turn, gives superior performance in the middle to upper frequency range. Greater care is needed, in the circuit design, however to avoid high frequency parasitic oscillation, which can cause rapid failure of the devices.

The principal technical problem in the evolution of power transistors derived from these charge operated devices, as distinct from the existing small signal MOSFETs, lies in obtaining an adequate capacity for current flow through the device. This problem is solved in two ways; by making the conduction path through which the current must flow as short as possible, and by fabricating a multiplicity of conducting elements on the surface of the silicon slice, which can be connected in parallel to lower the conducting resistance of the device.

V and U MOSFETs

In the 'V' or 'U' groove MOSFET, taking the case of the 'N' channel devices shown in Fig. 14.8, the required narrow conduction channel, in which electrostatically formed negative charges (electrons) may be induced in the depleted 'P' type layer, is obtained by etching a channel through a previously diffused, or epitaxially grown, pair of differently doped layers. With modern technology the effective thickness of such layers can be controlled with great precision.

In the case of the D-MOS device of Fig. 14.8(c), the required narrow channel length is achieved by the very precise geometry of the diffusion masks used to position the doped regions. It is customary in the D-MOS devices to use a polycrystalline silicon conducting layer, rather than aluminum, to provide the gate electrode, since this offers a lower likelihood of contamination of the thin gate insulating layer.

In all of these MOS devices, the method of operation is that a voltage applied to the gate electrode will cause a charge to be induced in the semiconductor layer immediately below it, which, since this layer of charge is mobile, will cause the device to conduct.

Both for the 'V/U' MOS and the 'D' MOS devices, the voltage breakdown threshold of the gate insulating layer, which must be thin if the device is to operate at all, will be in the range 15–35 V. It can be seen that part of this gate insulating layer lies between the gate electrode and the relatively high voltage 'drain' electrode. Avoidance of gate/drain voltage breakdown depends therefore on there being an adequate voltage drop across the N-doped intervening drain region. This in turn depends on the total current flow through the device.

This has the effect that the actual gate/drain breakdown voltage is output current-dependent, and that some protective circuitry may need to be used, as in the case of bipolar output transistors, if damage is to be avoided.

Output Transistor Protection

An inconvenient characteristic of all junction transistors is that the forward voltage of the P-N junction decreases as its temperature is increased. This leads to the problem that if the current through the device is high enough to cause significant heating of the base-emitter region, the forward voltage drop will decrease. If, due to fortuitous variations in the thickness of this layer or in its doping level, some regions of this junction area heat up more than others, then the forward voltage drop of these areas will be less, and current will tend to be funnelled through these areas causing yet further heating.

This causes the problem known as 'thermal runaway' and the phenomenon of secondary breakdown, if certain products of operating current and voltage are exceeded. The permitted regions of operation for any particular bipolar transistor type will be specified by the manufacturers in a 'safe operating area' (SOA) curve, of the type shown in Fig. 14.10.

The circuit designer must take care that these safe operating conditions are not exceeded in use, for example by the inadvertent operation of the amplifier into

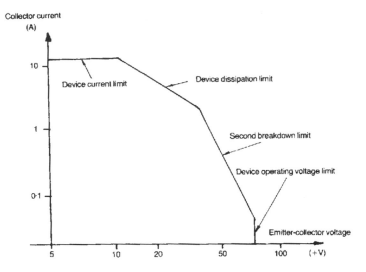

Figure 14.10 Typical safe operating area (SOA) curve for junction power transistor.

a short-circuited output load. Provided that the SOA limits are not greatly exceeded, a simple fuse on the output circuit will probably be adequate, but a more effective type of protection is that given by the clamp transistors, Q_7 and Q_8 in the power amplifier design shown in Fig. 14.7.

In this circuit arrangement, designed by A R Bailey, the clamp transistors monitor simultaneously the voltage present across the output transistors, by means, for example, of R_{18} and R_{23}, and also the output current, in the case of the upper output transistor, by monitoring the voltage developed across R_{29}. If the combination of these two voltage contributions exceeds the 0.55 V turn-on voltage of the clamp transistor, (Q_7), it will conduct and shunt the input signal applied to the first transistor of the output pair (Q_{10}) and avoid output transistor damage.

In the case of power MOSFETs, a simple zener diode, connected to limit the maximum forward voltage which can be applied to the output device, may be quite adequate. An example of this type of output stage protection is shown in the circuit for the output stages of a 45 W power MOSFET amplifier designed by the author, shown in Fig. 14.11.

Power Output and Power Dissipation

One of the most significant changes in audio technology during the past 25 years has been in the design of loudspeaker units, in which the electro-acoustic efficiency, in terms of the output sound level for a given electrical power input, has been progressively traded off against flatness of frequency response and reduced coloration.

This is particularly notable in respect of the low-frequency extension of the LS response, in closed box 'infinite baffle' systems. Amongst other changes here, the use of more massive bass driver diaphragms to lower the fundamental resonant frequency of the system – below which a –12 dB/octave fall-off in sound output will occur – requires that more input power is required to produce the same diaphragm acceleration.

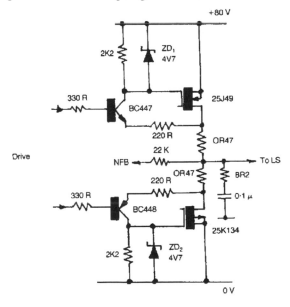

Figure 14.11 Zener diode protection for power MOSFET output stage.

On the credit side, the power handling capacity of modern LS systems has also been increased, so that equivalent or greater sound output levels are still obtainable, but at the cost of much greater input power levels.

Power levels

As a specific example of the power levels which may be needed for realistic reproduction of live sounds using typical modern LS units, recent measurements made in a recording studio showed that the peak output power required from the amplifier to match the peak sound levels produced by a grand piano played by a professional musician in the same room, was in excess of 300 W per channel. This test was made using a pair of high-quality monitor speakers, of a kind also widely used in domestic hi-fi set-ups, whose overall efficiency was typical of modern designs.

No conventional valve amplifier, operating in 'Class A', could meet this power output requirement, in a stereo system, without the penalties incurred in terms of size, cost, and heat output being intolerable. In contrast the solid-state class 'AB' power amplifier actually used in this test fitted comfortably into a 7 in. high slot in a standard 19 in. rack, and ran cool in use.

This reduction in LS efficiency has led to compensatory increases in power amplifier capability, so that the typical output power of a contemporary domestic audio amplifier will be in the range 50–100 W, measured usually with a 4 or 8 ohm resistive load, with 25 or 30 W units, which would at one time have been thought to be very high power units, now being restricted to budget priced 'mini' systems.

Design requirements

The design requirements for such power output levels may be seen by considering the case of a 100 W power amplifier using junction transistors in an output stage circuit of the form shown in Fig. 14.12, the current demand is given by the formula.

$$I = \sqrt{P/R}$$

where $P = 100$, and $R = 8$. This gives an output current value of 3.53 A (RMS), equivalent to a peak current of 5 A for each half cycle of a sinusoidal output signal. At this current the required base-emitter potential for Q_3 will be about 3 V, and allowing for the collector-emitter voltage drop of Q_1, the expected emitter-collector voltage in Q_3 will be of the order of 5 V.

In an 8 ohm load, a peak current of 5 A will lead to a peak voltage, in each half cycle, of 40 V. Adding the 1.1 V drop across R_3, it can be seen that the minimum supply voltage for the positive line supply must be at least 46.1

V. Since the circuit is symmetrical, the same calculations will apply to the potential of the negative supply line. However, in any practical design, the supply voltages chosen will be somewhat greater than the calculated minimum, so line voltages of ±50 V would probably be used.

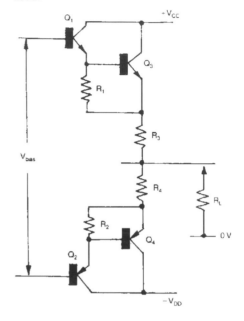

Figure 14.12   Typical transistor output stage.

For a true 'Class A' system, in which both of the output transistors remained in conduction for the whole output cycle, an operating current of at least 5 A, DC, would be required, leading to an output stage dissipation, for each channel, of 500 W.

In the case of a typical 'Class AB' output stage, with a quiescent current of 100 mA, a much more modest no-signal dissipation of 10 W would be required. At full power, the device dissipation may be calculated from the consideration that the RMS current into the load, in each half cycle, will be 3.53 A giving an RMS voltage swing of 28.24 V. The mean voltage across each device during the half cycle of its conduction would therefore be 50 minus 28.24, and the mean dissipation, for each half of the power output stage would be 76.8 W.

The worst case dissipation for such output stages, depending on supply line voltage margins, will be at about half power, where the dissipation in each half of the output stage could rise to 80–85 W, and the heat sinking arrangements provided must be adequate to meet this need. It is still, however, only a third of that required by a 'Class A' system, and even then only required on power peaks, which, with typical programme material, occur infrequently so that the average output stage heat dissipation, even under conditions where the amplifier is used at near its rated maximum power output, may well only be 20–30 W.

General Power Amplifier Design Considerations

Some of the techniques used for obtaining high gain and good linearity from a voltage amplifier stage, using semi-conductors, were discussed in Chapter 13.4. However, in the case of low-power voltage amplifier stages, all of the gain elements will be operated in 'Class A', so that there will not be a need to employ large amounts of overall negative feedback (NFB) to assist in linearising the operating characteristics of the stage.

In power amplifier circuits, on the other hand, the output devices will almost certainly be operated in a non-linear part of their characteristics, where a higher level of NFB will be needed to reduce the associated distortion components to an acceptable level.

Other approaches have been adopted in the case of power amplifier circuits, such as the feed-forward of the error signal, or the division of the output stage into low power 'Class A' and high power 'Class B' sections. These techniques will be discussed later. Nevertheless, even in these more sophisticated designs, overall NFB will still be employed, and the problems inherent in its use must be solved if a satisfactory performance is to be obtained.

Of these problems, the first and most immediate is that of feedback loop stability. The Nyquist criterion for stability in any system employing negative feedback is that, allowing for any attenuation in the feedback path, the loop gain of the system must be less than unity at any frequency at which the loop phase shift reaches 180°.

Bode Plot

This requirement is most easily shown in the 'Bode Plot' of gain versus frequency illustrated in Fig. 14.13. In the case of a direct-coupled amplifier circuit, and most modern transistor systems will be of this type, the low-frequency phase lead is unlikely to exceed 90°, even at very low frequencies, but at the higher frequency end of the spectrum the phase lag of the output voltage in relation to the input signal will certainly be greater than 180° at the critical unity gain frequency, in any amplifier employing more than two operational stages, unless some remedial action is taken.

In the case of the simple power amplifier circuit shown schematically in Fig. 14.14, the input long-tailed pair, Q_1 and Q_2, with its associated constant current source, CC_1, and the second stage, 'Class A' amplier, Q_3 with its constant current load, CC_2, form the two voltage gain stages, with Q_4 and Q_5 acting as emitter-followers to transform the circuit output impedance down to a level which is low enough to drive a loudspeaker (Z_1).

In any amplifier having an output power in excess of a few watts, the output devices, Q_4 and Q_5, will almost certainly be elaborated to one of the forms shown in Figs. 14.3 or 14.6, and, without some corrective action, the overall phase shift of the amplifier will certainly be of

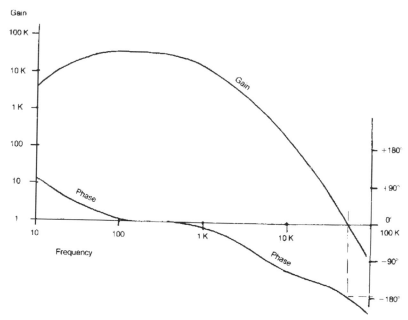

Figure 14.13 Bode plot relating gain and phase to frequency in notional marginally stable amplifier.

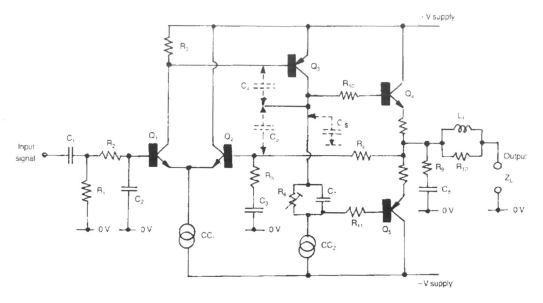

Figure 14.14   Basic layout of simple audio amplifier.

the order of 270°, at the critical unity gain point. This means that if overall negative feedback were to be employed, via R_5 and R_4, the circuit would oscillate vigorously at some lower, 180° phase-shift frequency.

Slew-Rate Limiting and Transient Intermodulation Distortion

Until a few years ago, the most common way of stabilising a negative feedback amplifier of this type was by means of a small capacitor, C_4, connected across the second amplifier transistor, Q_3. This particularly appealed to the circuit designers because it allowed the circuit to be stabilised while retaining a high level of closed-loop

gain up to a high operating frequency, and this facilitated the attainment of low overall harmonic distortion figures at the upper end of the audio frequency band.

Unfortunately, the use of an HF loop stabilisation capacitor in this position gave rise to the problem of slew-rate limiting, because there is only a finite rate at which such a capacitor could charge through the constant current source, CC_2, or discharge through the current available at the collector of Q_1.

Slew limiting

This leads to the type of phenomenon shown in Fig. 14.15, in which, if a continuous signal occurs at the same time as one which leads to a sudden step in the mean voltage

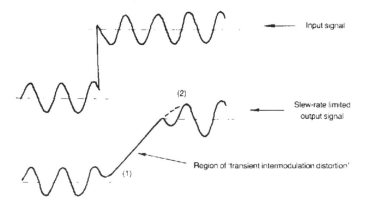

Figure 14.15   Effect of slew-rate limiting on amplifier handling combined signals.

level – which could happen readily in certain types of programme material – then the continuous signal will either be mutilated or totally lost during the period during which the amplifier output voltage traverses, (slews), under slew-rate limited conditions, between output levels '1' and '2'.

This type of problem had been well known among amplifier designers for many years, and the deleterious effect upon the sound of the amplifier had also been appreciated by those who were concerned about this point. However, the commercial pressures upon designers to offer circuits which met the reviewers magic specification of 'less that 0.02 per cent THD from 20 Hz to 20 kHz', led to many of the less careful, or more cynical, to use the slew-rate limiting system of HF stabilisation, regardless of the final sound quality.

Public attention was drawn to this phenomenon by M Otala, in a paper in September 1970, (*IEE Trans.* AU–18, No 3), and the subsequent discussions and correspondence on this point were valuable in convincing the reviewers that low levels of harmonic distortion, on their own, were an insufficient guarantee of audio quality.

HF stabilisation

The alternative method of HF stabilisation which could be employed in the circuit of Fig. 14.14, and which does not lead to the same problems of slew-rate limitation – in that it causes the HF gain to decrease throughout the amplifier as a whole, rather than leaving the input stage operating without feedback at all during the slew-rate limited period – is by the use of capacitor between Q_3 collector, and Q_2 base, as indicated by C_5, and as used in the author's 15 20 W 'class AB' design (*Wireless World*, July 1970).

Unfortunately, although this type of stabilisation leads to designs with pleasing sound quality, it does not lead to the same degree of distortion reduction at the HF end of the spectrum, unless a higher level of loop gain is employed. The 75 W design shown in Fig. 14.7, which achieves a typical distortion performance of less than 0.015 per cent over the whole audio spectrum, is a good example of this latter approach.

A useful analysis of the problem of slew-rate induced distortion, in non-mathematical terms, was given by W G Jung, in *Hi-Fi News*, (Nov. 1977), with some additional remedial circuit possibilities shown by the author in the same journal in January 1978.

Zobel network

Other components used for correcting the phase characteristics of the feedback loop are the so-called Zobel network (C_8, R_9) connected across the output, which will help avoid instability with an open-circuit load, and the inductor/resistor network (L_1, R_{10}) in series with the load, which is essential in some designs to avoid instability on a capacitative load.

Since L_1 is a very small inductor, typically having a value of 4.7 microhenries, its acoustic effects within the 20–20 kHz band will usually be inaudible. Its presence will, however, spoil the shape of any square-wave test waveform, as displayed on an oscilloscope, especially with a capacitative simulated LS load.

Bearing in mind that all of these slew-rate and transient waveform defects are worse at higher rates of change of the input signal waveform, it is prudent to include an input integrating network, such as R_2/C_2 in Fig. 14.14, to limit the possible rate of change of the input signal, and such an input network is common on all contemporary designs. Since the ear is seldom sensitive beyond 20 kHz, except in young children, and most adults' hearing fails below this frequency, there seems little point in trying to exceed this HF bandwidth, except for the purpose of producing an impressive paper specification.

A further significant cause of slew-rate limitation is that of the stray capacitances (C_8) associated with the collector circuitry of Q_3, in Fig. 14.14, and the base input capacitance of Q_4/Q_5. This stray capacitance can be charged rapidly through Q_3, but can only be discharged again at a rate determined by the output current provided by the current source CC_2. This current must therefore be as large as thermal considerations will allow.

The possible avoidance of this source of slew-rate limitation by fully symmetrical amplifier systems, as advocated by Hafler, has attracted many designers.

Among the other defects associated with the power output stage, is that of 'hang up' following clipping, of the form shown graphically in Fig. 14.16. This has the effect of prolonging, and thereby making more audible, the effects of clipping. It can usually be avoided by the inclusion of a pair of drive current limiting resistors in the input circuitry of the output transistors, as shown by R_{10} and R_{11} in Fig. 14.14.

Advanced Amplifier Designs

In normal amplifier systems, based on a straightforward application of negative feedback, the effective circuit can be considered as comprising four parts:

- the output power stages, which will normally be simple or compound emitter followers, using power MOS-FETs or bipolar transistors, using one of the forms shown in Fig. 14.17:(a)–(f):

- the voltage amplifying circuitry;
- the power supply system; and
- the protection circuitry needed to prevent damage to the output devices or the LS load.

The choice made in respect of any one of these will influence the requirements for the others.

Power supply systems

It has, for example, been common practice in the past to use a simple mains transformer, rectifier and reservoir capacitor combination of the kind shown in Fig. 14.18(a) and (b), to provide the DC supply to the amplifier. This

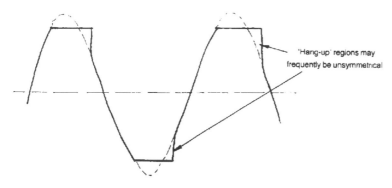

Figure 14.16   Effect of 'hang-up' following clipping.

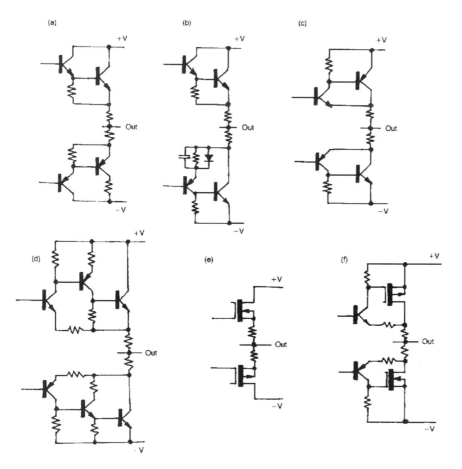

Figure 14.17   Output transistor circuits.

has an output voltage which will decrease under increasing load, and it is necessary to know the characteristics of this supply in order to be able to specify the no-load output voltage which will be necessary so that the output voltage on maximum load will still be adequate to provide the rated output power.

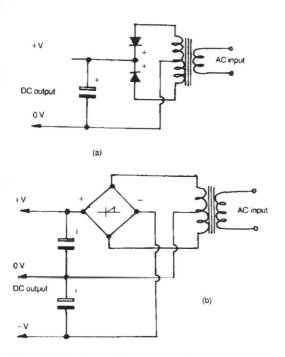

(a)

(b)

Figure 14.18 Simple DC power supplies.

Again, it is probable that the power supply for any medium to high power amplifier will provide more than enough current to burn out an expensive LS system in the event of an output transistor failure in the power amplifier, so some output fuse will be required. Unfortunately, most fuseholders tend to have a variable contact resistance, which introduces an unwelcome uncertainty into the critical LS output circuit.

The presence of the mains frequency-related AC ripple on the supply lines from such a simple power supply circuit, which will become worse as the output power to the load is increased, means that the voltage amplification circuitry must be chosen to have an adequate rejection of the supply line ripple.

Also, due to the relatively high circuit impedance, signal components present on the supply lines will worsen the distortion characteristics of the circuit if they break through into the signal path, since in a Class B or AB system these components will be highly distorted.

Such signal intrusion can also cause unwanted interchannel breakthrough, and encourage the use of entirely separate power supply systems. Such simple power supplies may carry hidden cost penalties, too, since the working voltage rating of the power transistors must be adequate to withstand the no-load supply potential, and this may demand the use of more expensive, high-voltage transistors having relatively less good current gain or high frequency characteristics.

Stabilised power supplies

If, on the other hand, an electronically stabilised power supply is used, the output voltage can be controlled so that it is the same at full output power as under quiescent current conditions, and largely independent of mains voltage. Moreover, the supply can have a 're-entrant' overload characteristic, so that, under either supply line or LS output short-circuit conditions, supply voltage can collapse to a safely low level, and a monitor circuit can switch off the supplies to the amplifier if an abnormal DC potential, indicating output transistor failure, is detected at the LS terminals. Such protection should, of course, be fail-safe.

Other advantages which accrue from the use of stabilised power supplies for audio amplifiers are the low inherent power supply line 'hum' level, making the need for a high degree of supply line ripple rejection a less important characteristic in the design of the small signal stages, and the very good channel-to-channel isolation, because of the low inherent output impedance of such systems.

The principal sonic advantages which this characteristic brings are a more 'solid' bass response, and a more clearly defined stereo image, because the supply line source impedance can be very low, typically of the order of 0.1 ohms. In a conventional rectifier/capacitor type of power supply, a reservoir capacitor of 80,000 µF would be required to achieve this value at 20 Hz. The inherent inductance of an electrolytic capacitor of this size would give it a relatively high impedance at higher frequencies, whereas the output impedance of the stabilised supply would be substantially constant.

The design of stabilised power supplies is beyond the scope of this chapter, but a typical unit, giving separate outputs for the power output and preceding voltage amplifier stages of a power amplifier, and offering both re-entrant and LS voltage offset protection, is shown in Fig. 14.19.

LS protection circuitry

This function is most commonly filled by the use of a relay, whose contacts in the output line from the amplifier to the LS unit are normally open, and only closed by

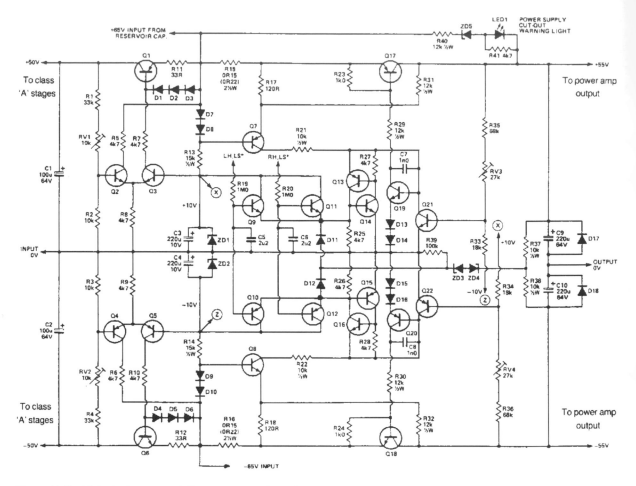

Figure 14.19　Twin dual-output stabilised power supply, with re-entrant overload protection and LS offset shut-down, by Linsley Hood.

the associated relay operating circuitry if all the appropriate conditions, including LS terminal DC offset and output load impedance, are within satisfactory limits.

Such circuitry is effective, but the relay contacts must have adequate contact area, and the relay must either be hermetically sealed or the contacts must be plated with gold or equivalent noble metal to preserve a low value of contact resistance in the presence of atmospheric contamination.

Output stage emitter follower configurations

Various circuit forms have been adopted for this purpose, of which the more useful ones are shown in Fig. 14.17. The actual form is known to have a bearing on the perceived sound quality of the design, and this was investigated by Otala and Lammasneimi, (*Wireless*

World, Dec. 1980), who found that, with bipolar transistors, the symmetrical compound emitter follower circuit of Fig. 14.17(c) was significantly better than the complementary Darlington configuration of Fig. 14.17(a), in terms of the amplifier's ability to reject distortion components originating in load non-linearities from, or electromagnetic signal voltages originating in, the LS unit.

The actual desirable criterion in this case is the lowness of the output impedance of the output emitter follower configuration before overall negative feedback is applied to the amplifier, the point being that the feedback path within the amplifier is also a path whereby signals originating outside the amplifier can intrude within the signal path.

The major advantages offered by the use of power MOSFETs, typically in one or other of the configurations shown in Fig. 14.17(e) or (f), are their greater intrinsic linearity in comparison with bipolar junction transistors, and their much higher transition frequency, which simplifies the design of a stable feedback amplifier having a low harmonic distortion. On the other hand, the output impedance of the simple source follower is rather high, and this demands a higher gain from the preceding voltage amplifier stage if an equivalent overall performance is to be obtained.

The inclusion of a low value resistor in the output circuit of the amplifier, between the emitter follower output and the LS load, as shown in the (1972) 75 W amplifier of Fig. 14.7, greatly reduces this type of load induced distortion and is particularly worth while in the case of the circuit layouts shown in Fig. 14.17(a) and (e).

Power amplifier voltage gain stages

The general design systems employed in transistor gain stages have been examined in Chapter 13. However, for high quality audio power amplifiers higher open-loop stage gains, and lower inherent phase shift characteristics, will be required – to facilitate the use of large amounts of overall NFB to linearise output stage irregularities – than is necessary for the preceding small signal gain stages.

Indeed, with very many modern audio amplifier designs, the whole of the small signal pre-amplifier circuitry relies on the use of good quality integrated circuit operational amplifiers, of which there are a growing number which are pin compatible with the popular TL071 and TL072 single and dual FET-input op. amps. For power amplifier voltage stages, neither the output voltage nor the phase shift and large signal transient characteristics of such op. amps. are adequate, so there has been much development of linear voltage gain circuitry, for the 'Class A' stages of power amplifiers, in which the principal design requirements have been good symmetry, a high gain/bandwidth product, a good transient response, and low-phase shift values within the audio range.

A wide range of circuit devices, such as constant current sources, current mirrors, active loads and 'long-tailed pairs' have been employed for this purpose, in many ingenious layouts. As a typical example, the circuit layout shown in Fig. 14.20, originally employed by National Semi-conductors Inc. in its LH0001 operational amplifier, and adopted by Hitachi in a circuit recommended for use with its power MOSFETs, offers a high degree of symmetry, since Q_3/Q_4, acting as a current mirror, provide an active load equivalent to a symmetrically operating transistor amplifier, for the final amplifier transistor, Q_6.

This circuit offers a voltage gain of about 200000 at low frequencies, with a stable phase characteristic and a high degree of symmetry. The derivation and development of

this circuit was analysed by the author in *Wireless World* (July 1982).

An alternative circuit layout, of the type developed by Hafler, has been described by E Borbely (*Wireless World*, March 1983), and is shown in Fig. 14.21. This is deliberately chosen to be fully symmetrical, so far as the transistor characteristics will allow, to minimise any tendency to slew-rate limiting of the kind arising from stray capacitances charging or discharging through constant current sources. The open/loop gain is, however, rather lower than of the NS/Hitachi layout of Fig. 14.20.

Both unbypassed emitter resistors and base circuit impedance swamping resistors have been freely used in the Borbely design to linearise the transfer and improve the phase characteristics of the bipolar transistors used in this design, and a further improvement in the linearity of the output push-pull Darlington pairs $(Q_5/Q_6/Q_8/Q_9)$ is obtained by the use of the 'cascode' connected buffer transistors Q_7 and Q_{10}.

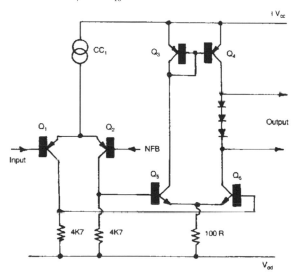

Figure 14.20 Symmetrical high gain stage.

The particular merit of the cascode layout in audio circuitry is that the current flow through the cascode transistor is almost entirely controlled by the driver transistor in series with its emitter. In contrast, the collector potential of the driver transistor remains virtually constant, thus removing the deleterious effect of non-linear internal voltage dependent leakage resistances or collector-base capacitances from the driver device.

The very high degree of elaboration employed in recent high-quality Japanese amplifiers in the pursuit of improvements in amplifier performance, is shown in the circuit of the Technics SE–A100 voltage gain stage, illustrated in a somewhat simplified form in Fig. 14.22.

In this, an input long-tailed pair configuration, based

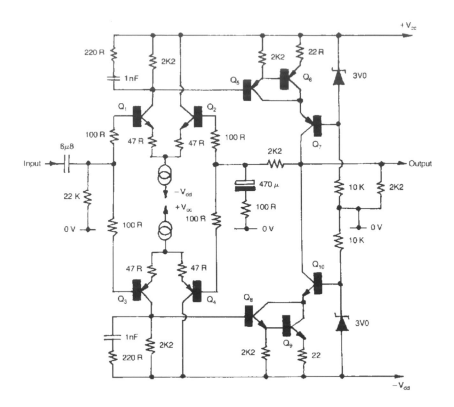

Figure 14.21   Symmetrical push-pull stage by Borbely.

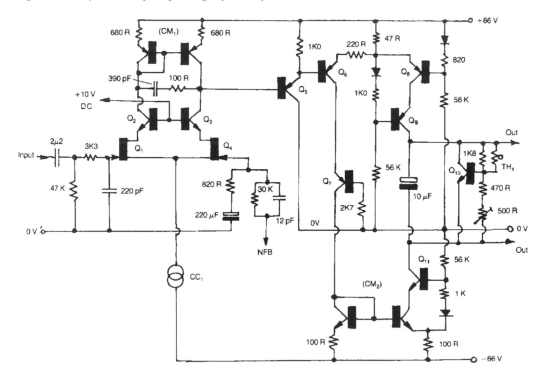

Figure 14.22   Technics voltage gain stage.

on junction FETs (Q_1, Q_4 with CC_1), to take advantage of the high linearity of these devices, is cascode isolated (by Q_2, Q_3) from a current mirror circuit, (CM_1), which combines the output of the input devices in order to maximise the gain and symmetry of this stage, and drives a PNP Darlington pair amplifier stage, (Q_5, Q_6).

The output transistor, Q_6, drives a current mirror (CM_2) through a cascode isolating transistor (Q_7) from Q_6 collector, and a further cascode isolated amplifier stage (Q_8, Q_9) from its emitter, for which the current mirror CM_2 serves as an active load. The amplified diode transistor, Q_{10}, serves to generate a DC offset potential, stabilised by a thermistor, (TH_1), to forward bias a succeeding push-pull pair of emitter followers.

As a measure of the effectiveness of this circuit elaboration, the quoted harmonic distortion figures, for the whole amplifier, are typically of the order of 0.0002 per cent.

Alternative Design Approaches

The fundamental problem in any 'Class B' or 'Class AB' transistor amplifier is that some non-linearity inevitably exists at the region where the current flow through one output transistor turns on and the other turns off.

This non-linearity can be minimised by the careful choice of output stage quiescent current, but the optimum performance of the amplifier depends on this current value being set correctly in the first place, and on its remaining constant at the set value throughout the working life of the amplifier.

One answer is, of course, to abandon 'Class AB' operation, and return to 'Class A', where both output transistors conduct during the whole AC output cycle, and where the only penalty for an inadvertent decrease in the operating current is a decrease in the maximum output power. The author's original four-transistor, 10 W 'Class A' design (*Wireless World*, April 1969) enjoys the distinction of being the simplest transistor operated power amplifier which was capable of matching the sound quality of contemporary valve designs. The problem, of course, is its limited power output.

The Blomley non-switching output circuit

The possibility of achieving a higher power 'Class AB' or even 'Class B' amplifier circuit, in which some circuit device is used to remove the fundamental non-linearity of the output transistor crossover region, in such circuits, is one which has tantalised amplifier designers for the past two decades, and various approaches have been explored. One of these which attracted a lot of interest at the time was that due to P Blomley, (*Wireless World*, February/March 1971), and which is shown, in simplified form, in Fig. 14.23.

In this, Q_1, Q_2, and Q_3 form a simple three-stage voltage amplifier, stabilised at HF by the inclusion of capacitor C_1 between Q_2 collector and Q_1 emitter. The use of a constant current load (CC_1) ensures good linearity from this stage. The crux of the design is the use of a pair of grounded base or cascode connected transistors, (Q_4, Q_5) whose bases are held, with a suitable DC offset between them, at some convenient mid-point DC level, which route the output current from the gain stage, Q_3, to one or other of the push-pull output triples (Q_6, Q_7, Q_8 and Q_9, Q_{10}, Q_{11}) which are arranged to have a significant

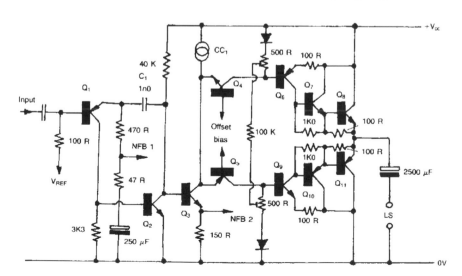

Figure 14.23 The Blomley non-switching push-pull output stage.

current gain and also to be forward-biased, and therefore conducting, during the whole of the output signal cycle.

Although acclaimed as a non-switching 'Class AB' output configuration, in reality, the switching of the output half cycles of a 'Class B' system still takes place, but through the small signal transistors Q_4 and Q_5 which, since they are freed from the vagaries of the output loudspeaker load, and the need to pass a substantial output current, may be assumed to do the switching more cleanly and rapidly. Nevertheless, the need to maintain an accurate DC offset between the bases of these switching transistors still remains, and errors in this will worsen residual crossover distortion defects.

The Quad current dumping amplifier design.

This unique and innovative circuit, subsequently employed commercially in the Quad 405 power amplifier, was first disclosed by P J Walker and M P Albinson at the fiftieth convention of the Audio Engineering Society, in the summer of 1975, and a technical description was given by Walker later in the year, (*Wireless World*, December 1975).

This design claims to eliminate distortion due to the discontinuous switching characteristics of the unbiased, 'Class B', push-pull output transistor pair, by the use of a novel dual path feedback system, and thereby eliminate the need for precise setting-up of the amplifier circuit. It has been the subject of a very considerable subsequent analysis and debate, mainly hingeing upon the actual method by which it works, and the question as to whether it does, or even whether it can, offer superior results to the same components used in a more conventional design.

What is not in doubt is that the circuit does indeed work, and that the requirement for correct adjustment of the output transistor quiescent current is indeed eliminated.

Of the subsequent discussion, (P J Baxandall, *Wireless World*, July 1976; Divan and Ghate, *Wireless World*, April 1977; Vanderkooy and Lipshitz, *Wireless World*, June 1978: M McLoughlin, *Wireless World*, September/October 1983), the explanation offered by Baxandall is the most intellectually appealing, and is summarised below.

Consider a simple amplifier arrangement of the kind shown in Fig. 14.24(a), comprising a high-gain linear amplifier (A_1) driving an unbiased ('Class B') pair of power transistors (Q_1, Q_2), and feeding a load, Z_L. Without any overall feedback the input/output transfer curve of this circuit would have the shape shown in the curve 'x' of Fig. 14.25, in which the characteristic would be steep from M' to N' while Q_2 was conducting, much flatter

between N' and N while only the amplifier A_1 was contributing to the output current through the load, by way of the series resistance R_3, and then steeper again from N to M, while Q_1 was conducting.

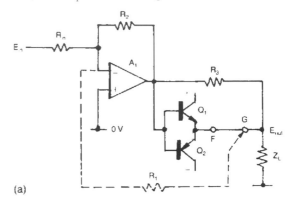

(a)

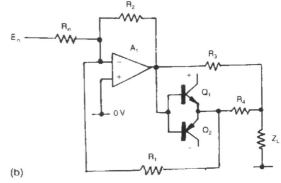

(b)

Figure 14.24   The basic current dumping system.

If overall negative feedback is applied to the system via R_1, the extent of the discontinuity in the transfer curve can be made less, especially if the closed loop gain of A_1 is sufficiently high, leading to a more linear characteristic of the type shown in 'y'. However, it would still be unsatisfactory.

What is needed is some way of increasing the negative feedback applied to the system during the period in which Q_1 and Q_2 are conducting, to reduce the overall gain of the system so that the slope of the transfer characteristic of Fig. 14.25 is identical in the regions M' to N' and N to M to that between N' and N.

This can be done by inserting a small resistor, (R_4), between points F and G, in the output circuit of the push-pull emitter followers Q_1 and Q_2, so that there will be a voltage drop across this resistor, when Q_1 and Q_2 are feeding current into the load, and then deriving extra negative feedback from this point, (F), which will be related to the increased current flow into the load.

If the values of R_1, R_2, R_3 and R_4 are correctly chosen, in relation to the open loop gain of A_1, the distortion due to the unbiased output transistors will very nearly vanish, any residual errors being due solely to the imperfect switching characteristics of Q_1 and Q_2 and the phase errors at higher frequencies of the amplifier A_1.

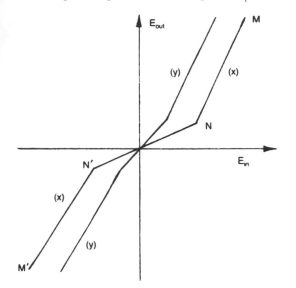

Figure 14.25 Current dumping amplifier transfer characteristics.

Unfortunately, the output resistor, R_4, in the LS circuit would be wasteful of power, so Walker and Albinson substitute a small inductor for this component in the actual 404 circuit, and compensate for the frequency dependent impedance characteristics of this by replacing R_2 with a small capacitor.

While the amplifier circuit still works within the performance limits imposed by inevitable tolerance errors in the values of the components, this L and C substitution serves to complicate the theoretical analysis very considerably, and has led to a lot of the subsequent debate and controversy.

Feed-forward systems

The correction of amplifier distortion by the use of a comparator circuit, which would disclose the error existing between the input and output signals, so that the error could be separately amplified and added to or subtracted from the voltage developed across the load, was envisaged by H S Black, the inventor of negative feedback, in the earlier of his two US Patents (1686792/1928). Unfortunately, the idea was impracticable at that time because sufficiently stable voltage amplifiers were not obtainable.

However, if such a system could be made to work, it would allow more complete removal of residual errors and distortions than any more conventional negative feedback system, since with NFB there must always be some residual distortion at the output which can be detected by the input comparator and amplified to reduce the defect. In principle, the 'feed-forward' addition of a precisely measured error signal could be made to completely cancel the error, provided that the addition was made at some point, such as the remote end of the load, where it would not be sensed by the error detection system.

Two practical embodiments of this type of system by A M Sandman, (*Wireless World*, October 1974), are shown in Fig. 14.26 (a) and (b). In the second, the iterative addition of the residual distortion components would allow the distortion in the output to be reduced to any level desired, while still allowing the use of a load circuit in which one end was connected to the common earth return line.

'Class S' amplifier systems

This ingenious approach, again due to A M Sandman, (*Wireless World*, September 1982), has elements in common with the current dumping system, though its philosophy and implementation are quite different. The method employed is shown schematically in Fig. 14.27. In this a low-power, linear, 'Class A' amplifier, (A_1) is used to drive the load (Z_1) via the series resistor, R_4. The second linear amplifier, driving the 'Class B' push-pull output transistor pair, Q_1/Q_2, monitors the voltage developed across R_4, by way of the resistive attenuator, R_5/R_6, and adjusts the current fed into the load from Q_1 or Q_2 to ensure that the load current demand imposed on A_1 remains at a low level.

During the period in which neither Q_1 nor Q_2 is conducting, which will only be at regions close to the zero output level, the amplifier A_1 will supply the load directly through R_4. Ideally, the feedback resistor, R_2, should be taken to the top end of Z_1, rather than to the output of A_1, as shown by Sandman.

A modified version of this circuit, shown in Fig. 14.28, is used by Technics in all its current range of power amplifiers, including the one for which the gain stage was shown in Fig. 14.22. In this circuit, a high-gain differential amplifier, (A_2), driving the current amplifier output transistors (Q_1, Q_2), is fed from the difference voltage existing between the outputs of A_1 and A_2, and the bridge balance control, R_y, is adjusted to make this value as close to zero as practicable.

Under this condition, the amplifier A_1 operates into a nearly infinite load impedance, as specified by Sandeman, a condition in which its performance will be very good. However, because all the circuit resistances associated with R_3, R_4 and R_x are very low, if the current drive transistors are unable to accurately follow the input waveform, the amplifier A_1 will supply the small residual error current. This possible error in the operation of Q_1 and Q_2 is lessened, in the Technics circuit, by the use of a small amount of forward bias (and quiescent current) in transistors Q_1 and Q_2.

Contemporary Amplifier Design Practice

This will vary according to the end use envisaged for the design. In the case of low cost 'music centre' types of system, the main emphasis will be upon reducing the system cost and overall component count. In such equipment, the bulk of the circuit functions, including the power

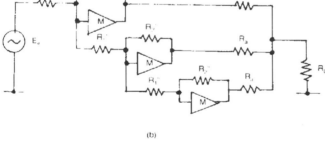

Figure 14.26   Feed-forward systems for reducing distortion.

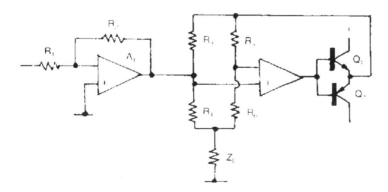

Figure 14.27   Sandman's 'Class S' system.

output stages, will be handled by purpose built integrated circuits.

From the point of view of the manufacturers, these ICs and other specialised component groupings will be in-house items, only available from the manufacturer, and a more substantial profit margin on the sale of these to accredited repair agents will assist in augmenting the meagre profit levels imposed by competitive pressures on the original sale of the equipment.

In more prestigious equipment, intended to be assessed against similar units in the hi-fi market, the choice of circuit will lie between designs which are basically of the form shown in Fig. 14.14, but using more elaborate first stage amplifier circuitry, and with either bipolar or power MOSFET transistor output devices, or more elaborate systems derived from the Blomley, Sandman, or Current Dumping designs, or on systems in which the amplifier quiescent current is automatically adjusted during the output cycle with the aim of keeping the output stages operating, effectively, in 'Class A', but without the thermal dissipation penalty normally incurred by this.

Many, but not all, of the better quality units will employ stabilised DC power supplies, and virtually all of the high quality designs will be of the so-called direct-coupled form, in which the LS output is taken directly from the mid-point of the output emitter followers, without the interposition of a DC blocking output capacitor. (The use of true DC coupling from input to LS output is seldom found because of the problems of avoiding DC offset drift.)

Such direct-coupled amplifiers will, inevitably, employ symmetrical positive and negative supply lines, and in more up-market systems, the power supplies to the output stages will be separated from those for the preceding low power driver stages, and from any power supply to the preceding pre-amp. circuitry. This assists in keeping cross-channel breakthrough down to a low level, which is helpful in preserving the stability of the stereo image.

Great care will also be exercised in the best of contemporary designs in the choice of components, particularly capacitors, since the type of construction employed in these components can have a significant effect on sound quality. For similar reasons, circuitry may be chosen to minimise the need for capacitors, in any case.

Capacitors

Although there is a great deal of unscientific and ill-founded folklore about the influence of a wide variety of circuit components, from connecting wire to the nature of the fastening screws, on the final sound quality of an audio amplifying system, in the case of capacitors there is some technical basis for believing that imperfections in the operational characteristics of these components may be important, especially if such capacitors are used as an integral part of a negative feedback comparator loop.

The associated characteristics which are of importance include the inherent inductance of wound foil components, whether electrolytic or non-polar types, the piezo-electric or other electromechanical effects in the dielectric layer, particularly in ceramic components, the stored charge effects in some polymeric materials, of the kind associated with 'electret' formation, (the electrostatic equivalent of a permanent magnet, in which the material retains a permanent electrostatic charge), and non-linearities in the leakage currents or the capacitance as a function of applied voltage.

Polypropylene film capacitors, which are particularly valued by the subjective sound fraternity, because of their very low dielectric loss characteristics, are particularly

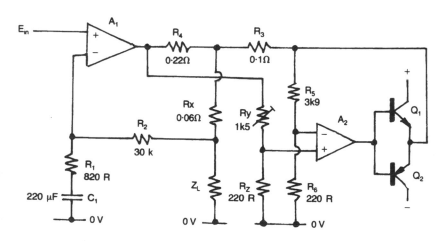

Figure 14.28 Technics power amplifier circuit (SE-A100).

prone to electret formation, leading to an asymmetry of capacitance as a function of polarising voltage. This effect is small, but significant in relation to the orders of harmonic distortion to which contemporary designs aspire.

Care in the decoupling of supply lines to the common earth return line is also of importance in the attainment of high performance, as is care in the siting and choice of earth line current paths. Such aspects of design care are not disclosed in the electronics circuit drawings.

Sound Quality and Specifications

Most of the performance specifications which relate to audio systems – such as the power output (preferably measured as a continuous power into a specified resistive load), the frequency response, the input impedance and sensitivity, or the background noise level in comparison to some specified signal level – are reasonably intelligible to the non-technical user, and capable of verification on test.

However, the consideration which remains of major interest to the would-be user of this equipment is what it will sound like, and this is an area where it is difficult to provide adequate information from test measurements.

For example, it has been confidently asserted by well-known engineers that all competently designed power amplifiers operated within their ratings will sound alike. This may be true in respect of units from the same design stable, where the same balance of compromises has been adopted by the designer, but it is certainly untrue in respect of units having different design philosophies, and different origins.

As a particular case in point, until the mid-1970s a large majority of commercial designs employed a second-stage slew-rate limiting capacitor, in the mode discussed above, as a means of attaining stable operation without sacrifice of THD characteristics at the upper end of the audio band.

The type of sonic defect produced by slew-rate limiting is unattractive, and clearly audible by any skilled listener who has trained his ears to recognise the characteristic degradation of sound quality due to this. Since the publicity given to transient intermodulation distortion by Otala, this type of stabilisation is now seldom used in feedback amplifiers and other, technically more correct, methods are now more generally employed.

Since this type of shortcoming is now recognised, are we to accept that those of the preceding generation of designs which suffered from this defect (and which, in many cases, originated from the drawing boards of those same engineers who denied the existence of any differences) were, indeed, incompetently designed?

Design compromises

Unfortunately, the list of desirable parameters relating to the sound quality of audio amplifiers is a long one, and some of the necessary specifications are imperfectly understood. What is beyond doubt is that most of the designers operating in this field are well aware of their inability to attain perfection in all respects simultaneously, so that they must seek a compromise which will necessarily involve the partial sacrifice of perfection in one respect in order to obtain some improvement in some other mutually incompatible region.

The compromises which result, and which have an influence on the amplifier sound, are based on the personal judgment or preferences of the designer, and will vary from one designer to another.

An example of this is the case of low harmonic distortion figures at higher audio frequencies, and good transient performance and freedom from load induced instability, in a feedback amplifier. These characteristics are partially incompatible. However, 'THD' figures are prominently quoted and form an important part of the sales promotion literature and the reviewers report. Considerable commercial pressure therefore exists to attain a high performance in this respect.

Transient characteristics and feedback loop stability margins are not quoted, but shortcomings in either of these can give an amplifier a 'hard' or 'edgy' sound quality and it is not uncommon for poor amplifier transient performance to lead to a redistribution of energy in the time domain or the frequency spectrum which may amount to as much as a quarter of the total transient energy.

Bearing in mind the importance of good behaviour in this respect, it is to be regretted that if the transient performance of an amplifier is shown at all, it is likely to be shown only as a response to symmetrical square-waves, rather than to the more complex asymmetrical transients found in programme material.

Measurement systems

A measurement system which attempts to provide a more readily quantisable technique for assessing amplifier performance, in the hope of lessening the gap which exists between existing performance specifications – which mainly relate to steady state (i.e., sinusoidal) test signals – and the perceived differences in amplifier sound, has been devised by Y. Hirata, (*Wireless World*, October 1981).

This technique uses asymmetrical, pulse type, input signals which approach more closely in form to the kinds of transient pressure waveforms generated by, for example, a bursting balloon, a hand clap, or cracker. The changes in these test waveforms caused by a variety of amplifier faults is shown by Hirata, but the interpretation of the results is too complex for it to be likely to replace the ubiquitous, if misleading, harmonic distortion figure as a criterion of amplifier goodness.

A further type of measurement, being explored by the BBC, has been described by R. A. Belcher, (*Wireless World*, May 1978) using pseudo-random noise signals, derived by frequency shifting a 'comb filter' spectrum. This is claimed to give a good correlation with perceived sound quality, but is, again, too complex at the moment to offer an easily understood measurement by which a potential customer could assess the likely quality of an intended purchase.

Conclusions

The conclusion which can be drawn from this discussion is that harmonic distortion figures, on their own, offer little guidance about sound quality, except in a negative sense – that poor THD figures, in the 'worse than 0.5 per cent' category, are likely to lead to poor sound quality. Fortunately, the understanding by design engineers of the requirements for good sound quality is increasing with the passage of time, and the overall quality of sound produced, even by 'budget' systems, is similarly improving.

In particular, there is now a growing appreciation of the relationship between the phase/frequency characteristics of the amplifier and the sound-stage developed by a stereo system of which it is a part.

There still seems to be a lot to be said for using the simplest and most direct approach, in engineering terms, which will achieve the desired end result – in that components which are not included will not fail, nor will they introduce any subtle degradation of the signal because of the presence of minor imperfections in their mode of operation. Also, simple systems are likely to have a less complex phase/frequency pattern than more highly elaborated circuitry.

For the non-technical user, the best guarantee of satisfaction is still a combination of trustworthy recommendation with personal experience, and the slow progress towards some simple and valid group of performance specifications, which would have a direct and unambiguous relationship to perceived sound quality, is not helped either by the advertisers frequent claims of perfection or the prejudices, sycophancy and favouritism of some reviewers.

On the credit side, the presence of such capricious critics, however unwelcome their views may be to those manufacturers not favoured with their approval, does provide a continuing stimulus to further development, and a useful counter to the easy assumption that because some aspect of the specification is beyond reproach the overall sound quality will similarly be flawless.

15 Loudspeakers

Stan Kelly

The conversion from electronic signals to sound is the formidable task of the loudspeaker. In this chapter, Stan Kelly examines principles and practice of modern loudspeaker design.

A loudspeaker is a device which is actuated by electrical signal energy and radiates acoustic energy into a room or open air. The selection and installation of a speaker, as well as its design, should be guided by the problem of coupling an electrical signal source as efficiently as possible to an acoustical load. This involves the determination of the acoustical load or radiation impedances and selection of a diaphragm, motor, and means for coupling the loaded loudspeaker to an electrical signal source. The performance of the speaker is intimately connected with the nature of its acoustic load and should not be considered apart from it.

Radiation of Sound

Sound travels through the air with a constant velocity depending upon the density of the air; this is determined by its temperature and the static air pressure. At normal room temperature of 22°C and static pressure p_o of 751 mm Hg (10^5 N/m^2) the density of the ambient air is 1.18 Kg/m^3. Under these conditions, the velocity of sound is 344.8 m/s, but 340 m/s is a practical value. The wavelength of a sound (λ) is equal to the velocity of propagation described by its frequency:

$$\lambda = \frac{340 \text{ m/s}}{f} \qquad (15.1)$$

The sensation of sound is caused by compressions and rarefactions of the air in the form of a longitudinal oscillatory motion. The energy transmitted per unit area varies as the square of the distance from the source. The rate with which this energy is transmitted expresses the **intensity** of the sound, which directly relates to the sensation of **loudness**. This incremental variation of the air pressure is known as **sound pressure** and for practical purposes it is this which is measured in determining the loudness of sound.

Sound pressure level (SPL) is defined as 20 times the logarithm to base 10 of the ratio of the effective sound pressure (P) to the reference sound pressure ($P_{\text{ref.}}$):

$$SPL = 20 \log \frac{P}{P_{\text{ref.}}} \text{ dB} \qquad (15.2)$$

$P_{\text{ref.}}$ approximates to the threshold of hearing and numerically is 0.0002 microbar (2×10^{-5} N/m^2).

The intensity (I) of a sound wave in the direction of propagation is:

$$I = \frac{P^2}{p_o C} \text{ W/m}^2 \qquad (15.3)$$

$$p_o = 1.18 \text{ Kg/m2}$$

$$C = 340 \text{ m/s}$$

The intensity **level** of a sound in decibels is:

$$IL = 10 \log \frac{I}{I_{\text{ref.}}} \text{ dB} \qquad (15.4)$$

$$I_{\text{ref.}} = 10^{-12} \text{ W/m}^2 = 2 \times 10^{-5} \text{ N/m}^2$$

The relation between intensity level (IL) and sound pressure level (SPL) is found by substituting Equation (15.2) for intensity (I) in Equation (15.4). Inserting values for $P_{\text{ref.}}$ and $I_{\text{ref.}}$ gives:

$$IL = SPL + 10 \log \frac{400}{p_o C} \text{ dB} \qquad (15.5)$$

It is apparent that the intensity level IL will equal the sound pressure level SPL only if $p_o C = 400$ Rayls. For particular combinations of temperature and static pressure this will be true, but under 'standard measuring conditions' of:

$$T = 22°C \text{ and } p_o = 751 \text{ mm Hg}, p_o C = 407 \text{ Rayls} \quad (15.6)$$

The error of -0.1 dB can be neglected for practical purposes.

Characteristic Impedance

The characteristic impedance is the ratio of the effective sound pressure to the particle velocity at that point in a free, plane, progressive sound wave. It is equal to the product of the density of the medium times the speed of sound in the medium (p_oC). It is analogous to the characteristic impedance of an infinitely long, dissipation-less, transmission line. The unit is the Rayl, or Newton s/m³.

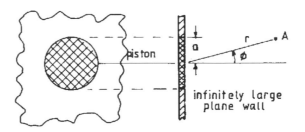

Figure 15.1 Piston in infinitely plane wall.

Radiation Impedance

When a vibrating diaphragm is placed in contact with air, its impedance to motion is altered, the added impedance seen by the surfaces which emit useful sound energy is termed 'radiation impedance'. The radiation reactance is usually positive, corresponding to an apparent mass. Both reflective mass and resistance as seen by the diaphragm depend on its size, shape, frequency, and the acoustical environment in which it radiates.

Radiation from a Piston

Many radiating sources can be represented by the simple concept of a vibrating piston located in an infinitely large rigid wall. The piston is assumed to be rigid so that all parts of its surface vibrate in phase and its velocity amplitude is independent of the mechanical or acoustic loading on its radiating surface.

Figure 15.1 shows the problem: we wish to know the sound pressure at a point A located distance r and angle $\emptyset$ from the centre of the piston. To do this, we divide the surface of the piston into a number of small elements, each of which is a simple source vibrating in phase with all the other elements. The pressure A is, then, the sum in magnitude and phase of the pressures from these elementary elements. For r large compared with the radius of the piston a the equation will be:

$$P \text{ (sound pressure N/m}^2) = \frac{\sqrt{2} jfp_o u_o \pi a^2}{v} \left[\frac{2J_1 (K_a \sin \emptyset)}{K_a \sin \emptyset} \right] e^{j\omega (t-r)} \quad (15.7)$$

where u_o = RMS velocity of the piston

$J_1 ( )$ = Bessel Function of the first order.

Note the portion of Equation (7) in square brackets yields the directivity pattern.

Directivity

At frequencies where the wavelength of sound (λ) is large compared with the diameter of the piston, the radiation is spherical. As the frequency is increased, the wavelength becomes comparable or less than the piston diameter and the radiation becomes concentrated into a progressively narrowed angle.

The ratio of pressure $P\emptyset$ at a point set at an angle $\emptyset$ off the axis, to the on axis pressure P_A at the same radial distance is given by:

$$\frac{P_\emptyset}{P_A} = \frac{2J_1 \left(\frac{2\pi a}{\lambda} K_a \sin \emptyset \right)}{\frac{2\pi a}{\lambda} K_a \sin \emptyset} \quad (15.8)$$

Figure 15.2 shows radiation patterns for different ratios of λ/D. The radiation from a piston is directly related to its velocity, and we can compute the acoustic power radiated and the sound pressure produced at any given distance in the far field.

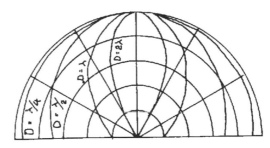

Figure 15.2 Directivity of piston as function of piston diameter and wavelength.

Sound Pressure Produced at Distance r

Low frequencies: When the piston diameter is less than one-third wavelength ($K_a \leq 1.0$) it is essentially non-directional and can be approximated by a hemisphere whose RMS volume velocity u_1 equals:

$$u_1 \text{ (diaphragm velocity)} = Sd \times u_c = \pi a^2 u_c \quad (15.9)$$

and the RMS sound pressure at distance r is:

$$p(r) = \frac{Sdfp_o}{r} = \frac{\pi a^2 fp_o}{r} \text{ N/m}^2 \quad (15.10)$$

total power radiated W_t:

$$W_t = \frac{4\pi p_o}{c}(Sd \times f \times u_c)^2 \text{ W} \quad (15.11)$$

Medium frequencies: At frequencies where the radiation from the piston becomes directional but still vibrates as one unit, the pressure produced at a distance r depends on the power radiated and the directivity factor Q:

$$p(r) = \sqrt{\left(\frac{W_t Q P_o C}{4\pi r^2}\right)} \quad (15.12)$$

$$Q = \frac{4\pi P a x^2}{2\pi \int_0^x P^2 \phi \sin \phi \, d\phi} \quad (15.13)$$

The mechanical impedance in MKS mechanical ohms (Newton-seconds/metre) of the air load upon one side of a plane piston mounted in an infinite baffle and vibrating sinusoidally is:

$$Z_m = R_{mR} + jX_m = a^2\pi c\rho_o\left[1 - \frac{J_1(2K_a)}{K_a}\right] + \frac{j\pi\rho_o c}{2K^2}K_1(2K_a) \quad (15.14)$$

where Z_m = mechanical impedance in Newton seconds/metre

a = radius of piston in metres
ρ_o = density of gas in Kg/cubic metre
c = velocity of sound in metres/second
R_{mR} = mechanical resistance in Newton seconds/metre
  note: this component varies with frequency
X_m = mechanical reactance in Newton seconds/metre
K = $\infty/c = 2\pi/\lambda$ = wave number
$J_1 K_1$ = two types of Bessel function given by the series:

$$J_1(W) = \frac{W}{2} - \frac{W^3}{2^2 \times 4} + \frac{W^5}{2^2 \times 4^2 \times 6} - \frac{W^7}{2^2 \times 4^2 \times 6^2 \times 8} \quad (15.15)$$

$$K_1(W) = \frac{2}{\pi}\left(\frac{W^3}{3} - \frac{W^5}{3^2 \times 5} + \frac{W^7}{3^2 \times 5^2 \times 7}\right) \quad (15.16)$$

where $W = 2Ka$

Figure 15.3 shows graphs of the real and imaginary parts of this equation:

$$Z_m = R_{mR} + jx_m \text{ as a function of } Ka.$$

It will be seen that for values of $Ka<1$, the reactance X_m varies as the first power of frequency, whilst the resistive component varies as the second power of frequency. At high frequencies, (i.e., $Ka>5$) the reactance becomes small compared with resistance which approaches a constant value. The graph can be closely approximated by the analogue Fig. 15.4, where:

$R_{m1} = 1.386 \, a^2 \, \rho_o c$ MKS mechanical ohms
$R_{m2} = \pi a^2 \, \rho_o c$ MKS mechanical ohms
$C_{m1} = 0.6/a \, \rho_o c^2$ metres/Newton
$M_{m1} = 8a^3 \, \rho_o/3$ Kg

Numerically, for $Ka<1$:

$R_{mR} = 1.5^{\omega2} \, a^4 \, \rho_o/c$ ohms
$M_m = 2.67 \, a^3 \, \rho_o$ Kg

It will be seen that the reactive component behaves as a mass loading on the diaphragm and is a function of diaphragm area only.

The term Ka has special significance: it relates the diaphragm radius to the wavelength of sound at any particular frequency. It is numerically equal to:

$$K_a = \frac{2\pi a}{\lambda} \quad (15.17)$$

where a = radius of diaphragm and λ = wavelength

When the wavelength λ is greater than the circumference of the diaphragm, the loudspeaker behaves substantially as a point source and the sound field pattern is essentially omnidirectional. At the same time the radiation resistance increases with frequency. Thus, at frequencies below $Ka = 1$, the increase in radiation resistance with frequency is exactly balanced by the reduction in velocity of the diaphragm with frequency due to its mass reactance (assuming there are no resonances in the diaphragm) and the sound pressure will be constant. At values above $Ka = 1$, the radiation resistance (neglecting the minor 'wiggles') becomes constant, but because of focusing due to the diaphragm dimensions being greater than λ, the sound pressure on the axis remains more or less constant. The velocity of sound in air is approximately 340 m/s., therefore a 150 mm (6 in) diameter diaphragm will behave as a point source to a limiting frequency of about 720 Hz; thereafter it begins to

A.
2b

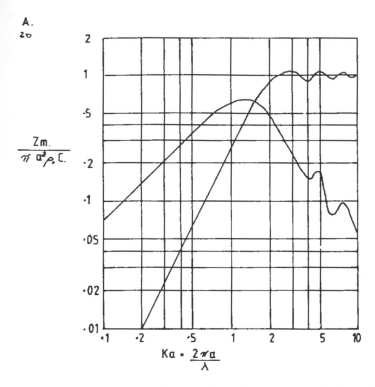

Figure 15.3 Air load on plane piston; mechanical impedance ref.; driving point.

focus. Various artifices (such as corrugations) are used with paper cones to extend this range, with more or less success. This was the classic premise which Rice and Kelogg postulated in 1925 when they re-invented the moving coil loudspeaker, and it is still fundamental today.

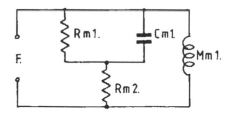

Figure 15.4 Impedance analogue of Fig. 15.3.

To summarise, the loudspeaker should operate under mass controlled conditions and (neglecting directional effects due to focusing of the diaphragm) sound pressure will be constant and independent of frequency; for a given magnet and coil system it will be inversely proportional to the total mass of the diaphragm and moving coil system.

Electrical Analogue

The analysis of mechanical and acoustical circuits is made very much easier by the application of analogues in

which mass is equivalent to inductance, compliance to capacitance, and friction to resistance. Using SI units, direct conversion between acoustical, mechanical and electrical elements can be performed.

The three basic elements (RCL) of acoustical electrical and mechanical systems is shown schematically in Fig. 15.5. The inertance M of an acoustic system is repre-

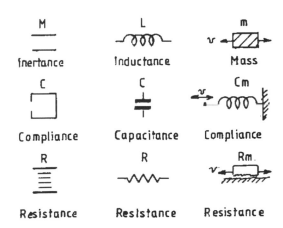

Figure 15.5 Acoustical, electrical and mechanical analogues.

sented by the mass of gas contained in a constriction which is short enough so that all particles are assumed to move in phase when actuated by a sound pressure. The compliance C of the system is represented by an enclosed volume, with its associated stiffness. It should be noted that the mechanical analogue of acoustic compliance is not mechanical stiffness, but rather its reciprocal, mechanical compliance $C_m = 1/s$. Although resistance of an acoustic system may be due to a combination of a number of different factors, irrespective of its origin, it is conveniently represented by narrow slits in a pipe, for the viscous forces that arise when gas is forced to flow through these slits always results in the dissipation of energy.

The Helmholtz resonator may be graphically represented by Fig. 15.6, but converting it to its electrical analogue shows it to be a simple resonant circuit which can be analysed easily using general circuit theorems.

The beauty of the analogue method of analysis is that it is possible by using various transformation equations to refer the acoustic and electrical parameters to the mechanical side, or conversely, the mechanical and acoustic parameters to the electrical side, etc. For the purpose of this analysis the electrical and acoustical parameters are referred to the mechanical side. The diaphragm can be thought of as an acoustic/mechanical transducer – that is, a device for transforming acoustic energy to mechanical energy, and vice versa. Under these circumstances it will also act as an impedance transformer, i.e., it will convert acoustic inertance into mechanical mass and acoustic compliance into mechanical compliance and acoustic resistance into mechanical resistance. The equivalent mechanical values of the acoustical quantities may be obtained from the following relationships:

Mechanical	Acoustic
Force	= Pressure × Area
	$F_m = p \times Sd$ Newtons (15.18)
Velocity	$= \dfrac{\text{Volume velocity}}{\text{Area}}$
	$u = \dfrac{U}{Sd}$ (15.19)

$$\text{Displacement} = \frac{\text{Volume displacement}}{\text{Area}}$$

$$x = \frac{x_v}{Sd} \qquad (15.20)$$

Resistance = Acoustical resistance × Area squared

$$R_m = R_a \times Sd^2 \qquad (15.21)$$

Mass = Inertance × Area squared

$$m = M \times Sd^2 \qquad (15.22)$$

$$\text{Compliance} = \frac{\text{Acoustical capacitance}}{\text{Area squared}}$$

$$C_m = \frac{C_a}{Sd^2} \qquad (15.23)$$

Diaphragm/Suspension Assembly

Assuming the diaphragm behaves as a rigid piston and is mass controlled, the power response is shown in Fig. 15.7 where f_o is the system fundamental resonant frequency. Above this the system is mass controlled and provides a level response up to f_1; this corresponds to $Ka = 2$, see Fig. 15.3. Above this frequency the radiation resistance is independent of frequency, and the response would fall at 12 dB/octave, but because of the 'directivity' effect the sound field is concentrated into a progressively narrower beam. The maximum theoretical rate of rise due to this effect is 12 dB/octave, thus the on axis HF response should be flat. In real life this is only approximated.

Diaphragm Size

It has been found experimentally that the effective area of the cone is its projected or base area. This should not be confused with the advertised diameter of the loudspeaker, which is anything from 25 mm to 50 mm greater than the effective cone diameter. In direct radiator loudspeakers and at low frequencies radiation resistance is proportional to the fourth power of the radius (square of the area) and the mass reactance to the cube of the radius. The resistance/reactance ratio (or power factor) of the radiation impedance is therefore proportional to piston radius, thus the electro-acoustic efficiency, other factors being constant, at low frequencies increases with diaphragm area. For constant radiated power the piston displacement varies inversely with area, hence 'long throw' type of small diaphragm area loudspeakers. With fixed amplitude the radiated power is proportional to the square of the area at a given frequency, or a frequency

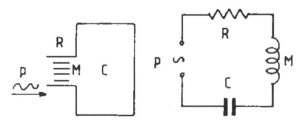

Figure 15.6   Schematic representations of a Helmholz resonator.

one octave lower may be reproduced if the area is increased by a factor of four. The upper limit of diaphragm size is set by increased weight per unit area required to give a sufficiently rigid structure.

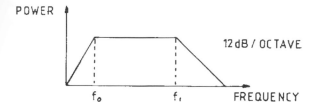

Figure 15.7 Power response of an infinitely rigid piston.

Figure 15.8 shows the necessary peak amplitude of a piston mounted in an infinite baffle to radiate one acoustic watt of sound power at various frequencies (one side only of the piston radiating). Peak amplitudes in mm are marked on the family of curves. For any other value of acoustic power output (P) multiply peak amplitude by $\sqrt{P}$. With an average room of 2000 ft^3, a reverberation time of one second, and a sound pressure level of +94 dB, the total sound output power is of the order of 30 mW. To radiate this power at 50 Hz the peak amplitude of a 250 mm radiator will be about 2 mm, whilst a 100 mm piston to radiate the same power would require a peak displacement of just over 13 mm. Even with 'long throw' loudspeakers it is not possible to obtain a peak to peak displacement of 26 mm, thus the sound power capabilities must be severely limited at low frequencies. One will often see response curves of these small speakers taken to apparently extraordinarily low frequency limits, but these are always undertaken at low power input levels.

The directional radiation characteristics of a diaphragm are determined by the ratio of the wavelength of the emitted sound to the diaphragm diameter. Increasing the ratio of diaphragm diameter to wavelength decreases the angle of radiation. At frequencies in which the wavelength is greater than four times the diaphragm diameter the radiation can be considered substantially hemispherical, but as this ratio decreases so the radiation pattern narrows. Figure 15.9 shows the polar response of a piston in terms of the ratio of diameter over wavelength. This shows the degrees off the normal axis at which the attenuation is 3, 6, 10 and 20 dB (as marked on the curves) as a function of the ratio of the piston diameter over the wavelength of the generated sound wave.

Diaphragm Profile

A practically flat disc is far removed from the theoretical 'rigid piston'. With the exception of foamed plastic, the mass, for a given rigidity will be excessive, resulting in very low efficiency, and if the cross section is reduced the system becomes very flexible and inefficient.

Decreasing the angle from 180° increases the stiffness enormously; concomitantly the thickness can be reduced, resulting in a lighter cone for the same degree of self support. The flexure amplitudes will be reduced, but the bell modes will make an appearance. As the angle is reduced it reaches an optimum value for level response at the transition frequency. There will be another angle for maximum high frequency response, resulting ultimately

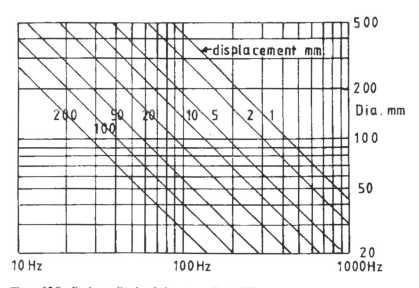

Figure 15.8 Peak amplitude of piston to radiate 1 W.

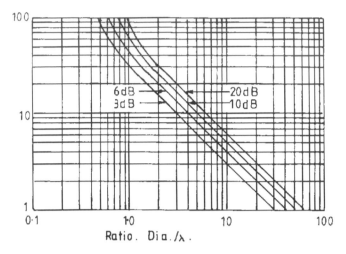

Figure 15.9 Directional radiation pattern with circular piston in infinite baffle.

in peaking in the upper treble region. Continuing the reduction in angle, the high frequency peak will be reduced, but the response above the peak will fall rapidly.

If instead of a straight sided cone the profile is curved, the 'smoothness' of the overall response can be improved considerably: bell modes are discouraged, and the on axis high frequency response improved. The price charged for this facility is reduced low frequency power handling capacity because, for a given weight, the curved cone is just not as stiff (and as strong) as the straight sided version.

The most efficient shape at low frequencies is circular. Theoretical and experimental investigations have shown that an ellipse with a major-minor axis of 2 has an average of 7 per cent lower radiation resistance in the useful low frequency range than a circle of the same area; the loss becomes progressively greater as the shape departs still further from circular. The shape of the cross section or profile of the cone depends on the power handling and response desired.

For domestic loudspeaker systems, which must be cost-conscious, the loudspeaker size is limited to 150 mm to 200 mm and a frequency response of 100 Hz to about 7 kHz with, possibly, a 25 mm soft dome to accommodate the high frequencies. Straight sided cones are usually employed when good 2–5 kHz response is required and when reproduction above, say, 7 kHz may be undesirable. Curved cones improve the response above 6–7 kHz by providing an impedance viewed from the voice coil which has a more uniformly high negative reactance and therefore absorbs more power from the high positive reactance due to voice coil mass seen looking back into the voice coil. This improvement is obtained at the expense of response in the 2–5 kHz region, a weaker

cone structure, and reduced power handling in the extreme bass.

Straight-Sided Cones

The most important parameter affecting the performance of a loudspeaker is 'cone flexure'. Because real materials are not infinitely rigid and have mass, the velocity of propagation through the material is finite. The cone is driven at the apex and the impulse travels outwards towards the periphery where it is reflected back to the source. At particular frequencies when the distance to the edge are odd quarter wavelengths, the returning impulse will be 180° out of phase and tend to cancel; conversely, when the distance is multiples of half wavelengths they will augment – under these conditions the system can be considered as a transmission line, and theoretically (and to some extent, practically) if the outer annulus were made resistive and of the correct value the line would be terminated and no reflections would occur, see Fig. 15.10 (a).

The conical diaphragm also has radial or 'bell' modes of flexure. These are similar to the resonances in a bell and occur when the circumference is an integral number of wavelengths, see Fig. 15.10 (b).

Obviously, both modes occur simultaneously, and at some frequencies reinforce each other and at others tend to cancel. Their main effect on performance are the 'wiggles' on the response curve and transient and delay distortions. It is instructive to apply a short 'tone burst'; it will be seen that at particular frequencies during the duration of the input signal the diaphragm is stationary

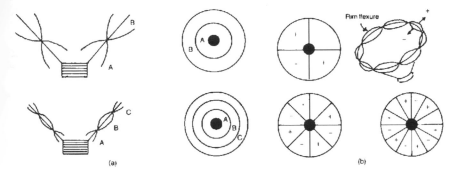

Figure 15.10 (a) concentric modes (b) bell modes.

and on cessation of the pulse it will burst into oscillation at some frequency unrelated to the driving current.

The art of diaphragm design is to minimise these deleterious effects. One method is to introduce concentric corrugations; the effect is to increase the stiffness seen by the bell modes and decrease stiffness for the concentric modes. By correctly proportioning the number, width, and wall thickness of these corrugations the outer edges of the cone are progressively decoupled as frequency increases. This results in the 'working' diameter of the diaphragm being reduced at high frequencies, thus improving the high frequency performance.

Material

Hard impregnated or filled pressed calendered papers are used when loudness efficiency and apparent high frequency response are important. The impregnant is usually a hard thermo-setting resin. Radiation response provides very little dissipation in direct radiator cones, hence by using paper having low internal flexural losses the transmission line is made to have strong resonances. The transient response of this type is necessarily poor since non-centre moving modes of the cone are unappreciably damped by the motor unit. Soft, loosely packed, felted cones are used when some loss in high-frequency response can be tolerated and a smooth response curve with reduced transient distortion is required. The apparent loudness efficiency of high loss cones of this type are anything up to 6 dB lower than that of low loss cones of similar weight.

In an effort to overcome the intransigencies of paper cones, resort has been made to other materials. Lightweight metal (aluminium alloys, etc.) immediately springs to mind because of its stability, homogeneity and repeatability but, because of the very low internal frictional losses, strong multiple resonances occur in the upper frequencies. A diaphragm of, say, 250 mm in diameter made from 0.1 mm thick aluminium alloy with a total mass of 40 g will show a 'ruler' level response up to approximately 2 kHz when multiple resonances occur. These are extremely narrow band (in some cases only 1 or 2 Hz wide) with an amplitude of anything up to 40 dB and an effective Q of several hundred. Putting a low pass filter cutting very sharply at, say, 1 k Hz does not eliminate shock excitation of these resonances at low frequencies and the result is a 'tinny' sound. Reducing the cone diameter and making the flare exponential reduces this effect and also places the resonant frequency a few octaves higher, but does not entirely eliminate the problem. Using foamed plastic materials (and sometimes coating the surfaces with a metal to form an effective girder structure) has met with some success. There are problems associated with the solid diaphragm in that the different finite times taken for the sound wave to travel directly from the voice coil through the material to the front and along the back edge of the diaphragm to the anulus and then across the front causes interference patterns which result in some cancellation of the emitted sound in the mid upper frequencies, say 800–1100 Hz. This effect can be mitigated by using a highly damped anulus, with the object of absorbing as much as possible of the 'back wave'. Expanded polystyrene is the favourite material for these diaphragms although expanded polyurethane has met with some success. An extension of this principle is exemplified where the diaphragm is almost the full size of the front of the cabinet (say 24 × 18 in). In this case the diaphragm, even at low frequencies, does not behave as a rigid piston. The overall performance is impossible of any mathematical solution and must be largely determined experimentally, but the lower bass (because of multiple resonances) is, in the opinion of its advocates, 'fruity' and 'full'! It has been developed to use two or even three voice coils at strategic places on the diaphragm. For synthesized noise it is possible, but in the writer's opinion, for 'serious' music listening it adds nuances to the music never envisaged or intended by the composer.

Vacuum-formed sheet thermoplastic resins have become very popular. Their mechanical stability is excellent, they are non-hygroscopic, and repeatability (a very important facet when mass producing units in hundreds of thousands) is several orders of magnitude better than paper cones. However, there is a price to pay: most of them contain a plasticiser which increases the internal mechanical losses in the structure, and hence the magnitude of diaphragm resonances are reduced. However, under user conditions, dependent upon electrical power input and thus operating temperature, they tend to migrate. This results in a changed cone (or dome) shape, and because the internal mechanical loss is reduced, the frequency response is changed. In extreme cases, especially with small thin diaphragms, cracking has occurred, but it must be emphasised that with a correctly designed unit operating within its specified power and frequency limits, these 'plastic' diaphragms (especially those using specified grades of polypropylene) give a cost effective efficient system.

Soft Domes

For use at medium and high frequencies, the 'soft dome' system has found favour. It consists of a preformed fabric dome with integral surround and usual voice coil assembly. It is very light and its rigidity can be controlled by the amount of impregnant, but the beauty of the concept is that the damping can be adjusted by the quantity and viscosity of the 'dope' applied to the dome. Responses flat ±1 dB to 20 kHz are standard, even on cheap mass-produced units!

Suspensions

The purpose of the suspensions is to provide a known restoring force to the diaphragm/voice coil assembly and at the same time to have sufficient lateral rigidity to prevent any side to side movement of the system. This latter requisite is most important when it is remembered that the clearance between the voice coil and the magnet pole pieces is of the order of 0.15 mm for tweeters and 0.4 mm for 150 W woofers. The average domestic 200 mm (8 in) loudspeaker is about 0.25 mm.

The combined stiffness of the front and rear suspensions are formulated to resonate with the total moving mass of the diaphragm/voice coil assembly and air load to the designed LF resonance. The front suspension radial width is usually about half that of the rear (in order to maximise cone diameter for a given cradle size) and it

is this factor which limits the peak to peak displacement. Figure 15.11 shows displacement/force for a roll surround. It will be seen that the maximum linear movement is limited: it follows the familiar hysteresis curve of non linear dissipative systems.

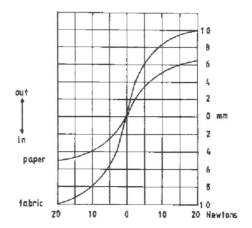

Figure 15.11   Displacement relative applied force.

The annulus of the diaphragm can either be an extension of the cone material itself, or, as is more usual with high fidelity loudspeakers, a highly compliant surround produced from cotton or man made fibres, Neoprene, or plasticised PVC. In the case of woven materials these must be sealed and the sealant is usually used to provide some mechanical termination of the cone.

The front suspension represents a discontinuity in the diaphragm system and because it has its own mass and compliance it is capable of a separate resonance. When this takes place it presents a very high impedance to the edge of the cone, reducing its output and causing a dip in the response. Because of its non-linearity it radiates considerable distortion, especially at low frequencies where the amplitude is greatest. The requirements are high flexibility and high interval losses. Probably the most successful material is plasticised PVC, using a very stable non-migrant plasticiser such as dibutyl sebacate.

The rear suspension is the major restoring force, the radial width is usually at least twice that of the front suspension and is a multi roll concentrically corrugated fabric disc, impregnated with a phenolic resin. The weave of the material, number of corrugations, diameter and amount of impregnant determining the stiffness. It should provide a substantially linear restoring force over the designed maximum amplitude of displacement. The whole structure behaves mechanically as a series resonant circuit. The mass being determined by the weight of the cone, voice coil and former, and the stiffness by the combined effects of the rear suspension and the annulus, the Q of the circuit being determined almost wholly by the losses of the restoring force.

Voice Coil

The dimensions of a voice coil are determined primarily by the rated power handling of the loudspeaker. It must be emphasised that with direct radiators 95–99 per cent of the input electrical power is dissipated in the form of heat; even with the most efficient horn loaded units a minimum of 50 per cent is used for heating purposes only.

Figure 15.12 shows voice coil temperature versus input power. The limiting temperature is set by:

(a) maximum temperature rating of the former: 100° C for paper based materials; 150° C for 'Nomex', which is an aromatic polyamide; 250° C for Polyimide;
(b) temperature rating of the wire enamel: maximum 220° C for ML insulation, down to 110° C for self-bonding and self-fluxing wires;
(c) adhesive; from 110° C for cyanoacrylic, to 250° C for those with polyimide base:
(d) mechanical expansion of the voice coil diameter at elevated temperatures.

Fortunately, there is an in-built semi protection for the assembly, namely temperature coefficient of resistance of the wire, which is +0.4 per cent for 1° C rise in temperature! Thus at a temperature of +250° C above ambient the voice coil resistance has doubled, and for a constant voltage input (which is the norm for modern amplifiers) the indicated power E^2/R_{nom} is twice the actual power. Note, R_{nom} is the manufacturer's specified resistance and is, or should be, the value at the series resonant frequency $R\omega$ where the input impedance is minimum and resistive.

Moving Coil Loudspeaker

Figure 15.13 shows the structural features of a moving coil direct radiator loudspeaker, and for purposes of analysis it will be convenient to divide it into two parts: the 'motor' or 'drive' unit, and the acoustic radiator, or diaphragm, described above.

The driving system consists of a solenoid situated in a radial magnetic field. It is free (within certain restraints) to move axially (see Fig. 15.14). When a current is passed through the coil a magnetic field will be generated, the magnitude being directly proportional to il, and this will react with the steady field of the permanent magnet and a mechanical force will be developed which will tend to

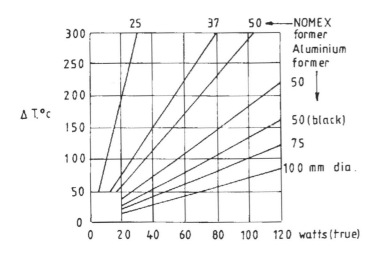

Figure 15.12   Voice coil temperature versus input power (300 Hz).

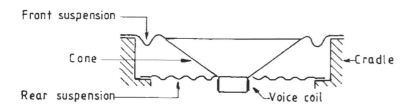

Figure 15.13   Moving coil loudspeaker cone, suspensions, voice coil assembly.

move the coil axially, either inwards or outwards depending on the direction of the current. The magnitude of this force is:

$$F = Bli \qquad \text{(Newtons)} \qquad (15.24)$$

where B is flux density in Webers/m^2, l is conductor length in metres, i is current in Amperes.

It should be noted that 'l' refers only to that portion of the coil situated in the magnetic field. As will be shown later, to reduce distortion the coil is often longer than the working magnetic field defined by the gap dimensions.

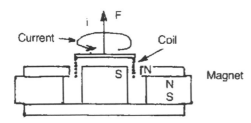

Figure 15.14   Moving coil motor element.

If the coil is free to move, its velocity will be determined by the applied force F and the mechanical impedance Z_m. The mechanical impedance Z_m will be a function of total mass (L_m) of the system, i.e. voice coil and former, diaphragm, air loading, etc. resistance R_m due to losses in the suspension and radiation, and total stiffness ($1/C_m$) due to the restoring force.

Using normal circuit theory, the impedance will be:

$$Z_m = R_m + j\left[\omega L_m - \left(\frac{1}{\omega C_m}\right)\right] \qquad (15.25)$$

This represents a simple series resonant circuit shown in Fig. 15.15. Whilst one can predict resonant frequency from lumped mechanical constants, the analysis must be carried several stages further to arrive at the correct transfer characteristic from electrical input to sound output in practical loudspeaker design.

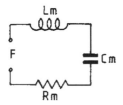

Figure 15.15   Analogue of lumped mechanical constants.

Figure 15.16 takes the analogue a stage further:

- L_{md}, C_{md} and R_{md} are the mechanical components of diaphragm, voice coil and suspension

- L_{ma}, R_{ma}: mechanical impedance components of air load
- Z_{me}: mechanical impedance due to electrical system
- Z_{ma}: mechanical impedance due to air load on rear of cone
- Z_{mx}: normally zero, but see motional impedance.

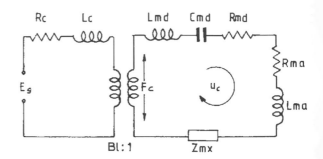

Figure 15.16   Analogue referred to mechanical side.

Motional Impedance

The moving coil system is a reversible transducer, a current through the coil will produce a force, and the resultant velocity of the coil will produce an EMF. This voltage will be a function of velocity, conductor length, and magnetic field strength; thus if an external EMF is applied to the coil the resultant motion will generate a back EMF. (180° out of phase) which will tend to counteract the forward current flow, thus increasing the electrical impedance. This is 'motional impedance'.

If the motion of the system can be prevented, by applying an infinite mechanical impedance (Z_{mx} in Fig. 15.16) there will be no back EMF and the electrical impedance will be only the voice coil resistance and inductance (blocked impedance). Reducing the mechanical impedance (reducing mass and resistance and increasing compliance) will result in an increase in velocity and the motional impedance will be increased. Intuitively, this indicates that motional impedance will be proportional to Bl and an inverse function of the mechanical impedance. The common factor is the velocity of motion.

$$\text{Back EMF} \quad E_b = Blv \quad \text{V} \qquad (15.26)$$

$$\text{from (15.26) velocity v} = \frac{\text{Force}}{Z_m} = \frac{Bli}{Z_m} \quad \text{m/s} \qquad (15.27)$$

Thus motional impedance $Z_{cm} = \dfrac{E_b}{i} = \dfrac{B^2 l^2}{Z_{m1}}$ (15.28)

$$Z_m = Kg/s \quad Z_{em} \text{ Ohms}$$

from (15.27) and (15.28)

$$Z_{em} = \dfrac{B^2 l^2}{R_m + j\left(\omega L_m - \dfrac{1}{\omega C_m}\right)}$$ (15.29)

$$Z_{em} = \dfrac{1}{Y_{em}} = \dfrac{1}{G_{em} + j B_{em}}$$ (15.30)

where Y_{em} = electrical admittance due to mechanical circuit

G_{em} = electrical conductance $\left(\dfrac{1}{R_{em}}\right)$

B_{em} = electrical susceptance $\left(\dfrac{1}{X_{em}} = \omega C_{em} - \dfrac{1}{\omega L_{em}}\right)$

thus $\quad Z_{em} = \dfrac{B^2 l^2}{R_m + j\left(\omega L_m - \dfrac{1}{\omega C_m}\right)}$ (15.31)

from $\quad R_{em} = \dfrac{B^2 l^2}{R_m} \quad$ Ohms (15.32)

$C_{em} = \dfrac{L_m}{B^2 l^2} \quad$ Farads (15.33)

$L_{em} = C_m B^2 l^2 \quad$ Henries (15.34)

where $\quad R_m = Kg/s \quad$ (mechanical Ohms)

$L_m = Kg$

$C_m = $ metres/Newton

and by inversion:

$$R_{me} = \dfrac{B^2 l^2}{R_e} \quad \text{Mechanical Ohms} \quad (15.35)$$

$$C_{me} = \dfrac{L_e}{B^2 l^2} \quad \text{Metres/Newton} \quad (15.36)$$

$$L_{me} = C_e B^2 l^2 \quad Kg \quad (15.37)$$

It will be noted that in the analogue inductance (mass) in the mechanical circuit will become a capacitance in the electrical circuit, etc.

At mechanical resonance

$$\left(\omega L_m - \dfrac{1}{\omega C_m}\right) = 0 \quad (15.38)$$

and the mechanical impedance Z_m will have a minimum value $= R_m$. The velocity will be maximum, thus the back EMF and motional impedance will be maximum, indicating a parallel resonance. It will be seen that series components in the mechanical circuit appear as parallel components in the electrical side and vice versa.

Analogue models

We can now assemble the various parameters to produce a basic analogue for a loudspeaker. Figure 15.17 shows the low frequency analogue referred to the mechanical side. The quantity fc represents the total force acting in the equivalent circuit to produce the voice coil velocity u_c:

$$u_c = \dfrac{E_g Bl}{R_c + (R_m + jX_m)} \quad (15.39)$$

Let us divide the frequency region into five parts and treat each part separately by simplifying the circuit in Fig. 15.18 to correspond to that part alone. In region A, where the loudspeaker is stiffness controlled, the power output increases as the fourth power of frequency, or 12 dB/octave. In region B, at resonance frequency ω_0 the power output is determined by the total resistance because X_m passes through zero. For large values of Bl and small values of R_c the total circuit resistance becomes sufficiently large so that the resonance is more than critically damped. The sound pressure will increase linearly with frequency (+6 dB/octave). In region C, the power output (and sound pressure) approaches a constant value, provided that the circuit impedance approximates a pure mass reactance. That is to say RMR and X_m^2 both increase as the square of the frequency.

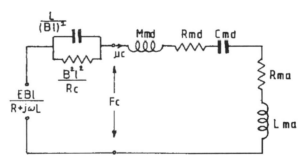

Figure 15.17 Low-frequency mechanical analogue.

From Fig. 15.17 it will be seen that the inductance of the voice coil is reflected into the mechanical circuit as a compliance (very much smaller than C_{ms} and this will resonate with the total mass at a mid frequency (usually 150–700 Hz), see Fig. 15.18. At this frequency the total electrical impedance is resistive, has the lowest absolute value, and is the value which is (or should be) quoted as the Rated Impedance in the manufacturer's specification and corresponds to d in Fig. 15.18.

Instead of referring all the parameters to mechanical

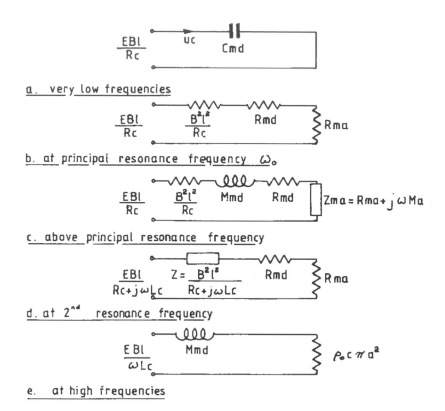

a. very low frequencies

b. at principal resonance frequency ω_o

c. above principal resonance frequency

d. at 2nd resonance frequency

e. at high frequencies

Figure 15.18 Simplified circuit, valid over restricted frequency ranges (a) very low frequencies (b) at principal resonance frequency ω_o (c) above principal resonance frequency (d) at second resonance frequency (e) at high frequencies.

mesh it is sometimes more convenient to refer to the electrical input, see Fig. 15.19, in which

R_c = voice coil resistance
L_c = voice coil inductance

$$C_{em} = \frac{m_{md}}{B^2 l^2} \qquad \text{Farads} \qquad (15.40)$$

$$L_{cm} = C_{md} B^2 l^2 \qquad \text{Henries} \qquad (15.41)$$

$$R_{em} = \frac{B^2 l^2}{R_{md}} \qquad \text{Ohms} \qquad (15.42)$$

R_c and L_c are the 'blocked' impedance values and not the DC resistance and inductance measured in 'air'.

$$C_A = \frac{\rho}{B^2 l^2} \times \frac{8a}{3\pi} \times Sd \qquad \text{Farads} \qquad (15.43)$$

$$R_A = \frac{B^2 l^2}{\rho_o C \, Sd} \qquad \text{Ohms} \qquad (15.44)$$

where a = Diaphragm radius
Sd = Diaphragm area

It should be noted that the factor $8a/\pi$ in Equation (15.43) is actually the 'end correction' used to describe

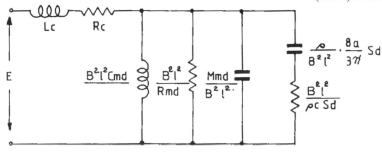

Figure 15.19 Analogue referred to electrical input.

the accession to inertia acting on one side only of a rigid piston of radius 'a' vibrating in an infinite baffle. The air loading on the back side of the diaphragm is determined by the loading presented by the enclosure.

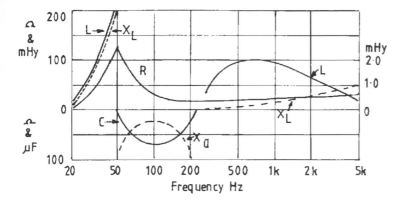

Figure 15.20   Complex impedance of moving coil loudspeaker.

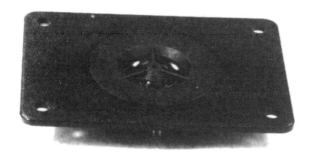

Figure 15.21   A 25 mm OEM soft-dome tweeter.

Figure 15.22   A 200 mm OEM bass Mid-range loudspeaker drive unit.

Figure 15.20 shows the impedance of a 300 mm (12 in) loudspeaker in an 85 litre enclosure (3 ft³). It will be seen that the modulus of impedance rises to 125 Ohms at the mechanical resonant frequency of 55 Hz, drops to 8 Ohms at the series resonance, and rises to 40 Ohms at 10 kHz. Of equal interest is the reactive component: below the first resonance an inductive reactance is presented to generator (rising to infinity at resonance), whilst between the two resonances a capacitive reactance is presented. At 100 Hz the effective capacitance is about 90 µF, and at that frequency the phase angle is 45°. Above the second resonance the impedance rises slowly.

It will be seen that the design of a successful moving coil loudspeaker owes as much to art as science. CAD can and does simplify much of the detail work, but after the basic design parameters (diaphragm size, magnetic field strength, conductor length, etc.) have been calculated, the nub of the problem is what diaphragm material? adhesives? cradle material and shape? etc. etc. etc. – the art of loudspeaker design is 5 per cent inspiration, 95 per cent perspiration, plus the essential compromise.

References

Beranek, L.L., *Acoustics*, McGraw-Hill.

Beranek, L.L., *Acoustic Measurements*, McGraw-Hill.

Borwick, J., *Loudspeaker and Headphone Handbook*. Butterworth-Heinemann (1988).

Colloms, M., *High Performance Loudspeakers*, 4th edition. Wiley (1991).

Olson, H.F., *Acoustical Engineering*, D Van Nostrand.

Olson, H.F., *Dynamical Analogies*, D Van Nostrand.

Walker, P.J., *New Developments in Electrostatic Loudspeakers*, J.A.E.S. **28** (11) (Nov. 1980).

Weens, D.B., *Designing, Building and Testing Your Own Speaker System*, 3rd edition. TAB (1990).

Weens, D.B., *Great Sound Stereo Speaker Manual*. TAB (1989).

16 Loudspeaker Enclosures

Stan Kelly

Loudspeakers and their enclosures are inseparable, and Stan Kelly here continues his discussion of the loudspeaker system with the effect of enclosures and the various types that are employed.

Fundamentals

To recreate in a domestic environment the same sound level as obtained under concert hall conditions, a maximum instantaneous sound level of +105 dB referred 2×10^{-5} N/m^2 is desirable, and this is equivalent to a dissipation of approximately 0.5 W in a room of 2000 ft^3 with a reverberation time of one second. With the normal type 12 in. moving coil loudspeaker, the effective diameter of the diaphragm usually employed is about 250 mm, giving a projected area of 0.05 m^2, and in order to dissipate 0.5 W from one side of the cone a peak to peak displacement of 18 mm at 40 Hz is required.

The ultimate low frequency response of a loudspeaker is determined by the 'fundamental resonance' of the diaphragm system, this being controlled by the total mass of diaphragm, voice coil, air loading etc., and the compliance of the suspension. This frequency is given by the expression:

$$f_o = \frac{1}{2\pi\sqrt{MC_m}} \qquad (16.1)$$

where f_o = frequency of resonance
M = mass
C_m = compliance

The effective mass can be determined by adding a known mass to the base of the cone (a length of solder of, say 50 gm, bent into a circle and taped on to the diaphragm is suitable) and the new resonant frequency determined. The dynamic mass is then given by:

$$M = \frac{M_1}{\left(\dfrac{f_o^{\,2}}{f_{o1}^{\,2}}\right)} \qquad (16.2)$$

where M_1 = Added mass
f_{o1} = New resonant frequency
and the compliance computed from:

$$C_m = \frac{1}{39.4\,f_o^{\,2}\,M} \qquad (16.3)$$

Thus to extend the low frequency response either the mass or compliance (or both) must be increased. These measurements are made with the loudspeaker in 'free air' – that is, without any acoustic loading due to baffle, cabinet etc. When fitted into an enclosure the bass response of the loudspeaker combination is determined by the lowest system resonant frequency, which is a function of the total effective mass of the cone plus the air load, the stiffness of the surround and centering device of the loudspeaker, and the stiffness of the volume of air obtained in the enclosure, and because of this added stiffness the system resonant frequency will be higher than the 'free air' resonance of the loudspeaker.

Infinite Baffle

Figure 16.1 shows the analogue (in the low-frequency spectrum) of an infinite baffle type of enclosure and its loudspeaker. It is seen that the speaker parameters are the total mass of the cone/voice coil etc., the total stiffness of the suspensions, and the loss resistance associated with these suspensions. Additionally, there are the radiation resistance and reactance due to the air load on the front of the diaphragm and the stiffness and loss resistance associated with the volume of the enclosure. For the purpose of this argument it is assumed that the walls of the enclosure are rigid. This can be achieved to a first approximation by using chipboard or plywood of thicknesses in excess of 16 mm for cabinets with a volume of less than 55 ltrs. If the cabinet volume is in excess of 55 ltrs the wall thicknesses should be a minimum of 20 mm. Cementing damping material in the form of rubberised felt 12 mm thick will reduce wall vibrations and their effect on the overall response.

The analogue in Fig. 16.1 can be reduced to the simple circuit of Fig. 16.2 in which all masses, compliances and losses are lumped into their individual components. This is a simple series resonant circuit in which the Q is determined by the total amount of loss resistance. Q, or the magnification factor is given by:

$$Q = \frac{2\pi f_o M}{R} \qquad (16.4)$$

where R = total loss resistance

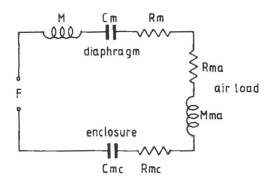

Figure 16.1 LF analogue of IB system.

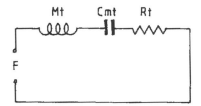

Figure 16.2 Lumped constants of Fig. 16.1.

Figure 16.3 shows frequency response of a series resonant circuit for various values of Q, and it will be seen that in the interest of smooth response the Q should be kept minimal. Under normal conditions, Qs of 1 to 1.4 are quite acceptable and 'ringing' due to this parameter is then negligible. The ultimate low frequency response is strictly a function of this resonant frequency and if it is required to extend the low frequency response by, say, on octave the resonant frequency of this system must be lowered by the same amount. The basic formula for resonant frequency is given in equation (16.1) above, but it must be remembered that the effective mass will now be different from the 'free air' conditions and the compliance will be reduced due to the reflected compliance

from the enclosure being in series with loudspeaker compliance, the effective compliance being:

$$C_{me} = \frac{C_m C_{ma}}{(C_m + C_{ma})} \qquad (16.5)$$

where C_{ma} = reflected acoustic compliance
C_{me} = total circuit compliance

Thus the frequency of resonance is determined by mass M and compliance C_{me} only; the resistive component not entering into the equation.

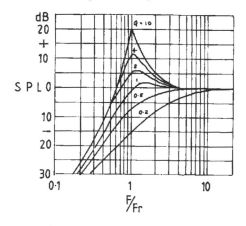

Figure 16.3 LF sound pressure and Q.

Damping

Addition of damping material will reduce the velocity of sound in the enclosure, increasing its effective compliance and reducing the ultimate resonant frequency of the system; but this does not invalidate the present argument. Therefore, assuming that the total compliance of a given system is constant, the low frequency response will be strictly a function of the mass of the diaphragm and the air load. From the analogue Fig. 16.1, it will be seen that at frequencies below the resonant frequency the system is stiffness controlled and the response will decrease at 12 dB/octave below it. At frequencies above the res onant frequency the system will be mass controlled, and it is normal practice for the speaker to be operated under these mass controlled conditions.

As stated above, operating the loudspeaker unit in an enclosure will be determined by the ratio of the reflected acoustic compliance and the loudspeaker compliance:

$$\frac{f_{o2}}{f_o} = \sqrt{\frac{C_{ma}}{(C_{ma} + C_m)}} \qquad (16.6)$$

where f_{o2} = resonant frequency in enclosure

and:

$$C_{\text{ma}} = \frac{V}{\rho\,c^2\,A^2} \qquad (16.7)$$

where V = volume of enclosure in m³
         ρ = air density 1.18 kg/m³
         c = sound velocity 334 m/s.
         A = diaphragm area in m²

The reflected acoustic compliance is inversely proportional to the square of the diaphragm area, thus reducing the diaphragm area will increase the reflected compliance and minimise the increase in resonant frequency due to the acoustic stiffness. Intimately connected with the reflected compliance is the required effective mass of the diaphragm to determine the low resonant frequency. In the absence of any diaphragm restoring force (loudspeaker compliance infinitely large) the required mass M_t is:

$$M_{\text{t}} = \frac{1}{\omega_2{}^2 C_{\text{ma}}} = \frac{c^2\,A^2}{\omega_2{}^2\,V} = \frac{3.55 \times 10^3\,A^2}{V f_o 2}2 \qquad (16.8)$$

where $\omega_2 = 2\pi f_{o2}$

With practical loudspeakers with some compliance, this mass M_t will have to be increased by an amount determined by the ratio of the two compliances mentioned in Equation (16.6) above.

Efficiency

The overall efficiency 'ε' is the ratio of acoustic power delivered by the loudspeaker system to the electrical input power supplied to the input terminals:

$$\varepsilon = \frac{B^2\,L^2\,A^2}{R_c\,M_t{}^2}5.45 \times 10^{-4} \qquad (16.9)$$

where B = flux density in Webers/m
         L = coil length in m
         A = diaphragm area in m²
         R_c = coil resistance in ohms
         M_t = total diaphragm mass in kg

This equation is fundamental, depending only on speaker parameters. It is seen that the efficiency is proportional to the square of the diaphragm area and inversely proportional to the square of the diaphragm mass. This latter value also includes the mass due to air loading, voice coil and former mass, and incidental masses such as suspension, cements, etc. It also assumes that the unit is working above the major low frequency resonance and the diaphragm is moving as a piston without any odd resonances, etc.

From Equation (16.9) it is seen that overall efficiency is proportional to the diaphragm area squared and to the inverse square of total effective mass. These two constants are the ones involved in determining the low frequency resonance (and hence the ultimate low frequency response). It can be shown that there is an optimum size of diaphragm area for any given enclosure volume, this is independent of the system resonant frequency and numerically equal to:

$$D = 10\left(\frac{V}{C_{\text{m}}}\right)^{-4} \qquad (16.10)$$

where D = diaphragm diameter in mm
       V = enclosure volume in ltrs
       C_{m} = loudspeaker compliance in metres/Newton

Frequency response

It is the aim of the designer to maximise frequency response, efficiency and power handling for a given product cost; 200 mm loudspeakers have found wide acceptance because they optimise the above requirements for domestic reproducers, but for extended bass response they do present some major snags for acceptability by the distaff side of the menage, namely the size of the enclosure necessary to keep the resonant frequency (and hence the ultimate low frequency response) to the lowest possible value.

The low frequency resonance can be determined by Equation (16.1) where M is the total mass and C_m the effective compliance of the enclosure and loudspeaker suspension in series. From Equation (16.10) it will be seen that for a resonant frequency of 50 Hz and a 200 mm loudspeaker with a compliance of 10^{-3} metres/Newton, the optimum enclosure volume will be over 200 ltrs, which is somewhat large even for the most ardent hi-fi enthusiast, especially when two enclosures are required for stereo! Therefore, under practical conditions, either the diameter of the diaphragm must be reduced, or the overall efficiency must be sacrificed.

Reflex Cabinets

One method of minimising this difficulty is to use a reflex cabinet where some of the latent energy available from the rear of the diaphragm can be made available at low frequencies. The reflex enclosure is a closed box with a hole or tunnel in one (usually the front) wall. The area of the port is equal to or smaller than the effective area of the driving unit. Figure 16.4 shows the usual construction, and Fig. 16.5 the analogue. It is seen that basically we have placed a mass in parallel with the enclosure compliance, and fundamentally it is nothing more nor less than a Helmholtz resonator. The addition of a port of area A_p behaves as a second diaphragm, since an effective mass of air oscillates in the opening. The mass of air M_p in the port includes the radiator mass on each side of the port as well as the mass of air inside the port. In the area of interest where $2\pi R/\lambda$ is less than one half, the radiation mass of the port is:

$$M_p = \frac{16\,A_p\rho}{3\pi} \qquad (16.11)$$

where M_p = radiation mass of port
A_p = area of port

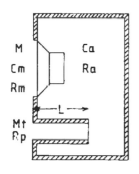

Figure 16.4   Vented enclosure.

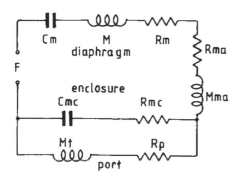

Figure 16.5   Analogue of Fig. 16.4.

and the radiation resistance of the port is:

$$R_p = \frac{A_p{}^2\omega^2\rho}{2\pi c} \qquad (16.12)$$

where R_p = radiation resistance
The mass contribution due to the length L of the port is:

$$M_p = A_p L\rho \qquad (16.13)$$

where M_p = mass of air in port
The total port mass is:

$$M_t = A_p\rho\,(L + 0.96\,\sqrt{A_p}) \qquad (16.14)$$

where M_t = total mass due to port
This mass will resonate with the acoustic capacitance C_a:

$$C_a = \frac{V}{\rho c^2} \qquad (16.15)$$

where V = volume of enclosure
c = velocity of sound in air
It is good design to make the acoustic resonance of the loudspeaker enclosure equal to that of the driving unit. It is even better to make the reflected mass of the port equal to the diaphragm total mass and the enclosure reflected compliance equal to the suspension compliance of the loudspeaker. These values can be computed from:

$$M_m = M_t A^2 \qquad (16.16)$$

where M_m = mass reflected to loudspeaker driving point
M_t = acoustic inertance of port (see Equation (16.14))
A = diaphragm area

$$C_m = \frac{C_a}{A^2} \qquad (16.17)$$

where C_m = compliance reflected to loudspeaker driving point
C_a = acoustic capacitance

Helmholtz resonator

Lord Rayleigh investigated the Helmholtz resonator and derived the following formula:

$$f = \frac{c}{2\pi}\sqrt{\frac{A_p}{V L\,\sqrt[1/2]{\pi\,A_p}}} \qquad (16.18)$$

where f = resonant frequency
        c = velocity of sound
        A_p = area of port
        L = length of port

From the analogue of Fig. 16.5 it will be seen that the system has two resonant and one anti-resonant frequencies. The upper resonant frequency f_{o1} is the total mass of the speaker diaphragm resonating with the compliance of the diaphragm suspension and the reflected acoustic compliance in series; it should be noted that this is the same frequency as for an infinite baffle. The second resonant frequency f_{o2} is the total mass of the speaker diaphragm plus the effective mass of the port and the diaphragm suspension compliance, thus the resonant frequency will be **below** the fundamental resonance of the loudspeaker. The anti-resonance f_{o3} is that of the enclosure capacitance and the port mass.

At this frequency, f_{o3}, the impedance is resistive and is maximum, so the loudspeaker diaphragm velocity is minimum and radiation from the port predominates and is in quadrature with the diaphragm radiation. At the other two resonant frequencies (one above and one below f_{o3}) the mechanical impedance is minimum and will produce an impedance maximum on the electrical side. Below f_{o2} the phase shifts rapidly so that the radiation from port and diaphragm are 180° out of phase, whilst above f_{o2} the two radiating surfaces are in phase. What has been achieved is that the radiation efficiency of the system has been increased at frequencies below the normal resonance of an infinite baffle to a limiting lower frequency, determined only by the acoustic resonance of the enclosure volume and the port mass. This can amount to at least half an octave with correct design.

Because of increased acoustic loading, the distortion relative to an infinite baffle is reduced at frequencies above the enclosure resonance (f_{o2}). Below f_{o2} the distortion will be increased due to absence of the stiffness loading of an infinite baffle (as the mass reactance of the port decreases with descending frequency).

It will be seen that increasing the area will increase the port length and hence the port volume, which must be added to the effective enclosure volume to determine the overall cabinet dimensions. At low resonant frequencies with large port areas the port volume becomes increasingly large and one soon encounters the law of diminishing returns. For example, with a design centre of 85 ltrs, 250 mm loudspeaker (diaphragm area 0.06 m²) and resonant frequency of 40 Hz, for a port area of 0.06 m² the tunnel length is 925 mm and the port volume 50 ltrs! Therefore one either reduces the port area, increase the resonant frequency, or increase the enclosure volume!

Maximising bass response

A way out of this difficulty is to replace the air mass in the tunnel with a rigid piston; the mass of the piston can be varied independently of the area. The volume of the assembly is negligible compared with that of a tunnel, it is therefore possible to maximise the bass response for a given set of conditions. The analogy between the rigid piston of given mass and an equivalent tunnel is exact, except that viscosity effects of the tunnel volume are absent and the Q of the piston is therefore somewhat higher. But the piston does have to be supported, and this introduces another component into the circuit: a compliance, due to the restoring force of the piston surround. Figure 16.6 shows the analogue. It is seen now to be two series resonant circuits in series, one (the passive diaphragm) shunted with the enclosure capacitance, this latter mesh can result in an anti-resonance at a frequency higher than the speaker resonance. It is good practice and perfectly feasible to introduce sufficient acoustic damping in the enclosure to reduce the effect to negligible proportions. It will be seen that the low-frequency response can now approach that of the basic loudspeaker resonance.

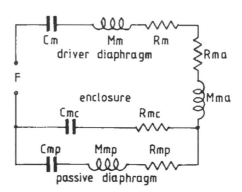

Figure 16.6   Enclosure with passive radiator.

The beauty of the passive diaphragm is that one can vary three parameters independently, i.e. the mass, compliance, and area. It is possible to use more than one passive diaphragm, but the law of diminishing returns soon operates, and in practice two units are about the maximum which can be used.

The equations quoted above for infinite and vented enclosures apply to passive diaphragms equally well. If the passive unit diaphragm and suspension is made identical to the driving loudspeaker the low frequency response will be extended by one half octave relative to an equivalent infinite enclosure. Unlike the port of the vented enclosure, the impedance of the passive diaphragm will increase at frequencies below resonance due to the compliant reactance of the surround, thus the impedance seen by the driving diaphragm will increase with decreasing frequency (actually the effective reactance of the enclosure compliance in parallel with the passive diaphragm surround compliance). Thus the distortion in reflex enclosures at frequencies below f_{o2} due to reduced acoustic loading is virtually eliminated.

Figure 16.7 shows the frequency response of a 250 mm diameter unit in an 85 ltrs infinite box. 'A' is the response of the system on an infinite baffle, and 'B' with the addition of the passive diaphragm. This diaphragm consists of a second speaker unit without the magnet and a 7 gm mass of Araldite in place of the voice coil. The bass response extends to 40 Hz (–3 dB) and total distortion is less than 10 per cent for inputs of 30 W. Under these conditions the sound pressure is 103 dB in a room of 2500 ft³, and reverberation time of 1.2 s. Correctly used, the passive diaphragm can extend the lower frequency response considerably; it can be (and is) used to good effect on miniature bookcase loudspeakers, but it must be remembered that although the frequency response can be extended in a downward direction to possibly 40 Hz, the power handling capacity is limited by maximum permissible displacement of the loudspeaker diaphragm. This is

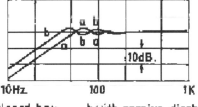

a. closed box.   b. with passive diaphragm.

Figure 16.7 Resonant enclosure (a) closed box (b) with passive diaphragm.

usually about 10 mm maximum, thus with a diaphragm of 150 mm diameter the output is limited to a level of 100 dB referred $2 \times 10^{-5} \text{N/m}^2$ at 120 Hz, below which the power available decreases at –6 dB/octave. It may be noted that 100 dB is a high sound level, equal to the

crescendo of a fairly large orchestra 30 ft away in, say, the Festival Hall.

Labyrinth Enclosures

The major problem of all enclosure design is the back radiation from the loudspeaker diaphragms. Ideally, it would be reversed in phase (180°) and re-radiated, this would double the efficiency of the loudspeaker; but because of the finite velocity of sound this is impossible. The vented enclosure goes some way towards solving this problem, giving at least an extra half-octave in the bass for the effort. The other problem of back radiation is the internal acoustic resonances due to reflections in the enclosure. These can be mitigated by adding damping, but complete eradication is impossible. One solution is to use a long tube; if it were infinitely long there would be no reflections and no colouration. A practical solution is to make it a half-wavelength long at the bass resonant frequency and add damping. The cross-sectional area should match the diaphragm area, but in practice one can reduce it by up to 50 per cent (although the damping becomes more critical).

Figure 16.8 shows the analogue. R_1 represents the 'frictional' content and R_2 the absorption of the damping material; M is the mass of air, and C its compliance per unit length. R_1 will increase and R_2 decrease with frequency, thus for optimum low frequency performance the damping material lining the labyrinth should be fairly thick. However, if the damping material is too thick the air loading on the back of the diaphragm will become excessive and thus limit the high-frequency radiated power.

Obviously if all the energy radiated from the rear of the diaphragm is absorbed it does not matter if the end of the labyrinth is open or closed. If the end is left open and the damping is reduced there will come a point when a substantial amount of low frequency energy can be radiated; the amount will be inversely proportional to

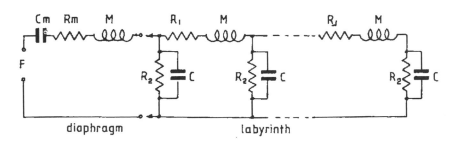

diaphragm   labyrinth

Figure 16.8 Labyrinth enclosure.

frequency, so at low mid frequencies (say, 100 Hz upwards) absorption will be 100 per cent and colouration due to back radiation eliminated. Correctly designed, the labyrinth offers the most effective way of getting rid of the unwanted back radiation from the diaphragm and is amongst the least 'coloured' of all enclosures.

Thus we are still presented with the fundamental proposition that to move large volumes of air at low frequencies large radiators are required. Reflex enclosures and passive diaphragms help to extend the low frequency response of the loudspeaker system by using some of the available energy from the rear of the loudspeaker diaphragm, and even though this amounts to only a few dB it must be remembered that an increase of 3 dB in sensitivity is equal to doubling the power of the amplifier output stage. The labyrinth is beautifully smooth but large, complicated, and expensive.

We thus return to the economics of the situation – does the cost of the loudspeaker enclosure complication offset the otherwise increase in cost of a power amplifier with, say, twice the power output for extended bass response?

Professional Systems

The specification for professional studio monitors is much tighter than that for domestic units because:

(a) the maximum sound level is high, 114 dB SPL is minimum, and with systems used in 'pop' recording 124 dB SPL is often specified:

(b) the acoustic environment is much larger than the domestic one;

(c) distortion levels even at the maximum output are significantly less (especially in the lower bass) than the domestic counterpart.

Assuming an auditorium $50 \times 40 \times 15$ ft (10^3m^3) with a reverberation time of 1.5 s, a power of 8 acoustic W would be required for an SPL of +114 dB; 32 W for +120 db; and 80 W for +124 dB. A speaker system of sensitivity +92 dB SPL for 1 W input has a conversion efficiency of 1 per cent, thus to produce +114 dB SPL assuming hemispherical radiation would require an electrical input of 800 W. Fortunately, due to reflections from the boundary surfaces the **effective** sound pressure is considerably increased, so the power input can be reduced by about –3 dB (halved!). Additionally, the speakers are normally used in pairs so the total input power per speaker for 114 dB SPL under normal operating conditions would be 200 W.

Assuming a 3-way system, it would be divided:

(a) bass power 80 W;
(b) mid-range 70 W;
(c) treble 50 W;

and could consist of a 15 in (380 mm) unit for bass, two 4 in (100 mm) mid-range, and a 1 in (25 mm) soft dome for treble. The treble unit should be of a Ferrofluid type to raise the power handling capacity. Normal practice is to use a 3-way passive network, but for optimum results it is preferable to drive each unit separately from its own amplifier with low level crossover; in this way the sound level from each of the speakers can be matched accurately and the problems of impedance mismatch between the crossover networks and the loudspeakers eliminated.

For larger systems, say for high level 'pop' studio monitoring where levels of +124 dB SPL are the norm, and levels of +130 db SPL peak can and do happen (what effect this has on the hearing of the artists is another matter) a total input power of 1500 W per speaker system is the accepted standard.

A typical system will use four 12 in (300 mm) bass units using 4 in (100 mm) diameter voice coils in a vented enclosure of 1 m³ tuned to 40 Hz with a Q of 1.4, the –3 dB point being 35 Hz. They are connected in parallel and fed from a 500 W RMS amplifier. The mid-range is a 100 mm soft dome unit, the gap flux density is 1.85 Tesla and is fed from a 500 W amplifier. The treble unit is a 40 mm soft dome, gap flux density 2.0 Tesla and is also fed from a 500 W amplifier. The crossover frequencies are 250 Hz and 5000 Hz (split at low level at the input to each amplifier). Additionally, a 24-way one-third octave equaliser is fitted to the input and the system is then adjusted in situ so that under operating conditions the system is flat ±1.5 dB 35 Hz to 17 kHz, and –3 dB 23 kHz at 120 dB SPL.

Networks

The requirements for a single loudspeaker to cover the 'full' frequency range of 20–20 000 Hz are in practice self-defeating; if the diaphragm is sufficiently large to radiate 20 Hz efficiently, at 20 kHz entirely apart from cone resonances etc., the focusing would result in a very narrow beam. Thus it is necessary, for wide band reproduction to split the frequency range into a number of sections. A good practical compromise would be two sections with a crossover frequency of about 2 kHz; a more exotic system would use three bands with crossover frequencies of, say, 500 Hz and 5 kHz.

By definition a 'low' pass filter will 'pass' all frequencies up to a specified frequency, and 'stop' all frequencies above it; a high-pass filter will be its inverse. This definition is impossible in the real world, and the crossover frequency is the –3 dB point of the attenuation curve.

The simplest network uses a series inductance for the

low-pass and a series capacitance for the high-pass. Assuming the load is resistive, the attenuation will be –6 dB/octave and because it introduces minimal phase shift it is preferred by some designers. The next filter is the 'two-pole', i.e. for a low-pass network a series inductance and shunt capacitor across the load with the elements interchanged for high-pass. The attenuation slope is –12 dB/octave. By adding a third element (in series) and forming a 'T' network, a 'three-pole' system results and the attenuation slope is –18 dB/octave. A fourth element (in shunt across the load) will give a 'four-pole' system with a loss of –24 dB/octave, and so on, see Fig. 16.9. In commercial practice it is rare to use more than a three-pole network, and the two-pole is by far the most popular.

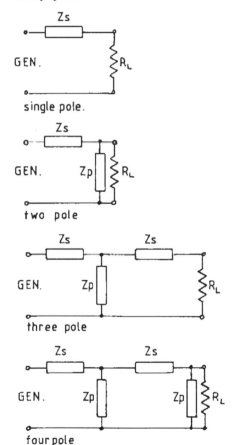

Figure 16.9 Basic network systems.

Before design of the network can commence four independent parameters must be known:

(a) generator impedance
(b) load impedance
(c) cut-off frequency
(d) attenuation slope.

Modern power amplifiers have massive amounts of overall negative feedback and are basically constant voltage generators, so the impedance 'looking into' the output terminals is a short circuit. The load impedance presented by the loudspeaker varies widely with frequency, and bears only passing resemblance to the manufacturer's impedance rating. Therefore, the first requirement is to measure the impedance at the crossover frequency. It will, in most cases, be a complex load, a resistance in series with an inductance for LF drivers and a resistance in parallel with a capacitance for tweeters.

To terminate the network correctly the reactive component must be removed and the easiest way is to use a 'Zobel' network, which is connected in parallel with the loudspeaker. For the low pass case this consists of a resistance R in series with a capacitance C. R and L equal the measured resistance and inductance of the loudspeaker at the cross-over frequency, and $C = L/R^2$. For the tweeter (high-pass case) the resistances are again equal but the reactance is generally capacitative, therefore the other element in the Zobel network is an inductance: $L = C/R^2$ although in some cases, where the tweeter is used well above its resonant frequency (frequencies of 5 kHz and above) it may present an inductive reactance, in which case the Zobel network will use a capacitance as in low frequency drive units.

Figure 16.10 shows the schematic for a two-pole network, the values of the two inductors and two capacitors are equal:

$$L = \frac{R_L}{\sqrt{2}\pi f_c} \qquad C = \frac{1}{\sqrt{2}\pi f_c R_L} \qquad (16.19)$$

Note: R_L is the measured AC resistance of the loudspeaker (plus Zobel network if used).

Figure 16.11 3 is for a three-pole network:

$$L_1 = \frac{1.6R_L}{2\pi f_c} \qquad C_1 = \frac{1}{\pi f_c R_L}$$

$$L_2 = \frac{R_L}{2\pi f_c} \qquad C_2 = \frac{1}{3.2\pi f_c R_L}$$

$$\qquad\qquad\qquad\qquad\qquad\qquad\qquad (16.20)$$

$$L_3 = \frac{R_L}{4\pi f_c} \qquad C_3 = \frac{1}{2\pi f_c R_L}$$

The network Fig. 16.10 would normally be used on medium power systems, say maximum 100 W, whilst network Fig. 16.11 would be used for higher power systems. For a three-way system, the low-pass section of Fig. 16.10 would be used for the bass and the high-pass for the treble. A band pass network for mid range is shown in Fig. 16.12. The element values calculated from Equation (16.19), L_aC_a would be for the high frequency cut-off (say 5000 Hz) and L_bC_b for the low frequency cut-off (say 500 Hz).

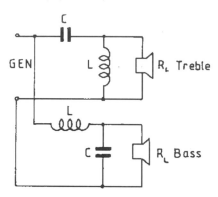

Figure 16.10   Two-pole network.

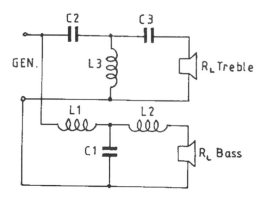

Figure 16.11   Three-pole network.

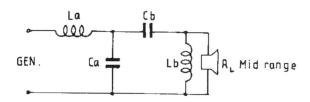

Figure 16.12   Band-pass network.

Components

For medium-priced equipment, bipolar electrolytic capacitors are normally used, but it must be remembered that the normal tolerance is ±20 per cent or ±10 per cent

to special order. This means that unless one is prepared to select capacitors to closer tolerances, say ±5 per cent, the cut-off frequency will vary out of design centre tolerances. The best capacitors are of the non-polar dielectrics such as polypropylene, PTFE, and polystyrene.

Inductors can be either air cored or ferrite cored. The air cored inductor has the advantage of no hysteresis losses or distortion but is expensive, especially with high power (which require thicker wire), high inductance coils. With ferrites there is non-linear distortion and because of low saturation induction there is a limit to their power handling capacity, being at about 350 Amp. tns for a 10 mm diameter core (proportional to cross sectional area) for 5 per cent THD. The advantage of the ferrite is the saving in copper. Coil resistance is important, with an 8 ohm loudspeaker a series coil resistance of 1 ohm will result in a power loss of about –1 dB.

Ribbon Loudspeaker

The ribbon loudspeaker is one of the most efficient high frequency units in production. Its construction (Fig. 16.13) is simplicity itself. The diaphragm consists only of an aluminium foil ribbon which is suspended between the poles of a magnet. The nominal flux density in the gap is 0.6 Tesla and high energy columnar magnets are used.

A practical ribbon is 55 mm long, 10 mm wide, 3 μm thick, and its mass is 3.5 mgm. The resistance of the ribbon is of the order of 0.1 ohm, necessitating a matching transformer when used with commercial amplifiers designed for a 4–8 ohm load.

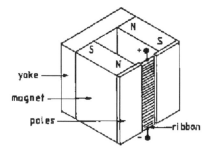

Figure 16.13   Ribbon transducer.

If the ribbon were flat, it would have to be stretched slightly to prevent it twisting and 'sagging' in the middle, and possibly moving out of the gap. This would also result in multiple mechanical resonances of the ribbon which would adversely affect the frequency response.

The solution is to corrugate it laterally, as this not only removes the mechanical problem but also prevents cross resonances. Additionally, it is very difficult to obtain a uniform magnetic field over a gap width of 10 mm (the flux density is about 15 per cent higher at the pole faces compared with the centre), therefore the force is not constant across the ribbon face and it would flex longitudinally and eventually break; the lateral corrugations effectively prevent this.

Because of the very low mass the ratio of applied force (*Bli*) to mass (3.5×10^{-6} Kg) is extremely high, giving a good efficiency and excellent high frequency and transient response, the cut-off point (–3 dB) being 55 kHz. However, the small radiation area puts the 'low' frequency cut-off at 5 kHz. This problem is overcome by providing horn loading, and with a suitable design the 'low' frequency response can be extended down to 1 kHz; this demands a very steep cutting high pass filter with an attenuation slope in excess of –60 dB/octave to prevent the high power mid frequencies destroying the ribbon. Figure 16.14 shows a suitable design in which the mutual inductance of the two shunt arms resonates with C_2 giving infinite rejection at 700 Hz.

The rear surface of the ribbon is acoustically coupled to the cavity between the magnets, which is filled with an absorbing material to prevent resonances.

The excellent high frequency and transient response is primarily dependent upon the low mass of the diaphragm which is also the conductor. This means that the thermal capacity is low and, concomitantly, the power handling. With present designs this is limited to 30 W and a 50 per cent overload for 5 ms. would melt the ribbon, so some form of overload protection should be built into the amplifier.

Wide Range Ribbon Systems

The 'wide range' ribbon loudspeakers currently available have claimed frequency response 25 Hz to over 30 kHz, but being di-pole radiators (similar to electrostatics) the final acoustic performance is dependent on the listening room geometry, furnishings, and speaker placement. Maximum SPL is about +115 dB at 4 m in a room $27 \times 14 \times 10$ ft and reverberation time of 1.5 s, using two 100 W amplifiers per speaker. The quoted SPL is for a pair of units in normal stereo configuration. The system consists of bass, mid-range, and treble units, the bass being fed by one 100 W amplifier, and the mid-range and treble by a similar 100 W amplifier.

These speakers are not small, each weighing about 136 Kg (300 lbs) and are 2000 mm high, 890 mm wide, and 10 mm deep. The drive units are mounted side by side (see Fig. 16.15) and are 1850 mm long. The bass is trapezoid in shape, 250 mm wide at the top and 350 mm at the bottom; the mid-range is 50 mm wide, and the treble unit 20 mm wide. The diaphragms are 10 μm Kapton with an aluminium conductor 20 μm thick cemented to it.

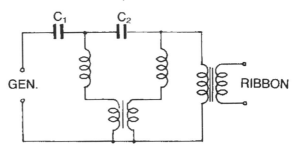

Figure 16.14 High-pass network.

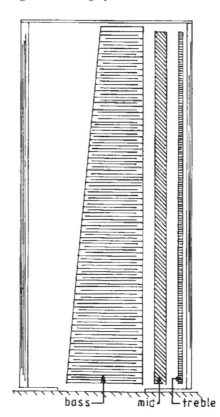

Figure 16.15 Wide range ribbon loudspeaker.

Figure 16.16 shows the construction of the bass unit. The conductor is arranged in 'zig-zag' fashion and the magnets alternately N-S-N-S etc., the conductor overlapping a pair of magnets, thus the resultant force from the interaction of field due to the current and that from the magnets will move the diaphragm, the displacement being proportional to magnitude and direction of the current. It is stretched slightly to prevent sagging, and the system although not as 'transparent' as the electrostatic has excellent transient response.

The mid-range driver is under slight tension and is supported along its edge by plastic foam and the corrugations are at 45° to reduce transverse vibrational modes.

The treble unit is smaller to the mid-range, except for width and transverse corrugations. A simple two-pole crossover network is used for the two units which, like the bass, has a substantially resistive input impedance of 4 ohms. The crossover frequencies are 400 Hz and 5 kHz.

The distortion at rated output is less than 1 per cent over the whole frequency range, and intermodulation products are negligible.

Pressure Drive Units

From time immemorial a horn has been used to 'amplify' sound; when used as a musical instrument and applied to the mouth the tiny vibrations of the lips can produce an ear-shattering sound (especially the brass family). With the invention of the telephone it was rational to use a horn in an attempt to amplify the low level of sound from the early receivers, indeed the first 'loudspeakers' used for wireless reception in the 1920s were nothing more nor less than a headphone with an attached horn. From these humble beginnings the most powerful and efficient form of sound reproduction has evolved.

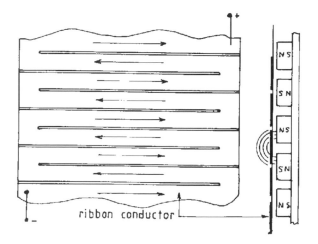

Figure 16.16   Bass unit.

The modern unit consists of a moving coil drive unit and matching horn. For optimum results the two units should be designed as a system.

Horns

A horn is basically an acoustic transformer. It transforms a small area diaphragm into an effective large area diaphragm without the disadvantages of increased mass, cone resonances, etc. The radiation resistance of a large area diaphragm is much greater than one of small area, thus more power is radiated for a given velocity volume of air. The basic parameters for designing a horn are maximum acoustic power, frequency range, and tolerable distortion. Knowing these, the driving unit can be specified, then the throat and mouth diameters and the form (or shape) and length of the horn can be calculated.

Mouth size

The mouth (large end) should have a circumference large enough so that the radiation impedance is substantially resistive over the designed frequency range. i.e., $Ka > 1$; that is, $C/\lambda > 1$, where C is the mouth circumference $(2\pi\,a)$, and λ is the wavelength of the lowest design frequency. Thus, for $f_c = 50$ Hz, the horn mouth diameter should be ≈ 2.2 m (7.25 ft) which for domestic purposes is rather large.

Rate of flare

Having determined the mouth dimensions and the throat diameter which in turn are controlled by the driving unit constants, the length of the horn is determined by the rate of flare, which in turn is a function of the cut-off frequency. The actual relationship depends upon the particular flare law used.

The conical horn

This is the oldest and simplest form. The area A_x at any distance x from the throat is:

$$Ax = A_t x^2 \tag{16.21}$$

where A_t = Throat area
The low frequency limit f_L taken –3 dB is:

$$f_L = \frac{9x}{Ax} \qquad \text{or} \qquad x = \frac{f_L Ax}{9} \tag{16.22}$$

Exponential Horn

The exponential horn will extend the low frequency response by about three octaves compared with a conical horn of the same overall dimensions. The secret is the rate of flare. The area Ax at any distance x from the throat is:

$$Ax = A_t \varepsilon^{mx} \qquad (16.23)$$

where m = Flare constant
         $\varepsilon = 2.718$
The low frequency limit (-3 dB) is:

$$f_L = \frac{mc}{\sqrt{2\pi}} \qquad (16.24)$$

but the frequency of complete cut-off normally quoted is:

$$f_c = \frac{mc}{4\pi} \qquad (16.25)$$

Hyperbolic horn

For the same overall dimensions this horn will show an improvement of one-third octave downwards compared with the exponential horn. The law is:

$$Ax = A_t (\cosh x/x_o + T \sinh x/x_o)^2 \qquad (16.26)$$

where x_o = Reference axial distance from throat
         x = Axial distance from throat
         T = Parameter which never exceeds unity and is usually about 0.6

$$f_c = c/2\pi x_o \qquad x_o = c/2\pi f_c \qquad (16.27)$$

The hyperbolic has a nearly vertical cut-off curve and the -3 dB point can be considered the same as f_c.

Figure 16.17 shows the relative impedance of these three horns; R_{ma} = Resistance, and X_{ma} = Reactance.

- Curve No. 1 = Conical
- Curve No. 2 = Exponential
- Curve No. 3 = Hyperbolic

These curves are for horns of infinite length. Practical horns will have the same general shape but there will be 'wiggles' on the curve due to reflections from mouths of finite size. See Fig. 16.18 – the shorter the horn the larger the discrepancy.

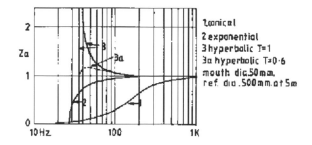

Figure 16.17   Normalised throat resistance infinite horns.

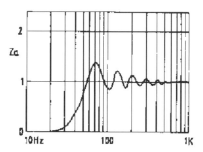

Figure 16.18   Exponential horn Fig. 16.17 but length = 2.5 m.

Distortion

For optimum coupling the horn throat is usually of smaller area than the diaphragm of the driver and the horn throat forms a constricted volume of air.
     The variation of pressure in air is a function of volume and is given by:

$$p = 0.726/V^{1.4} \qquad (16.28)$$

where p = Absolute pressure in bars (1 bar = 10^5 N/m²)
         V = Differential change of volume of air in throat due to movement of diaphragm (in m³/Kg)
     For very small changes in volume, this equation is substantially linear, but where the relative volume displacement due to diaphragm excursion may be large, distortion is generated. Assuming a sinusoidal motion of the diaphragm, the second harmonic distortion is:

% second harmonic distortion $= L \times 73\, f/f_c \sqrt{I_t} \times 10^{-2}$ (16.29)

where f = Driving frequency
         f_c = Horn cut-off frequency
         I_t = Intensity in W/m² at horn throat

Drive units

Figure 16.19 shows the layout of a horn/driver unit. Over the working frequency range the mechanical impedance Z_{mt} presented by the throat of the horn is a pure resistance:

$$Z_{mt} = p_o C S_t \qquad \text{where } p_o C = 407 \text{ Ohms} \qquad (16.30)$$
$$S_t = \text{Throat area in m}^2$$

The mechanical impedance presented to the diaphragm is proportional to the ratio of square of the diaphragm area to the throat area:

$$Z_D = Sd^2/Z_{mt} = Sd^2/p_o C S_t \qquad (16.31)$$

The effect of the horn is to produce a large increase in Z_D at low frequencies (above the horn cut-off frequency, of course); such a system is therefore much more efficient at these frequencies than the small drive unit alone. Additionally, because the moving mass of a small drive is low, the inherent efficiency is proportionally greater when compared with a direct radiator loudspeaker of comparable cone size to the horn mouth.

Horn type loudspeakers can easily be built with efficiencies in excess of 20 per cent (best examples reach 50 per cent) over a wide range of frequencies above cut-off frequency of the horn. The primary difficulty is that in order to produce high outputs at low frequencies the displacement of the small diameter diaphragm must be very large, so that it requires considerable ingenuity and skill on the part of the designer to construct a suspension system which will not introduce non-linear distortion.

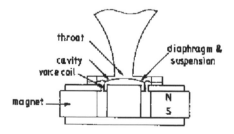

Figure 16.19   Horn driver unit.

It will be seen from Equation (16.31) that increasing the ratio of diaphragm to throat area increases Z_D and thus the efficiency at low and medium frequencies, but because of the increased mass of the driving system will reduce the ultimate high frequency response – what one gains on the swings!

The analogue of the mechanical circuit is shown in Fig. 16.20. C_D is the compliance of the suspension, M_t total mass of voice coil, diaphragm, etc., R_t total mechanical losses (suspensions, damping, etc.), $Z_t = Ra + jXa$ the reflected acoustic resistance and mass reactance (which increases with frequency) at the throat. and C_c reflected compliance of the air chamber between the diaphragm and the throat:

$$C_c = V_c/1.4 \times 10^{-5} Sd^2 \qquad (16.32)$$

and this reactance acts in parallel with the radiation impedance Z_t at the throat of the horn, thus at some high frequency the reactance of C_c will be less than Z_t and the high frequency response will fall.

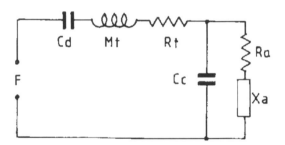

Figure 16.20   Analogue of Fig. 16.19.

The motional impedance, speaker efficiency, and power output, can be calculated from this mechanical circuit and the known electrical characteristics of the motor system. Figure 16.21 shows the complete analogue referred to the electrical side, which over the working frequency range can be simplified to Fig. 16.22. It will be seen that it is purely resistive, $i^2 R_{mt}$ being the actual power radiated by the horn.

Figure 16.23 shows a response curve; at frequencies below A the response will drop because of reduced loading due to horn 'cut-off'. The mid-frequency range B is flat; in this region the diaphragm velocity is constant with frequency and the acoustic load is resistive, therefore the acoustic output will be flat. At high frequencies the response is limited, principally by the combined mass of the diaphragm and voice coil assembly. If there were no cavity between the diaphragm and throat the response would fall at –6 dB/octave, C in Fig. 16.23. In real life there is always a cavity and it is possible to choose C_c so that it will resonate with the total mass M_m and extend the upper frequency response by about an octave – note the response then falls at –12 dB/octave rather than the –6 dB/octave when C_c is zero.

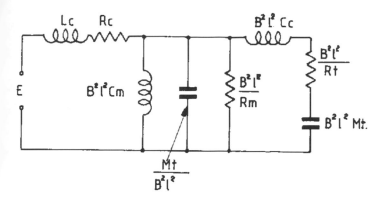

Figure 16.21 Analogue of Fig. 16.19 referred to electrical side.

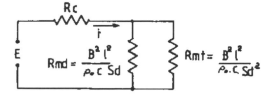

Figure 16.22 Simplified circuit of Fig. 16.21.

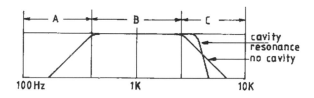

Figure 16.23 Response of Fig. 16.21.

The principal diaphragm/suspension resonance is usually chosen at the centre frequency of the desired range and is highly damped by the reflected acoustic resistance. With optimum design a flat response can be obtained over a range of four octaves.

Electrostatic Loudspeakers (ESL)

The electrostatic transducer is by far the oldest piece of electrical equipment to transform electrical energy into another form, either mechanical or acoustic. The electric machines of the 18th century with their whirling globes of glass or brimstone and with their discs plastered with silver paper and 'what-have-you' provided the experimenters with ample opportunity to become familiar with the loud impulsive sounds which accompany spark discharges in air. Accumulating the charge on a condenser

or, as it was known in those days, a Leyden Jar, still further enhanced these noisy discharges. The main spark provided a noise by direct ionisation of the air, but also the Leyden Jar itself acted as an electro-acoustic transducer and provided a small amount of acoustic energy. The effect can now be identified as a shock excitation of the compressional waves in the material of the Jar, produced by the sudden release of the electro-strictive stress established by the electric charge. Some of the Leyden Jars actually emitted a musical tone corresponding to the frequencies of the mechanical resonance of the Jar itself. However, these resonance vibrations died out very quickly, presumably due to the damping introduced into the mechanical circuit by the adhesive used to stick the foil onto the Jar. This was later identified in 1863 by Lord Kelvin who inferred that the cause of the effect must be closely related to the electrostatic stresses about which Faraday had speculated.

Since that time, at periodic intervals, the use of the electrostatic forces have been harnessed (generally unsuccessfully) into making an electro-acoustic transducer. The most successful of the early attempts was in 1917, when Wente produced the first successful electrostatic or condenser microphone. It was the first transducer design in which the sensitivity was deliberately offset for uniformity of response; and it was certainly the first in which electronic amplification was relied upon to gain back the ground lost by eschewing resonance.

The principal changes introduced into the condenser microphone itself during the next three decades consisted of the virtual elimination of the cavity in front of the diaphragm and a drastic reduction in the size of the instrument. As a matter of course the action was reversed and an electric potential applied to the condenser microphone, which behaved as a miniature loudspeaker. However, the electrostatic loudspeaker failed to gain any commercial acceptance in spite of extensive activity devoted to it between about 1925 and 1935. Principally, there were several serious shortcomings which still adhered to the basic design. Either the diaphragm or the air cavity itself had usually been relied upon to provide the protective insulation against electrical breakdown, but this protection was often inadequate and limits were therefore imposed on the voltages that could be used and, concomitantly, the maximum power output.

Close spacings, a film of trapped air, stiff diaphragm materials, and vulnerability to harmonic distortion, combined to restrict to very small amplitudes both the allowable and the attainable diaphragm motion. As a consequence large active areas had to be employed to radiate useful amounts of sound power, especially at low frequencies. When large areas were employed, the sound radiation was much too highly directional at high frequencies, and several of the early Patents bear on one or other of these features. It is now apparent that an integration of such improvements would have made it possible to overcome, almost but not quite, every one of those performance handicaps. The problem was that no one sat down and coherently thought about it.

The essential components of a simple electrostatic loudspeaker are shown diagrammatically in Fig. 16.24. One plate, P_1 is fixed, the other plate, P_2 being movable and therefore responsible for sound radiation. If a steady DC voltage is applied between the plates the attracted force will cause P_2 to approach P_1 and ultimately stay there unless prevented from doing so. This action follows from the law of inverse squares, namely, force $(E/D)^2$. It is imperative, therefore, to introduce some form of elastic constraint between P_1 and P_2. For analytical purposes it is immaterial how this is effected so long as the force displacement law is known. To make the analysis practicable, we must perforce deal with a linear differential equation, the force displacement law should be linear. Accordingly, we postulate a thin dielectric substance in which the law of compression is $F = Sx$, where x is the displacement in the direction of F.

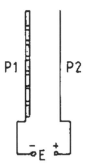

Figure 16.24   Basic single-sided electrostatic loudspeaker.

If an alternating voltage be applied across P_1P_2, then seeing that the attraction is independent of polarity, the diaphragm moves towards the plate during each half

cycle. The relationships are indicated in Fig. 16.25, where curve (a) corresponds to the electric input and curve (b) corresponds to the acoustic output. This output can be resolved by Fourier analysis into an infinite series of sequences. The sine wave input is therefore reproduced with an infinite retinue of harmonics and electro-mechanical rectification ensues. If a polarising voltage large by comparison with the signal voltage is imposed thereon, the action of the device undergoes a remarkable transformation. The steady voltage causes the dielectric to be compressed by an amount αE^2; the signal voltage causes a fractional variation in the compression according to harmonic law when the signal voltage is additive; the dielectric is compressed $(E + \Delta E)^2$, whilst in the middle of the next half-cycle it is compressed $(E - \Delta E)^2$. The force varies as $1/D^2$ but if ΔD is a small fraction of the dielectric thickness the variation with distance can be disregarded and the action is substantially linear. Thus, if the dielectric is massless the device reduces mechanically to a harmonically driven mass on a simple coiled spring; the diaphragm is the mass and the dielectric provides the spring effect; whilst the alternating electric force does the driving. Under such conditions the acoustic output will be a replica of the voltage input, providing the amplitude of vibration is small.

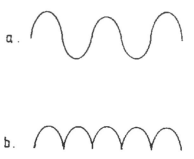

Figure 16.25   Rectified acoustic output, single-side transducer (a) electric input (b) acoustic output.

This simple system has been used with some success in the production of high frequency tweeters. If the diaphragm is to be mechanically stable it is obvious that there must be a limit to the polarising potential that may be applied. Irrespective of corona discharge the electrostatic force acts as a negative compliance and if this increases above the static compliance of the diaphragm it will collapse onto the other electrode.

The beauty of the electrostatic loudspeaker is that (in theory at least) the driving force is applied equally over the whole surface of the diaphragm, which thus moves in phase as a coherent unit, whereas the cone of a moving coil unit is driven at the apex only and problems arise due to finite velocity of the acoustic wave through the diaphragm material and resonant modes dependent on shape, size, and material of the diaphragm present the designer with a permanent headache.

There are problems, of course. It is customary to regard electrostatic devices as basically non-linear since the tractive force between condenser plates at a constant potential difference varies inversely as the square of distance between them, and Fig. 16.26 shows the schematic of a single-sided transducer. There are several basic parameters: (a) the force of attraction between oppositely charged condenser plates; (b) the capacitance of a parallel plate condenser; (c) relation between charge and EMF across the plates.

$$F_e = \frac{-q^2}{2\varepsilon_o Sd}; \qquad C = \frac{\varepsilon_o Sd}{D + x}; \qquad q = C. E_o \quad (16.33)$$

F_e is the mechanical force originated by the charge q, and it will be noticed that the stress is a tension and acts in a negative x direction. The presence of q^2 term in Equation (16.33) indicates that the system is non-linear. It follows an inverse square law, thus the input AC voltage must be a small proportion of the polarising voltage if distortion is to be kept within reasonable limits.

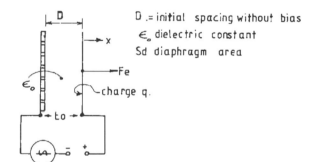

D .= initial spacing without bias
ε_o dielectric constant
Sd diaphragm area

Figure 16.26   Single-sided system with DC bias.

Another important factor is the negative compliance generated by q, which at some value of E_o will exceed the diaphragm stiffness which will then collapse onto the stationary back plate with disastrous results.

Push-pull electrostatic loudspeaker

One way of overcoming this problem is to place a second electrode on the other side of the diaphragm. This, so far as DC is concerned, is connected to the first electrode, thus there is now an equal and opposite force applied to the diaphragm and, if the potential is not too high, the diaphragm will remain central. Thus if the two outer

electrodes although at zero DC potential are connected together through the secondary of a transformer and an alternating potential applied to the primary the diaphragm will be alternately attracted to one and then the other of the outer plates according to whether the instantaneous potential is increasing or decreasing, (See Fig. 16.27).

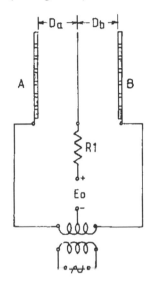

Figure 16.27   Constant charge push-pull system.

It will be seen that the two outer plates are each at ground potential except for the applied AC signal voltage; this is assumed to be divided equally between the two meshes. R_1 is of small value and is purely for protection against short circuiting the power supply if the diaphragm touches either outer electrode. As shown in the single-sided transducer, the device is basically non-linear because the tractive force between the diaphragm and back plate is essentially non-linear since for a constant potential difference F_e varies as the inverse square of their separation.

Constant charge push-pull system

If, however, R_1 is made very large, then the time constant of $R_1 Ca$ and $R_1 Cb$ is very long (several seconds in practice) and can be considered, so far as the signal voltage is concerned, infinite; this will be the equivalent of a constant charge on the diaphragm and will be independent of its position in the space between the electrodes. Thus the force acting on the diaphragm will be determined only by the magnitude of the unvarying charge on the diaphragm and the electric field established by the signal voltage, most importantly it will be independent of the position of the diaphragm in the space between the electrodes. Thus the major drawbacks of the electrostatic loudspeaker are removed by this simple expedient. The non-linearity due to diaphragm spacing is eliminated as is the requirement to restrict the ratio of signal to polarising voltage, and in this form the electrostatic transducer comes out as a clear winner in the distortion stakes when compared with electrodynamic units.

The efficiency is directly proportional to the (polarising voltage)^2 and inversely as the spacing between the diaphragm and the outer electrodes, and is limited by voltage breakdown. To prevent catastrophic breakdown the outer electrodes are covered with a plastic coating which not only has a high voltage breakdown capacity but also a high dielectric constant; with modern materials over 95 per cent of the polarising voltage appears in the air space.

The other problem is ionising of the air in the gap. It is usual to arrange for the polarising voltage and half the peak-to-peak signal voltage to be 90 per cent of the breakdown voltage. This leaves very little safety margin and a momentary overload will cause ionisation and consequent audio distortion and in extreme cases puncture the diaphragm. However, with the present generation of loudspeaker systems protection circuitry is built into the system.

Diaphragm size

The ultimate diaphragm dimensions are controlled by a number of conflicting requirements: to radiate low frequencies the area must be large – the lower the cut-off frequency the larger the diaphragm – for good dispersion at high frequencies the converse is true. Fortunately, there is a compromise solution – that is, to split the diaphragm into a number of discrete electrical sections whilst still maintaining a coherent whole.

Low frequencies

The first limitation is the spacing between the plates and diaphragm which limits the excursion at low frequencies. The fundamental resonant frequency of a rectangular diaphragm f_o is:

$$ f_o = \left[\frac{1}{2} \sqrt{\frac{TA}{Lm_t} \left(\frac{1}{a^2} + \frac{1}{b^2} \right)} \right] \qquad (16.34) $$

where T = Tension
A = Area
Lm_t = Total dynamic mass
a = Short dimension
b = Long dimension

The static force is $(E_o/D)^2$; it will be linear with displacement, will behave as a negative stiffness, and provided this does not exceed the positive mechanical stiffness of the diaphragm it will remain stable. This sets the limit for maximum displacement, and for a given low frequency resonance the diaphragm size – thus one trades low frequency response against diaphragm size and efficiency.

Compared with the dynamic loudspeaker the total mass of the moving system is very small, in theory and to some extent realised in practice, this leads to high efficiency; being thin and relatively large it is flexible and under working conditions is lightly stretched, but the major restoring force is a function of the electric field. It can be shown that a stretched membrane will break up into an infinite number of frequencies, not necessarily harmonically related: the expression Equation (16.34) gives the first (or fundamental), others can be calculated from:

$$ f_n = \frac{1}{2} \left\{ \frac{TA}{Lm_t} \left[\left(\frac{m}{a} \right)^2 + \left(\frac{n}{b} \right)^2 \right] \right\} \qquad (16.35) $$

where m = 1,2,3, etc.
n = 1,2,3, etc.

Fortunately, these resonances are highly damped by the resistive component of the air load and the perforated outer electrodes, and do not present any difficulties with current designs.

Of all practical loudspeakers, the ESL is the nearest approach to an ideal piston. It provides direct conversion from a constant voltage to a constant force applied to the air load. This air load is the major portion of the mechanical impedance of the system and with modern designs approaches 70 per cent of the total loading; thus the diaphragm velocity (and hence radiated power) is directly controlled by the radiation impedance.

Because of the low acoustic impedance of the diaphragm, it is virtually 'transparent' to a sound wave. If the rear of the speaker is contained in a box, because of this 'transparency' all reflections from the walls and resonances of the box will pass through the diaphragm unattenuated and will, according to the phase of the reflected signal, augment or cancel the original signal, and the resultant sound field will be a considerable degradation of the otherwise flat response. Additionally, the diaphragm has a much greater area than the traditional moving coil loudspeaker and the dynamic mass is an order of magnitude lower. The low-frequency resonance is controlled by the total mass of the moving system and the total compliance. Because of the low mass of the diaphragm the effective mass is that of the air load, and the compliance will be inversely proportional to the square of the diaphragm area. Thus the reflected compliance would be extremely low and together with the low mass, even with a large enclosure would place the 'low' frequency resonance in the upper mid frequencies. For this reason all practical ESLs are 'unbaffled' – this is known as a 'doublet' source, because the maximum radiated energy levels are found on the front and rear axis with the minimum levels in the plane of the diaphragm.

Directivity

Focusing at the higher frequencies is inevitable because of the relatively large size of the diaphragm. There is, however, a simple solution which was used in the first commercial ESL. This consisted of splitting the outer electrodes into three vertical sections, the centre section covers the frequency range from about 500 Hz upwards, and the two outer sections handle the bass frequencies. Since a step-up transformer is necessary, it is possible to form effective crossover network components by deliberately introduced leakage inductance in the transformer to match the inherent capacitance of the loudspeaker.

The latest developments carry this philosophy to its logical conclusion. It was developed by PJ Walker in his Quad ESL 63 using the 'constant charge' concept and taking advantage of the 'transparent' diaphragm. These techniques allow one to short circuit the somewhat cumbersome methods of equivalent analogue circuit analysis and use a much simpler reciprocity system. The 'transparent' diaphragm means that if it is immersed in any sound field it will not introduce any discontinuities and will not disturb the field. When this is the case radiation impedances and diaphragm velocities can be eliminated from the equations and instead a much simpler relationship between electrical input current and acoustic sound pressure is revealed, even for complicated surface shapes.

The basic transduction mechanism must be defined and then by reciprocity the current/pressure relationships are derived. Figure 16.27 represents a section through an ESL of as yet undefined area. Of these electrodes, the two outer ones are assumed to be stationary plates, parallel to each other and perforated to allow free passage of air. The centre electrode, a thin stretched membrane, is assumed to be sufficiently light and unrestrained so that if placed in any sound field its normal vibrations will follow that of the air particles prior to its introduction. There are no baffles or boxes or other structural impediments.

The electrodes are open circuited and the system charged such that the equal voltages E appear between the electrodes as shown. The polarising forces on the centre electrode are in equilibrium and no voltage appears across the outer electrodes.

Let the central electrode be moved to, say, the left. The voltage on the left-hand side will now reduce since with no charge migration it must remain in direct proportion to the spacing. The voltage on the right-hand side will increase, again in proportion to the spacing. The electrical forces on the central electrodes are still in equilibrium but now a voltage has appeared across the outer electrodes. The fact that the electrical forces on the central electrode still cancel to zero means that no work has been done in moving the electrode from one position to another, and yet a voltage has appeared across the outer electrodes. This is not the creation of energy but merely a manifestation of negative compliance. The mechanical system 'sees' the capacity between the outer electrodes in series with a negative capacity of equal value. Thus the two-terminal system is linear and the voltage appearing across the outer electrodes is directly proportional to the displacement of the central electrode. Additionally, it is necessary to prevent the charge distributing itself unevenly over the diaphragm during its excursions, which would produce variations of sensitivity in different areas. To avoid this the conductive coating is very highly resistive, thus localising the instantaneous charge.

Imagine a point source of volume velocity U in the far field at right angles to the plane of the membrane and at a distance r: near the loudspeaker unit the waves from this point source will be essentially plane with a pressure of $U.f_p/2r$ and have a particle displacement of $U/4\pi rc$ which is independent of frequency. The centre electrode will vibrate in accordance with this displacement and will produce a voltage between the outer electrodes which will be independent of frequency. Thus a volume velocity at point P produces a voltage at a pair of open circuit terminals and by reciprocity a current into those same terminals will produce a pressure at point P equal to:

$$p = E/D \times I/r \times 1/2\pi C \quad \text{Newtons/m}^2 \quad (16.36)$$

where E is the initial polarising voltage on the membrane when central, and D is the distance between the two fixed electrodes. This, then, is the expression for the far field axis pressure. The result, dependent only upon two simple electrical and dimensional measurements, independent of frequency, area, or shape of the loudspeaker elements, is in sharp contrast to the laborious postulation connected with the design of, say, a moving coil loudspeaker.

The next problem is directivity – ideally, a small surface for high frequencies and a progressively larger area as the frequency is reduced. Theoretically, a thin flexible membrane with vibration patterns arranged to radiate waves as if they had come from a point source some distance behind the membrane surface. It would be expected that the membrane would vibrate in the form of expanding rings, falling in amplitude as they expand in accordance with simple calculations.

Figure 16.28 shows a point source S producing spherically expanding sound waves at a distance r (in practice, about 300 mm) to the transparent membrane. It will be seen that for the membrane when driven electrically to synthesize the spherical wave, the input signal must be progressively delayed proportional to distance from its centre. Assume that the wave front M with radius r is just arriving at the membrane surface; after time t_1 it will have radius d_1, and $t_2 = d_2$ etc., it will be seen from the diagrams that if the differences between successive delay lines are

equal the radii of corresponding circles will get progressively closer as they move out. If the membrane is now placed between two perforated plates on which conductors are arranged in concentric circles, the signal is fed to the centre one and also through successive steps in a delay line to the outer ones in turn, each step of the delay line being equal to $t–t_1$, $t_1–t_2$, etc., then the membrane movement will be that of a plane through which the sound field is passing, see Fig. 16.29.

For simplicity the electrodes are divided into six sections of equal area and hence equal capacity. It is assumed equal delay time and attenuation to each section. Figure 16.30 shows the simplified electrical circuit, and Fig. 16.31 one section of the delay line. The self capacity of the inductors is balanced out by the cross connected capacitors, so the delay is to a large extent independent of frequency.

By optimising the delay line constants it is possible to produce smooth axis and polar curves with an almost complete freedom of choice for directivity index versus frequency. The solution is to examine the motional

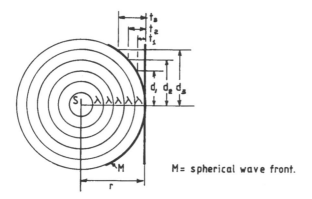

Figure 16.28   Radial delay from point source.

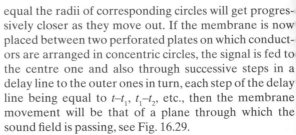

Figure 16.31   Delay-circuit section.

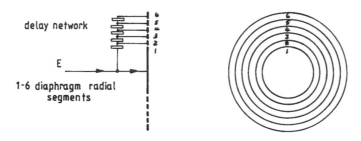

Figure 16.29   Radial delay and attenuation of signal currents.

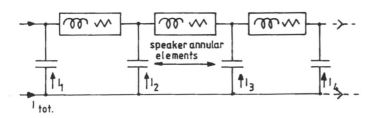

Figure 16.30   Simplified electrical circuit.

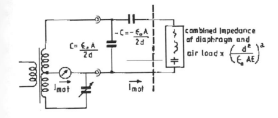

Figure 16.32 Bridge for direct measurement of membrane velocity.

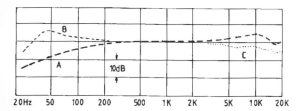

Figure 16.33 Results of bridge measurements.

current because this is a direct measurement of the membrane velocity:

$$\mu = \frac{I_{mo} T d^2}{E_o E A} \qquad (16.37)$$

Figure 16.32 shows a suitable bridge circuit to extract

the motional current – the so-called 'blocked impedance' is obtained by switching off the polarising supply since this has the effect of increasing to infinity the apparent impedances to the right of the dotted line. With the bridge carefully balanced, the polarising supply is switched on and the membrane velocity (obtained from the motional current into the system) can be plotted directly on a standard curve tracer. This motional current also gives direct access for accurate distortion measurements.

Figure 16.33 shows a practical example of these procedures: (a) is the vector sum of electrode currents; (b) is deviation due to suspension stiffness from motional current measurements; and (c) is calculated deviation due to diaphragm mass and plate obstruction. The actual response measured outdoors at 2m is shown in Fig. 16.34, and shows excellent correlation when compared with the predicted response of Fig. 16.34.

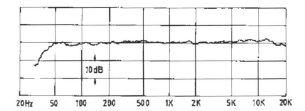

Figure 16.34 Outdoor axis response at 2 m.

17   Headphones

Dave Berriman

Not all sound reaches us through loudspeakers, and the advent of the Sony Walkman has led many listeners back to headphone stereo. In this Chapter, Dave Berriman shows the very different type of problems and their solutions as applied to this alternative form of listening.

A Brief History

The history of modern headphones can be traced back to the distant first days of the telephone and telegraph. Then, headphone transducers for both telephone and radio, worked on the same moving-iron principle – a crude technique compared to today's sophistication, but one which served so well that they are still to be found in modern telephone receivers.

Moving-iron headphones suffered from a severely restricted frequency response by today's standards, but they were ideal. These sensitive headphones could run directly from a 'cat's whisker' of wire judiciously placed on a chunk of galena crystal. Radio could not have become such a success so early in its development without them.

As 'cat's whiskers' bristled around the world, these sensitive headphones crackled to the sound of the early radio stations and headphones became as much a part of affluent living as the gramophone. But the invention of sensitive moving-iron loudspeakers was the thin end of the wedge for headphones. Though little more than a telephone earpiece with a horn or radiating cone tagged on (with developments like the balanced armature to increase sensitivity and reduce distortion) they freed individuals or even whole families from the inconvenience of having to sit immobile.

Later, in the 1930s, with the invention of the then revolutionary but insensitive moving-coil loudspeaker and the development of more powerful amplifiers to drive them, headphones were largely relegated to the world of communications, where they stayed until the mid 1950s, when they underwent something of a revival.

Why the change? In a nutshell, stereo. Though stereo was developed in the thirties by Alan Blumlein, it was not to see commercial introduction, via the stereo microgroove record until the 1950s. Hi-fi, as we now know it had been developing even before the introduction of the LP, with the 78 rpm record as its source, but the introduction of the microgroove record and then stereo were definitely two large shots in the arm for this emergent science of more accurate sound reproduction. Stereo was also a major stimulus to headphones, and though it may seem an obvious thing now, it took an American Henry Koss to think of the idea of selling stereo headphones as 'stereophones' and creating a whole new market for quality stereo headphone listening. Needless to say, Koss has not looked back since and neither has Sony since the introduction of its first Walkman personal cassette player. Ironic, perhaps, because stereo was originally developed by Alan Blumlein for loudspeaker reproduction!

Pros and Cons of Headphone Listening

Good headphones, having no box resonances, can produce a less coloured sound than loudspeakers – even today. Headphones have the further acoustic advantage of not exciting room resonances and thus giving the listener a more accurate sense of the recorded acoustics.

On the other hand, headphones do not seem capable of producing the sheer impact available from loudspeakers, and unfortunately, stereo images are formed unrealistically inside the head due to the way stereo is recorded for loudspeaker reproduction.

None of these disadvantages matter much if your prime requirements are privacy or, in the case of Walkmans, portability. You can cheerfully blast away at your eardrums without inflicting your musical tastes on others. The closed back type of headphone in particular is very good at containing the sound away from others and insulating the listener from outside sounds.

Dummy heads

One way of overcoming the sound-in-the-head phenomenon which occurs when listening to normal speaker-oriented stereo through headphones, is to record the signal completely differently – using microphones embedded in artificial ears in a dummy human head. This 'dummy head' recording technique effectively ensures that each listener's ear receives a replica of the sound at the dummy head's ear positions. The result is much greater realism, with sounds perceivable behind, to the side, and in front of the listener – but not inside the head.

The exact location of sound images depends to some extent on how similar the listeners' ears are to the dummy's, but even using a plastic disc for a head-like baffle between two microphones gives very passable results.

Why then, has this not been taken up more widely? In a word, the answer is loudspeakers. Dummy head recordings do not sound at all convincing through loudspeakers due to the blending of sounds between the microphones which occurs around the dummy head during recording. The only way to listen to dummy-head recordings with loudspeakers is to sit immediately between them and listen as if they were headphones, which is not exactly practical.

Nevertheless, listening to good dummy-head recordings through fine headphones is an experience which is not easily forgotten. The sound can be unnervingly real.

Cross-blending

Other ideas used to try to reduce or eliminate the unwanted images in the head with headphone stereo have centred around cross-blending between the two stereo channels to emulate the influence the listener's head would normally have. In other words, to reproduce and imprint electronically the effect of the listener's head on the stereo signals.

The head has two main effects. Firstly, it acts as a baffle, curtailing the high frequencies heard by the ear furthest from the sound. Secondly, it introduces a time delay (or rotating phase shift with increasing frequency) to the sound heard by the furthest ear from the source.

Naturally such a simulation can only be approximate. Basic circuits simply cross-blend filtered signals between channels to emulate the absorptive factor. More complex circuits have been devised to introduce a time delay, in addition to high-frequency absorption, into the cross-blended signals. This is reported to be more effective, lending a more natural out-of-the-head sound to headphone listening with normal stereo recordings.

Biphonics

There were attempts in the 1970s to emulate dummy-head listening when using loudspeakers. The technique was known as biphonics and the idea was to cross-blend signals feeding two front-positioned loudspeakers to give an all-round effect, with sounds appearing to come from behind the listener as well as in front (Fig. 17.1). Surprisingly, the system worked, but only when the listener's head was held within a very limited space in front of the speakers. Predictably, the idea did not catch on.

Headphone Types

Moving iron

Early headphones relied on many turns of very fine wire wound on to a magnetic yoke held close to a stiff disc

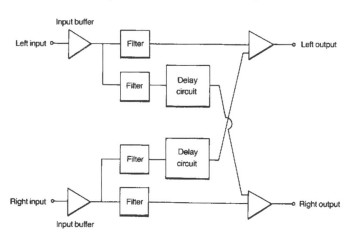

Figure 17.1 Block diagram of headphone cross-blend circuit with delay.

made of 'soft' magnetic alloy such as Stalloy (Fig. 17.2). A permanent magnet pulled the thin disc towards the yoke with a constant force and audio signals fed to the coil caused this force to vary in sympathy with the input. They were very sensitive, needing hardly any power to drive them, and very poor in sound quality due to the high mass and stiffness of the diaphragm, which caused major resonances and colorations – not to mention distortions due to magnetic non-linearities.

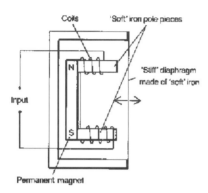

Figure 17.2   Moving-iron headphone. The current flowing in the coil either strengthens or weakens the force on the soft iron diaphragm. AC audio signals thus vibrate the diaphragm in sympathy.

Currently used in telephone receivers, they are not found today in any application requiring high-quality sound. That is reserved for other more advanced techniques.

Moving-coil

Moving-coil headphones (Fig. 17.3) work in exactly the same way as moving-coil loudspeakers. A coil of wire, suspended in a radial magnetic field in an annular magnetic gap, is connected to a small radiating cone.

When an alternating audio signal is applied to the coil, the coil vibrates axially in sympathy with the signal, recreating an analogue of the original wave shape. The cone converts this into corresponding fluctuations in air pressure, which are perceived as sound by the listener's nearby ears. Figure 17.4 shows a cut-away view of a Sennheiser unit.

The major difference, of course, between moving-coil headphones and loudspeakers is that the former are much smaller, with lighter and more responsive diaphragms. They can consequently sound much more open and detailed than loudspeakers using the moving-coil principle. They are usually also much more sensitive, which can mean that, in addition to reproducing detail in the signal which is inaudible through loudspeakers, they

can also reproduce any background noise more clearly, particularly power amplifier hiss which is not reduced when the volume is turned down. This is not peculiar to moving-coil headphones and can occur with any sensitive headphone.

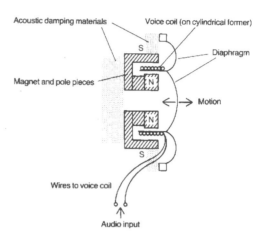

Figure 17.3   Typical moving-coil headphone transducer. The current through the voice coil creates a force on the diaphragm which vibrates it in sympathy with the audio input.

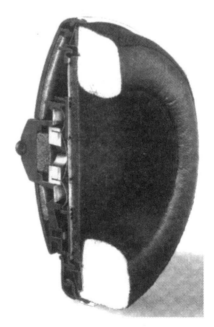

Figure 17.4   Cut-away view of a Sennheiser headphone earpiece showing the diaphragm, magnets and acoustic damping materials.

Moving-coil headphones are essentially medium to low impedance devices with impedances between eight and a few hundred ohms. They are usually operated via the normal amplifier headphone socket, which simply takes the loudspeaker signal and diverts it through a resistor network to reduce it to a level more suitable for the high sensitivity of the headphones. Alternatively, some high-quality amplifiers provide a separate amplifier to drive the headphones directly.

Many cassette decks also provide a headphone outlet and some headphones of lower-than average sensitivity sometimes do not work very well in this situation due to the limited output available.

Electrodynamic/orthodynamic

This type of headphone is essentially in the same family as the moving-coil type, except that the coil has, in effect, been unwound and fixed to a thin, light, plastics diaphragm.

The annular magnetic gap has been replaced by opposing bar magnets which cause the magnetic field to be squashed more or less parallel to the diaphragm. The 'coil' is in fact now a thin conductor zig-zagging or spiralling its way across the surface of the diaphragm, oriented at right angles to the magnetic field so that sending a constant direct current through the conductor results in a more or less equal unidirectional force which displaces the diaphragm from its rest position. An alternating music signal therefore causes the diaphragm to vibrate in sympathy with it, creating a sound-wave analogue of the music.

The great advantage of the electrodynamic, or flat diaphragm type of headphone is that the conductor moves the air almost directly, without requiring a cone to carry the vibrations from a coil at one point, to the air at another, with the very real risk of introducing colorations, break-up, distortion and uneven frequency response.

Unlike a cone, the film diaphragm does not have to be very stiff, though it is sometimes pleated to give it a little more rigidity because the force on it is not entirely uniform. So it can be very thin and light, which results in a lack of stored energy and a very great reduction of the other problems inherent in cones as outlined above.

Consequently, good headphones of this type tend to sound more like electrostatics, with an openness and naturalness which eludes even the best moving-coil types.

Electrodynamic headphones tend not to be quite so sensitive as their moving-coil counterparts and are often best used at the output of amplifiers, rather than with cassette decks. Impedance is usually medium to low and is almost entirely resistive.

Electrostatic

The electrostatic headphone, like the electrodynamic, uses a thin plastics diaphragm, but instead of a copper

track it requires only to be treated to make it very slightly conductive so that the surface can hold an electrostatic charge. It can consequently be very light.

The diaphragm (Fig. 17.5) is stretched under low mechanical tension between two perforated conductive plates to which the audio signals are fed via a step-up transformer.

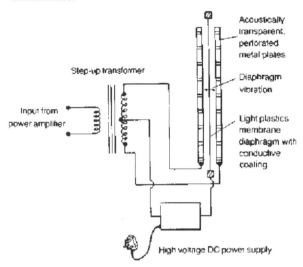

Figure 17.5 Electrostatic headphone. The transformer steps up the audio signal for feeding to the outer metal plates. The central diaphragm is given a high DC charge with a power supply. An audio signal causes the diaphragm to be attracted alternately to the outer plates.

The central diaphragm is kept charged to a very high voltage with respect to the outer plates using a special type of power supply, capable of delivering only a non lethal, low-current, high voltage from the house mains, or alternatively, by an energiser which uses some of the audio signal to charge the diaphragm to a similarly high but safe voltage.

The diaphragm experiences electrostatic attraction towards both outer plates. The spacing between the plates and diaphragm, the voltage between them and the tension on the diaphragm are all carefully chosen so that the film does not collapse on to either plate. Instead it stays in a stable position between the outer plates, attracted to each one equally during no-signal conditions. When an audio signal is fed to the transformer, it is stepped up at the secondary from a few volts to around a thousand volts. This unbalances the forces on the diaphragm in sympathy with the audio signal, causing it to be attracted alternately to each plate and of course reproducing an analogue of the original sound.

The push-pull action of the transformer and plates effectively produces a linear force on the diaphragm regardless of its position between the plates – unlike normal single-ended electrostatic attraction, which follows an inverse square law and would create large amounts of distortion in a transducer.

The electrostatic headphone is therefore the most linear of all the types available and with its super-light diaphragm, weighing less than several millimetres of adjacent air, it is not surprising that good headphones of this type can offer superb quality sound – the best available. However, this essentially simple technique is the most complex to execute. Not surprisingly, electrostatic headphones are the most costly to manufacture and buy. Figure 17.6 shows details of the Jecklin Float PS-2 type.

They are generally less sensitive than moving-coil types and are usually operated directly from the amplifier's loudspeaker terminals. Due to the capacitive nature of the electrostatic element and the complex inductive/capacitive nature of the transformer, they tend to have a more reactive impedance than other types, but this does not usually pose any problem for good amplifiers. Air ionisation between the diaphragm and the plates limits the maximum signal and polarising voltages. Likewise, the signal voltage on the plates must not exceed the polarising voltage. These factors impose limitations on the maximum sound pressure level which can be achieved.

Electrets

Basically the electret headphone is an electrostatic type but using a material which permanently retains electrostatic charge – the electrostatic equivalent of a permanent magnet. The electret has the advantages of the conventional electrostatic, but does not require an additional external power supply. It is similarly restricted in maximum sound pressure level, though both types produce perfectly adequate sound pressure levels with conventional power amplifiers.

High polymer

High polymer is basically a generic name to cover piezo-electric plastics films such as polyvinylidene fluoride film.

Piezo-electric materials have been known for many years. They change their dimensions when subjected to an electric field, or, conversely generate a voltage when subjected to mechanical strain.

The ceramic barium titanate and crystals like Rochelle salt and quartz are two materials which have been used for many years in devices like crystal phono cartridges, ultrasonic transducers and quartz oscillators but their stiffness is too high and mechanical loss too low for wide-frequency-range audio applications.

High polymer films, on the other hand are very thin,

some 8 to 300 μm, and have very low mechanical stiffness, which makes them ideal for transducer diaphragms. The basic film is made piezo-electric by stretching it to up to four times its original length, depositing aluminium on each side for electrodes, and polarising with a high dc electric field at 80–100°C for about an hour.

When a voltage is later applied across the film it vibrates in a transverse direction, becoming alternately longer and shorter. If the material is shaped into an arc, this lengthening and shortening is translated into a pulsating movement which will generate sound waves in sympathy with the electrical input signal.

It is a relatively simple matter to stretch the high polymer diaphragm across a piece of polyurethane foam pressing against a suspension board to create an arc-shaped diaphragm and make a very simple form of headphone.

High-polymer transducers were first developed by Pioneer. Advantages are claimed to be a very low moving mass, similar to electrostatics but without the complexity and, of course, no power supply. The high polymer headphone is also claimed to be much more sensitive than the electrostatic type and be unaffected by humidity, which can reduce the sensitivity of electrostatics. Harmonic distortion is also said to be very low at under one percent.

Basic Headphone Types

Apart from the many different operating principles described above, there are two basic categories into which headphones will fall, though some designs will include features of both and will therefore not function purely as one type or the other.

Velocity

The open or free-air headphone, known as the velocity type, sits just away from the ear flap, often resting on a pad of light, acoustically-transparent reticulated foam. This type of headphone cannot exclude outside sounds, which can intrude on the reproduced music, but the best models can produce a very light, open, airy sound. This type of headphone has been very successful in recent years, particularly since the growth in personal stereos – 'Walkman' clones – where their light weight and compact dimensions have made them ideal.

A theoretically perfectly-stiff diaphragm operating in free air, like a velocity headphone, has a frequency response which falls away at low frequencies at 6 dB per octave and so you would expect the system not to work

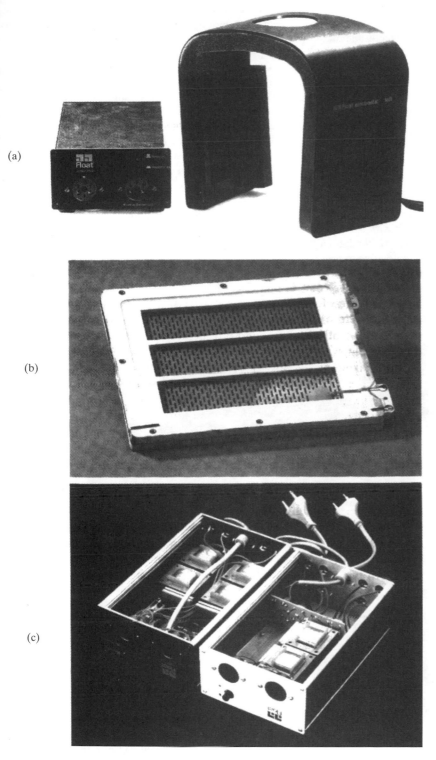

(a)

(b)

(c)

Figure 17.6 facing page (a) Jecklin Float PS-2 electrostatic headphones and energiser unit (b) Jecklin Float electrostatic headphone module (c) Jecklin Float energiser units.

particularly well. However, by juggling with the parameters of mechanical diaphragm stiffness, mass and acoustic damping it is possible to get a perfectly adequate low-frequency performance from velocity headphones, which is not dependent on the exact position of the headphones on the ears. The rear face of the diaphragm in velocity headphones is essentially open to the air, which helps impart a more open airy sound, but does not exclude outside noises.

Pressure

The other category is the closed or circumaural headphone, known as the pressure type, which totally encloses the ear flap and seals around it with a soft ear cup. The principal advantages of this type of headphone are that sealing around the ear makes it possible to pressure couple the diaphragm to the ear drum from about 700 Hz down to very low frequencies, with a linear response down to 20 Hz easily achievable – so long as the seal is effective (Fig. 17.7). The frequency response of the

Figure 17.7   Supra-aural headphones (RE 2390) from Ross Electronics.

pressure type, unlike the velocity type, is essentially flat down to a low-frequency which is dependent only on the degree of sealing. A poor seal due to inadequate headband pressure, or the wearing of spectacles for instance, can cause a marked deep bass loss.

The principle disadvantages are that closed headsets tend to be heavier, require greater headband pressure and can make the ears hot and uncomfortable. Pressure-type headphones can either be closed backed or open-backed. Closed-backed headphones offer the exclusion of outside sounds as a distinct advantage, in situations which require this, like recording studios for instance. On the other hand there is a body of opinion that judges the sound of closed-back headphones to be closed-in compared to open-backed types.

Intra-aural

Another category, which has been recently commercialised is the intra-aural type which actually fits into the ear to provide very lightweight personal listening (Fig. 17.8). Obviously the transducer size limits the bass performance, though this is helped by channelling the sound almost directly into the ear canal.

Measuring Headphones

Both types of headphone work into an entirely artificial environment in which the close proximity of the diaphragm and, in the case of the closed headphone, a trapped volume of air, modifies the frequency response as perceived by the ear. This situation is additionally confused by the fact that each person's ears are different in shape, and at high frequencies will introduce their own pattern of reflections, causing reinforcements and cancellations at different frequencies. Measuring headphones with artificial ears reveals some pretty horrifying curves, but the results on the heads of real listeners are equally alarming.

The fact of the matter is that, even if a headphone were made to measure 'flat' on a person's head as measured inside the cavity at the entrance to the ear, the headphone would not sound right. This is quite simply because, even when listening without headphones, the sound pressure response at this point, in the ear canal, at the eardrum, or wherever you care to measure it, is simply not flat. We are *not* dealing with an amplifier or a electromechanical transducer but the human ear. The ear and brain have of course figured out that the normally complex pattern of reflections and cancellations

Figure 17.8   Sanyo 'turbo' intra-aural earphones with Sanyo's GP600 D personal stereo.

with which it has to continually deal are perfectly natural and they sound that way.

Unfortunately each individual's ears make their own different imprint, but there are some rough trends which can be observed. For instance, there is generally a 2 kHz to 5 kHz boost, a dip around 8 kHz and all sorts of peaks and dips above 8 kHz. For a natural sound balance with headphones, they should produce measurements which follow this rough trend when tested on real ears. However, measurements carried out using headphones on real ears, or artificial ears (designed to mimic real ears for the purposes of testing) reveal differing results which can only be interpreted by the experienced observer. It is clear, though, that published response graphs of headphones should be taken with a very large pinch of salt.

The Future

Guessing on the future is always impossible – one can only be guided by current trends, which are unavoidably based on the past! It is very difficult to see how headphones will develop. New principles of operation are unlikely to spring out of nowhere. Just about all the likely candidates of moving air have been exploited. However,

as with the high polymer headphone, the invention of new, better, materials can often turn a previously impossible type of headphone into a reality. It is consequently unlikely that tomorrow's headphones will be anything other than developments of today's (unless it becomes possible to inject the audio signals, suitably coded by digital techniques, directly into the auditory nerve, or brain).

That may be highly unlikely, not to say impractical and unnecessary, but digital technology could play a part in equalising the response of headphones. For instance, it is possible to equalise the headphone signal so that the response at the ears when using headphones, more closely mimics the headphone-less characteristic. It would be possible to undertake this equalisation digitally to exactly compensate for each individual and give improved sound quality. At present this would be fairly expensive, but with processing power falling it should not be long before it is commercially viable. The technology already exists.

Likewise it is perfectly feasible to build digital filters which compensate for each individual's head-baffle effect, thus converting stereo into dummy-head binaural sound for headphone listening. This could all be done at the same time as the earphone-response correction outlined above. The only question marks are the very limited market for such equipment and the high cost of developing it. The two conflicting sides (development costs versus economies of manufacturing scale) may not add up to a very balanced equation now, but who knows about the future?

Finally, the diaphragm itself may be driven, not by analogue, but directly, using digital signals. This is not nearly so far-fetched as it sounds. With compact disc as the source, and given the transducer technology, it would be perfectly feasible to retain the signals in digital form from microphone to headphone, with any filtering or correction (as outlined above) carried out without the signal having to leave the digital domain.

Guesswork may be way off beam, but there is one thing which is certain. There is a great future ahead for headphones.

18 Public Address and Sound Reinforcement

Peter Mapp

Audio in the home is only one part of the audio industry. A very large part of the audio industry is concerned with public address problems of all types. In this Chapter, Peter Mapp deals with the special nature of public address systems, and reveals a technology that is very different from the domestic audio scene.

Introduction

Although the function of a commercial public address or sound reinforcement loudspeaker is essentially the same as its hi-fi counterpart, i.e. to reproduce sound, the commercial loudspeaker will typically be employed in a very different way. Furthermore, the unit may have to withstand and continue to operate under extremes of environmental conditions indoors or out of doors, e.g. temperature, humidity, dust or even corrosive atmospheres.

Whereas a hi-fi loudspeaker is designed to reproduce sound as accurately as possible or perhaps be engineered to sound good on typical programme material, its commercial counterpart will be designed for robustness, longevity, ease of maintenance and servicing and ease of installation – not to mention the price. The sound contracting business is highly competitive – the lowest price tender is the one which usually gets the job and not necessarily the best technical solution.

The loudspeaker drive unit may be in the form of a traditional cone unit mounted in a backbox or alternatively the assembly may be installed directly into a ceiling without an enclosure. Compression drivers mounted to exponential, multi-cellular or, more commonly these days, 'constant directivity' horn flares are frequently used in large sound system installations where advantage is taken of their increased sensitivity and in the case of the CD horns of their predictable coverage pattern.

Typically a PA cone drive unit would have a sensitivity of around 90–93 dB 1 W 1 m and a compression driver coupled to a 'long throw' CD horn flare around 112–114 dB 1 W 1 m as compared to a hi-fi loudspeaker of typically 87–90 dB 1 W 1 m. Low-frequency drive unit/cabinet combinations used in conjunction with the mid/high frequency horn units would typically comprise of either a single or double cone driver of 12 or 15 in producing around 100–103 dB 1 W 1 m.

Large theatre or cinema systems may be augmented with additional sub-woofers working below 100 Hz and super-high-frequency units – though with modern drive units and horn technology this latter requirement is rapidly reducing as even large horns can operate up to 16 kHz without significant beaming – a disadvantage of earlier designs.

Just as with hi-fi loudspeakers, the quality and performance of the drivers or complete systems varies significantly. When a well engineered, good quality PA/SR system is correctly installed, it should be capable of achieving a quality of reproduction similar to that of most hi-fi systems, though the level the system has to operate at in order to achieve the desired listening level may be many times higher than the conventional hi-fi installed in a relatively small room. Furthermore, the PA system may also be working to overcome an adverse acoustic environment with perhaps a reverberation time of one or two seconds or even four to eight seconds if the system is installed in a large church, ice rink, sports stadium or exhibition hall, etc.

Today it is possible to design good quality, intelligible sound systems capable of catering for most environments – the components, acoustic theory and engineering techniques are all available, so that there is no technical excuse for the well-known 'station PA system' quality and lack of intelligibility. However, too many such systems are still being installed. Whilst inadequate budgets are often to blame, there is still a remarkable ignorance in many quarters concerning modern sound system engineering and design.

Signal Distribution

The way in which the loudspeaker signal is fed and distributed to a commercial PA or SR system again differs from its hi-fi counterpart. The signal may be distributed at either high or low level or more accurately high or low impedance, depending on the application or length of associated cable run.

High-impedance distribution

The high-impedance distribution system allows loudspeaker signals to be distributed over relatively long distances without suffering undue losses due to the resistance of the transmission cable. It also allows multiple loudspeakers to be simply connected to the system by simple parallel connections. It is therefore very much akin to the conventional electrical mains distribution.

High-impedance distribution systems are often referred to as 100 V (UK and Europe) or 70 V (USA) systems, so termed because of the nominal voltage at which transmission occurs for a fully driven system. Figure 18.1 illustrates the basic principle of operation.

The PA amplifier is fitted with a line-matching transformer which converts the amplifier output to a higher voltage and impedance. The PA system loudspeakers are correspondingly fitted with a line matching transformer which matches the low impedance loudspeaker (e.g. 8 ohms) to the line. The transformer generally has a number of power tappings allowing the power input and hence output sound level of each loudspeaker to be adjusted.

For example, consider the case of a 200 W amplifier connected to a 100 V line distribution system.

To dissipate its rated output of 200 W at 100 V the amplifier would need to see a load impedance of 50 ohms, i.e. from $W = V^2/R$:

$$W = V^2/R = 10\,000/200 = 50 \text{ ohms}$$

Therefore the neglecting any line losses the amplifier may be loaded up to a total of 50 ohms at 200 W. The amplifier loading can most conveniently be thought of in watts with any combination of loudspeakers being nominally permissible up to the amplifier's rated output. It is perhaps useful to remember that 1 W would be dissipated into a load impedance of 10 Kohms i.e. a 1 W load is equivalent to 10 Kohms, 10 W equivalent to 1 Kohm and 100 W equivalent to 100 ohms (conversely for a 70 V distribution system or more accurately 70×7 V 1 W = 5 Kohms, 10 W = 500 ohms, 100 W = 50 ohms).

Although by transforming and transmitting the loudspeaker signal at a higher voltage in order to minimise line losses, with long cable runs, the resistive cable loss can still become significant. As a general rule, line losses should be kept below 10 per cent i.e. 1 dB by selecting an appropriate grade of cable. For medium size installations, e.g. a theatre paging system, a typical cable conductor size would be 1.5 mm², whereas for a large industrial complex or railway terminus or airport etc, main feeder cables with 4 mm² conductors may be employed with local distribution to the loudspeakers or groups of loudspeakers at 1.0 or 1.5 mm² cabling.

The size of conductors required will depend not only on the total length of cable (do not forget to include the return signal cable which will double the actual cable length/resistance) but also on the anticipated loudspeaker load which the amplifier has to drive i.e. the higher the power load, the lower the total resistance of the load and hence the potentially greater effect of the transmission cable. The total cable resistance should not exceed 10 per cent of the total load impedance if power losses are to be maintained within 10 per cent.

Do not forget to add the cable power loss factor to the total load when sizing the power amplifier. Table 18.1 provides a useful range of maximum cable lengths versus cable size and circuit loading.

When sizing the amplifier for a high impedance distribution system, sufficient allowance should also be made for future expansion requirements. Because it is so easy to connect up further loudspeakers it is also very easy to load an amplifier to beyond its rated capacity – leading to distortion and possible failure of the unit. An allowance of 10 to 20 per cent should be made.

Headroom

A further point to note when calculating power requirements and amplifier loadings is the allowance of a suffi-

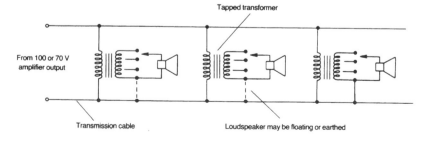

Figure 18.1   100 V transmission system.

Table 18.1

Conductor size (areas in mm²)	AWG	Resistance ohms 1000 (300m)	Low-impedance			High-impedance systems					
			4 Ω	3 Ω	16 Ω	200 W 100 V or 100 W 70 V (50 Ω)	100 W 100 V 50 W 70 V (100 Ω)	50 W 100 V 25 W 70 V (200 Ω)	10 W 100 V 5 W 70 V (1000 Ω)	2 W 100 V 1W 70 V (5000 Ω)	1 W 100 V ½ W 70 V (10000 Ω)
5.26	10	1.00	120	240	480	1500	3000	6000	30000	150000	300000
3.3	12	1.59	75	150	300	940	1800	3800	18000	94000	180000
2.08	14	2.50	48	96	190	600	1200	2400	12000	60000	120000
1.3	16	4.02	30	60	90	370	740	1500	7400	37000	74000
0.82	18	6.39	19	38	76	230	460	920	4600	23000	46000
	20	10.1	12	24	48	150	300	600	3000	15000	30000

2 wire copper cable lengths for 0.5 dB loss in SPL

cient headroom factor. Whilst this is widely appreciated within hi-fi circles, it is often inexplicably forgotten in commercial PA systems. A typical speech signal will have rms peaks of up to 12 dB over the mean. Furthermore, the natural voice may vary by 20–25 dB in normal speech. The system design must therefore take due allowance of this. The resultant effect in amplifier power requirement terms is significant. For example, employing a fairly moderate headroom factor of 6 dB results in an amplifier power requirement factor of four times i.e. a 100 W amplifier becomes 400 W or a 1 KW system becomes 4 KW. Needless to say, this design factor requirement significantly affects the cost of an installation. It is therefore not surprising that in competitive bidding this factor is reduced or often completely neglected altogether – leading to the all to familiar fuzzy distorted sound all too frequently associated with PA systems. Whereas within a hi-fi system the odd couple of dB in loudspeaker sensitivity is not particularly significant – in commercial PA systems it can become very important – for example, consider the effect of a 3 dB sensitivity factor on the total power requirement for a large PA system containing several hundred loudspeakers!

In the past 100 V or 70 V PA systems have often tended to be associated with poor quality sound. Whereas technically there is no real reason why this should be the case it is often true. The reasons essentially stem from economic considerations. One of the prime reasons for the inferior quality is due to the quality of the matching transformers used – particularly for the loudspeakers. Models are available which maintain a wide frequency response/stable impedance characteristic and do not saturate (i.e. distort) prematurely – though at a cost. Furthermore, the quality of the drive units employed in budget ceiling loudspeakers such as typically found in shops, supermarkets, and other similar commercial or industrial installations is generally pretty poor, though a good frequency response can be achieved – even within the budget end of the market.

Unfortunately, all too frequently the poorer response units are also installed where an improved response is required, again resulting in poor quality PA and doing little for the reputation of PA systems in general.

A further factor also all too frequently neglected is the correct loading of such ceiling type loudspeakers. In most commercial situations the unit is mounted directly into the ceiling and retained by spring clips. Whilst this makes for easy installation, it also further degrades the potential quality of the reproduction, as does fitting too small an enclosure to the rear of the loudspeaker, which can have a disastrous effect on the low frequency response of a unit. An enclosure is often required for fire regulation purposes as well as protection to the loudspeaker itself.

A number of manufacturers, however, are aware of the problem, and whereas large backboxes can be fitted, installing a 1 ft³ enclosure on site is very often totally impracticable and expensive. A unit which overcomes the problem and achieves good sound dispersion (another factor of sound system design discussed later) is the Bose 102, which correctly loads the 4-in drive unit with a properly ported small enclosure which has also been made easy to install.

Low-impedance distribution

Low-impedance distribution systems are very much more akin to the wiring of their hi-fi counterparts.

Low-impedance systems are used where high-quality sound is required and where only a few loudspeakers are to be connected to an amplifier. (Installing complex series-parallel arrangements to match load impedances is often not practical in commercial sound systems.)

Low-impedance sound systems are typically to be found as the main systems often operating in the form of a central cluster or side of stage and on stage units in theatres, conference venues or cinemas, or in discos and in rock band type installations. However, large arenas or stadia served by either one or two 'centralised' loudspeaker systems are often run at low impedance with multiple auxillary loudspeakers, e.g. covering walkways or bars or other rooms etc being catered for via a standard high impedance 100 V line systems.

The same rule for cable losses applies to low impedance systems as to 100 V or high impedance systems, i.e. voltage losses should be kept below 10 per cent – frequently this is held within 0.5 dB. As the load resistance is low to begin with e.g. 4 or 8 ohms, cable resistance must also be correspondingly low – requiring substantial conductors to be employed (see Table 18.1). For this reason in many installations employing a centralised loudspeaker system, the power amplifiers are mounted adjacent to the loudspeakers themselves.

Loudspeakers for Public Address and Sound Reinforcement

Before looking specifically at the different types of loudspeaker employed in commercial sound systems, it is worthwhile perhaps defining what we mean here by public address and sound reinforcement systems.

A sound reinforcement system, as its name implies, is a system used to aid or reinforce a natural sound, e.g. speech in a conference room or church etc. The system should therefore be designed to sound as natural as possible. In fact you should hardly be aware that the system is operating at all – until it is switched off!

In a public address system, there is no natural sound to reinforce, e.g. paging system at an airport, foyer or railway station etc. Here the prime goal of the system is to provide a highly intelligible sound. In the adverse acoustic environments of airports, stations or shopping malls etc often a compromise has to be reached between naturalness and intelligibility – as frequently under such difficult conditions the most natural sound is not always the most intelligible. (See later for further discussion on speech intelligibility.)

Loudspeakers employed in commercial sound systems can roughly be split into five types or categories:

1. Full range cone driver units
2. Cone driver. Bass reproducers. Ported or horn loaded.
3. Column or line source LS (based on use of multiple cone drivers)
4. Re-entrant horn/compression driver loudspeaker units
5. Compression drivers fitted with multi-cellular exponential or constant directivity or other horn flares.

Cone Driver Units/Cabinet Loudspeakers

We have already briefly mentioned the ubiquitous cone driver in the preceding section. Drive units for full range working are generally either 4-, 6- or 8-in. cones – the latter frequently being fitted with a supplementary co-axial or parasitic cone Hf tweeter unit. 10–12-in. units are similarly sometimes employed. 12–15-in. cones are generally the most popular units used for bass or bass/mid frequency cabinets whilst 18-in. units are employed for bass and sub-woofer applications.

A wide variation both in terms of the quality and performance of the units exists.

Column loudspeakers

The column loudspeaker, also sometimes referred to as a line source, is much more widely used in Europe than in the USA. The majority of column loudspeakers generally tend to exhibit a pretty poor performance, both in terms of their frequency response, off-axis response and irregularity of coverage/dispersion angles (particularly in the vertical plane). There are notable exceptions, however. The concept behind the column loudspeaker is based on the combining properties of an array of multiple but similar sources which interact to produce a beaming effect in the vertical plane, whilst maintaining a wide dispersion in the horizontal plane. The vertical beam is created by multiple constructive and destructive wave interference effects produced by the phase differences between drivers; therefore unless the column is well designed to take this underlying factor correctly into account, a very uneven dispersion narrowing sharply at high frequencies and exhibiting severe lobing can result. Sadly, the vast majority of units commercially available do not attempt to deal with this problem. Furthermore, the quality of drive units generally employed in such units is extremely poor. The popularity of the column loudspeaker would appear to be based on its neat appearance, low cost, presumed directional control and ease of manufacture. The sound received from the majority of commercial column loudspeakers therefore very much depends on where a listener is standing or sitting relative to the main axis of the unit.

The original popularity of the column loudspeaker probably stems back to its use in the early 1950s in St Paul's Cathedral and Westminster Abbey. The designer of these units fully recognised the physics which underlie the array configuration and employed both power tapering and multiple or sub column array techniques.

The original loudspeakers used in St Paul's Cathedral were 11 ft long and fitted with a supplementary high frequency column unit to cater for frequencies above 1 kHz. The units were highly successful in achieving intelligible speech in the highly reverberant cathedral [13 s RT60 within the dome area] with throws of over 100 ft. Supplementary 9 ft column loudspeakers were used to cover the remainder of the Cathedral fed from an early signal delay unit in order to synchronise sound arrivals and maintain and enhance speech intelligibility. The installation was one of the first sound systems recorded as being capable of achieving satisfactory intelligibility in such a reverberant environment.

Figure 18.2 shows frequency response and dispersion data for a number of current column loudspeaker models.

The column loudspeaker should not be dismissed out of hand, as in many situations it can provide both excellent intelligibility and coverage provided that the system

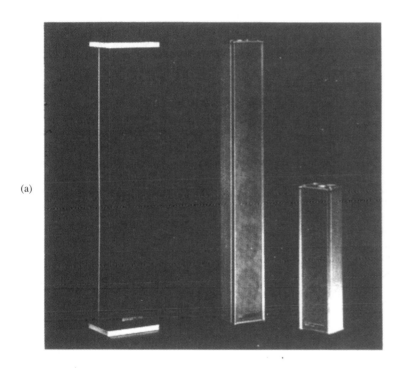

(a)

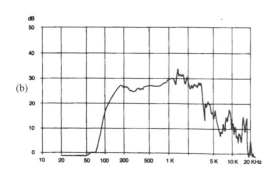

(b)

(c)

Frequency (Hz)	Vertical coverage angle in degrees				Typical horizontal coverage
	col. A	col. B	col. C	col. D	
250	105	76	90	–	> 180
500	54	40	22	52	180
1K	33	20	45	30	150
2K	20	10	21	30	100
4K	54	20	72	33	80
8K	24	18	70	30	60
Column length	0.95 m	1.2 m	1.2 m	1.2 m	WIDTH 200 mm

This table shows, the vertical coverage angle is hardly consistent, resulting both in large variations in the resulting frequency response within the audience area and within the excitation of the reverberant field.

Figure 18.2 (a) Typical commercial column loudspeakers (b) column LS data (c) comparison of coverage angles.

is correctly designed and an appropriate model of unit employed.

Re-entrant horn loudspeakers

The re-entrant horn loudspeaker is used solely for lower quality paging or public address systems. Its main advantage is its high sensitivity e.g. approx 110 dB 1 W 1 m and its compact size. Re-entrant horns, however, generally suffer from a limited frequency range and often exhibit strong resonances within their response. The majority of units are limited to the range between approximately 300 Hz and 4 or 5 kHz giving rise to their characteristic metallic/nasal sound. There are exceptions however. The horn's limited low frequency response can, however, be an advantage under some difficult acoustic or noisy conditions. The re-entrant horn can, however, be readily manufactured to be resistant to adverse environmental or potential explosive atmospheres. It is therefore ideal for use in emergency paging systems or paging systems in industrial or marine environments etc.

However, all too frequently the unit is used where a small column or some other form of cone driver unit with its superior frequency response would be more appropriate.

Figures 18.3 and 18.4 show the basis of operation of the re-entrant horn and some measurements made by the author on a number of typical re-entrant horn units. (The responses have been deliberately smoothed to illustrate the frequency response range rather than any irregularities or resonances within the response.)

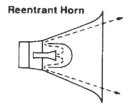

Reentrant Horn

Figure 18.3   Re-entrant horn loudspeaker: principles of operation.

Constant directivity horn loudspeakers

Although large horns date back to the 1920s, today most large horn flares take the form of the so called 'constant directivity horn' first introduced by Electro-Voice in 1971 (See Fig. 18.5).

The Constant Directivity horn was a major break-

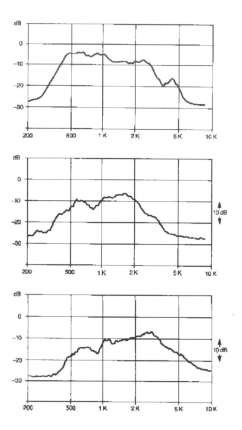

Figure 18.4   Re-entrant horn frequency response curves.

through in sound system design, as for the first time it allowed the system designer to uniformly cover areas over a wide frequency range – typically 600 Hz to 16 kHz with today's current range of devices. Apart from exhibiting a near constant coverage angle – and eliminating the high-frequency beaming associated with other horn flares, the constant directivity approach also allows the sound system designer to control and direct

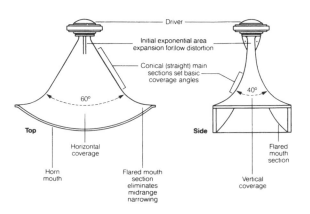

Figure 18.5   Constant directivity horns.

the sound so that it does not strike undesirable reflective surfaces, whilst maintaining appropriate audience coverage, so reducing the formation of echoes and helping to maintain a good ratio of direct to reflected sound – important for speech intelligibility and clarity.

The typical performance characteristics of a modern CD horn are shown in Fig. 18.6, where, for example, the device is shown to exhibit a coverage angle of $60° \times 40° \pm 10°$ over the range 500 Hz to 20 kHz. CD horns are generally manufactured to provide coverage angles of $40° \times 20°$, $60° \times 40°$ or $90° \times 40°$, which may be considered as long, medium and short throw. A $120° \times 40°$ format may also be available.

The CD horn format is comparatively large and bulky as compared to the column loudspeaker for example. This is purely as a result of the physics of the horn and the size of the wavelength of sound at mid to low frequencies. Typically CD horns are used down to around 800 to 500 Hz depending on the application, directivity control required and desired power handling capability. Compression drivers may fit directly onto the throat of the horn, or via an adapter section and typically have a throat diameter of one or two inches.

The majority of CD horn/compression drivers can exhibit a very smooth frequency response, but typically this is deliberately rolled off at around 3 to 4 kHz. The

response of the horn may, however, be readily restored by appropriate pre-equalisation of the signal being fed to it. (Some manufacturers provide a cross-over/equaliser unit specifically designed to match their range of horns and bass units – where possible an electronic cross-over is preferred together with bi-amplification of the horn and bass unit.)

Loudspeaker Systems and Coverage

Having looked at the basic loudspeaker units and methods of signal distribution, this section deals with how these components are actually used in a sound system and how appropriate coverage is obtained.

System types and loudspeaker distribution

Sound systems may be divided into central or point source, distributed or semi-distributed – although in practice a complete installation may incorporate features of all types.

An example of a distributed system would be a low ceiling conference room or airport terminal etc., relying

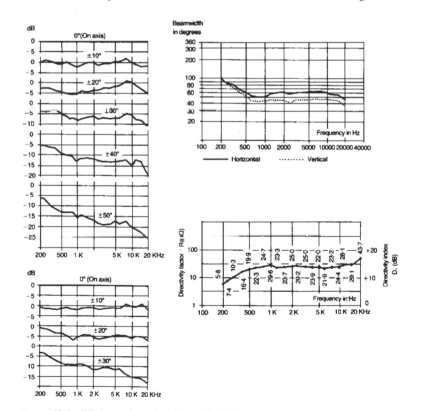

Figure 18.6 CD horn characteristics (HP 4060).

on coverage from the ceiling. An example of a semi-distributed system might be a large church using a relatively few number of column loudspeakers or a central cluster with repeater or satellite clusters. A central cluster or point source system, as its name implies, is a system whereby coverage of the area is achieved from one central position. This approach is frequently used in large reverberant spaces where localising the source of sound at one position and correctly aligning/directing the cluster can produce superior clarity or intelligibility over a distributed or even semi-distributed system. The central cluster cannot always provide the coverage required particularly in relatively low ceiling wide or long rooms – here a distributed or semi-distributed system may be required.

In low-ceiling rooms or under-theatre balconies etc., where a uniform and even distribution of sound is required, e.g. within ± 2 dB (why should it vary throughout the room?) a high density of loudspeakers will be required. Many systems fail purely due to there being an insufficient density of loudspeakers. The following provides some useful guidelines for correct coverage assessment.

An 8-in loudspeaker typically has a coverage angle of just 60° contrary to many manufacturers' data sheets. The coverage angle at 4 kHz should be taken, as using the angle at 1 or 2 kHz will deny system high frequency response and reduced intelligibility.

When calculating the coverage, the ear to ceiling (loudspeaker) height must be taken and not just the floor to ceiling height. How often will the listener be lying on the floor?

Depending on the uniformity of coverage required either edge to edge or edge to centre overlap should be used. In large industrial or commercial premises where the system is to be used primarily for paging (i.e. PA) a slightly greater spacing can be employed.

A quick and easy way to obtain the correct spacing is to space the loudspeakers at 0.6 h for optimum coverage or –1.2 h for a variation of approx 4–5 dB (h = ear-to-ceiling distance) with a half space at the boundary of the coverage area. Do not forget that the coverage angle of a loudspeaker is defined by its –6 dB points, i.e. the sound level will be 6 dB down relative to the on axis response – a marked change – particularly when the dispersion characteristics or off axis response is considered over the whole frequency range of interest.

When considering the coverage of column or horn loudspeaker systems the angle at which they subtend the vertical must be taken into account, requiring a brief incursion into 3D geometry – although for many purposes a 'flat plane' approach will give a good approximation.

When calculating the coverage of such systems, due account of the inverse law must be allowed for as described below.

Inverse square law

In general, the level of sound from a source falls off as the inverse square of the distance away from it. Put into decibel notation this means that the sound level will decrease by 6 dB every time the distance is doubled. For example, the sound level at 2 m is 6 dB less than at 1 m and conversely the level at 4 m is 6 dB less than at 2 m or 12 dB less than at 1 m. But the level at, say, 100 m is only 6 dB less than at 50 – but 40 dB less than at 1m.

The fall-off in level may simply be found from the following formulae.

$$20\left(\log\frac{D_1}{D_2}\right)$$

or when comparing levels with the standard reference of 1 m

– 20 log r will give the decrease in dB (where r = distance from LS)

(Note this is only strictly true with small sound sources, or where the wavelength is small as compared to 1 m. However, in practice the error is sufficiently small to be ignored for most loudspeaker types with the exception of loudspeaker stacks and arrays.)

From the manufacturers data of the sound pressure level (SPL) produced at 1 m for a 1 W input (or any other reference) it is a straightforward matter to calculate either the SPL at any given distance or the power required to produce a required SPL – bearing in mind that power follows a 10 log and not 20 log relationship.

The graphs in Fig. 18.7 and 18.8 give an easy method of establishing sound losses with distance and power requirements.

For example, let us assume that a given loudspeaker has a sensitivity of 90 dB 1 W 1 m. What would the sound level be at 20 m with 1 W input.

From the graph or from 20 log r one can immediately see that an attenuation of 26 dB will occur, i.e. the sound level will reduce to 90 – 26 = 64 dB.

If it is required that the level at this point should be 74 dB then the power would need to be increased from 1 to 10 W, or from 1 to 40 W in order to achieve 80 dB. (Note that at 1 m this would correspond to sound pressure levels of 100 and 106 dB.)

Whilst it is generally the case that the sound level will fall off by 6 dB for each doubling of distance – this is only true for sound travelling or propagating in a non reflective, e.g. open-air environment. Indoors or in an enclosed space the reflected or reverberant field should also be taken into account. (For initial estimate the reverberant component is generally ignored, but it can have a significant effect on the overall sound level and plays an important role in speech intelligibility considerations.)

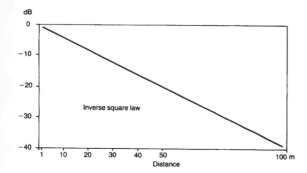

Figure 18.7 Inverse square law graph.

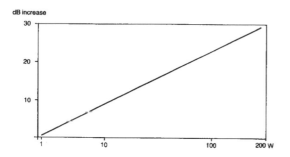

Figure 18.8 Power versus dB SPL increase graph.

Reverberant soundfields and loudspeaker Q factor

The following formulae relate the direct, reverberant and total sound field sound pressure levels.

SPL Direct $\quad L_p = LW + 10 \log (Q/4\pi r^2)$

SPL Reverb $\quad L_p = LW + 10 \log (4/R)$

SPL Total $\quad L_p = LW + 10 \log (Q/4\pi r^2 + 4/R)$

where $\quad L_p$ = Sound pressure level (SPL)
$\qquad LW$ = Sound power level (PWL)
$\qquad Q$ = Loudspeaker directivity factor
$\qquad r$ = Distance from source
$\qquad R$ = Room constant

The room constant R has to be calculated from a knowledge of the room surface areas and treatments

$$R = S\bar{a}/(1-\bar{a})$$

where $\quad S$ = total surface area
$\qquad \bar{a}$ = mean sound absorption coefficient
R can also be calculated from a knowledge of the room's

reverberation time and volume and is effectively a measure of the sound absorption present within the room

$$R = 0.161 \, V/T$$

where V = room volume in m³, T = RT60

Other methods also exist to calculate the reverberant level – particularly if the efficiency of the loudspeaker is accurately known e.g.

Reverberant SPL = 10 log (Total power input to LS × efficiency) + 126 dB

Two other useful formulae based on the above expressions for calculating direct and reverberant sound fields are as follows:

SPL (direct) $\quad$ = PWL – 20 log D + 10 log Q – 11
SPL (reverberant) = PWL – 10 log V + 10 log RT60 + 14

where PWL is the sound power level of the source.

The difference between the wanted (direct) sound and undesirable (reverberant) sound is given by:

$$\text{SPL}(D) - \text{SPL}(R) = 10 \log V - 10 \log \text{RT60} - 20 \log D + 10 \log Q - 25$$

It can be immediately seen that in order to attain a good direct to reverberant sound ratio in a given space with a fixed volume (V) and reverberation time (RT60) together with a fixed listening distance (D), it is important to keep the value of Q as high as possible.

Where more than one sound source is involved – as is normally the case, a 10 log N/L factor has to be included in the reverberant term, to take account of the additional sound power transmitted into the space. N is the number of like sources, whilst L is the number of sources which contribute to the direct sound field at any given listener position. In most cases it is generally assumed that $L = 1$.

Hence the direct to reverberant ratio =

$$10 \log V - 10 \log \text{RT60} - 20 \log D - 10 \log N + 10 \log Q - 25$$

Another important acoustical parameter to be aware of is DC, the critical distance. This is the distance away from a sound source where the direct sound field i.e. the component which follows the inverse square law is equal in magnitude to the reverberant component.

$$DC = 0.141 \sqrt{QR}$$

or

$$DC = 0.057 \sqrt{\frac{QV}{\text{RT60}}}$$

We can immediately see from this equation that the greater the Q or directivity of the source or the more absorbtive the room, the greater the critical distance – this is important as Q has been found to be an important factor in speech intelligibility in enclosed spaces or rooms.

To put the value of Q, the directivity factor or index of a sound source, into context, an omni-directional sound source has a Q of 1, a human talker a Q of 2.5, a typical column loudspeaker 8–10 and a $40° \times 20°$ CD horn a Q of around 50. A simplified definition of the Q of a loudspeaker would be 'the ratio of the sound pressure level measured on the main axis at a given distance away from the loudspeaker to the SPL at the same distance averaged over all directions from the loudspeaker.' The Q therefore depends not only on the coverage angle of the loudspeaker but upon its entire radiation pattern.

However, a simple estimate of Q may be obtained from the following formula:

$$Q = \frac{180}{\text{arc sin}\left[\sin\frac{a}{2} \times \sin\frac{b}{2}\right]}$$

where a and b are the vertical and horizontal dispersion angles of the device in question.

Speech Intelligibility

Effect of direct and reverberant sound components

It is well known that in reverberant environments speech intelligibility becomes difficult, e.g. a swimming pool or cathedral etc. In practice it has been found that little problem generally exists in situations where the reverberation time is 1.5 s or less. However, in environments with reverberation times greater than 1.5 s great care needs to be taken in order to ensure adequate speech intelligibility is achieved.

Speech intelligibility may be expressed in a number of ways. However, the most useful from a sound system design and assessment point of view is the percentage loss of consonants method (per cent Alcons) devised by Peutz in the early 1970s. The method is based on empirical data and provides a simple but effective method of predicting the probable intelligibility of a sound system. Over the years, the basic formula has been subtly modified to take into account second order effects and to improve its accuracy. Further more recent developments in acoustic instrumentation now mean that intelligibility can be directly measured.

Although the vowel sounds are responsible for the majority of the power or energy of the human voice, it is the consonants which are responsible for speech intelligibility. The frequencies between 200–4000 Hz contribute

91 per cent to the intelligibility of speech – which occurs over a total range of 60 Hz to 10 kHz. (The 2 kHz ⅓ octave alone is responsible for 11 per cent of the total intelligibility.) A measure of the loss of consonant speech information is therefore likely to be a very sensitive predictor of overall speech intelligibility – which is exactly what Peutz found.

The basic formula he derived is as follows:

$$\%\text{Alcons} = \frac{200\,D^2\,(\text{RT60})^2\,(n+1)}{VQM}$$

where D is the distance from the loudspeaker to the furthest listener

RT60 is the reverberation time of the space in seconds

V is the volume of the room or space in cubic metres

Q is the directivity ratio of the sound source

M is an acoustic or DC modifier – which takes into account the non uniform distribution of the sound absorption within the space – e.g. the additional absorptive effect of an audience in an otherwise hard surfaced space. M is usually chosen as 1 except in these special circumstances.

N = Number of like sources (loudspeakers or loudspeaker groups) not contributing to the direct sound field.

The basic Alcons equation is very straightforward to use and gives a good estimate of likely intelligibility whereby a loss of 10 per cent or less is regarded as good, 10–15 per cent as satisfactory except for complex information whilst 15 per cent is generally regarded as the practical working limit of acceptability.

It should be noted that the above % Alcons formula is valid only when the distance to the listener is less than three times the critical distance, i.e. when $D < 3\,DC$.

When $D > 3\,DC$, the formula simplifies to

$$\% \text{Alcons} = 9\,\text{RT60}$$

Figure 18.9 shows another approach to the prediction of Alcons – based on direct-to-reverberant ratio relationships.

The figure clearly shows that it is quite possible to achieve acceptable intelligibility even when a negative direct-to-reverberant ratio occurs.

A number of other acoustic factors also affect the intelligibility of speech and must be taken into consideration together with the % Alcons criterion. The factors include:

(a) Signal to noise ratio (speech level to background noise)

(b) Frequency response/range of the system

(c) Echoes or late arriving sound components/reflections > 50 ms

(d) Strong very early reflections < 3 ms

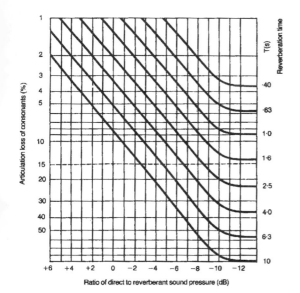

Figure 18.9   Articulation loss of consonants versus reverberation time and direct-to-reverberant sound ratio.

Speech signal to noise ratio

It is important to ensure that the sound system achieves an adequate speech signal to ambient or background noise level so that speech information is not masked by the background noise (just as excessive reverberation masks the impulsive nature of speech – whereby the reverberant hangover of one syllable masks the next).

It should be noted that the effects of reverberation and noise masking are directly additive on a log basis – the problem of achieving intelligible speech in a noisy, reverberant environment becomes at least twice as difficult.

There is considerable conjecture as to what the optimum signal to noise ratio should be. Many pragmatists believe a ratio of +6 or 10 dBA to be adequate. Whilst other research indicates that no further improvement in intelligibility will be gained after +15 dB, Peutz states that a 25 dB S/N ratio is required if speech intelligibility (as assessed by the % Alcons method) is not to be degraded. This requirement has later been qualified as a 25 dB S/N ratio in the 2 kHz ⅓ octave band.

In many instances it is simply not possible to achieve a signal to noise ratio approaching 15–25 dB, either because sufficient gain before feedback is not possible or because to apply such a large S/N factor in noisy areas

would require the speech signal to be uncomfortably loud.

In the author's experience most general announcement systems should be limited to an operational maximum level of 95 dBA. However, there are exceptions, e.g. sports stadia etc. where the crowd noise as a goal etc. is scored can easily reach 95 dBA. In these circumstances a higher PA signal level may be required if the system also forms part of the emergency evacuation system of the building or stadium (which it most likely will be).

In such circumstances an emergency override or automatic noise level sensing and control (or both) should be employed. An automatic noise level sensing and control system automatically tracks the ambient noise level and alters the gain of the sound system to maintain the desired signal to noise ratio. This is effected by placing special noise sensing microphones throughout the complex which feed into a VCA type controller. During an announcement the signal level is 'frozen' at the level occurring immediately before the commencement of the transmission.

Although a number of manufacturers make these units, the majority do not allow optimum performance to be attained, by either (a) having a fixed or too low a range of gain adjustment or (b) not incorporating appropriate signal averaging to the incoming sensing microphone control signal such that short transient sounds can trigger the system inappropriately.

The use of an automatic noise sensing and control system is not just restricted to noisy environments but such systems can also be used extremely effectively in generally low noise environments which are subject to variations in background noise level, e.g. lounges or departure gates at airports, where the ambient level can change significantly with occupancy. Many industrial, leisure or commercial paging/announcement systems can similarly benefit, enabling the broadcast signal to be sufficiently intelligible without becoming annoyingly loud.

System frequency response

As we saw previously, the range of the human voice extends from approximately 60 Hz to 10 kHz, but 91 per cent of the intelligibility information is contained within the frequency band 200 Hz to 4 kHz. This is, in fact, a remarkably restricted range – and although the resultant sound should be highly intelligible (assuming no other signal degradation occurs) the quality will be judged as being extremely poor. A modern good quality PA system should be capable of responding over the range 100 Hz to 6 kHz and preferably to 10 kHz, though in some circumstances, e.g. reverberant environments, it is often desirable to roll off the lower frequency range – from perhaps 150 or even 200 Hz. For general music reproduction the system should be capable of responding from around 80 Hz to 10 kHz and up to 15 kHz for high-quality theatre-type installations.

Although a frequency range of, say, 150 Hz to 6 kHz does not seem a particularly onerous requirement, it is surprising how many installations fail to meet even this most basic criterion. There are a number of reasons for this. The most obvious reason already touched upon is that the loudspeakers themselves are not able to adequately respond over the required frequency range. However, even in installations where the loudspeaker's response has been checked and found to be reasonably flat from 100 Hz right up to 10 kHz a large 'in-room' frequency imbalance can result. This is caused by more acoustic power being produced at the lower to mid frequencies than at the higher frequencies > 2 kHz, for example. (Remember that at 250 Hz a typical ceiling mounted loudspeaker will radiate uniformly over a 180° angle or hemisphere, but at 4 kHz for example the radiation will be reduced to a 60° cone and although the 'on axis' frequency response may be flat, the off axis radiation will not be and it is the summation of both the on axis and all the off axis components which account for the total power radiation into the space. The imbalance will be further increased if the floor is carpeted or the seating directly below the loudspeakers is sound absorptive, e.g. upholstered, as the 60° beamed Hf sound will be directly absorbed, whilst the lower frequencies radiating over a much wider total included angle will be multiply reflected off the wall and other room surfaces – an imbalance in the combined direct and reverberant sound fields will therefore result.

In many instances the frequency imbalance can be readily overcome by appropriate equalisation of the sound system.

Echoes and strong reflections

Although excessive reverberation can mask speech syllables by causing one to run into another, the effect of strong single (or multiple) reflections can be even more problematical – and often the effect of such reflections is incorrectly put down to reverberation – further confusing the issue.

Strong early reflections e.g. those which occur within approximately 3 ms of the direct sound, can cause serious aberrations in the frequency response of the system as a result of complex interactions occurring between the direct and reflected wavefronts producing severe comb filtering. Frequently deep notches within the response of up to an octave in bandwidth can occur with a resulting loss of information and hence intelligibility. The same effect also occurs between loudspeakers in array systems attempting to cover the same area.

These large frequency response aberrations can not be equalised out by conventional means, but require proper signal alignment or control at source with correct acoustic treatment of the offending surfaces. (See Chapter 3 on studio acoustics for further information.)

Strong reflections or late arriving sounds which occur approximately 40–50 ms after the direct sound can also serve to reduce speech intelligibility – reflections arriving after 60–70 ms causing distinct echoes.

The sound system should obviously be engineered to avoid such problems, e.g. by using well-controlled and aligned directional loudspeakers to stop high levels of direct sound from striking potentially harmful hard reflective surfaces. Alternatively, where this is not possible such surfaces should be treated with an appropriate sound absorbing material.

In large spaces where either a semi-distributed system or central cluster with satellite infills or local perimeter coverage is employed, a signal delay unit can and should be employed to synchronise sound arrivals and so defeat any potential echo sources before they can give rise to a problem.

Signal (time) Delay Systems

Electronic signal delay units, often referred to as time delay units, used to be a laboratory curiosity or only found in the most elaborate recording studios. However, advances in digital audio signal processing and memory techniques now mean that a good quality audio signal delay unit can fit the budget of most medium or low cost sound systems. Delay units may be used both to synchronise sound arrivals in overcoming potential echo problems and to improve the naturalness or realism of a sound reinforcement system by helping to maintain the correct impression of sound localisation – a technique which relies heavily on the psycho-acoustic properties of the human hearing system.

As we noted in the previous section, sounds occurring after approximately 50 ms can significantly degrade the intelligibility of speech. In high-quality sound reinforcement system, in a well-controlled and designed acoustic space, or in outdoor environments, a value of 35 ms is probably the practical limit of the ear's ability to integrate or fuse together multiple sound arrivals, i.e. below 35 ms the ear will effectively ignore any discrete reflections – using the information to merely add to the apparent loudness of the initial sound. The actual integration time is highly dependent on the exact circumstances of the particular situation or installation. Often reflection sequences which occur within the 35 ms can extend this period. In other situations full integration may only occur up to approximately 25 ms. Many authorities therefore regard 25 ms as the practical limit for total integration of speech signals.

Musical signals with a high transient content may require a still shorter integration period e.g., 15–20 ms.

From the foregoing discussion, we can see that a signal delay line may prove to be useful in situations where secondary loudspeakers are employed whereby the primary and secondary sounds reaching a listener will be separated by more than 35–50 ms. (This is equivalent to a path length difference of approximately 38–55.)

Apart from improving speech intelligibility by synchronising sound arrivals, a delay line can also be used to help maintain correct sound localisation whilst improving the overall quality and clarity or intelligibility of a sound, by making use of what has become known as the Haas effect.

The effect again relies on the underlying resolution of our hearing system – or rather lack of it. Haas, amongst others, established that a secondary sound when delayed in the region of 10–25 ms could be significantly louder (i.e. greater in amplitude) than the first sound, yet a listener would hear the sound as though still originating at the first source. In practice, this means that a weak primary signal, e.g. from a distant loudspeaker, can be reinforced by a louder local signal without localisation being lost. This has an enormous impact on modern sound rein-

forcement system design for both speech and musical material. In practice, in the author's experience, the secondary source can be as much as 4–6 dB louder than the primary signal before localisation is disrupted – though this depends heavily on the reverberation time and local reflection sequences naturally occurring within the space and on the frequency response or spectral content of the two signals.

Infill

A common application and good example of the use of signal delay lines is the delaying of signals used to feed under balcony infill loudspeakers in a theatre installation where the primary loudspeaker system cannot effectively reach. (See Fig. 18.10.)

The in-fill loudspeakers are generally used to help fill in the missing high-frequency information which does not penetrate beneath the balcony. Delaying the signal to the under balcony loudspeakers ensures that the time difference between the primary sound from the central cluster stage system and overhead signals will be such as to ensure full integration occurs so that intelligibility is improved and not degraded. Furthermore, by ensuring that the sound from the stage loudspeakers arrives first – with the under balcony loudspeaker sound arriving after a delay of around 15–20 ms, correct localisation on the stage will be maintained. (Note that the optimum

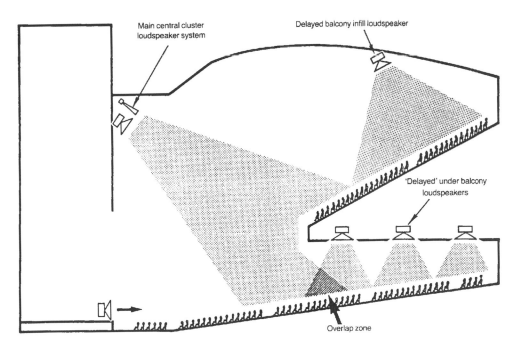

Figure 18.10   Under balcony loudspeaker delay system.

localisation effect is realised when the secondary signal is delayed by approximately 15–20 ms relative to the primary or first arrival.)

Modern digital delay lines enable accurate delays to be set with 1 ms second or better resolution up to typically 500 ms or, in some cases, resolutions of 20 µs can be achieved over a delay range of over a second. (This facility is very useful in correctly aligning horn clusters so that a single coherent wavefront is radiated.) The frequency response of current signal delay lines usually extends from around 20 Hz up to 15 or even 20 kHz, with a dynamic range capability of over 90 dB.

Equalisers and Sound System Equalisation

Even when a sound system is engineered using the best possible components which measure ruler flat (on axis in an anechoic chamber) it can still benefit from being correctly equalised (e.g. to overcome the frequency imbalance which frequently occurs when the system is installed within the room or space).

Of course equalisation only affects the direct sound component but it can improve the subjective (and objective) tonal balance of a system as the ear will respond to a combination of both direct and reflected reverberant sound.

As we have already seen, frequency equalisation will not overcome complex acoustic or phase interaction effects.

The object of equalisation is to smooth and contour the overall response shape of the sound system and overcome any sharp peaks in the response which may cause premature feedback when the microphone and loudspeakers are located in the same space. (See Fig. 18.11.)

System equalisation is usually performed using a narrow band e.g. ⅓ octave equaliser – generally a graphic equaliser in conjunction with a ⅓ octave real time spectrum analyser.

A 'flat response' measuring microphone is placed in a typical seating position and 'pink noise' (i.e. filtered random noise) is fed into the system. The coverage of the system is first checked together with any major response anomalies. Once these have been satisfactorily resolved, then equalisation – response smoothing and contouring – can begin. (See Fig. 18.12.)

Sound systems are not generally equalised 'flat' but a gradual high frequency roll off, e.g. 3 dB per octave above 2 or 4 kHz is introduced as this has been found subjectively to sound best. (A flat response can sound extremely harsh and hard under these circumstances.) During the equalisation process, the system is regularly 'talked' using a microphone or a well known piece of music is played through the system off high quality tape or compact disc.

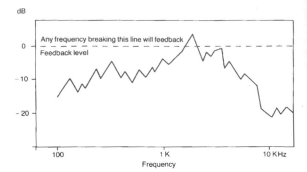

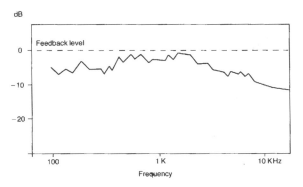

Figure 18.11 Equalisation curve – before and after equalisation.

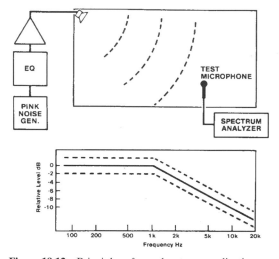

Figure 18.12 Principles of sound system equalisation.

Generally speaking, each different loudspeaker type or group or different acoustic area within a system will require its own equaliser and separate equalisation curve.

If the system is being equalised to optimise the acoustic gain of the system, then regenerative equalisation may be performed, whereby a system microphone located at a typical working position, e.g. the pulpit in a church, is connected into the system and allowed to 'feed back' in a controlled manner – enabling the primary feedback frequencies to be established and brought under control by selective attenuation either at the 'ring' frequency itself, or by adjustment to the bands on either side of the particular frequency of interest. A ⅓ octave or tuneable parametric notch filter are essential for optimum results, the ⅔ octave or full octave band equalisers having too broad a filter response curve. (See Fig. 18.13.)

Proper equalisation of a system is a very time-consuming job, but the difference in system performance can on occasion be quite dramatic. A thorough understanding of the system, the loudspeakers employed and their coverage patterns is an essential prerequisite to correct system equalisation. Figure 18.14 shows a basic schematic diagram for a simple theatre system showing how equalisers and delay lines are typically used in practice.

Further detailed information on system equalisation and the use and psycho-acoustic background of signal delay lines may be found in Davis and Davies and Mapp (see references).

Compressor-Limiters and Other Signal Processing Equipment

Another signal processor frequently found in permanent or temporary sound system installations along with signal delays and equalisers is the signal compressor or compressor-limiter.

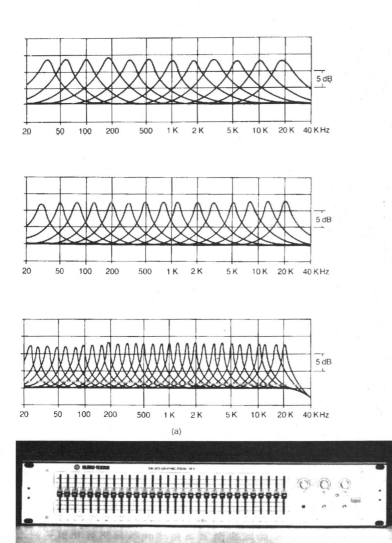

Figure 18.13 (a) Equaliser filter curve. (b) Typical third octave graphic equaliser.

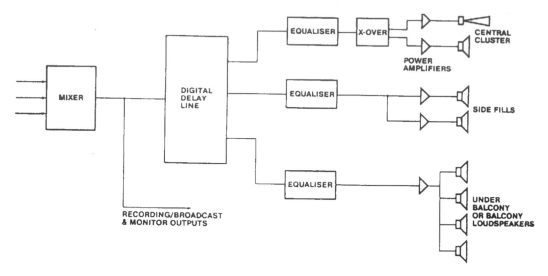

Figure 18.14   Typical sound system set up with equaliser and delay time.

The compressor or compressor-limiter does just as its name suggests – that is it compresses or reduces the dynamic range of a signal and/or limits further output once a predetermined threshold has been reached. For example, a compressor set to have a compression ratio of 2 would mean that for every 2 dB increase in level that a signal undergoes, the signal leaving the activated compressor would only be increasing by 1 dB in level.

Compressors are used in sound systems to control the dynamic range of the transmitted or broadcast signal, e.g. to keep the signal level more constant to improve intelligibility and also to reduce the peak power requirements on the system. As the compressed signal will have a higher average energy level, it also tends to sound slightly louder – another useful factor in helping to get the message through. (Note some compressors also incorporate an automatic level control function whereby the average output signal level, e.g. from a microphone, is held constant over a given range, regardless of the input signal level. This is extremely useful in paging systems, for example, where different announcers with different voice levels use the system.)

The limiting function is generally used for protection of the system, i.e. so that the following stages are not driven into overload. For example, a limiter may be employed immediately in front of a power amplifier to ensure that the amplifier cannot be driven into overload or clipping – a sure way of burning out the loudspeaker drivers connected to the other side of the amplifier – particularly in a low-impedance system – where there is no matching transformer to help reduce the high levels of Hf energy which are produced by a clipped signal.

Other signal processing elements which are commonly found in sound system installations may include:

- *De-essers* to reduce voice sibilance
- *Phase/frequency shifter* to help improve the feedback margin
- *Electronic-crossovers* often with inbuilt equalisation or delay facilities for loudspeaker alignment
- *Parametric equalisers* allowing both the frequency of operation and the bandwidth of the filter to be adjusted
- *Effects units* such as digital reverberators, time delay/echo effects and phasing and flanging etc.

Amplifiers and Mixers

So far as we have worked backwards up the sound system chain from room acoustics to loudspeaker to equaliser etc we have missed out the amplifier – though its requirements were partially discussed earlier when dealing with loudspeaker signal distribution systems.

Today the power amplifier can very much be taken for granted. Although there is some debate as to whether a Mosfet power amplifier sounds any different to a bi-polar, under the normal working conditions of most sound systems the difference is very much of secondary importance. The primary consideration is that the amplifier should be able to reliably deliver its rated output into the load presented to it. Amplifiers for commercial sound systems must be adequately protected against open and short circuit conditions – situations which occur remarkably frequently in the real world.

When installing power amplifiers, it is essential that adequate provision is made to ventilate them – many amplifiers run remarkably hot – with typically 15 per cent of their power rating being converted directly into heat. In a recent installation the author was involved in, over 15 kW of heat output from the amplifiers had to be catered for.

Another important point to watch out for – particularly in multiple amplifier installations, is the initial switch-on current demand – although an amplifier may only draw 3 A or so under normal working conditions, for example, it may well draw up to 10 A or so at switch-on, particularly if fitted with a toroidal transformer when the initial switch on surge on large amplifiers can be very high indeed. Modern installations should therefore incorporate a method of sequenced switch on to stagger and control the load. Wherever possible amplifiers with delay relays which disconnect the loudspeaker load until the amplifier has stabilised should be specified in order to protect the loudspeaker voice coils – particularly in low impedance systems.

Mixers used for sound reinforcement purposes vary considerably, ranging from a simple 4 into 1 combining unit without tone controls or level indication to the sophistication of a full 32-channel into 8×8 matrix mixer as used in today's larger theatres and auditoria.

The type of mixer required very much depends on the task in hand and the sophistication of the control required. Computerised control of channel inputs to output groups is now becoming an established technique in live sound mixing.

Essentially a mixer takes a microphone signal – perhaps only a millivolt or so and pre-amplifies it up to nominal line level i.e. 775 mV or 0 dBm. At this stage it can be equalised – usually by a combination of high and low frequency shelving filters and one or two bands of tuneable parametric equalisation, routed or sent off for further processing before reaching the main channel fader which controls how much of the signal will be combined into the mix. The signal may be mixed into a number of possible group options – which may themselves undergo further signal processing, e.g. reverberation or more specific equalisation or compression etc. Line inputs are treated in the same way but via either a different initial pre-amplification stage or possibly via an appropriate attenuator network.

Microphone input stages should be designed for balanced low impedance operation (e.g. to match 200 ohm microphone input). Line inputs are generally not balanced – the signal level being appreciably higher, but balancing may be required under certain circumstances. Note unbalanced high-impedance microphones should only be used for very short cable runs e.g. 20 ft or less as otherwise they can become subject to RF interference and high-frequency cable losses.

Balanced operation ensures that any interference picked up by the conductors on the transmission pair is cancelled out at the input stage by mutual phase cancellation between the signal carrying conductors which are of opposite polarity.

Great care must be taken with earthing of grounding of sound system installations to avoid undesirable earth loops and hum or RF pick-up.

Interconnection between equipment, e.g. equalisers and other signal processing equipment, may or may not be balanced. However in Rf or other electrical interference prone environments, e.g. where large amounts of thyristor dimming equipment is installed or where long runs between equipment occurs, such as mixer to amplifier racks etc., fully balanced operation is recommended.

Cinema Systems and Miscellaneous Applications

Cinema systems

Cinema systems are really no different from basic high quality reinforcement systems, except that the input to the system is either the optical film sound reader – the solar cell or a magnetic pick up if the soundtrack is on magnetic tape. (Optical sound pick-up is by far the most common.)

Film soundtracks have in the past gained a reputation for poor performance. This is in part due to the massive high frequency roll off the old 'Academy' curve produced e.g. 25 dB down at 9 kHz and inferior and old reproduction equipment.

Modern Dolby stereo standards require the sound level within the auditorium to achieve 85 dBC without undue distortion etc whilst maintaining a flat frequency response curve within ± 2 dB over the range 80 Hz to 2 kHz with a well-controlled roll-off and response extension to 12.5 kHz. This response cannot be achieved without proper system equalisation. Separate equalisers are used to compensate for the high frequency slit loss produced at the optical sensing head and ⅓ octave equalisation to compensate for loudspeaker response and room response anomalies in addition to the significant high frequency transmission loss which the film screen itself causes – even though perforated.

A number of sound track formats are employed but essentially a left, centre and right signal are recorded and produced through three behind the screen loudspeakers which generally comprise of an MF/HF horn and bass bin

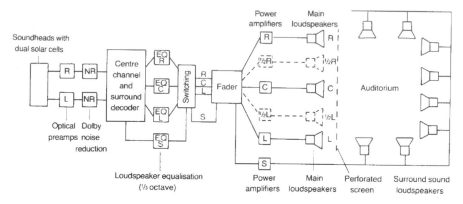

Figure 18.15   Cinema system basic diagram.

combination. Each channel is independently equalised and aligned. When applicable surround sound is fed to a separate amplifier circuit with compact full range surround sound loudspeakers located round the auditorium. The surround sound channel generally incorporates a delay line to ensure that the correct impression of localisation is maintained.

Sub-woofer systems are also sometimes included together with an appropriate decoder and amplifier channel.

If a Dolby system is employed, all the appropriate signal processing is included within the Dolby stereo decoder unit. Figure 18.15 shows a basic block diagram of a Dolby cinema system.

Sound systems

Sound systems are used for a variety of purposes other than straight-forward speech or music reproduction.

Other typical applications include sound masking systems used in large open plan offices to improve speech privacy by creating a more uniform background ambient noise, artificial reverberation enhancement is used in multi-purpose halls, theatres or studios, where the normal natural acoustics tuned with a short reverberation time for satisfactory orchestral music performance. A number of such systems now exist which alter not only the perceived reverberation time, but by careful use of artificial reflections, alter the whole feeling, size and envelopment created by the space.

References and Bibliography

'Articulation loss of Consonents as a Criterion for Speech Transmission in a room' *JAES* **19** (11) (1971).

Barnett, P.W. and Dobbs, V., *Sound System Design – Intelligibility* Public Address (Feb. 1985).

Barnett, P.W. and Mapp, P., *Sound System Design – Distance and Power* Public Address (October/December 1984).

Capel, V., *Public Address Systems*: Focal Press, Butterworth-Heinemann (1992).

Cremer, L. and Muller, H.A., *Principles and Applocations of Room Acoustics Vols 1 & 2* (translated by T.J. Shultz), Applied Science (1982).

Davis, D. and Davis, C., *Sound System Engineering*, Howard Sams.

Haas, H., 'The Influence of a Single Echo on the Audibility of Speech' *Acustica* **1** (49) (1951).

Hodges, R., 'Sound for the Cinema' *dB Magazine* (March 1980).

Klepper, D.L., and Steele, D.W., 'Constant Directional Characteristics from a Line Source Array' *JAES* **11** (3) (July 1963).

Lim, J.S., (Ed), *Speech Enhancement*, Prentice Hall.

Mapp, P., *Audio System Design and Engineering Part 1 Equalisers and Audio Equalisation. Part 2 Digital Audio Delay*, Klark Teknik (1985).

Parkin, P.H., and Taylor, J.H., 'Speech Reinforcement in St Pauls Cathedral' *Wireless World* (Feb. 1952).

Steenken, H.J.M. and Houtgast, T., 'RASTI: A tool for evaluating Auditoria' *Bruel and Kjaer Technical Review* (3) (1985).

Steenken, H.J.M. and Houtgast, T., 'Some Applications of Speech Transmission Index (STI)' *Acustica* **51** (1982) 229–234.

Taylor, P.H., 'The line source loudspeaker and its applications' *British Kinematography* (March 1964).

Uzzle, T., *Movie Picture Theatre Sound* Sound and Video Contractor (June 1985).

19   In-Car Audio

Dave Berriman

Another growth area in audio has been in-car entertainment. The car is not exactly an ideal listening space, and in the chapter, Dave Berriman shows how the complex problems of obtaining acceptable reproduction have been tackled.

Modern Car Audio

Back in the days of the humble car radio, that medium was the only source of in-car sound. Originally, of course, it was simply AM, with permeability tuning using variable ferrite-cored chokes to avoid the vibration problems in the car environment and provide an easy means of pre-setting stations.

Nowadays we have the luxury of FM radio reception with its much wider hi-fi quality bandwidth, and the bonus of stereo – previously a luxury but now expected as standard on all but the most basic car radios.

Radio reception has been pretty much revolutionised since those early permeability-tuned sets. Modern receivers feature the extensive use of integrated circuits to pack increasing sophistication and features into the diminutive car-radio slot.

Synthesized tuning is virtually standard today and because the local oscillator frequency is locked to that of a quartz crystal, tuned frequencies are extremely stable and effectively drift free. Displays are digital as a matter of course – an easy-to-apply bit of technology when added to frequency synthesis.

The detailed circuitry of modern car radios is very complex, but integrated circuits reduce it all down to a few basic interconnected circuit blocks. Modern circuitry has done away with rows of IF tuning 'cans' and FM decoders have gone 'phase-lock loop' which makes for ease of manufacture and even higher performance, reliability and stability. Circuit diagrams of car radios now look more like block diagrams and service engineers need not concern themselves with the minutae of electronic circuitry. Gone are many of the symbols of individual electronic components, to be replaced by circuit-processing 'blocks', several of which are to be found within each integrated circuit.

FM Car Reception

FM has the potential for a much higher sound quality level than the AM wavebands. That much we all know. The main bugbear with in-car FM reception is due to the much shorter wavelength of the carrier frequency used on this waveband. Reflections from nearby buildings and objects result in the carrier being simultaneously received by more than one path. This is known as multi-path reception.

Due to its short wavelength, the FM carrier is either reinforced or weakened by the received reflection depending on the position of the car's aerial. As the aerial moves through the nulls, severe audible distortion, or complete loss of signal can result, producing an annoying click-clicking sound. The increased multiplicity and strength of reflections in urban environments makes this a particular problem in towns and cities, and it is a problem wherever FM is received on the move.

One particularly novel way of overcoming this major problem has been introduced by Clarion in its 999MX diversity reception system. This utilises two aerials situated at different locations on the vehicle: one is a conventional whip aerial, the other a window strip aerial. A comparator monitors the state of the tuned signal at each and selects whichever has the strongest signal at that instant.

Since it is much less likely that both aerials will be at a reception null than with a single aerial, there will be very many fewer instances of distortion or signal loss due to multi-path reception.

Auto selection

An altogether different approach to the problem of changing reception conditions has been adopted by

Philips, who has a system which it has dubbed Micro Computer Control, or MCC (though it reported to be working on a diversity system with four aerials).

MCC can store the frequencies of transmissions carrying identical programme modulation and then monitors their signal strength. If one drops too low in level, the receiver will automatically switch to the strongest station, from a selection of ten user-selected frequencies.

It enables the receiver to select whichever is the strongest station during long journeys, avoiding the annoyance of retuning, and could reduce the number of audible clicks and buzzes while on the move, but to achieve similar results to the Clarion system it would require duplication of programmes on a number of different carrier frequencies covering the same location, which only occurs in overlapping 'fringe' reception areas in the UK.

The technique could be aided by the BBC's proposal to transmit station identifying codes. If this were to become commonplace in commercial and BBC radio, receivers could use this code to identify and select any desired transmission, without the receiver having to be pre-programmed with this information. The strongest transmission containing the correct programme could therefore be selected automatically using the transmission's own identifying signals, detected and decoded by the receiver.

Another way around the problem of multipath drop outs is the use of automatic noise reduction systems, going under various titles, such as, Mitsubishi's SRC (Fig. 19.1) and Nakamichi's ANC, for example. In essence, these monitor the signal level and progressively introduce a degree of stereo-to-mono blending at high frequencies, thus minimising the audability of noise while sacrificing stereo performance. Eventually, for very poor reception, the tuner switches itself to mono or even to full mute, which is hardly an ideal solution. Noise Killer circuits, which detect noise spikes and mute them from the audio output are another technique used by most manufacturers to minimise the effects of ignition noise and other interference. A typical example, Trio's ANRC III, is illustrated in Fig. 19.2.

Power Amplifiers

Not so many years ago, you were lucky to get five watts of mono sound. That was considered powerful. Today, car radios boast many more watts per channel in an effort to drown out the high levels of in-car road, engine and wind noise. Using the standard car battery's 14 V or so of fully-charged potential and using a four ohm loudspeaker to extract more power from the amplifier, there is, unfortunately, a technical limit to the power which can be delivered. Absolute maximum is about five or six watts for a normal output stage.

Bridging is a system whereby there are effectively two output stages per channel, each driven in anti-phase. This virtually doubles the output voltage available for

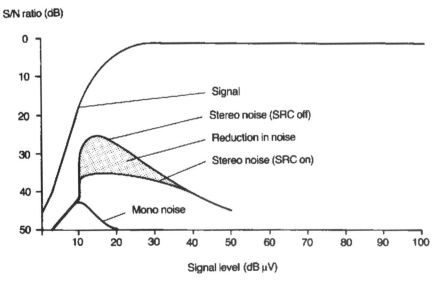

Figure 19.1   Mitsubishi's stereo reception control (SRC) is its answer to multi-path noise due to weak signals. As the field strength weakens, SRC automatically changes from stereo to mono mode and smoothly attenuates the higher audio frequencies.

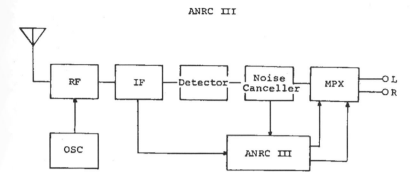

Figure 19.2 In Trio's ANRCIII a pulse noise blanking system circuit placed between the IF and multiplex (MPX) sections activates the high-blend mode. When the multi-path noise drops below a predetermined threshold the high blend is taken out, gradually returning to stereo and avoiding sudden changes to the sound character.

the loudspeakers and hence pushes the power up to just over 20 W. Most radio/cassette units offer 4–20 W per channel.

Separate Power Amps

Today's enthusiasts want enough power to deliver high sound levels over and above the ambient noise level. The availability of higher-quality car loudspeakers with lower sensitivity and power-hungry sub-woofers (not to mention the temptation to wind up the bass control, which seems to have become *de riguer* for some ICE reviewers these days) means that 20 W is not really enough for 'serious' in-car hi-fi enthusiasts. For them, more and more manufacturers are producing separate power amplifiers. As much as 50 or 100 W is quite commonplace these days and to achieve these power levels it is not possible to rely on the battery voltage – it is simply

not enough. A much higher voltage rail is necessary and is generated by using a DC-to-DC convertor (Fig. 19.3). This basically comprises a high-frequency oscillator driven by the car battery, coupled to a step-up transformer which increases the AC voltage and finally a rectification and smoothing stage to bring the AC back to DC again, but at a higher voltage suitable for the power amplifier. The principle is very similar to that of switched-mode power supplies.

Manufacturers say little or nothing about their DC-to-DC convertors, perhaps realising that their significance will not be grasped. Trio goes so far as to use two DC-to-DC convertors, one at ±25 V and one at ±45 V. It is a variation on techniques used by Hitachi, Acoustic Research, Proton and Carver for example. The Trio KAC-9020 of Fig. 19.4 runs using the 25-V rail power supply for most of the time, switching in a pair of extra transistors to take the output up to the 45 V maximum, just for those music peaks that need it. It minimises heat dissipation and is a very cost-effective route to high power outputs.

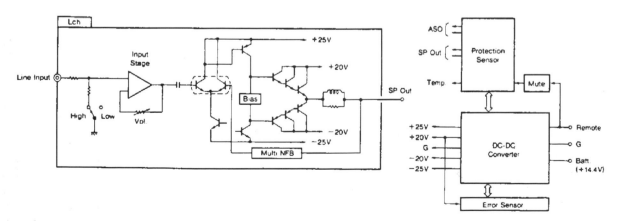

Figure 19.3 Nakamichi's PA-200 power amplifier uses output triples and provides the driver transistors with a higher rail voltage.

A 'Class A' power amplifier unit from Soundstream is illustrated in Fig. 19.5.

Multi-speaker Replay

Another trend is the move towards providing four power amplifiers, even in some radio cassette units, which feed four loudspeaker systems distributed around the car. These are even becoming popular as manufacturer-fitted systems in new cars, though they merely reproduce the same stereo twice, through two sets of loudspeakers.

They do produce more of a wrap-round sound than front or rear mounted units on their own, though this approach is hardly ideal.

Ambisonics

Ambisonics, in contrast, is a system which was designed from first principles to more accurately recreate the original sound field, using a number of sound channels ranging between two and four. The most basic system uses just two channels as in stereo but encodes the signals differently. After decoding, four signals can be fed to four loudspeakers to recreate the original sound field more accurately than stereo, or the many ill-conceived quadraphonic systems which failed so miserably. The effect is to reproduce sounds accurately in space horizontally around the listener. With more channels than two and more than four loudspeakers, height information can also be reproduced for even greater realism.

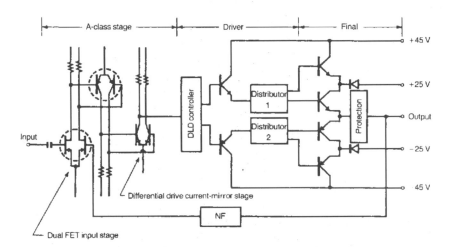

Figure 19.4   In the KAC-9020, Trio's 'Supercharger' system uses two voltage supplies. One powers a medium-to-low power amplifier, and the other provides a higher voltage to another set of output transistors for handling higher power musical peaks.

Figure 19.5   Pure 'Class A' power from Soundstream.

It is a system which has not caught on in the home due, in part, to the need for more than one set of loudspeakers and also the lack of suitably encoded software. In the car, four loudspeakers present no problem and so one company Troy Ambisonic has decided to launch an Ambisonics decoder (Fig. 19.6) specifically for in-car use. The system has a good pedigree, with much early research on encoding and decoding parameters done by the BBC being taken into account by Ambisonics designers. Both the BBC and the IBA are in favour of the system, though there have been relatively few transmissions of Ambisonic programme material. It is a bit of a catch-22 situation, as we are unlikely to have more Ambisonic software until more people can replay it and only a few people will buy the equipment without the software to play through it. Fortunately, ordinary stereo can be processed very simply by the Troy Ambisonics decoder to give an enhanced stereo spread which is claimed to be very pleasing. Some of the smaller record companies are issuing Ambisonic CDs and if the broadcasting autho2r-ities see a demand, Ambisonics may eventually be lifted out of its current obscurity.

Cassette Players

Philips invented the compact cassette, let manufacturers have free licences to manufacture tapes and players (provided they conformed to strict specifications) and so created a worldwide standard.

The medium was originally intended to be used for dictation machines and mid-to low-fi music replay, but very soon after its inception hi-fi compact cassette players became available. They were bulky and sound quality was not very inspiring, but it was a start. The main problems were, and still are, the narrow track width and slow tape speed. The former limited the amount of signal available off tape and therefore presented serious noise problems and the latter meant that the high-frequency performance from compact cassette tape was restricted.

Background noise was overcome by a variety of competing noise-reduction systems, of which Dolby B (and latterly Dolby C) emerged clear victors. Dolby B was the first complimentary noise reduction system which was taken up by tape manufacturers and is still, today, the most prevalent. It is easy to forget that Dolby B played a significant part in turning compact cassette into a hi-fi medium because it did its job so dramatically well.

Large strides were made by tape and tape head manufacturers to further improve fidelity. Tape manufacturers have developed very short ferric particles, chrome, so-called pseudo-chrome, and more recently pure iron particles, for tape coatings. Cassette deck manufacturers have made great advances with very hard tape heads which have ultra-fine gaps. Both have greatly improved

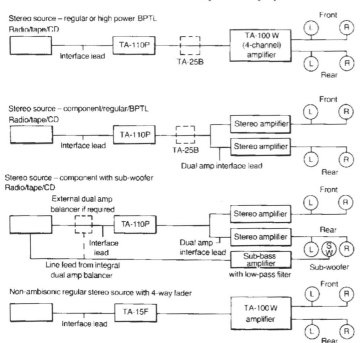

Figure 19.6 Troy's TA-110P Ambisonic decoder can be used with Troy's own four-way channel amplifier or with other equipment for Ambisonic replay in the car.

the high-frequency performance of cassette tape reproduction. Cassettes were of course an obvious choice for in-car use, with their ease of handling, reasonable cost and the capability for making recordings. Progress in in-car cassette units has mirrored those of domestic hi-fi. Initially car cassette players seemed well behind their domestic counterparts in performance, but the demand for quality tape reproduction on the move has lead to the emergence of cassette players and cassette/receiver units (Fig. 19.7) which deserve comparison with quality domestic hi-fi units.

Modern units

Now you can buy in-car cassette units (at a price) with superb specifications and a sound to match, built and marketed along the same lines as top hi-fi units. Conversely, there are a great many average performing cassette units.

Powerful electronics may be mass produceable down to a price, but stable tape transports require mechanical integrity and precision – and they don't come cheaply.

Many units now offer auto-reverse, which requires much better tape head alignment, because any angular error is effectively doubled when the tape reverses. For the recorded signal to be replayed faithfully at high frequencies, the tape and head should be accurately aligned. With compact cassettes, accurate alignment is absolutely crucial, as very small angular errors result in a dramatic losses in output at high frequencies.

In the Trio KRC 747D, for example, the tape head is actually automatically re-aligned when the tape reverses (Fig. 19.8), to compensate for any angular differences, while other manufacturers, like Philips for example, utilise the lower-cost alternative of guides built into the head. These are claimed to align the tape at right angles with the gap for optimum high-frequency response regardless of tape direction.

Figure 19.7   The Hitachi CSK 350 auto-reverse tuner/cassette unit.

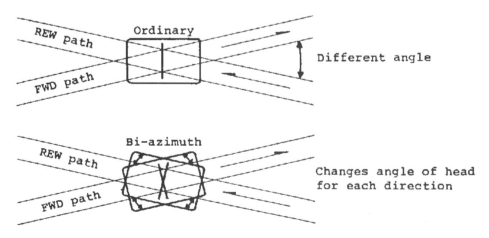

Figure 19.8   Trio's Bi-azimuth tape head adjusts itself when the tape reverses to avoid high-frequency losses due to tape/head misalignment.

Compact Disc

Compact disc is another Philips invention. It was spawned by the technology developed for the video LP – a video disc scanned by laser light which reproduces stunningly clear pictures and excellent sound, as anyone who has viewed one will testify.

Video LP uses frequency modulation to encode and decode the picture and sound signals in a fashion not too dissimilar to FM radio, hence the excellent quality. The signal is held by the disc as a spiral of microscopic bumps in a reflective layer, protected by a transparent plastics coating.

Philips engineers realised that the disc could be scaled down for audio use and made the intellectual and technological leap of going for digital signal encoding and decoding which was ideally suited to the on/off nature of the recording system (i.e. a bump or no bump representing a logical '1' or a logical '0').

While digital recording had been used for its high quality by broadcasting authorities, it had not been used for mass market domestic recording and replay. The technology was extremely young and needed considerable work, particularly on the error correction (which 'fills in' any information not picked up by the laser from elsewhere on the disc) and also on reducing the highly complex digital processing circuitry on to large scale integrated circuits.

The advantages of digital recording, providing there are enough binary digits, or 'bits' in each digital 'word' used to encode the audio signal, are imunity to background noise and low distortion. If enough redundant information is included in the digital signal, the decoder can use this to make up for any signals lost due to dust or scratches, for example.

Consequently, digital recording and replay can be very tolerant of lost information, provided the error correction works well. Scratches, for instance, which would be only too audible with an LP, can be inaudible with compact disc.

Error correction was not a field in which Philips had much experience, but the Japanese company Sony had just the expertise that Philips required, having designed and manufactured the industry standard PCM 1600 analogue-to-digital, digital-to-analogue (A/D, D/A) converter and latterly the PCMF-1 portable A/D, D/A converter. In fact, it leads the world in digital error correction, while Philips was significantly advanced in laser disc technology.

The two companies could have gone their own ways, but fortunately they pooled their resources to produce compact disc, a source of music which is now establishing itself firmly as a serious alternative to the vinyl LP in the home.

In-Car CD

The same qualities which make CD attractive in the home hold just as good in the car, with only vibration and the risk of scratching posing significant threats to disc replay see Fig. 19.9. Within certain limits, the servo system which keeps the laser light focused on the rapidly spinning disc can cope with vibration, with error correction making up for the rest. However, beyond that the system will either conceal the error (not audible) or mute (audible) while the errors persist. Skipping from one section of the information spiral to another is commonly caused by vibration.

There are now an increasing number of in-car CD

Figure 19.9 Hitachi's CD-D4 compact disc player uses anti-vibration techniques for improved sound quality.

Cross-section of KDC-9 shock absorber   KDC-9 vibration tolerance characteristics

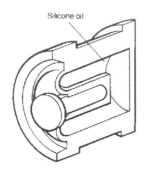

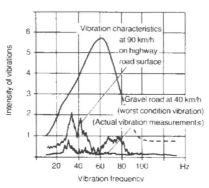

Figure 19.10   Mechanical filtering action of shock absorber used in the Trio KDC-9 car CD player.

players on the market. They tackle the vibration problem by means of carefully-designed suspension systems, as illustrated in Fig. 19.10 which isolate the player's mechanism from vibration and shock.

Though scratches, up to a point, are usually coped with well by the servo mechanism, error correction and concealment, it is still not a good idea to damage the surface of compact discs. Badly scratched discs can cause skipping and there is little doubt that some players sound noticeably under stress when their error correction circuits are working 'hard', though theory might indicate otherwise.

If you take pride in your home hi-fi system, then you will need to be careful not to scratch your CDs in the car – just the sort of environment where they might slide about and get damaged.

Some manufacturers have taken this seriously enough to design transparent 'caddies' which will hold an individual CD. The whole 'caddy' is inserted into the player and when not in use the caddy protects the disc from dust and scratches.

There is a safety aspect to all this too, because that way, the driver does not have to fiddle about with putting the disc back in the rather unergonomic Philips-type 'jewel case' while hurtling down the *autostrada, autobahn* or motorway.

Another approach to this problem has been introduced by Pioneer, which make a CD 'juke box' type autochanger which can be mounted in the boot and operated from the front end. Most users will not wish to go for caddies or juke boxes and, provided users are careful with storage, no harm will come to their CDs. Compact disc in the car is only just taking off and in-car players are expensive, but with the enthusiasm which CD has received in the home it should establish itself fairly

quickly, especially as prices of players fall. Figure 19.11 shows a block diagram of a car CD player.

Digital Audio Tape

Digital tape recording in the home is not new. Sony's PCMF-1 A/D. D/A convertor was made for just this purpose, with recordings going on to ordinary domestic video tape.

Digital audio tape is basically an updated domestic version of this idea but using the very latest miniaturised 8 mm vapourised-metal film video tapes and with the A/D and D/A convertors all built in to the tape machine.

Though the digital technology is essentially the same as for CD, the standards for digital encoding and decoding have been deliberately made different to those of CD to prevent direct copying of the digital pulse train from CD to DAT. This would have made perfect recordings possible, and that did not go down too well with those bodies trying to prevent copyright infringement through home taping.

People will still copy tapes anyway, of course, and put up with the minor irritation of marginally less than perfect recordings.

The major advantage of tape is, of course, that you can make your own recordings and this will be an attraction to those who currently break the law by recording cassettes off disc or tape for use in the car. Other advantages for in-car use are DAT's relative immunity to vibration and when the tape emerges from the machine it is protected by a flap and so it is less likely to be damaged than the polished surface of compact discs.

In-Car DAT

When it comes to which medium will succeed in the car, DAT will be hampered by lack of software. It will be some time before software manufacturers catch up with the already large catalogue available on CD and compact cassette. The success of compact cassette in the car has been due in no small part to its massive acceptance in the home and there is no reason to suppose that CD or DAT will be very different.

Home DAT will also be held back due to the high cost of machines, which are mechanically similar to video cassette players and require precise well-engineered mechanisms. It is debatable whether these will ever be as cheap to manufacture as CD mechanisms. The digital-to-analogue convertors and analogue sections of the machines will be very similar in concept to CD, if different in exact execution, and so will be manufacturable every bit as cheaply as those for CD.

The signal-processing technology used for DAT is very similar to that used for CD, though dedicated integrated circuits will be needed for DAT. At the time of writing, manufacturers have been experiencing problems with making the vaporised metal tapes in sufficient quantities. It is a new technology with specific problems

which need to be overcome and informed projections are that DAT tapes will be almost as expensive as CDs. However, there is every hope that these problems will be overcome and then prices may fall.

Sound quality of DAT will be virtually identical to CD, so contrary to some claims, there will be no advantage here. The slightly higher sampling frequency of DAT will make no audible difference to sound quality in practice.

Eventually we will probably see both media gaining acceptance with car users, depending on the system they use at home. Both will offer a considerable improvement in quality over compact cassette, though the latter can achieve a very high standard in the best car players. It may take some compact cassette fans a lot of persuading to change to either of the new systems.

Loudspeakers

Car loudspeakers essentially comprise the same ubiquitous moving-coil drive system found in so much other sound-reproducing equipment from transistor radios to top-flight hi-fi. Moving-coil loudspeakers are basically

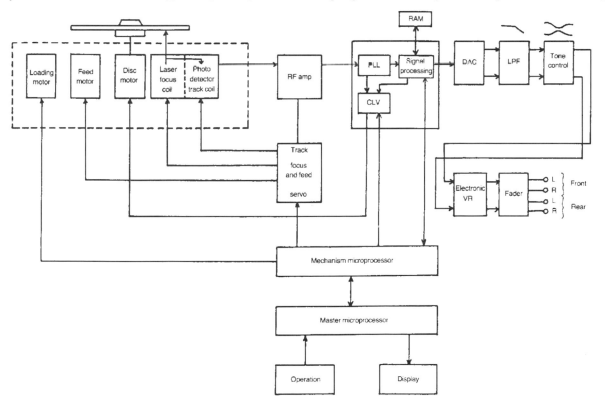

Figure 19.11 Block diagram of Trio KDC-9 car compact disc player.

simple and are consequently easy and relatively cheap to manufacture compared to other more esoteric sound reproducing methods, and so it is little wonder that transducers of this type have a virtual monopoly wherever you go.

In the moving-coil loudspeaker the magnetic field is made to flow radially across an annular gap between cylindrical pole pieces, with the coil suspended freely in the gap. Current flow in the coil produces a corresponding axial force which drives the coil in the gap. A direct current would apply a constant force, of course, and merely a constant displacement of the cone, but audio signals vibrate the coil in sympathy.

On its own, the coil would make very little noise of course, so it is connected to a diaphragm to couple it to the air. The diaphragm should be both light for good sensitivity and rigid for low distortion and extended frequency response. The most cost-effective (and some would say the best) way of achieving this is to use a cone for the diaphragm. This has the advantage that it can be shaped to produce a radiating area which shrinks with increasing frequency, giving a more extended frequency response than the theoretical 'perfect piston' action of a perfectly rigid diaphragm would allow.

Diaphragm materials

The traditional material used for this purpose is paper pulp, which can be deposited on a former to produce the desired contour. Paper, however, changes its properties with variations in humidity and of course will disintegrate if it becomes wet for long enough. It is consequently not the best of materials for door panels in cars

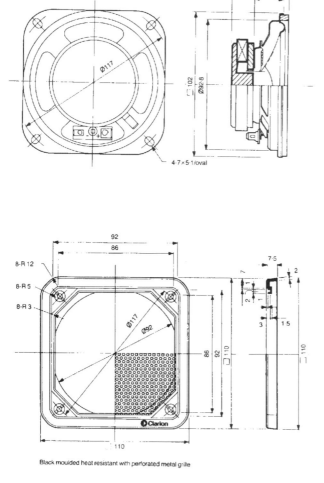

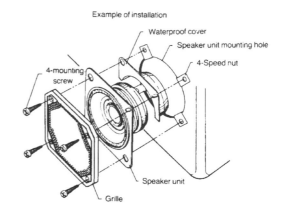

Figure 19.12   A typical car loudspeaker with mounting details.

where rain water or condensation could collect and for this application has been an increase in the use of plastics materials such as polypropylene. Figure 19.12 shows the construction and installation of a typical low-cost car loudspeaker.

Polypropylene has many advantages. It is resistant to moisture, easy to vacuum form, consistent, requires no additional damping coatings and gives cleaner, less coloured sound reproduction. Not surprisingly, it is now being used by more and more companies, Acoustic Research for example, in their quality car loudspeaker systems. Kevlar is a woven fibre material which, when bonded together, forms a very light, rigid material ideal which is claimed to be excellent for loudspeaker cones. One company which has used this material for their domestic loudspeakers and has now applied it to their in-car loudspeakers is Bowers and Wilkins, though it now uses polypropylene for its domestic products.

An alternative type of diaphragm construction has been borrowed from aeronautical engineering, where light, rigid structures are of course essential. This is the flat honeycomb sandwich, which has been used, for example, by Trio and Panasonic, to name just two in their top-of-the-range of car loudspeakers. Normally a flat, thin diaphragm would have very little rigidity, but by sandwiching an open honeycomb of thin sheet material set on edge between two thin outer skins, a very rigid yet light structure is formed. The main drawback of this type of diaphragm is that, while it is undoubtedly very rigid, its flexure cannot be controlled in the same way as a cone to modify the effective radiating area. Instead of progressive flexure, the highly rigid structure tends to have a pronounced resonance at the top of the unit's pass band, but this can be minimised by connecting the coil by means of a short cone to the honeycomb at the nodal points for the first resonance – the so-called 'nodal' drive (Fig. 19.13).

Low frequencies – in the bass

The low-frequency performance of moving-coil loudspeakers is governed by the complex relationship between the drive unit's mechanical, electrical and magnetic characteristics, together with the air compliance and losses in the enclosure. These have been so well researched and documented for the formulae not to need repetition here and this solid core of theoretical and practical work has made the low-frequency design of loudspeakers in enclosures highly predictable.

This is fine where there is a precisely defined enclosure for the drive unit to work into. Unfortunately, with car loudspeakers mounted in doors, boots or dashboards, things are somewhat less predictable and so the loudspeaker designer has inevitably to allow for a range of possible mounting positions and leave the rest to fate.

Consequently, to get really accurate low-frequency performance in cars, it is necessary to have proper enclosures for the woofer units, but often the luggage space lost from the boot is considered too high a price to pay.

KEF unit

One manufacturer who has met this problem head on is KEF with its GT200 loudspeaker system (Fig. 19.14). This uses a variation on techniques first applied in its domestic hi-fi loudspeakers in which a sealed box woofer is coupled to the air by means of a port which increases

Principal of nodal drive

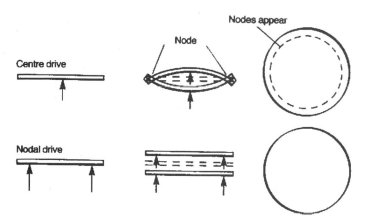

Figure 19.13 Driving a flat diaphragm at the centre point produces a strong resonance with a ring-shaped node. By connecting the voice coil via a short sub-cone at the nodal position, this node is suppressed, giving a flatter frequency response and less harmonic distortion.

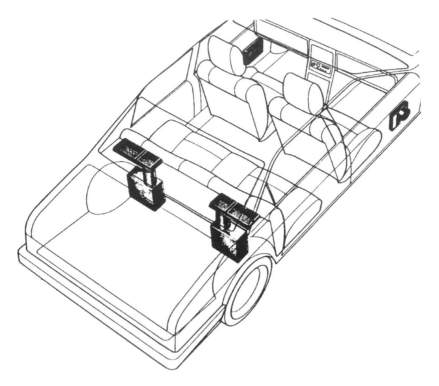

Figure 19.14   Rear parcel shelf mounting for KEF's GT200 sub-woofer with matching mid-range tweeter satellite units.

the maximum sound pressure level by increasing the loudspeaker's 'volume velocity'.

Ideally, the port should be exactly dimensioned for its purpose, as it is in KEF's domestic models. This has not been possible in the KEF car derivative, because boots come in all shapes and sizes and the flexible plastics ducts used to couple the bass sound from the drive unit to the car interior must be trimmed to fit, but KEF claims distinct advantages in bass extension and power handling as a result of using this technique. Additional drive units are, of course, required to handle midrange and treble and these are mounted alongside the port openings on the rear parcel shelf.

Multi-unit systems

Not so long ago, in-car-audio loudspeakers comprised simply a couple of full-range drive units in the dashboard or the rear parcel shelf, but the demand for improved quality has lead to loudspeakers containing more than one drive unit to handle the audio frequency band. One relatively expensive example has already been mentioned, but there are many less complex systems.

The simplest form of cross-over (Fig. 19.15) is just a

single capacitor feeding the tweeter. Its high capacitive reactance at low frequencies filters out signals which could otherwise damage the delicate tweeter by overheating its moving coil or fatiguing the leadout wires. The cross-over also reduces distortion and coloration, which would otherwise be seriously detrimental to sound quality. More complex cross-overs involve the use of inductors and capacitors which provide a higher rate of 'cut off' outside the desired operating band of the unit, where greater attenuation is required.

Bass units can be rolled off and tweeters 'rolled in' in this way, just as in conventional domestic hi-fi loudspeakers. Woofers roll off naturally at high frequencies anyway, and in many in-car applications this has been used by the designer as a cost-effective alternative to additional electrical components. Figure 19.16 illustrates some currently available units.

Active cross-overs

All of these techniques so far discussed are of the socalled 'passive' cross-over type. They offer the advantages of cost effectiveness and compactness, but there is a section of the market which seems to be on a relentless

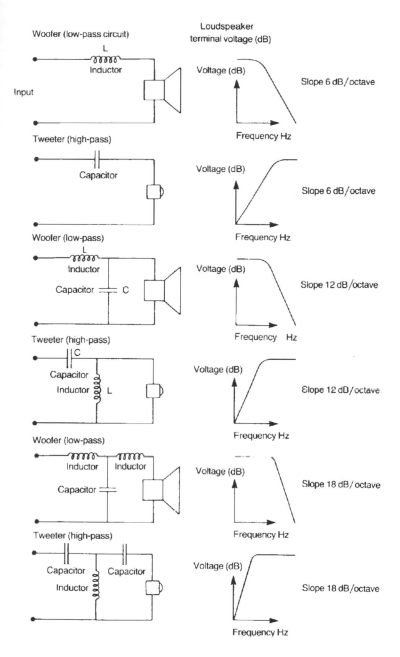

Figure 19.15 Passive cross-over systems can be simple or complex. Car audio cross-overs tend to be of the simple 6 dB/octave type in basic installations, but more sophisticated systems incorporate more complex networks for sharper cross-over cut-offs.

quest for better and better quality. Seekers after in-car perfection are willing to pay considerably more for the active type of crossover as offered by companies such as Nakamichi, Soundstream, and Trio (Fig. 19.17), for example.

In this type of cross-over, the signal filtering is carried out electronically by 'active' circuits in which their responses can be tailored more accurately. Unlike 'pas-

sive' crossovers which intercede between power amplifier and drive units, 'active' cross-overs connect between the signal source and the power amplifier and work at a lower voltage level (Fig. 19.18).

Each drive unit thus requires its own power amplifier in addition to its own filter circuit. The active filters for an in-car hi-fi system need to be purpose-designed for the drive units and so it is usual to buy the cross-over (which

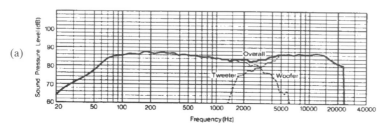

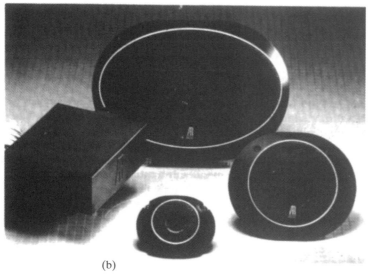

(b)

Figure 19.16   (a) Woofer and tweeter are blended by the cross-over to create a homogeneous overall frequency response. The curves shown here are for Nakamichi's SP200 loudspeaker (b) CCS-100 three-way component speaker system from Acoustic Research showing the cross-over module.

comprises all the electronic filters required in one box), speaker units, and even the power amplifiers from the same manufacturer.

The resulting systems are by no means inexpensive, but there are an increasing number of buyers who are satisfied by nothing but the best from their car audio systems, spending as much, or more than their home hi-fi on their mobile sound system.

Acoustics

It is well known in hi-fi and professional sound circles that loudspeaker positioning has a marked bearing on sound quality. A loudspeaker placed near a wall or a corner will sound different to one mounted away. Likewise, floor-mounting a loudspeaker which should be placed on an open stand can ruin the performance of a normally excellent design. So, what about the car? How does positioning and acoustics come into in-car sound?

The greatest variable at home is the distance between the loudspeaker and the room boundaries, because each

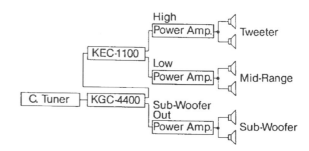

Figure 19.17   A three-way system from Trio. Its KGC 4400 low-bass cross-over feeds one power amplifier for the sub-woofer, while the KEC 1100 splits low and high frequencies for the rest of the frequency range. Similar multi-way systems are available from most quality manufacturers.

nearby boundary causes dips and peaks in the frequency response at frequencies where the reflection either reinforces or partially cancels the direct sound.

In a domestic speaker the greatest influences will be the rear wall, floor and side walls, depending on how close they are. The rear wall is often a major problem because the speaker stands in front of it and thus the

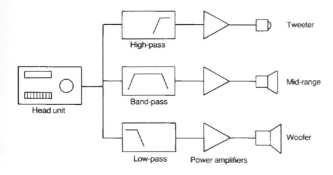

Figure 19.18 Active loudspeaker system.

reflection cannot be eliminated. At least in the car, the speaker units are flush mounted, which overcomes this effect.

However, there are more sources of reflections in the car, notably the windows. Ironically, beaming sound up from the parcel shelf and reflecting it off the windows is often promoted by manufacturers as an advantage, but does introduce a strong direct/reflected mix with unavoidable dips and peaks in the frequency response. Figure 19.19 shows the elements of the reflection problem.

Reflections inside the car environment are very complex, and the only way to adequately deal with them would be to design each speaker system for a particular car, or at least adjust the speaker to suit the car. One advantage of multi-drive-unit active systems is that they allow a degree of adjustment to individual speaker sound outputs and flexibility in speaker positioning, though this is not really perceived by the purchaser.

It would be tempting to assume that graphic equalisers were the solution for the problem of in-car acoustics, because these can boost and cut specific broad ranges of frequencies, but while they can correct for general trends in frequency response they cannot eliminate dips due to strong reflections, because pumping more power into the speaker merely reinforces the reflection as well!

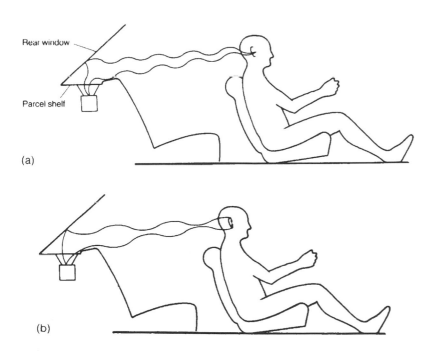

Figure 19.19 (a) When path difference equals one wavelength (λ) sound waves reinforce at the listener (b) when path difference equals one half wavelength (λ2) sound waves will cancel at the listener.

Reflection Compensation

One system which could correct for reflections is the Adaptive Digital Signal Processor, or ADSP, which was originally developed for domestic use by Teledyne Acoustic Research.

This compensates for reflections in the time domain by means of digital processing. To set up the system, a microphone picks up the waveform produced at the main listening position when a special signal is replayed through the loudspeakers. The microphone is then removed – it is only there for setting up.

The digital circuitry then computes a digital filter which exactly compensates for reflections. Each channel of the audio signal can then be converted into digital form and can be corrected in real time by this digital filter, which in effect creates in the audio waveform the inverse of errors due to reflections. A digital-to-analogue converter then turns it back to analogue.

The sonic effect of the ADSP is to remove colorations due to reflections, but so far the technology has proved too expensive for domestic use, let alone for inclusion in the car. However, with costs of complex digital circuits falling and with the potential cost savings which could be brought about by mass production, the system may possibly find its way into cars of the future.

Installation

With the advent of separate power amplifiers, head units, equalisers, multi-way loudspeakers and cross-overs, the installation of in-car hi-fi has become a highly specialised business.

What used to be a simple bolt-in and connect-up job can now involve much more than the typical do-it-yourselfer, or even some garages, can adequately cope with. Recent advances in technology and manufacture which have provided very real gains in sound quality have also encouraged users to be fussier than in days past. When some of the top systems available can cost a few thousand pounds, the user has a right to expect his in-car system to give of its best and improper installation can ruin an otherwise superb system.

In-car audio has perhaps not surprisingly taken a lead from specialist hi-fi dealers who like to install hi-fi in the home, rather than let buyers walk out of the shop with cardboard boxes. Correct installation can make all the difference in the home and in the car and since you have got to have your car system installed, you might as well have it done properly. Some systems come with the car, of course, and so there is no problem (except when they are pulled out to make way for something better), and

some are still just a bolt-in job, but more complex systems take a bit more effort.

Specialised in-car installers will fit just about anything, from sub-woofers with cross-overs to air conditioning and mobile telephones. Some take in-car hi-fi very seriously, others less so and it is up to the buyer to beware and go on recommendation if possible. It is certainly becoming apparent that, while the technically competent individual can cope, there are enough pitfalls for most mortals to make it well worthwhile employing a specialist for the job.

Apart from the question of avoiding radio interference (which we will refer to later), with systems involving separate parts (i.e. with a front end, amplifier and even a cross-over), there is always the risk of the audio sections picking up interference from various sources within the car which has been transmitted through the wiring or encouraged by poor earthing practices. A competent installer will know how to avoid or cure such problems.

Then there is the purely audio hi-fi side of things. Specialist hi-fi retailers and enthusiasts have known for years that connecting cables can have a significant affect on the sound of a system and the same applies in the car. It is no use buying a 100 W-per-channel power amp, and then wiring up with poor cable. The full performance of the system will never be realised.

At home, enthusiasts are used to paying extra for loudspeakers with better quality, more rigid cabinets and solid loudspeaker stands. It does not take a lot of brainpower to realise that mounting sub-woofers, mid-range and tweeter units on a thin cardboard and perforated steel parcel shelf will not be a particularly good recipe for success. Bolting the same units firmly on a solid baffle of 19 mm plywood (cut to fit and carefully covered to blend with the car interior) can make a noticeable difference which will be well worth the extra expense.

There are more down-to-earth considerations, like avoiding fouling the window mechanism when mounting door-panel loudspeakers and deciding on the mounting positions for mid-and tweeter units. They are not always mounted on the rear shelf with the woofers – the technically 'correct' position – to direct more mid-and treble towards the listener. Cars are not the perfect audio environment and sometimes demand unconventional solutions.

Some domestic amplifiers employ separate power supplies for parts of their circuitry to eliminate mutual crosstalk and distortion. There are some in-car installers who echo this philosophy and run separate leads back to the battery rather than use common leads for power amps and head or cross-over units. They claim that this noticably improves sound quality. Some of the more powerful amplifiers can draw several amperes of current – enough to produce a few volts of rather nasty power line ripple with only half an ohm or so of common power cable impedance, so the idea is not really so fanciful, particularly in complex systems with multiple amplifiers and cross-over modules which might be sensitive to ripple. Needless to say, however these high-power amps are connected up, thick low-DC-resistance cables should be the order of the day.

Another pitfall for the unwary is the little matter of signal levels – output voltages and input sensitivities – a problem which has been banished as far as domestic hi-fi users have been concerned ever since all manufacturers agreed on a *de facto* standard. Unfortunately, behind physically compatible in-car cables and connectors there often lurk unseen incompatibilities. Units which look as though they should work together, do not, and specifications can be confusing enough to make this best left to specialists. Ironically, items which work well together sometimes have different connectors and so require special leads, which either the manufacturer provides or the installer makes us. Isn't it about time there was some standardisation?

Interference

In general, interference is generated by electrical equipment, in particular spark-generating contacts, within outside the car (Fig. 19.20). There are also external sources of interference, such as high-tension cables, other cars and buildings. There is little that can be done about interference from outside the car, other than minimising its effect by installing the aerial properly and buying a good-quality radio.

Interference from within the car is another matter and can usually be cured by a combination of suppression by strategically placed capacitors, filters and studious attention to earthing. Knowing where interference gets into the system is invaluable when deciding how to install a system or how to cure problems in an installation.

Radio interference becomes a menace in areas of poor signal strength. Where radio signals are strong, the wanted signal will effectively drown out the interference. However, when signal strength drops, interference can intrude on listening pleasure.

Interference entry points

Aerial

The aerial is most likely to pick up interference from nearby conductors which are carrying strong-high frequency currents.

It can pick up this radiated interference either directly, or the cable connecting it to the radio can pick it up from nearby conductors.

That is why most car aerials are boot-mounted these days – to avoid interference from the engine either via the aerial or cable.

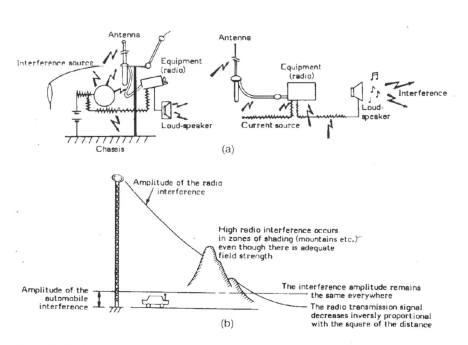

Figure 19.20 (a) Points at which interference can enter the in-car audio system (b) radio interference suppression measures for automobiles.

Power Supply Leads

If the power supply leads from the battery are placed close to leads carrying high-frequency currents, then the interference can become superimposed on the battery voltage and enter the equipment that way.

Signal Interconnects

In situations where there are separate pieces of equipment (for instance a console mounted head unit and remote power amp) the interconnecting cables can pick up interference directly from the interference source or by electrical induction from nearby cables which are carrying interference. This can be minimised by relocating the interconnect cables (Fig. 19.21).

The Equipment

Most equipment is built with a metal case which offers a degree of screening, but if there are cut-out slots or gaps between metal seams, these can act like slot aerials at interference frequencies, which let the unwanted signals in almost unattenuated. Changing the location or mounting angle of the unit can reduce the effects of this type of interference.

Where is it coming from?

If there is an interference problem in an installation the best way to solve it is to locate the sources of interference. It will then be possible to minimise it at point of entry and source.

First, site the car in an area free from outside interference with the bonnet closed and the aerial fully extended. Detune the car radio away from radio transmissions and listen at three or five points over the tuning range while briefly revving the engine at about 1500 rpm. Do not maintain the engine at this speed for long, as you could damage the engine.

Also listen using a cassette player switched to replay with no tape or a blank unrecorded tape to assess the level of interference picked up by the cassette section. The same thing goes for a CD player, only you will have to use a quietly recorded CD for this purpose. Turning up the treble control helps in these tests by making the interference more obvious.

The Aerial

Start with the aerial. If this is disconnected from the radio, it is obviously eliminated as a collecting point for interference. Life is not quite so simple, though, as this reduces the set's sensitivity and will appear to reduce the interference, even if it is not gaining access through the aerial. To prevent this from happening it is necessary to replace the aerial with a dummy aerial (Fig. 19.22).

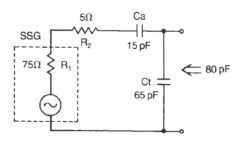

R₁ . . . SSG output impedance
R_1 . . . SSG output impedance
R_2 . . . 80Ω-R_1
Ca . . . Effective aerial capacity
Ct . . . Aerial affective capacitance + aerial stray capacitance
SSG . . . Measuring transmitter

Figure 19.22 Supplementary circuit-dummy aerial.

If the interference disappears with the aerial replaced with the dummy, then the aerial's installation needs to be checked.

The most common cause of problems here is poor earthing between the aerial's metal fixing base and the car chassis. In many cases, the metal shell's inner coating does not provide perfect grounding at high frequencies. If the paint has not been pierced by the fixing washers and nuts, this will create an open circuit, resulting in interference. In either case, the aerial cable's inner conductor, instead of merely conveying the radio signal, becomes part of the aerial itself and will pick up any

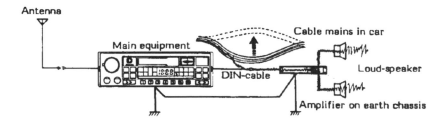

Figure 19.21   The inter-connect cable should be kept clear of other cables which could induce interference.

interference around under the dashboard where it will run close to other cables.

To minimise interference make sure that the aerial is perfectly grounded and relocate the signal cable clear of other leads.

If the interference does not disappear when the aerial is replaced by the dummy, then either the radio is picking up interference directly, the chassis lead of the radio has not been sufficiently well grounded to the car's metal body, or there are problems with the supply-line filter, or its connection between the battery and the set.

Connecting the set to a separate car battery and screening the car radio with a temporary earthed metal netting will help to pinpoint which route of entry the interference is taking.

Interconnect Leads

If interference is not entering through the voltage supply, directly into the set or through the aerial, then it may be entering by way of the interconnect leads. If they are picking it up, moving them about will affect the level of interference. In rare cases, the loudspeaker leads can also collect interference and guide it into the head unit or amplifier.

If this is the case, repositioning the leads away from sources of interference such as other cables will be effective, and care is needed with earthing (Fig. 19.23).

Internally-Generated Interference

If no source of interference can be located, it is just possible that the radio or head unit could be at fault and

unless there is an obvious cause, such as loose components or circuit boards, repairs are best left to specialists. In any case, if the guarantee is in force, an individual will invalidate it if he opens up the set and attempts his own repairs.

Interference Suppression

Following the above guidelines will help to locate where the interference is entering the in-car audio system and minimise its effect. It will also be necessary to suppress the sources of interference.

Ignition System

Interference generated by spark plugs is the strongest source in the car and is obviously synchronous with engine speed. Curative measures include connecting a 5 Kohm to 20 Kohm resistor in the high-voltage lead between the ignition coil and the distributor and between the distributor and the spark plugs, fitting suppressed plugs or a suppressing distributor cap and plug leads.

On the low-voltage side of the ignition system, connect a 0.5 μF capacitor between the battery connection at the ignition coil and the chassis (Fig. 19.24). The carbon brushes which connect with the rotor in the distributor coil should also be checked for good contact.

Another cure for ignition interference, which may be finding its way via either the radio or inter-connect leads, is to connect a noise suppressor (not just a capacitor) between the battery connection to the ignition coil and ground, as shown in Fig. 19.24.

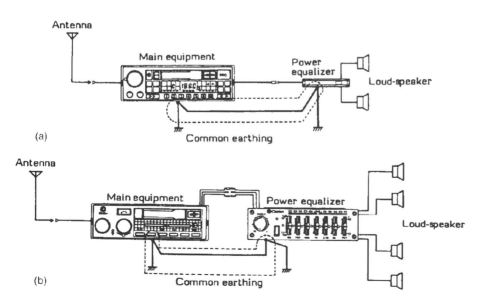

Figure 19.23 Common earthing by using an inter-connect lead when wiring systems comprising more than one component.

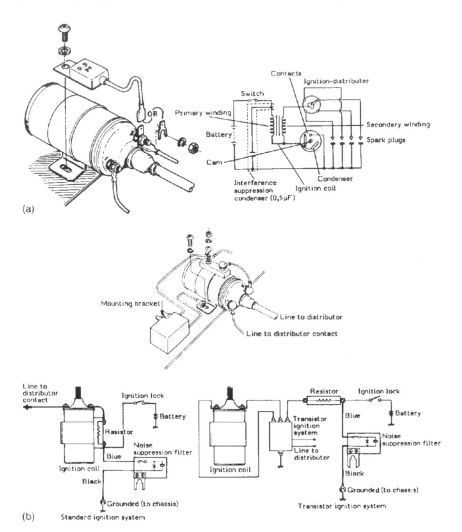

Figure 19.24 (a) Suppressing the ignition system using a 0.5 μF capacitor (b) suppressing the ignition system by connecting a filter to the low-voltage (primary) side of the ignition coil.

Generator

The alternator generator produces AC which is rectified to produce DC for the battery and there can be some residual AC voltage in the output, which can cause a humming or whining noise. This can be cured by connecting a 2.2 μF capacitor between the battery terminal and the chassis (Fig. 19.25 (a)).

Regulator

This can produce a scratching noise when the battery is charging which can be cured by a 0.5 μF capacitor mounted close to the regulator between the battery supply lead and the chassis (Fig. 19.25 (b)).

Wiper Motors

These can cause annoying interference which can be cured

by a special filter comprising two coils and a capacitor which is wired into the battery leads to the wiper motor terminals, or by a 0.5 μF capacitor connected between the wiper motor terminal and the chassis. It should be mounted as close to the motor as possible (Fig. 19.25 (c) and (d)).

Windscreen Washer Pump

This produces a shrill noise which can be remedied by the same type of filter used for wiper motors, connected into the leads to the pump motor or across its supply terminal and the chassis (Fig. 19.26 (a)).

Heater Motor

This produces a deep noise which can be cured as for wiper motor and windscreen washer interference (Fig. 19.26 (b)).

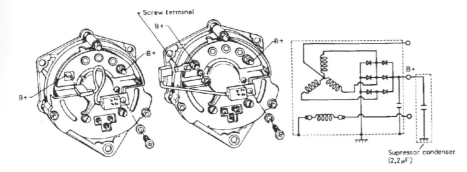

Figure 19.25 (a) Suppressing the generator using a capacitor (2.2 µF or 0.5 µF values are used).

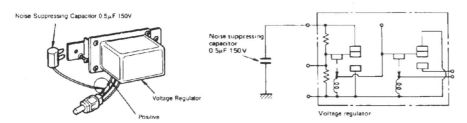

Figure 19.25 (b) Suppressing regulator noise by fitting a 0.5 µF capacitor between the positive terminal and ground.

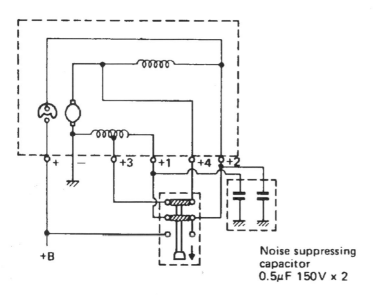

Figure 19.25 (c) Suppressing the wiper motor using capacitors. These can be used for brake lights, indicators, fuel pump etc. Mount close to the source of interference, connected between the battery feed to the device and ground.

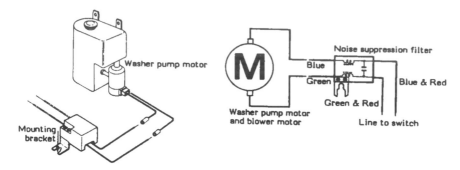

Figure 19.25   (d) Suppressing the wiper motors using an LC filter.

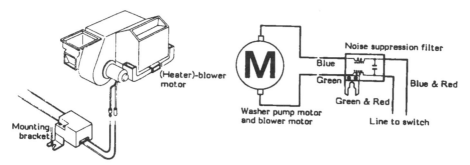

Figure 19.26   (a) Suppressing the windshield washer pump using an LC filter.

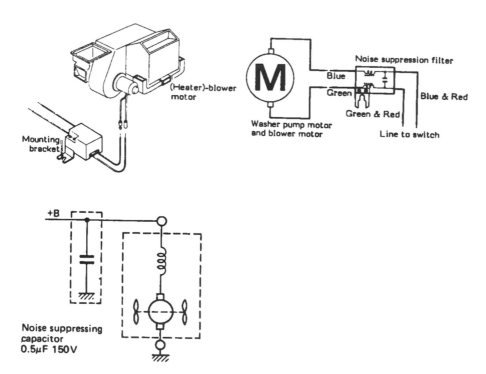

Figure 19.26   (b) Suppressing the fresh air/heater motor using an LC filter; suppressing the fresh air/heater or heater motor using a capacitor between the battery supply point and ground.

Brake Lights, Traffic Indicator and Fuel Pump

Interference from these sources can be remedied by LC filters connected into the battery supply and to the chassis for each respective item (Fig. 19.26 (c)), or by 0.5 μF capacitors connected as for wiper motors, etc. (Fig. 19.26 (d)).

Voltage Supply

If tests indicate that interference is being picked up through the voltage supply from the ignition system or generator, this can be cured by connecting an LC filter with an earth connection in series with the battery supply to the items of equipment in question. One filter may be sufficient for more than one item.

Grounding and shielding

Careful attention to grounding will help reduce interference at source.

Unfortunately, grounding is a topic which is often not understood. Even though the DC resistance of grounding leads may be virtually zero ohms, this is not necessarily as low at radio frequencies, where the impedance can be sufficiently high for the apparent short-circuit to behave like an aerial, picking up strong interference. In such cases radio reception can be accompanied by loud background noise even when all the anti-interference measures have been taken. The cure is to use very short, heavy earthing leads well connected to the car's metal body. The same goes for head units and power amps, though some manufacturers provide inter-unit earth connections with their screened interconnect leads.

Because of the many conducting paths around the car engine and body, there are extremely complex patterns of interference-producing currents. Impedances around these paths are sufficiently high, due to all the bolted and screwed connections and the short wavelengths involved, to produce significant voltage differences at radio frequencies. Consequently, moving the position of

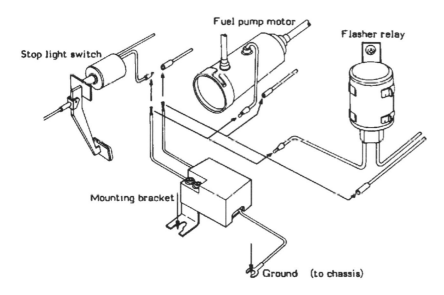

Figure 19.26   (c) Suppressing the brake lights, traffic indicator and fuel pump.

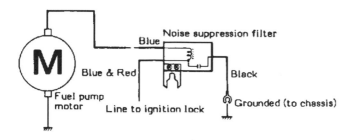

Figure 19.26   (d) Suppressing the fuel pump.

a grounding point only a few centimetres can make a very large difference to the interference levels.

Grounding can be highly beneficial, particularly between the motor block and the chassis (Fig. 19.27 (a)), the radiator grille and the bodywork or motor block, gear housing and bodywork, bonnet and bodywork and silencer and bodywork. Each car will have its own optimum positions for connecting grounding leads, which in these cases should be heavy braided flexible copper conductors made specifically for this purpose. Grounding items like the bonnet (Fig. 19.27 (b)), and grille will also have a screening effect which will help shield the aerial from interference.

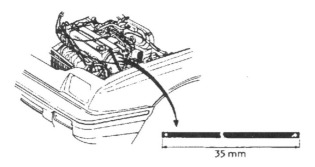

Figure 19.27   (a) Flexible ground lead between motor and bodywork.

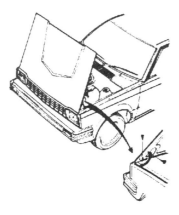

Figure 19.27   (b) Flexible ground lead between the bonnet and the bodywork.

Again, grounding and shielding are definitely jobs for the specialist who knows which bolts can be undone safely and which positions are likely to be optimal.

That really covers just about every interference problem one is likely to experience with in-car audio installations, whether fitting a system from scratch or trying to cure a problem in existing systems. Knowing the possible pitfalls from the outset should at least avoid building difficult-to-cure faults into a car audio system.

The Future for In-Car Audio

This is where the author gets the chance to partake in a little crystal ball gazing and he makes no apologies if some of these guesses are entirely wrong.

Going on current trends, in-car audio or ICE (in-car-entertainment) as it is often called, is likely to become ever more sophisticated as micro-electronics makes it more and more possible. Diversity reception is one new development which will probably become more commonplace to reduce the annoying audible effects of FM multi path signal nulls. Compact disc (and possibly digital audio tape if initial problems can be overcome) will be more widespread and may eventually threaten to usurp compact cassette in the car. By this time, however, we may have a solid-state music storage system, using little plug-in-music modules, which will be ideal for both in-car and domestic use.

Normal run-of-the-mill car radio or radio cassette units will become ever more decked with features, while there is little doubt that there will be a growing number of buyers willing to pay the extra for high-quality systems which perform a sonic quantum leap over typical car units.

For most people, who are probably not so concerned with sound quality, the car radio or car radio/cassette/CD/DAT player will become much more than just an entertainment system. Already there are systems like Car and Information, or CARIN, the experimental car navigation system under development by Philips, which uses a compact disc to store visual information, maps and what have you for recall and display. It can even plot a journey and can be linked during the journey to up-to-date road and traffic information by radio data system (RDS) transmissions.

In some countries, traffic bulletin information is broadcast on certain frequencies. These can be automatically monitored and voice messages relayed to the driver, overriding whatever signal source he has selected.

It is likely that similar services will become more commonplace and it is only a relatively small step to transmit coded data in the way that CEFAX data is transmitted along with the TV signal for decoding by television receivers. Then what is currently regarded as a means of entertainment will become a comprehensive information centre, processing all sorts of traffic and weather data – even perhaps giving details of and taking hotel or flight bookings, though preferably not distracting the driver while on the move!

In-car telephones became even more popular as prices fell and they also make possible the transmission and reception of all sorts of data – Prestel for instance – which could be processed by an onboard computer. This is not really very far fetched, as some radios on the market already use micro-computer technology. The mobile office will truly have arrived.

Perhaps one day the problems of terrestrial FM reception will be done away with completely by direct broadcast by satellite (DBS) of digital stereo or multi-channel sound, requiring a small roof aerial pointing skywards and a reasonably powerful satellite transmitter. Currently not feasible perhaps, but given the investment in the required geostationary satellite transmission channel improved aerial and receiver technology, who knows?

20 Sound Synthesis

Mark Jenkins

Not all sound starts off as waves in air, and a substantial number of the sounds that we hear nowadays have their origins in electronic instruments. Mark Jenkins reviews here the rapidly changing scene of electronic music making and its relationship with computers.

Electronic Sound Sources

Virtually any electronic circuit which has a periodic function can be used as a sound source simply by connecting its output to a speaker, which converts electrical energy into mechanical energy in the form of sound waves. Early experiments in the electrical and audio fields indicated that an electronic 'musical instrument' was a possibility, simply because an electronic oscillator could be set to a variety of 'musical' pitches. Controlling the pitch so that an electrical output becomes a meaningful musical 'performance' has been the problem ever since.

One of the first electronic instruments ever devised, Elisha Gray's 'Musical Telegraph' of 1874, solved this problem by adding a keyboard of sorts to the electronic circuits involved. But later instruments such as Leon Termen's 'Theremin', which post-dated the invention of the valve in the 1920's, showed that this approach was not the only valid one – the Theremin used a vertical antenna with the proximity of the performer's hand determining the pitch, which was continuously variable rather than limited to certain notes.

The Theremin's sound was based on a simple sine wave, but one with enough side bands to retain some depth and musical expression. Other instruments combined simple waves to produce more complex sounds, and the term 'sound synthesis' to describe this method of sound creation was used by Thaddeus Cahill, inventor of the Telharmonium, as early as 1896.

The Telharmonium used an electromechanical system involving cogged wheels which closed electrical contacts at a variable rate. This system was later adapted by the Hammond Organ company, whose popular home elec-tronic keyboard introduced in 1929, used a similar 'tone wheel' system. But the second generation of Hammond organs, introduced in 1939, abandoned the tone wheel system in favour of valves, and the valve in turn gave way to the transistor and the integrated circuit, each in turn making more and more complex methods of sound generation easily available.

Electronic sound sources, then, treated as oscillators of one kind or another, have developed rapidly in many ways, the most recent being the digital storage of a waveform based on microcomputer wave creation (as on some of the Japanese Casio keyboards) or on direct 'sound sampling' from acoustic sources (as on microprocessor-based instruments such as the Fairlight Computer Musical Instrument, Emulator, Prophet 2000 and many others). However, the simplest way to understand the electronic creation of musical sounds is to consider the history of the music synthesizer.

Synthesizers, Simple and Complex

Although engineers such as Donald Buchla were experimenting around the same time, the first commercial music synthesizer is generally credited to Dr Robert Moog, whose early work was in constructing Theremins. Moog's invention was the Voltage Controlled Oscillator, the pitch of which was determined by an input voltage, making it easily controlled by a chain of resistors connected to a keyboard or other musical device. His 1964 paper on 'voltage controlled electronic music modules' for the Audio Engineering Society of America was rapidly followed by designs for a voltage-controlled oscillator (VCO), a voltage-controlled amplifier (VCA) and later a voltage-controlled filter (VCF).

Due to slight distortion in the circuit, Moog's patented VCO always produced a rich and full sound which could not (legally) be reproduced by competitors such as ARP. Moog's filter also has an outstanding reputation for helping to colour the sound produced by the various wave shapes available from the VCO.

The other major sound source available at the time was the noise generator, generally based simply on a noisy diode which required some time to warm up to operating temperature. While the diode has now been replaced by a digital pseudo-random shift register to simulate the distinctive hissing sound of totally random tones, the applications remain the same – filtering with a low-pass filter to cut off high-pitched tones for white noise, or with a band-pass filter to create the almost-pitched tones of red noise, each used for a variety of musical and non-musical purposes.

For almost 20 years the science of sound synthesis was based on these same sources – the oscillator and the noise generator – although the methods of control improved gradually, as we will see. It is only with the introduction of sound sampling that this situation has greatly changed.

Returning to the basic 'modular' synthesizers, we see that most of Moog's components could be controlled instantaneously by a keyboard, or over a period of time by a transient or envelope generator, or by another component such as an oscillator running too slowly to create an audible tone (low frequency oscillator, LFO, or modulation generator, MG).

Moog, a self-confessed non-musician, needed input from established performers such as Walter (now Wendy) Carlos and Keith Emerson to design practical performance synthesizers. Pedals, footswitches and sequential controllers (sequencers) for repeated phrases were all added to the system, which was eventually simplified into the still-popular MiniMoog, which has two VCOs, one VCF, two VCAs and one LFO.

While complex synthesizers still existed (some of Moog's larger systems had hundreds of modules), the trend as the synthesizer became more popular was toward more powerful but more compact and manageable instruments. ARP under Alan R Pearlman and Tom Oberheim's Oberheim Electronics in the USA, and Peter

Zinovieff's EMS in the UK, came out with models which abandoned the modular approach for simpler but still flexible 'patchable' designs, while Japanese companies such as Yamaha and Roland began to introduce synthesizers with no patching at all, simply a few variable parameters as on the MiniMoog, or even a fixed number of preset sounds.

Two major problems remained. Most synthesizers only produced one note at a time, having only one or two oscillators, and so it was impossible to play chords. In addition, it took some time to produce and tune a sound, and even longer to reproduce it at a later date if necessary (such as when playing studio work live on stage).

Synthesizers had to become both polyphonic and programmable, and as we shall see below, this began to become practical only when the microprocessor became available.

Radiophonics and Sound Workshops

Long before Dr Moog developed his synthesizer, composers from the classical *avant garde* such as Karlheinz Stockhausen and Luciano Berio were using a selection of techniques to create original musical pieces on tape.

Once magnetic tape recording became reliable it was quickly found that it could be exploited in various ways – cutting, splicing and reversing tape for instance – to create new and unfamiliar sounds. Natural and instrumental sounds were used to create this *musique concrète*, but some composers (such as Stockhausen) wished to go beyond these techniques to create totally electronic music. Sine wave oscillators and tape recorders were his only instruments for early pieces such as Studie 1 (1953) and Studie 2 (1954), the latter piece also using white noise.

In the UK the term used for this kind of music has generally been 'radiophonics', a monstrosity perpetrated by the BBC. Their Radiophonic Workshop under Delia Derbyshire and other pioneers introduced first concrète,

Figure 20.1 A small analogue modular synthesizer – Roland's System 100.

Figure 20.2   A modern processor-controlled polyphonic analogue synthesizer – the Roland JX-10.

Figure 20.3   Yamaha's TX816 comprising eight rack-mounted FM synthesizer modules.

and then synthesizer sounds, to millions of viewers and listeners. Early works were created using tape machines with special modifications, banks of oscillators and acoustic instruments, and basic synthesizers such as the EMS VCS3 were only introduced after some years.

Since the Workshop's output was intended mainly for TV and radio use, little of it has become available for general consumption. But one early project, 'White noise – an electronic storm in hell', by David Vorhaus with Delia Derbyshire and Brian Hodgson, perfectly illustrates the Workshop's combination of vocal and instrumental sounds, electronic sounds and effects processing.

Even the more sophisticated synthesizers used nowadays make great use of external processing, although in some cases such effects are built in. Modular synthesizers frequently offer a phase shifter, while the EMS VCS3 had a built-in spring line reverberation unit, for instance.

The radiophonic workshop, or indeed the more conventional recording studio or specialised synthesizer studio, has access to many kinds of effects processors.

Chorus

Chorus is the most common device found on commercial synthesizers and is used for thickening sounds. Many cheap 'home keyboards' and most of the cheaper poly-phonic synthesizers such as the Roland Juno 6 and 60 and the Korg Poly 6 and 800 use it to add life and movement to basically thin sounds.

When a group of violinists play the same note, they are never precisely in tune with each other and never play at exactly the same moment, and so the sounds produced are full and thick. Chorus units simulate several instruments playing together by delaying the incoming signal by a variable amount and mixing the delayed signal back in with the original; this causes any slight variations in the sound to give the impression of two or more instruments playing; sometimes it is possible to alter the feedback of the circuit to its own input to multiply this number.

Chorus units based on analog delay lines are very popular in pedal form and common models include those from Boss, Pearl, Tokai, Rozz and many others. Some models have a stereo pair of outputs, so the straight and delayed signals can be sent to two sides of a stereo amplifier or to two channels of a mixer. Stereo positioning and added thickness in the sound can give an impression of movement in space, an important element in adding life to electronic recordings.

Other delay techniques

Flangers and Phasers work on similar delay principles to give a regular sweep effect to sounds. Phasers (phase

Figure 20.4   A small synthesizer workstation based on the Yamaha DX-100.

shifters), as their name suggests, use a variable very short delay to catch the incoming signal at a different phase of its waveform and combine this delayed version with the original to produce a sweeping or 'skying' effect. Phasing was first produced by mixing the outputs of two tape machines playing identical sounds, but electronic means of production now dominate. Variable feedback or resonance in the filter circuit and stereo outputs are also common.

Flangers use an analogue delay line in the region of a few hundred milliseconds long and almost all have a resonance setting to alter their sweep effects from very subtle to highly coloured.

A graphic or parametric equaliser can change the tone of an oscillator output. A graphic equaliser, basically a sophisticated tone control with five or ten bands per channel rather than the usual bass and treble, makes it possible to boost just the deepest frequencies for powerful bass sounds, or just the centre frequencies for more precise snare drum or vocal sounds, or just the top frequencies for clearer cymbals or melody lines.

Parametric equalisers boost or cut just one frequency in which the user is interested, and usually have just two controls – boost/cut amount, and frequency. Some more sophisticated models also have a third control to alter the width of the frequency band affected.

Echo is a term used by recording engineers to describe what most of us would call reverberation – the effect of a sound coming from a deep well. Discrete repetitions of a sound (such as in a mountain range) are referred to as 'Repeat', and while digital systems to create either effect can cost tens of thousands of pounds, the cheapest electromechanical delay and reverb units cost only a few tens of pounds.

Around 300 ms of delay can be created by a cheap pedal unit based on a charged coupled device (CCD) circuit which uses a series of charging and discharging capacitors to delay the signal; this third of a second is about the level at which separate repeats as opposed to constant reverberation become distinguishable. Delay units generally have controls for repeat time, number of repeats (amount of feedback from output to input) and mix between original and echo signal.

There are several other methods of producing delay; magnetic tape in a loop or a cartridge is one popular method which offers a gradually degrading sound which can be attractive.

Digital methods

But fully digital audio delay systems have now become quite inexpensive, with models from Boss and other

Figure 20.5 Yamaha's chief demonstrator David Bristow with a DX7 synthesizer, QX1 sequencer, RX15 drum machine and CX5 music computer system.

companies selling for under £200. A digital delay 'samples' the incoming sound at a certain rate and with a certain level of resolution (bit number); both sampling frequency and bit resolution affect the quality of the delayed sound. Some delay units include a modulation oscillator to regularly vary the delay time to produce effects such as flanging and vibrato, and so represent very good value for money.

The most expensive digital delay units have very fast sampling rates (giving frequency response beyond the normal range of human hearing), 16-bit (Compact Disc quality) operation for high fidelity, programmable settings and delay times of up to a minute or more.

Many delay units also produce reverberation, although until recently it was necessary to use a dedicated device. The simplest reverb units is a spring line with a transducer at either end, one to convert the electrical signals representing the sound into vibrations, and one to reconvert it into electrical signals with all the reflections of the spring added. A pair of metal plates are often used in the same way (plate reverb).

Spring line reverbs are inexpensive but have relatively low fidelity, while digital reverbs using various computational algorithms to simulate reverberant spaces of different sizes, shapes and natures can cost from £300 upwards. The Yamaha SPX90, Roland DEP-5, Alesis MIDIfex and other devices are now capable of producing both delays and reverb.

Apart from instruments, the radiophonic workshop or synthesizer studio obviously needs recording and mixing facilities, and these are discussed elsewhere in the book. However, it is worth noting here that many synthesizer studios now dispense with the multi-track tape machine, since it is possible to record performances from many instruments in a computer or sequencer and play them back 'live' to a single master tape, a MIDI-based setup being most common for these applications. Also, the mixing process is becoming influenced by MIDI, so that changes can be made to levels, EQ and effects under the control of a MIDI sequencer or computer during mixdown. This is a new and relatively inexpensive way to automate the mixing process.

Problems of Working with Totally Artificial Waveforms

In the last few years the problems facing early synthesizers were solved when programmability and polyphony became inexpensive to fit. However, basic synthesizers still offered very simple sounds created perhaps by a single bank of oscillators offering sine, sawtooth or square waves.

As synthesizer pioneer David Vorhaus pointed out during a BBC interview in the 1970s. 'The problem with early synthesizers was that they produced very square waves in every sense of the word'. Vorhaus' concern was to use artificial ambience and other effects as an integral part of the sound, and since that time the cheaper availability of digital reverberation units has made this aim more easily realised.

The processing devices mentioned previously can help to make basic electronically-generated sounds more

Figure 20.6 The master of studio electronics – Karlheinz Stockhausen in rehearsal.

interesting, but in the last few years it has become easier for synthesizer manufacturers to offer more powerful-sounding basic waveforms.

One of the pioneers in this field was Wolfgang Palm of the German company PPG. Palm's 'Wave' keyboards store hundreds of waveforms digitally and create sounds by sweeping through a table of waves which can be arranged either in logical or totally random orders. Analogue treatment in the form of filters and envelopes are also available.

Some of Palm's waves were based on sounds from acoustic instruments, and the Waveterm unit introduced as an option for the PPG synths allowed more such sounds to be loaded. The art of sound sampling – digitising

Figure 20.7 The classic days of analogue sound synthesis – Chris Franke of Tangerine Dream.

acoustic sounds for playback from a keyboard – has become well-established since the introduction of the Wave, and non-MIDI-equipped samplers are now available for less than £100.

MIDI-equipped instruments such as the Ensoniq Mirage and Akai S900 offer sampling facilities plus filtering, envelope shaping and other musical parameters, and cost between £1000 and £2000. Since their sounds are based on those of acoustic instruments and are recorded with anything up to 16-bit (CD quality) fidelity, they can be as rich and complex as desired.

Another modern alternative to the problem of bland electronic sounds is the MIDI layering facility offered by the simple connection of two keyboards. Any piece played on one keyboard can be doubled on the other; if similar sounds are chosen, the two synthesizers can be detuned from one another to give a powerful ensemble effect, or alternatively two completely different sounds can be chosen. In the latter case, a brass sound with a sharp, sudden attack and many high overtones could be underpinned by a string sound with faded in with a smooth attack and a mellow, rich sound.

By and large the problems faced by the early pioneers of electronic music, such as Karlheinz Stockhausen in his experiments with sine wave generators, are no longer serious. Even inexpensive instruments and effects are now capable of producing full, rich sounds from electronic sources.

Computers and Synthesizers (MIDI and MSX)

As we have seen, the main problems of synthesizer manufacture – polyphony and programmability – were solved by the introduction of microprocessor control on models such as the Roland Jupiter 8 and Sequential Circuits Prophet 5.

One remaining problem – that of intercompatibility – was solved by the designer of the Prophet 5, Dave Smith. Smith's early experience, as the company's name suggests, was in building sequential controllers, and the difficulties of designing a reliable polyphonic sequencer to interface to the his Prophet 5 synth led him to propose a universal synthesizer interface – USI – which would make this simpler for all manufacturers.

Until that time, instruments from different manufacturers worked on largely different principles and interfaced to other units, such as sequencers and drum machines, but only from the same manufacturer. Smith's proposal was innovative as it made selective purchasing rather than enforced loyalty to one brand a much more realistic proposition.

The Japanese manufacturers, such as Roland (already using its own DCB or Digital Communication Bus system). Yamaha and Korg, suggested several additions to the system with coded 'flags' to identify each function, and it was decided to opto-isolate the system to prevent ground loops, and to set its operating speed at 31.25 KBaud. The resulting specification for a serial digital interface was renamed the Musical Instrument Digital Interface or MIDI.

The complete MIDI published by Sequential Circuits sets out the MIDI specification in full, but below we have covered the main operational points. Early objectors to the MIDI proposals such as New England Digital, Oberheim and Fender, pointed out rightly that a serial interface is limited in speed as compared to a parallel interface such as the short-lived Triad system; but MIDI is relatively cheap to fit, needing only inexpensive DIN

Figure 20.8 Inexpensive effects processors from Vesta Fire.

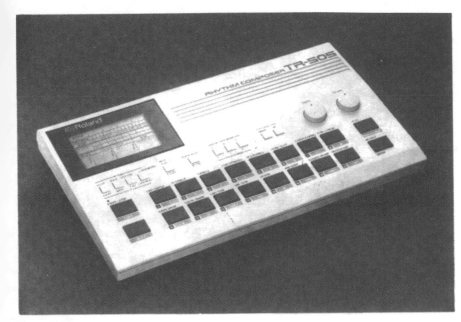

Figure 20.9   A MIDI-equipped sampled drum machine – the Roland TR-505.

leads and sockets to operate as opposed to expensive computer-grade multicore cables.

MIDI methods

MIDI is implemented in such a way that note information can be transmitted over a range of 10.5 octaves, changes of memory can be made, and specific codes for functions unique to one model or another are available in the 'System Exclusive' section of the MIDI standard. MIDI codes can also be applied to the timing of drum machines, sequencers and other devices apart from keyboard instruments, and an increasing number of effects units, guitar convertors, wind instruments, home computers and other non-keyboard units are being fitted with MIDI.

MIDI has received one update since its inception, at which time the original specification became known as

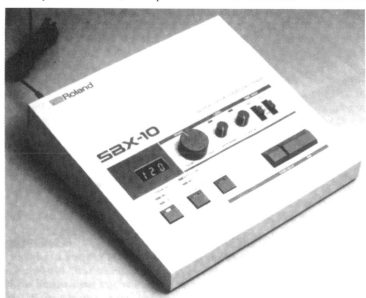

Figure 20.10   A MIDI-compatible synchronizer – the Roland SBX-10.

Figure 20.11   A MIDI-equipped FM piano – the Yamaha PF70.

OLD MIDI and the current version as MIDI 1.0. Another revision to encompass MSMPTE (MIDI-SMPTE, or MTC, MIDI time code) is at the time of writing being completed, and this will allow SMPTE time codes to be passed through a MIDI system.

The hardware aspect of MIDI is relatively simple; signals from the instrument microprocessor's UART (universal asynchronous receive/transmit) pass through an opto-isolator to prevent ground loops and damage caused by high or reverse polarity current. A +5 level indicates a high and a low is set when the LED in the opto-isolator turns on and switches on the output isolator.

Connection cables consist of 180° 5-pin DIN cords with Pin 2 as shield and pins 4 and 5 used for information. MIDI IN and MIDI OUT sockets are compulsory (this is not the case on keyboardless expander synthesizers which often have only MIDI IN, or on remote control keyboards which usually have only MIDI OUT); a MIDI THRU socket which passes on the input information unaltered to other units is optional. MIDI information is often shared between several units since 16 different 'channels' are available to perform different functions apparently at the same time.

Connecting instruments in a Chain via the MIDI THRU ports can cause unacceptable delays, but a splitter box with one MIDI IN and several MIDI THRU sockets, so that instruments can be connected in a parallel or star network, will solve this problem.

Modes

Instruments receiving MIDI information specific to a particular channel can be assigned one of three different modes – omni, poly and mono mode. Apart from note information, it's also possible to transmit volume or tone levels on individual notes, modulation, patch changes and changes of individual parameters within a patch, and pitch bend information. Messages which cannot be responded to, such as a code to sound a chord of C Major if sent to a drum machine, should simply be ignored.

In omni mode, MIDI transmitters send information on Channel 1, while receivers respond to events on all 16 input channels. This simplest mode allows two synthesizers to be connected together and played in parallel from one keyboard. For computerised control of several instruments with different parts to play, a more complex mode is needed however.

In poly mode an instrument will respond to only one MIDI channel selected by the user; with 16 channels available, up to 16 different instruments can play different parts 'simultaneously' if they are transmitted on the appropriate channels.

Mono mode is the most complex MIDI mode, allowing individual control of every voice on every synthesizer played (the number of 'voices' a synthesizer has is generally the maximum number of notes it can play simultaneously). Usually the user can set the 'base channel', which Voice 1 responds to, and subsequent voices respond to subsequent MIDI channels. For instance, the Casio CZ-101 synthesizer plays eight-note chords controlled by any MIDI channel from 1 to 16 in its normal mode; in poly mode it plays four different monophonic sounds controlled independently by MIDI channels 1–4, or 2–5, or 8–11, or whatever.

Apart from added instrumental complexity, mono mode has the advantage of reproducing variations of touch response on every individual note (as on the Prophet T8), and of reproducing a true polyphonic glissando (portamento or glide), which cannot be done in poly mode since every note has to be switched off before its value is changed.

MIDI information comes in five categories, of which four are independent of the channel used. Apart from channel information, the MIDI format recognises system common, system real time, system exclusive, and reset information, identifying the category by a system of 'flags' to allow instruments to respond appropriately. Once a category has been identified information can be sent until a change of category is needed, so it is not necessary to label absolutely every piece of information.

A number of different operations are possible under each category, so each has a number of 'status bytes' which define the exact operation to be carried out. Status and data bytes are differentiated by a 'Flag' – a status byte begins with a 1, whereas a data byte always begins with a 0. Since MIDI is a binary code designed for 8-bit processors, this leaves seven digits, with a maximum value of 1111111 (127), remaining.

For instance, data beginning 1001 (9 in decimal notation) is identified as 'Channel – note on' information, with the following half-byte specifying the channel from 1 to 16 in use. For example, the binary three-byte code

Byte;   1   2   3
1001/aaaa   0bbb/bbbb   0ccc/cccc

gives the following information:

- *Byte one*: 1 indicates status byte – 001 indicates 'note on' information coming – aaaa identifies channel to be used from 1 to 16.
- *Byte two*: 0 indicates data coming – bbb/bbbb defines key number from 0 to 127. This corresponds to a range of ten and a half octaves, with middle C as 60, a five-octave synthesizer ranging from 36 to 96 and an 88-note MIDI piano ranging from 21 to 108.
- *Byte three*: 0 indicates data coming – ccc/cccc defines key on velocity, 0–127 (softest to loudest). Instruments without velocity sensitivity default to 64. Setting 'key on' velocity to 0 is an alternative way to turn a note off.

So these three bytes signify 'note on' information, identify the MIDI Channel in use (the synthesizer to be played if using poly mode), select a note and switch it on with a certain key velocity.

'Note off' information must also be sent at some point, and 'key pressure information may also be sent; these also come under the category of channel information.

Below is a complete list of the MIDI codes in use according to *The complete SCI MIDI*, (first edition).

CHANNEL Codes are as follows:

- *Note on*: status byte 9 (1001)
  1001/aaaa, 0bbb/bbbb, 0ccc/cccc
  = Note on channel, key no. 0–127, key on velocity 0–127
- *Note off*: status byte 8 (1000)
  1000/aaaa, 0bbb/bbbb, 0ccc/cccc
  = Note off/channel, key no. 0–127, key off (release) velocity 0–127
- *Polyphonic key pressure*: status byte 10 (1010)
  1010/aaaa, 0bbb/bbbb, 0ccc/cccc
  = Poly pressure/channel, key no. 0–127, pressure after touch value 0–127 (omni mode)
- *Control change*: status byte 11 (1011) (see notes below)
  1011/aaaa, 0bbb/bbbb, 0ccc/cccc
  = Control change/channel, control address 0–127, value 0–127
- *Program change*: status byte 12 (1100)
  1100/aaaa, 0bbb/bbbb
  = Prog. change/channel, prog. no. 0–127
- *Channel pressure*: status byte 13 (1101)
  1101/aaaa, 0bbb/bbbb
  = Channel pressure/channel, after touch amount (for an individual channel in mono mode rather than an individual key in omni mode – compare with poly key pressure above).
- *Undefined*: status byte 14 (1110)
  Available for a different purpose for each manufacturer. Sequential circuits use this code for pitch wheel information on the Prophet 600, for instance.

Note on control changes

The fact that every synthesizer or other instrument has a different control layout is one of the largest problems of computer/MIDI control. Theoretically it is possible to take over every parameter of a sound with an external computer, but in practice this is frequently impossible because every synth has a different specification. MIDI was written with conventional analogue synths in mind and is flexible enough to cope with most designs, but some models such as the Yamaha DX7 use an FM synthesis system offering perhaps 150 programmable parameters. MIDI simply does not assign enough codes to cope with all these parameters.

Although MIDI control of FM parameters is rare at the time of writing, there are some packages capable of altering and storing DX7 patches and a couple which will work for almost any MIDI synth.

MIDI control addresses are from 0 to 127, with only the pitch bender specifically defined – this is assigned the address 0, and a value of 64 indicates no pitch bend. All other parameters have values varying from 0 (minimum) to 127 (maximum). Individual manufacturers can assign the front panel controllers to MIDI addresses in any way they desire as long as the allocation is given in the user's manual for each instrument.

The values of rotary, sliding or other continuous controllers are represented by the third byte of the control change code (see above). If only low resolution is needed then the first half only (MSB) is sent; if higher resolution is needed then both halves (MSB and LSB) are sent. If only the less significant half changes value, the most significant half need not be sent again.

MIDI control addresses are:

0	Pitch Bender MSB
1	Controller 1 MSB
2	Controller 2 MSB
3	Controller 3 MSB
4–31	Controllers 4–31 MSB
32	Pitch Bender MSB
33	Controller 1 LSB
34	Controller 2 LSB
35	Controller 3 LSB
36–63	Controllers 4–31 LSB
64–95	Switches (0 = off, 127 = on)
96–123	Undefined
124	Local/remote kybd
125	Omni mode/all notes off (value must = 0)
126	Mono mode/all notes off (value must = 0)
127	Poly mode/all notes off (value must = 0)

'System exclusive' codes are provided by the MIDI specification for computers or musical instruments to take control of functions unique to a particular design. Although MIDI is intended to be universal, many synths will have unique functions, and system exclusive commands need only be relevant to one particular product, and will be ignored by instruments otherwise equipped. System exclusive information opens with a two-byte identifier which is followed by the data itself, and then an 'end of block' code which terminates system exclusive status.

The format for system exclusive information is:

1111/0000, 0aaa/aaaa, data, 1111/0111

This is interpreted as:

Status byte, manufacturer's ID 0–127, data, end of block terminator.

An instrument which receives a system exclusive notice will ignore the following information if it contains the wrong manufacturer's ID; example IDs (in hex) include sequential circuits = 01, Kawai = 40, Roland = 41, Yamaha = 43, and so on.

Data sent can be in any format as long as the flag on each byte is a 0; the range of each piece of information can be from 0 to 127, and the end-of-block code, or alternatively the 'system reset code' (see below) terminates system exclusive status.

'System real time' codes are used to transmit timing information which can synchronise computers, sequencers, drum machines and other instruments together for complex performances. Twenty-four pulses

per quarter note (24ppqn) is generally accepted as being of high enough resolution for most synchronisation purposes; having a 24 ppqn standard within MIDI does not make the system directly compatible with older non-MIDI instruments using 24 ppqn clocks, since the MIDI clock is delivered in amongst other MIDI information and at TTL level. However, a range of convertors to detect MIDI clocks and convert them to normal pulses, and vice versa, is available.

MIDI sync information in the system real time category can be sent between any 'system common' (see below) or channel data consisting of two or more bytes, but like other codes should not be sent during system exclusive information (see Table 20.1).

Table 20.1

Status in hex	Function
F8	Clock in play (sent at 24ppqn while transmitter is in play mode)
F9	Measure end (sent at end of each measure instead of F8)
FA	Start from first measure (sent when play is hit on master – first F8 should follow within 5 ms).
FB	Continue start (restarts from point when last F8 was sent)
FC	Clock in stop (sent in stop to synchronise a phase locked loop for interpolating the timing clock)

'System common' codes are transmitted on all channels and cover various functions including setting sequencers and drum machines to the same part of a song (see Table 20.2).

Table 20.2

F1	Undefined
F2	Measure (three bytes – F2, 0aaa/aaaa, 0aaa/aaaa, the two data bytes comprising a 14-bit measure no.)
F3	Song (two bytes – F3, 0aaa/aaaa – the data byte comprising a 7-bit Song no.)
F4	Undefined
F5	Undefined
F6	Tune (initiates synth tune routine duplicating front panel control)

'System reset' is a single code which resets an instrument to its power-on condition, which usually leads to omni mode and other default values.

FF	System reset (should not be used at power-up, as two linked units could endlessly reset each other)

The five categories of MIDI information have a strict order of priority so that vital new signals can interrupt ongoing transmissions. The order is as follows; system reset, system exclusive information, system real time information (except reset), system common and channel information. Channel information has a four bit-channel number in the status byte, while all other information is intended for all channels. System real time information is interleaved with other data (except system exclusive) so that timing can be accurately kept.

We have seen how the MIDI standard can address the problems of transmitting note, velocity, patch and other information from one processor-controlled instrument to another. Now for a more detailed look at the mechanics of MIDI. To begin with, a short MIDI conversation between a micro-computer running a MIDI composition package, and a connected synthesizer.

10110011	All notes off
01111111	for MIDI channel 3
00000000	in poly mode
10010011	Note on channel 3;
00111100	note is C4;
01000000	velocity is 64
(10010011)	Note on channel 3;
00111110	note is D4;
01100000	velocity is 96
(10010011)	Note on channel 3,
00111100	note is C4;
00000000	velocity is 0
	(turns note off)
(10010011)	Note on channel 3;
01000101	note is A4;
01001000	velocity is 16
10110111	Selects mono mode
01111110	on channel 7;
00000000	turns notes off
10010111	Note on channel 7;
00111100	note is C4;
00100000	velocity is 32
(10010111)	Note on channel 7;
00111110	note is D4, legato, mono mode;
(11111000)	Timing clock sent at any time;
00110011	note velocity is 51
10000111	Note off channel 7;
00111110	note is D4;
00010000	release velocity is 16;
10110111	channel 7 silent
01111110	
00000000	Channel 7 all off

Channel 7 silent – Channel 3 D4 and A4 still sounding

In fact it would be more correct to describe these transmissions as a monologue since in this example the synthesizer has nothing to say in reply. Various notes are sounded with different velocities and terminated, the effect of increased velocities depending on the sound with which the synth is programmed. Initially the MIDI 'words' in this conversation come in groups of three bytes, but after a basic status such as 'note on' has been established, it is not strictly necessary to repeat it.

What does this conversation sound like in musical terms? The opto-isolators used in the standard MIDI input circuit can work very quickly, the standard demanding a rise and fall time of less than 2 µs, and MIDI has been set to work at 31.25 kBaud, or 31250 bits/s. Transmissions are organised in packages of ten bits – a start bit, eight data bits, and a stop bit, so these bits will take 320 µs (microseconds i.e. millionths of a second) to transmit. Therefore MIDI can function much faster than the human ear can detect, which is how a serial digital communication system can deal with several events which are supposed to occur at the same time such as all the notes of a chord sounding together.

Notes apparently played simultaneously via MIDI are not in fact sounding together, and are not even being controlled by a single piece of information each, since each one needs note on, note off polyphonic key pressure, velocity and all the other information associated with MIDI to be sent separately. It is true that if many synthesizers are used, each set to a different one of the 16 'channels' available for MIDI transmissions, then delays can become audible. However, one advantage of most micro-based MIDI composition packages is that they allow you to compensate for such delays. Slow scanning of a synth's keyboard is likely to create much longer delays, and so those who would seek to change the MIDI standard would do better to ask individual manufacturers to improve their keyboard response times.

But suppose we have ten synthesizers responding to 3 byte note on transmissions; each of these takes $3 \times 320 = 960$ µs, or around one thousandth of a second, so the delay for the synth on the end of the line can be ten times this long. One hundredth of a second delays are quite audible to some musicians, and if a lot of other information-on velocity, pitch bend or timing – is also being transmitted, we can look at delays of up to three or four times this long, so the problem does need some attention.

Mode Messages

As we have seen, the MIDI modes define how a MIDI-equipped unit responds to channel codes, but commands

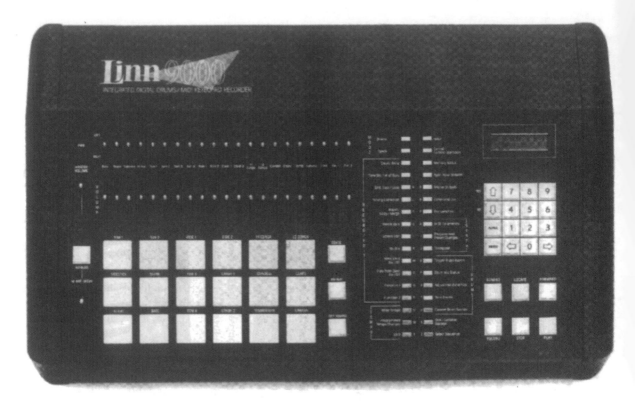

Figure 20.12   A typical MIDI sequencer the Linn 9000 also equipped with sampled drums.

Figure 20.13   A MIDI synthesizer capable of using the multi-timbral Mono Mode – the SCI Max.

Figure 20.14 Jay Stapley demonstrates Roland's GK-I interface from a conventional electric guitar of rack-mounted MIDI synthesizers.

to switch to a new mode can also be sent via MIDI. Mode messages consist of three bytes.

1011/aaaa, 0bbb/bbbb, 0ccc/cccc

but the exact meaning of these bytes has been changed since the MIDI standard was created (see Table 20.3).

Table 20.3

bbb/bbbb	OLD MIDI	MIDI 1.0
122	Undefined	Local keyboard on/off ccc/cccc=0; off ccc/cccc=127; on
123	Undefined	All notes off ccc/cccc=0
124	Undefined	Omni off + all notes Off ccc/cccc=0
125	Omni + all notes Off ccc/cccc=0	Omni on + all notes Off ccc/cccc=0
126	Mono + all notes Off ccc/cccc=0	Mono + all notes off ccc/cccc=M (M=MIDI Ch. 1–16)
127	Poly + all notes Off ccc/cccc=0	Poly + all notes Off ccc/cccc=0

The Tables 20.4 and 20.5 below summarise these changes.

Table 20.4 *OLD MIDI*

bbb/bbbb	Description	Functions		
		Omni	Mono	Poly
125	Omni	On	Off	Off
126	Mono	Off	On	Off
127	Poly	Off	Off	On

Table 20.5 *MIDI 1.0*

bbb/bbbb	Description	Functions		
		Omni	Mono	Poly
124	Omni off	Off	–	–
125	Omni on	On	–	–
126	Mono	–	On	Off
127	Poly	–	Off	On

Old MIDI products will ignore the codes 122, 123 and 124, so this could lead to problems unless instruments are updated. The new mode commands represent a slight simplification of the system, and allows one new feature to be introduced. This is 'local keyboard on/off', which stops MIDI information from a synthesizer's keyboard or other controller from reaching its voice generators. This means that you could play a synth from a sequencer, and simultaneously use its keyboard to control another synth without adding notes to the pattern that is playing. Sending Local Keyboard/127 reconnects the unit's own keyboard.

Real Time

There are also some units capable of reading the other main system real time code, the 'song pointer'. This code shows what pattern or measure number has been reached in a song, and when this figure is converted to a SMPTE code it is possible to lock MIDI units which give song pointers to tape machines, film and video equipment.

One very important point to note is that changes have been made to the system real time and MIDI Mode codes since MIDI was first defined. Second thoughts dictated a slight simplification of the system, and any units programmed with the original MIDI specification (OLD MIDI) may need an update to MIDI 1.0 to work efficiently with the latest equipment.

System Real Time codes consist of a single byte,

1aaa/aaaa

which can have a value from 128 to 255. Table 20.6 lists the original System Real Time codes with their updated versions.

So the new MIDI simplifies the system by throwing out 'clock in play', 'clock in stop' and 'clock plus measure end' in favour of just 'clock' and 'stop'. The 'active sensing' function is a new one – codes can be transmitted every 300 ms or so to make the receiver expect information and to go to all notes off if there is none. In other words, if the MIDI input cable is accidentally pulled out, any notes which have not been switched off will rapidly be silenced instead of droning on endlessly.

MIDI time code (formerly MSMPTE) is at present being introduced; (it was first seen on the Sequential Studio 440 sampler/sequencer/drum machine) and no doubt some other minor changes will be introduced at the same time.

But the fact that MIDI as a standard is growing does not mean that it is basically faulty. The beauty of MIDI is that it is relatively inexpensive to fit and relatively simple for the less experienced musician to use. At the same time, it is powerful enough to offer the professional a vast range of new tools which can allow him to compose and edit music and create new sounds and new techniques much more quickly and impressively.

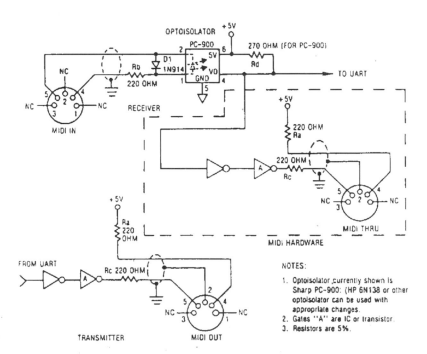

Figure 20.15   MIDI standard hardware specification.

Figure 20.16   The Fairlight Computer Musical Instrument – the original studio sampling keyboard.

Table 20.6

Value	OLD MIDI	MIDI 1.0
248	Clock in play (sent at 24 ppqn while transmitter is in play mode)	Clock
249	Measure end (sent at end of each measure instead of F8)	Unused
250	Start from first measure (sent when when play is hit on master – F8 should follow within 5 ms).	Start
251	Continue start (restarts from point when last F8 was sent)	Continue
252	Clock in stop (sent in stop to synchronise a phase locked loop for interpolating the timing clock)	Stop
253	Unused	Unused
254	Unused	Active sensing

In the immediate future it is likely that MIDI will be *de rigeur* on all new processor-controlled instruments (and will be added to many which need not be processor-controlled, such as effects units, wind instruments and guitars). The music industry can congratulate itself on the introduction of a standard which is much more universally respected than computer industry standards such as RS232 or MSX. In fact the computer industry's failure to establish the cheap MSX standard for home micros worldwide (Japan and Germany are still the main strongholds for this Z80-based system) did something of a disservice to the micro music market.

Yet the introduction of MIDI on a relatively afford-able home computer, the Atari ST, has led to its almost universal acceptance as the modern musician's micro-computer, though no other micros have been fitted with MIDI as a standard rather than an option. Music for MIDI and micros is currently a very exciting area and is likely to continue to be one for the foreseeable future.

References

Miles-Huber, D,. *The Midi Manual*. Howard Sams (1991).
Rumsey, F.J., *MIDI Systems and Control*. Focal Press (1990).
Thomas, T., *Sound Synthesis: Analogue and Digital Techniques*. TAB (1989).

21 Interconnections

Allen Mornington-West

The subject of cable connections is seldom dealt with adequately, and in some circles technology has been supplemented by what amounts to superstition. In this chapter, Allen Mornington-West strips away the fantasies and reveals the facts about cables and connectors. ·

Target and Scope of the Chapter

In this chapter we will covering some of the topics which lie behind signal interconnections. We will first brush up on the theory and properties of conductors and insulators and describe some of the typical properties and examples. We will take a brief look at the elementary physics of electronic components and discourse briefly on the topic of signal sources and their types before we arrive at a discussion on the implementation of interconnections. A look at commonly available wire and connector types and a brief discussion of their use with the cables on offer will be useful. Finally we will need to touch on the matter of wiring practice, safety and standards which may be applicable to audio equipments and installations.

We shall try to involve nothing more detailed than Ohms's law in order to make the topic readily accessible. It needs to be noted that it is certainly possible to be more rigorous (read this as 'it will involve a book-full of maths and physics') and to account for ever smaller secondary and tertiary effects. The intention is that all of the equations presented should be usable without an evaluation of j, the square root of –1, being required and they should also be able to account for as much of the audible world as that to which current research indicates we are sensible. We have also not attempted to show the derivation of the expression which have been used and the reader is thus encouraged to read more deeply in dedicated texts.

Basic Physical Background

The conduction of electricity is an electronic process when observed from a microscopic viewpoint. Earthly matter is composed of atoms whose main structure comprises a nucleus surrounded by a cloud of negatively charged lightweight particles known as electrons. The nucleus is largely made up of neutrally charged particles called neutrons bound together with positively charged particles, protons, of very similar mass. The organisation of the orbiting electrons into various orbital layers is well ordered and follows established rules whilst the distribution of the energies which individual electrons may have in each group of orbits can be described by statistical models. The outer orbit or **valence** electrons play a predominant role in electrical conduction since, while most of these outer electrons possess energies which lie in the valence energy band, some of the more energetic electrons possess energies which lie in the conduction energy band. Their behaviour in a metal is described by the Fermi-Dirac statistical distribution while the Maxwellian distribution functions describe the statistical distribution within semiconducting materials.

The distinction between conductors and insulators can be made simply by considering the ease with which charges may move within them. In conductors electrons in the conduction band of energies can be readily stripped from their attendant atoms whereas for insulators this stripping is very difficult. Substances normally considered to be conductors under nearly all circumstances include metals and strongly ionic aqueous solutions whilst examples of insulators include mica, porcelain, glass and most unmodified plastics. There is a small group of materials, known as semi-conductors, whose ability to conduct an electrical current lies between that of metals and of insulators.

A simple diagram, Fig. 21.1, shows this situation in terms of the occupancy and spacing of valence and conduction bands. Here we see that, within a conductor, the valence and conduction bands are adjacent and are not separated by an energy gap. Thus, in addition to the elec-

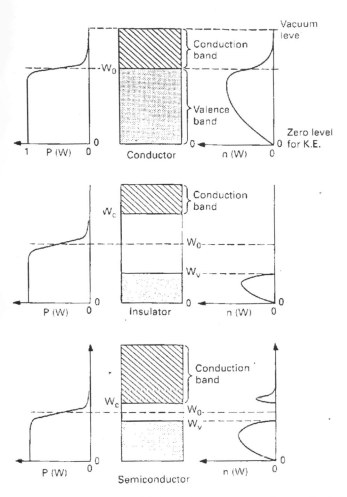

Figure 21.1 A diagram which shows the arrangement of energy levels in conductors, insulators and semiconductors. In addition the diagram shows the Fermi-Dirac probability distribution, P(W), and the electron density, n(W). Note that in a conductor there is no gap between the valence band and the conduction band. In an insulator the probability of electrons escaping the valence band to the conduction band is exceedingly small. To overcome this gap electrons must increase their mean energy. Semi-conductors show some occupation of the conduction band by electrons escaping the valence band.

trons occupying the valence band, there is a significant number of electrons with energies which lie within the conduction band. By comparison, insulators can be seen as materials in which the conduction band is separated by a large energy gap from the valence band. The probability that there will be any electrons with sufficient energy to escape across the forbidden gap and occupy the conduction band is very low, though it is finite.

Bridging these two extremes is the group of semiconductors in which it can be seen that the energy band gap between the valence and the conduction band is smaller than that of an insulator and that there is a significant probability that there will be electrons which can bridge the band gap. These electrons can take part in the conductive process. In particular this can be shown for copper, Fig. 21.2, where it can be supposed that, with many of the electron orbits overlapping, electrons are effectively able to occupy any of the intercrystalline space.

The electric charge associated with one electron is quite small (approximately $1.6 \ 10^{-19}$ coulomb) and the passage of one coulomb per second results in a current of one ampere. Some indication of the numbers involved may be useful here. A 1 metre length of copper wire 2 mm in diameter will contain of the order of 1.5×10^{22} atoms. Each copper atom will be associated with 29 electrons of which about a third may be found in the conduction band at room temperature. A current of one ampere thus involves a net drift velocity of electrons equivalent to around $100 \ \mu m \ s^{-1}$. This rate of electron movement or drift should not be confused with the rate at which an electrical disturbance is propagated since this is close to the speed of light.

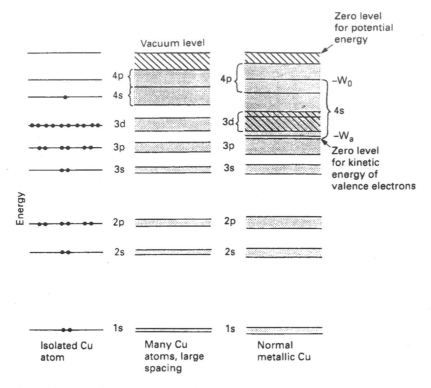

Figure 21.2   The electron energy level for copper whose atomic number, Z, is 29. The symbols 1s through to 4p refer to the names given to the different types of electron orbitals which can be occupied. At the far left the energy diagram is shown for an isolated copper atom. For a larger, though well spaced collection of copper atoms the laws of quantum physics are obeyed and the range of energies in each level increases. Electrons with an energy equivalent to the vacuum level are able to escape the copper atom nucleus. Finally, on the right, we see how the bulk properties of the metal provide a continuum of energy levels. The diagram is not to scale.

It should be recalled that within solids a current can also be carried by 'holes'. Holes are essentially the absence of electrons in the covalent bonds of crystalline substances such as germanium and silicon and they play a fundamental part in the description of the working of semi-conductors. Electrons within the body of a metal may be considered as a randomly moving mass of particles. The dynamics of a collection of such particles was originally derived through a consideration of the kinetic energies of gases and, with the proviso that the cloud of electrons is able to occupy energy levels freely, the statistics of gas dynamics can still be applied.

Thus within a conductor there will be electrons (or current carriers) which, due to thermodynamic activity, are travelling in a direction other than that of the direction of the current drift. They will collide principally with atoms in the crystal lattice and with those going in the direction of the current drift and this constant buffeting is a function of temperature. As the temperature rises there will be more of this buffeting and thus, in accord with experience, the resistance to motion will increase. Within some semiconducting materials the resistance to electron flow appears to fall since the increase in tem-

perature provides the energy to raise more electrons into the conduction band. As a consequence of this process of random activity a fundamental property of conductors is that they have a self-noise and the magnitude of this will be dependent on the temperature, the resistance of the conductor sample and the bandwidth over which the magnitude is to be measured. This is reflected in the following equation:

$$e_n = \sqrt{(4\,kTBR)}$$

where e_n = rms value of the noise
k = Boltzmann's constant, $1.38\ 10^{-23}$ W s K^{-1}
B = bandwidth in Hz
R = resistance in ohms, Ω
T = temperature in degrees Kelvin

As an example the 1 m sample of 2 mm diameter copper wire will have a resistance of around 5.4 mΩ and a self noise of around 1.3 nV rms (about −174 dBu) over a 20 kHz bandwidth.

The resistance of a metallic conductor to current flow is, provided that other environmental variables such as the temperature and pressure are held constant, not dependent on the magnitude of the current flowing through the conductor. This constancy is referred to through Ohm's law:

$$V = I R$$

where V volts is the voltage which causes a current
$\quad\quad I$ amperes to flow in a conductor of resistance
$\quad\quad R$ ohms, Ω

At the most elementary level of analysis an assumption is made that the metallic conductor is composed of a single crystal such that the dynamics of electrons and holes can be described adequately statistically. For semi-conducting materials the same assumption allows the general behaviour of intrinsic and doped (p and n type) semiconductors to be analysed. In order to appreciate the consequences of this assumption consider the size of the spacing between successive copper atoms in its face centred cubic crystal. The lattice is of the order of 361 pm (1 pm = 1 picometre = 1 millionth of 1 μm) and for successive silicon atoms 542 pm. The width of the finest conductive traces (usually aluminium which has a crystal lattice spacing of 404 pm) used in manufacturing integrated circuits (ICs) is currently of the order of 1 μm and thus each 1 μm length of such a trace comprises some 60 thousand million atoms.

Practically sized dimensions of wires tend to be much larger than these dimensions and they will be thus composed of a very large number of randomly sized and oriented crystals. The orientation will depend much on the processes which have been used to refine and form the wire. In addition practical materials will contain impurities some of which will have been added in order to modify the overall properties such as its ductility and strength. Impurities modify the crystal structure of the material even when present in very small quantities. A purity of 99.9999 per cent (or one impurity per million) will mean that even a 1 μm cube of conductor would contain 60 thousand impurities and it would be unrealistic to expect materials of this level of purity to be cheaply available. Commonly available electrolytically refined copper will be typically 99.99765 per cent pure and the main impurities will be silver, zinc, iron, oxygen and sulphur. Curiously, the most cogent reason for the use of very high purity copper and insulating materials is that after extended exposure to high temperatures (typically greater than 120° C) certain impurities, particularly the oxygen, in both the copper and the insulation material interact with the copper. Although there appears to be relatively little change in the electrical performance there is an increase in the brittleness of the conductor and if the conductor is subject to continual flexing or strong vibration there is consequently a risk of mechanical failure.

At the level of microscopic crystals there will be some degree of anisotropy and electron mobilities and scattering forces will depend on the relationship of crystal orientation with respect to the electron drift or current flow. The bulk electrical behaviour of the wire with which we are familiar is based on very dense mosaic of these randomly oriented crystals and current flow is thus effectively independent of crystal orientation. This situation is, of course modified for the materials used in semiconducting devices though it should be noted that it is difficult to manufacture reliably a semiconducting surface with an area larger than 100 mm² due to the unwanted presence of impurities.

Resistance and Electrical Effects of Current

The resistance to electron flow of a conductor arises through the collisions with other electrons and atoms in the crystal lattice. The rate of these collisions is a function of temperature. Thus most metals will increase their resistance with increasing temperature. The specific resistivity of some typical commercially pure materials is given in Table 21.1 along with the temperature coefficient. Note that the resistivity of copper changes by only 0.5 parts in 10^6 as the crystal size is reduced from 4 mm to 10 μm.

Table 21.1 *Resistivity of some common materials*

Material	Resistivity Ωm at 20°C	Temperature coefficient $\Omega\, m^{-3}$
(Elements)		
Aluminium	2.67×10^{-8}	45×10^{-4}
Copper	1.69×10^{-8}	39×10^{-4}
Gold	2.2×10^{-8}	40×10^{-4}
Iron	10.1×10^{-8}	65×10^{-4}
Lead	20.6×10^{-8}	42×10^{-4}
Silver	1.51×10^{-8}	41×10^{-4}
Tin	11.5×10^{-8}	46×10^{-4}
Silicon	600	
Germanium	0.6	
(Compounds)		
Nichrome wire	103×10^{-8}	1.0×10^{-4}
Constantan	48×10^{-8}	0.2×10^{-4}
Solder	15×10^{-8} approx	

We can calculate the resistance, in ohms, of a wire by using the following relationship:

$$R_{20} = \rho\, L/A$$

where $\quad R_{20}$ is the resistance at 20° C
$\quad\quad \rho \quad$ is the specific resistivity
$\quad\quad L \quad$ is the length of the wire in metres
$\quad\quad A \quad$ is the area of the wire in square metres

The resistance at a temperature other than the reference temperature of the resistivity can be approximated from:

$$R_t = R_{20}(1 + \alpha(T - 20))$$

where R_t is the resistance at the temperature
         T is in degrees C
         α is the temperature coefficient of resistance

The resistance change with temperature is not exactly a linear effect though over a limited range of 30° C the secondary effects can be certainly neglected for audio signals.

The passage of the electric current will create a heating effect in the wire which is the product of the current in the wire and the potential difference between its ends. The increase in the temperature of the wire and its rate of increase are dependent on several factors including the nature of the surrounding material or insulation and thermal conductivity to the surrounding air. A current of one amp in a copper wire with a diameter of 2 mm will cause a dissipation of 5.4 mW per metre. If a thermal resistance to the ambient air of 5° C per watt is assumed then the core of the wire will increase in temperature by about 27 m°C (27 thousandths of a degree Celsius) and a result in a change in the resistance of about 1000 ppm or 5.6 μΩ per metre. Since this rise may have a time constant of the order of 200 s we may safely conclude that the heating effect is unlikely to give rise to an audible effect due to the modulation of the resistance.

There are other effects of electrical current and metallic contact which we should not forget for the purposes of a discussion on interconnections and cables. Partly this is for completeness but also because, at various times, these effects have been invoked as being responsible for deteriorations in audio quality. It is instructive to assess the validity of the claims after some assessment of the likely magnitude of the claimed effect has been made.

The Hall effect occurs in both conductors and semiconductors and it is the creation of an electric field at right angles to both the current flow and a magnetic field. Figure 21.3 shows the outline of the phenomenon. For metals such as copper the effect is minute and a one amp current flowing in a wire of 2 mm diameter placed at right angles to the earth's magnetic field will develop a Hall effect field of around 315 pV. The effect is sufficiently pronounced in appropriately doped semiconductors (it is about 100 million times greater) for it to be useful in measuring magnetic fields and controlling motors. Its magnitude can be compared to the voltage levels required in loudspeaker cables which are in the region of ten thousand million times greater (approximately 190 dB greater).

Contact potential effects occur when two dissimilar materials are brought into intimate electrical contact. In atomic terms a free exchange of valence electrons can occur between the atoms of each material in the region of

the contact and an electrical potential difference will exist. This difference cannot be measured directly since attempts to do so will yield the thermal Seebeck potential at the junction. Contacts between conductors result in differences in the energies of electrons in the conduction band giving rise to a thermally driven voltage effect known as the Seebeck effect (and its companion Peltier effect). Within the microcrystalline environment of a metal wire there will be many such randomly oriented junctions producing the net effect of a homogeneous material.

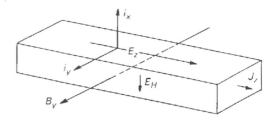

Figure 21.3   A diagram showing the arrangement of currents, voltages and magnetic fields for the purpose of understanding the Hall effect. B_y is the applied magnetic field, the applied current is shown as a current density vector, J_z, and resulting Hall effect voltage is E_H.

Even in contact between conductors and insulators the penetration of electrons extends over several nm (10^{-9} metre) and is responsible for the triboelectric effects which can be observed when one surface is moved over the other. The magnitude of the Seebeck effect potential depends on the materials which comprise the junction and their temperature difference. Between the materials commonly used as conductors the effect can be modelled as a simple very low voltage offset in which there is no directional dependency. It is important to recognise that this directional independency is valuable since it would otherwise prove arbitrarily difficult to produce distortionless connections and circuitry. The simple reminder of this point is that it is possible to carry out distortion measurements at audio frequencies using very simple, readily available test equipment down to levels of at least one million to one. Thus, despite the very large number of interconnecting metallic surfaces involved in making the measurement there is no measurable effect within the resolution of the test equipment. The one area where this effect may be of concern is in techniques for measuring very low DC voltage levels. This is more a matter of circuitry design technique for engineers involved in interfacing to low frequency sensors, such as seismic and strain gauges, with bandwidths from ZF (zero frequency) to as high as 100 Hz, than to interconnections for audio signals with bandwidths from 20 Hz to 20 kHz. For completeness Table 21.2 shows some of the common conductor combinations and their Seebeck potential effect values.

Table 21.2 *Thermoelectric coefficients of some common conductor combinations*

Conductor combination	Thermoelectric coefficient voltage per °C ref. cold junction at 0°C
Copper/nickel	+22.4 µV
Copper/gold	+200 µV
Copper/silver	−200 µV
Copper/lead	−3.2 µV
Copper/tin	−3.4 µV
Copper/brass	−1.6 µV
Chromel/alumel	40 µV

This last combination is used for temperature measurements.

Capacitive Effects

The electrons which carry the electric current are also responsible for electrostatic phenomena. You'll recall that the quantity of charge carried by each electron, or by a hole, is 1.6 10^{-19} coulomb. The study of electrostatics gives us analytical access to capacitance and electrostatic screening. From an electrostatic viewpoint the surface of a conductor has to be an equipotential surface and the whole of its interior will thus be at the same potential. The static charge on a conductor resides entirely on its outer surface. Capacitance is the ratio of electrical charge to voltage. The simplest form of a capacitor takes the form of two parallel plates of area A and uniform separation d, Fig. 21.4:

$$C = e_0 kA/d \text{ farads}$$

where
A is the area of plates in m^2
d is their separation in metres
k is the dielectric constant for the material separating the plates (no units)
e_0 is the primary electric constant also known as the permittivity of free space, equal to 8.854×10^{-12}, the units being farads per metre, F/m

Two further arrangements of conductors are of interest. The first is the arrangement known as the coaxial cable, Fig. 21.5:

$$C = e_0 2\pi k /(\ln(b/a))$$

where
b is the inner radius of the outer conductor in metres
a is the radius of the inner conductor also in metres
k is the dielectric constant of the insulating material between inner and outer conductors (no units)

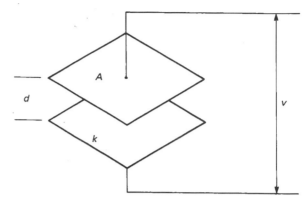

Figure 21.4 The general arrangement for a plate capacitor where the separation between two parallel plates is d, the area is A and the voltage applied to the plates is v.

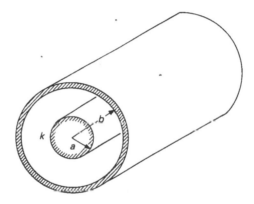

Figure 21.5 The general arrangement for a coaxial cable type of capacitor. The inner wire or conductor has a radius of a and the inner radius of the other conductor is b. The medium between the inner and outer conductors will have a permittivity of k.

The simple coaxial cable introduces the use of the outer conductor as a screen. This screen will be effective in removing the effects of unwanted electrostatic fields though not necessarily effective at screening all electromagnetic phenomena.

The second arrangement is the capacitance between two parallel wires, Fig. 21.6:

$$C = e_0 \pi k /(\ln(d/a))$$

where d is the separation between two conductors
a is the radius of each conductor in metres

The effect of the material between the conductors of a capacitance is to increase the charge which may be held on the conductors. The conventional model for this behaviour considers that the insulating material is put under an electrical strain by the field between the conductors. Insulators used in this way may not behave linearly with either the magnitude or with frequency of the applied voltage. In addition the phenomena of dielectric loss and dielectric storage or absorption occur. Dielectric loss can show a complex relationship with operating frequency and this is due to the work carried out on changing the direction of electric stress in the insulator.

Figure 21.6   The general arrangement for analysing the capacitance between two wires. The wires have a radius of *a* and are separated by *d*. The medium between the wires will have a permittivity of *k*.

Different dielectric materials behave differently with mica and PTFE having a low loss and materials used in electrolytic capacitors showing a higher loss. This loss is referred to as the power factor ($\tan \delta$) of the material at a given operating frequency. From an electronic point of view dielectric absorption can be modelled as a low level of electrical charge which is returned to the attached electrical circuit after the initial charge has been removed. Equivalent circuits of practical capacitors in which these deviations from the ideal capacitor can be reflected can be devised and they are of great practical use to the circuit designer. The relevance to audio interconnections lies mostly in the fact that conductors are kept apart from each other by insulators which will have significant dielectric constants and, potentially, dielectric loss. The performance of some common materials and the frequency at which they were measured are listed in Table 21.3.

Table 21.3   *Dielectric constants and loss angles for some common insulating materials*

Material	κ	at frequency	$\tan \delta$
Porcelain	7	1 MHz	70×10^{-4}
Quartz	3.8	100 MHz	2×10^{-4}
Mica	7	100 MHz	2×10^{-4}
Perspex	2.6	1 MHz	160×10^{-4}
PTFE	2.12	3 GHz	2×10^{-4}
Polythene	2.3	1 MHz	2×10^{-4}
PVC	4	1 MHz	600×10^{-4}
Silicone rubber	8.5	1 MHz	10×10^{-4}

It would be useful to estimate the effect of such losses on conventional audio cabling. The effect of the losses can be modelled by considering the nominally ideal capacitance as paralleled by a suitably valued resistance. Assume that the cable has a PVC covering and a nominal

capacitance of around 80 pF per metre, then at 20 kHz the effect of the dielectric loss is equivalent to driving a 1.65 GΩ resistor. This is the kind of loss which can be neglected readily in audio interconnections. However, where long lengths of cables are used for carrying digital audio signals, with their much higher bandwidth, the effect may start to be significant. One of the available cures is obviously to use a lower loss dielectric such as polythene.

It will be useful to recall that triboelectric effects can arise when conductors and insulators move relative to each other. The movement causes electrons in the insulator to be stripped from their position in the insulator and they will be able to join the pool of electrons which exist in the contacting conductor or, alternatively, they can be left behind in another insulator. In general the insulating materials involved have a high resistance and the currents caused by this effect can never be high. However the effect can be the cause of problems in circuits where the impedances are high (typically greater than 1 MΩ) and the use of special 'low-noise' cables are to be recommended. It might be thought that the cables usually used in interconnections would thus give rise to a form of audible distortion as a consequence of being inevitably placed in an audio sound field. However the magnitude of this effect is exceedingly small and even high sound pressure levels in the vicinity of a large amount of cable are unlikely to produce a triboelectric current greater that –180 dB of the current required to make the high sound pressures in the first place.

Magnetic Effects

We now need to bring to mind some of the magnetic consequences of current flow. The passage of a current produces a magnetic field and a magnetic field predicates the existence of a moving charge (philosophers may argue about the best way of stating this and we might be wise to permit them this liberty). In the coarse world of electronic components this consideration gives rise to the bulk visible effects of a magnetic field and the definition of inductance. The alternative viewpoint recognises that, together, the electric and magnetic fields constitute an electromagnetic wave. This second approach leads to the concept of the transmission line, a topic which is often not understood properly in audio applications.

For a long thin wire the magnetic field at a distance *r* from the wire is given by:

$$B = n_0 I / (2\pi r)$$

where B is the resulting flux in webers/m² (called tesla)

n_0 is the primary magnetic constant also known as the permeability of free space, equal to 1.257×10^{-6} in units of Henry per metre, H/m and

I   is the current in ampheres

The magnitude of this field or flux is dependent on the organisation of the wiring such that, for example, the field in a closely wound toroidal coil with N turns will be:

$$B = n_0 NI/(2\pi r)$$

where N is the number of turns on the ring and
        r is the mean radius of the ring

An important feature of such a ring is that, provided that the windings are closely spaced and even, the external field is very low. The field within the coil is much increased when the coil is wound around a magnetic material such as a ferrite or iron core and it is increased by the factor μ, the permeability of the core material. Magnetic core materials do not exhibit linearity over a wide range of operating fluxes and the effective value of μ for a given material depends on the flux level at which it is being operated (Fig. 21.7). Additionally all magnetic materials exhibit hysteresis and thus energy loss when they are subjected to alternating magnetic fields.

Materials whose relative permeability is essentially unity include the common highly conductive materials such as aluminium, copper and silver as well as the eutectic alloys employed in soldering. Common materials whose permeability is much greater than unity are based on iron and nickel. Mumetal has an initial permeability in the region of 20000 and is a mixture containing 76 per cent nickel, 17 per cent iron, 5 per cent copper and 2 per cent chromium. The steels used for transformer laminations have lower initial permeabilities (9000) but can withstand much higher flux levels and are based on a mixture of iron and between 4 per cent to 5 per cent of silicon. Thus mumetal, with its high initial permeability, is the preferred material for making magnetic screens for such items as small signal transformers.

The inductance of these simple arrangements is of interest since we can take the idea of the magnetic field or flux and recognise that it can be coupled into a second separate circuit and this is the basis of operation of transformers.

For a single conducting wire (we assume that the return current path is carried by another wire which is very much further away than the diameter of the wire) the internal self-inductance is 0.05 mH per metre of conductor and it is independent of the diameter of the conductor. More usually the return current is carried by a similarly sized conductor which runs close typically within five times the conductor diameter. Under this condition the self-inductance of the circuit is (Fig. 21.8):

$$L = n_0/\pi(\ln (b/a) + 0.25)$$

where L is the inductance in henries per metre
        a is the radius, in metres, of the conductors and
        b is their separation also in metres

Figure 21.8   Diagram showing the arrangement for analysing the magnetic field between two parallel wires of radius a and distance between centres of b.

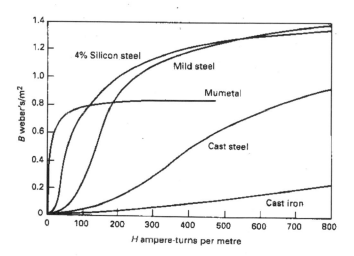

Figure 21.7   This set of graphs shows the effect of permeability of some common materials. The applied field is H and the resulting flux is B.

This relationship is derived without consideration of the effect of the flux at higher operating frequencies within the individual conductors but it will be valid for low frequencies and it does depend a little on the conductor shape. For example, the self-inductance of a parallel arrangement of strip conductors width 2a and separation 2b (in metres as ever) is given approximately by:

$$L_{self} = n_0/2\pi(\ln(b/a) + 1.5) \qquad \text{Henry per metre}$$

This is made a little more complicated when AC currents are involved because the magnetic flux which is created by the current will set up eddy currents in the conductive core of the cable. These eddy currents constrain the current to be restricted increasingly to the outer surface layers, or skin, of the conductor as the frequency is raised. Thus at high frequencies the resistance of a conductor will rise. We define the skin depth as that depth where the magnitude of the flux density has fallen to a fraction:

$$\sqrt{2}/e = 0.52026$$

(where e is the base of natural logarithms) and is equal to 2.71828 of its value at the surface or at ZF.

This depth is given by:

$$d = \sqrt{\frac{\rho}{\mu_0\mu_{rel}\omega}}$$

where d is the skin depth in metres
   ρ is the specific resistivity of the conductor
   μ_0 is the primary magnetic constant 1.257×10^{-6}
   μ_{rel} is the relative magnetic permeability of the conductor
   ω is the angular frequency ($\omega = 2\pi f$)

As an example this penetration depth is 6.6 mm for copper and 0.5 mm for iron at 50 Hz. The reader may recognise these dimensions as comparable to those used for copper bus bars in high current installations and for the thickness of laminations used in mains transformers and motors. In the case of wires which are essentially of circular cross-section we can define the net resistance at frequencies which are high enough such that the resistance performance is dominated by the skin effect as:

$$R = R_0/a/((2\sqrt{2})\,d)$$

where R_0 is the resistance of the sample at ZF
   a is the radius of the wire and
   d is the skin depth or flux penetration depth given above

Figure 21.9 shows how this relationship gives an acceptably accurate result when the ratio of a/d exceeds 3. Below this ratio the resistance of the wire is within 20 per cent

of its value at ZF. There are many circumstances where it is desirable to reduce the skin effect and this requires increasing the surface area available for conduction for a given amount of conductor. This can be achieved by using a larger diameter conductor recognising, ultimately, that the best form for this conductor will be hollow. The other common approach requires making up a bundle of very fine wires each of which are individually insulated and this is the basis for a wire known as Litz wire. A thin walled (walls of 0.13 mm) copper tube of 8 mm diameter will have the same area as the 2 mm diameter wire considered so far and it would have the same resistance at ZF but would spread the skin effect resistance over four times the surface area whilst the same performance would be obtained with Litz wire comprised of 16 individual cores each of 0.25 mm radius. Either approach is sometimes used in radio frequency (RF) circuit design but its application to baseband audio is arguably specious. Eddy current effects and the resulting skin effect resistance also occur within the pole systems of conventional loudspeakers and can be taken into consideration in exceptional driver designs.

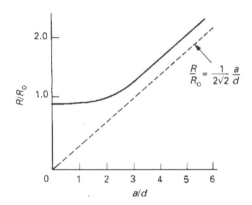

Figure 21.9 The relationship between skin effect depth and the wire diameter where a/d is the ratio of wire diameter to skin depth and the ratio R/R_0 is the resulting ratio of resistance at ZF to overall resistance at the frequency being considered. The derivation of this curve is quite a complex matter.

The total current lags the surface current by 45°, so that an expression for the internal self-inductance of a wire can be derived as:

$$L_i = \rho/2\sqrt{2}\pi ad\omega$$

in which d is proportional to the root of angular frequency, so that L_i tends to zero as the frequency is raised (and the skin that carries the current becomes thinner). Because of this, at high frequencies the reactance of the internal self-inductance approaches the AC value of resistance, and this value is independent of the material, making properties such as oxygen content, zinc content, boron doping *et al.* totally irrelevant.

Another aspect that can be ignored is the proximity effect, caused when the magnetic field of one wire affects the distribution of current in a nearby wire. Analysing this and estimating for 2.53 mm² copper wires at 20 kHz, separated by 10 mm gives a ratio of AC to DC resistance of 1.0086, so that the effect is of little significance in loudspeaker leads. The proximity effect is, however, significant in wound components such as transformers.

Characteristic Impedance

The electrical audio interconnection can be analysed either in terms of its lumped components or in terms of its behaviour as a transmission line. The lumped components approach uses the values of the cable's series resistance, inductance and shunt capacitance which have been given so far, groups them together to model the cable and, further, makes the assumption that the length of cable involved in carrying the signal is very much shorter than the wavelength of any signal being handled by the cable. This will be a very valid assumption in most installations since, for a sinewave of 20 kHz, a wavelength of the order of 15 km can be expected. An advantage of looking at the cable as a transmission line, where this is appropriate, is that the final calculations are simpler since the model can incorporate of the effect of field losses. The lumped component model offers the advantage that it is easier to appreciate the individual physical mechanisms which give rise to the losses, though evaluating it at any arbitrary frequency becomes a chore.

We can calculate the characteristic impedance of a cable, Z_{ch}, from its lumped component values:

$$Z_{ch} = \sqrt{(L_{self}/C_{self})}$$

where L_{self} is the cable's self-inductance per metre
           C_{self} is the cable's self-capacitance per metre

and its companion, the velocity of propagation (or the rate at which an electromagnetic disturbance travels) down the interconnect cable:

$$V_{ch} = 1/\sqrt{(L_{self} \times C_{self})}$$

In both cases, we will assume that, at high frequencies, the reactance of the self-inductance dominates the resistance component and that there is negligible shunt resistive loss in the cable. Thus there is an additional reason in not treating the loudspeaker cabling as a transmission line since, in general, the characteristic impedance of typical interconnects is usually at least one order of magnitude greater than the impedance offered by the combination of the loudspeaker, crossover and air-loading system. As a consequence for all practical purposes

loudspeaker cable will appear as an inductance in series with the loudspeaker system.

For operators of systems where the audio interconnect length is indeed long then there are real reasons for considering the transmission line approach to interconnecting. The operators of telephone systems have long known this and the approach involved has thus been historically linked firstly to broadcasting installations and thence the professional audio users where the phrase '600 Ω line' is encountered. However the 600 ohm line philosophy is not a panacea and, for many environments, it is an operational state worth avoiding.

Whilst baseband audio signals are still so common it is easy to forget the existence of both audio in a digital form and of digital control signals in complex systems such as mixing desks. Digitised audio signals occupy a band up to 3 MHz for a single AES/EBU channel. It is thus quite appropriate that the sending, receiving and transmission impedances have been well defined for both the domestic (IEC958) and the professional (AES-3-1992) standards. For domestic installations a 75 Ω system is defined whilst a 110 Ω system is used for professional systems. For the most reliable results connections should be made to input and output connectors using the proper connector style and a cable of the correct impedance. An example of a problem which can arise in audio equipment intended for domestic use is the apparent failure of a digital audio interlink. This may be blamed on the cable for the mythical reason that it does not sound 'musical'. A more honest appraisal of the situation may reveal that there was significant distortion of the digital waveform taking place due to the reflections of the digital signal at various impedance mismatches. A companion example can occur in professional audio mixing desks where the need to control upwards of 100 channels of analogue processing exists. In this particular application eighty faders had been unevenly distributed along a serial digital control bus some 30 m in length. Although standard ribbon multiway cable has a characteristic impedance of approximately 110 Ω, and could be correctly terminated thus giving a fair waveform, the failure to recognise the individual reflected contributions from each of the faders meant that the passage of data was not reliable.

In fact, the need to consider wire and cable connections crops up surprisingly often. If you are analysing system performance in the frequency domain, consider using the transmission line approach when the length of the circuit represents around a fifteenth of the wavelength of the highest significant frequency. The alternative analytical approach, working in the time domain, suggests that if the time required for the signal to travel the length of the conductor is more than an eighth of the signal rise time you should also consider the system as better described by transmission line theory. In addition to these simple rules consider those below which may help you to decide which approach to use for borderline cases. Bear in mind that different logic families switch at different speeds.

If $R_s > Z_{ch}$: consider the cable's capacitance from the driver's point of view.

If $R_1 < Z_{ch}$: consider the cable's inductance from the load's point of view

where R_s is the source or driver resistance
              R_1 is the load or sink resistance

For the sake of consistency, however, the 1 m length of 2 mm diameter copper wire which we have considered so far can be revalued. Consider two such lengths, a send and return current path, with their centres placed 6 mm apart, each covered with a PVC insulation and evaluated at 10 kHz.

wire diameter	2 mm
copper resistivity	1.7×10^{-8} Ωm^{-3}
vinyl coating dielectric constant	5
conductor square area	3.14 mm^2
resistance of 1 m at ZF	10.8 mΩ
self-inductance	439 nH
skin depth	0.463 mm
approx overall resistance at 10 kHz	13.04 mΩ
capacitance	126 pF
Z_{ch}	58.9 Ω
V_{ch}	1.3×10^8 ms^{-1}

It is worth noting that the existence of a very thin coating (approximately 0.05 mm) of tinning or solder over the wire does not materially affect the resistance nor does become significantly involved in the skin effect. The velocity of light in free space is around 3×10^8 ms^{-1}.

We have revised the background concerning the essential processes of conduction, electrostatic charges and the magnetic effects of current as they may affect the topic of interconnections. Before we can tackle some of the problem areas for audio signals we will need to recall some simple alternating current (AC) background.

Reactive Components

There are three linear passive electrical components with which we will be concerned and these are the resistor, the capacitor and the inductor. At low frequencies, for example 100 Hz, most components can be treated as ideal and we can represent them on circuit diagrams by the conventional simple symbols. Thus a resistor will have a behaviour which can be considered independent

of frequency whilst the capacitor and inductor will both exhibit a reactance:

$$X_c = 1/(2\pi f C) \qquad\qquad X_1 = 2\pi f L$$

where C is the value of the capacitance in farads
              L is the value of the inductance in henrys and
              X_c and X_1 will have the units of Ω
              f is the frequency in Hz

Non-ideal behaviour of resistive and capacitive components employed in the input and output stages of audio equipment is unlikely to be a cause of concern. To a first approximation typical 0.25 watt resistors may be simply modelled, Fig. 21.10, and capacitors can be modelled similarly, Fig. 21.11. Within the design of audio equipments it may be necessary to recognise these non-ideal behaviours and this may be particularly so for capacitors used in sample and hold amplifiers. The effect of dielectric absorption in the large values of capacitor used within conventional passive crossover networks can similarly be neglected within the band of audio signal frequencies. This can be appreciated through considering a more developed model for such types of capacitor in which the model of Fig. 21.11 is shunted by a series resistor and capacitor network. The values of these series components will depend on the materials used to form the capacitor but for polyester the series resistor will typically be more than 10 MΩ and the series capacitance value about 1/30 of the main value. For a capacitor with a nominal value of 10 µF this gives a time constant of 50 minutes, a value which may be safely considered irrelevant for audio purposes. In passing we may note that over-enthusiastic application of voltage may cause damage to resistors (primarily through over-heating) and to capacitors (by causing breakdown of the dielectric) and that before these effects produce some observable catastrophe the individual components will usually have ceased to perform linearly. We can thus appreciate that the simpler functions carried out by short lengths of cables are not going to be the cause of significant audible defect.

Combinations of R and the reactive components, L and C, will in general show a resistance which depends on frequency. This is given the name of impedance and the use of the term serves to remind us that the impedance is accompanied by a phase difference. We do not intend to cover the aspect of phase in this text but the reader might note that the small amounts of phase shift (less than 45°) associated with the small amplitude errors (typically up to 3 dB) which we have noted in this discussion have not been shown to have a directly audible consequence. Far greater amplitude and phase errors (more than 720°) can result from limitations in loudspeaker driver and crossover design and implementation. Where these errors occur in the mid-frequency band for audio signals,

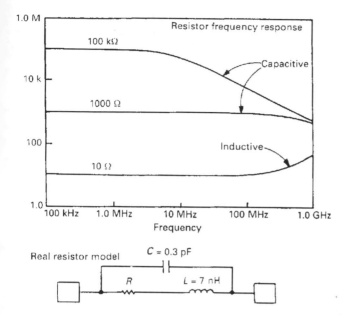

Figure 21.10 A model of a real resistor. The idealised resistor has no associated parasitic components. In this model the resistor is considered as being paralleled by a 0.3 pF capacitor and has approximately 7 nH of lead inductance. Real resistors are also subject to a very small extent tó the effect of applied voltage but extremely sensitive equipment is needed to measure the resulting harmonic distortion components since these are often 140 dB lower than the fundamental.

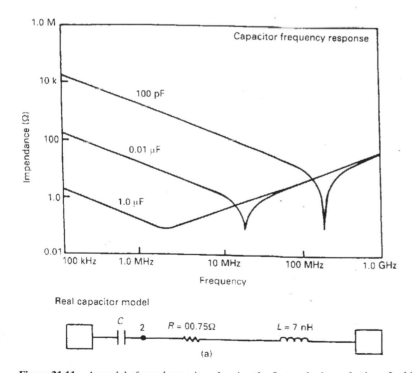

Figure 21.11 A model of a real capacitor showing the first-order imperfections. In this case the main imperfections can be modelled as a series resistance of approximately 75 m Ω and a lead inductance of 7 nH. Note that one effect of the inductance and capacitance is a series resonance. For very high values of capacitance this can fall in the audio bandwidth but this can only occur when the impedance presented by the capacitor is at an exceedingly low value. Certain types of capacitor, notably ceramic capacitors using high permittivity dielectrics can have their capacitance modified by the applied voltage and for this reason their use in the audio signal path is often avoided in quality audio designs.

around 4 kHz, they may produce an audible artefact for some listeners. Thus, while the amplitude errors associated with poor interconnections may give rise to some audible defect the accompanying phase changes are of no great significance.

The input impedances of audio equipment can usually be modelled simply as an input resistance with a small value of stray (or sometimes intentional) capacitance in parallel whilst the output impedances are often resistive accompanied·by a small amount of series inductance. Hardly any audio transducers are intrinsically entirely resistive (the use of resistive strain gauges as a vibration pick-up would be an example of one). Considerable skill is exercised by some transducer system designers in order to arrange for almost purely resistive impedance over most of the audio band and this is exhibited by many microphones and loudspeaker systems. Capacitive audio transducers do exist and examples include piezo pick-ups for musical instruments, capacitor microphones, piezo pick-up cartridges and a variety of positional and motional sensors.

For the purposes of interconnections the deviations from the ideal in individual resistive and capacitive components encountered at the input and output of equipments can be neglected since they are very small. The behaviour of inductors is slightly different because it is far from easy to design inductors which can be reliably modelled purely as simple inductors. Inductive signal sources include guitar pick-ups, dynamic microphones and vinyl record pick cartridges and the end target of audio power amplification is almost certainly the moving coil loudspeaker and its attendant crossover network. In addition connections between equipments used in professional environments are often accomplished using transformers for both input and output.

At its simplest an inductor has to be modelled as an ideal inductor associated with the series resistance of its windings and with an internal shunt capacitance which is very much a function of the individual construction (Table 21.4).

Table 21.4   *Characteristics of some common audio signal sources*

Type of inductance	Typical values for		
	L	*C*	*R*
Moving iron pick-up	600 mH	10 nF	1 kΩ
Moving coil pick-up	1 mH	200 pF	6 kΩ
Guitar pick-up	7 H	5 nF	10 kΩ
Mic transformer 1:7 ratio	1 H	1 nF	1 kΩ
Line input transformer	5 H	1 nF	1 kΩ
Line output transformer	500 mH	1 nF	75 Ω
Loudspeaker voice coil	500 mH	100 pF	8 Ω
Loudspeaker crossover coil	20 mH	300 pF	0.5 Ω

This arrangement of *L*, *C* and *R* will have a self-resonant frequency and an accompanying *Q*-factor which indicates the degree to which the arrangement is resonant. It will become clear that in certain cases poor interconnection practice may be justifiably understood to give rise to some auditive defect and that this will be largely due to a failure to take into consideration the relative nature of the source of the signal and its destination or load. In addition it might be due to the inappropriate use of connectors and the quality of the connection.

Interconnection Technologies

The problems of interconnections for audio purposes and the effects that should be considered can be broken down into three areas. In the first we will consider the connection of an input transducer to the first item of equipment, in the second group we will consider the interconnection of signal handling equipments and, in the third, we will consider the connection to the output transducer. The correct use of interconnection technology is as important within equipments for entirely similar reasons.

In the first we will be concerned with handling signals whose amplitude will range from around –120 dBu up to around 0 dBu. The term dBu is used to indicate a dB ratio with respect to 0.7745 V. Typical signal sources will include microphones, moving coil pickup systems and musical instrument pick-ups of various designs. Typical source impedances range from 30 Ω mostly resistive for ribbon microphone types, to inductive sources such as guitar pick-ups and conventional vinyl disc cartridges and, finally, to almost pure capacitive sources, which can be found in musical instrument transducers. It is fair to say the designers of microphone systems have, by incorporating appropriate electrical and acoustical loading of the raw transducer, arranged to provide for a largely resistive source. Designers of capacitor microphones, whether of the pre-polarised kind found in tie clip styles or of the bulkier studio microphone which requires an external polarising voltage, have nearly always incorporated a very high input impedance buffer amplifier very close to the diaphragm and thus can engineer a substantially resistive output impedance.

The second group will consider the transport of signals which lie in the region of –20 dBu to +20 dBu. Typically these signal sources will come from essentially resistive sources and will terminate in essentially resistive sinks but the power level will still be low and the maximum power involved is typically 10 mW. Note that it is common, particularly in professional systems, to operate with signals using balanced connections and that this is often achieved by using transformers whose non-ideal behaviour needs to be taken into account.

The final group is concerned with the transport of power and, in keeping a broad outlook, this group should include the range from the connection of mains power, the connection of power supplies and the connection of loudspeaker systems to power amplifiers. Power levels up to 5 kW can be involved and the small amount of the losses which associated with cables may become significant. With this wide range of signal and power levels to be handled it will not be surprising to note that there is a wide range of connector styles in current use. There is thus also a companion selection of cables which can be used with such connectors.

A simple classification of cables available to the audio engineer would cover unscreened, screened and power cable types. There is a very large range of such cables readily available and, in addition, manufacturers are usually able to make up specially designed cables in surprisingly short lengths.

Unscreened cables

The insulation materials most commonly used are based on one of the many PVC formulations. Higher temperature working is offered by plastics such as silicone rubber, kaptan and PTFE all at a much higher cost than PVC. Recent standards of fire safety are requiring the use of low smoke and fume (LSF) insulations especially where such cabling carries voice evacuation signals. Almost universally the conductor used will be copper in one form or another. Stranded copper is much easier to handle than is solid core copper of the same cross sectional area. Single copper cores larger than about 0.4 mm diameter also tend to be very intolerant to repeated flexing. Even a single sharp bend may create sufficient damage to restrict the current flow at the point of the bend.

Wire diameters are almost universally used in parts of the world where the metric system is in full use. In the USA wires are still reckoned by Standard American Wire Gage (AWG) which is similar to the old Imperial Standard Wire Gauge (SWG). It is usual to refer to wires by the number of strands and the diameter of each strand (see Table 21.5). In electrical installation work reference is usually made to the conductive cross-sectional area in mm². For electronic equipment practice it is usual for each strand to have been given a very thin coating of solder, a finish known as tinning. This coating has no significant effect on the electrical performance but it does make it possible for the wire to be readily soldered and it also protects, to a limited degree, the surface of the copper core from being attacked by atmospherically borne acids. Other variants on tinning include plating by

precious metals such as silver. This, as we have disclosed above, will have negligible effect on the conductivity at audio frequencies but it adds considerably to the cost. Braided forms of cabling are available and are usually used where some flexibility is required and high currents need to carried. They can often be found as part of the earth bonding system in equipment racks.

Note that the total circuit resistance must allow for the return path. The indicated maximum current rating assumes a single layer of insulation and it can be achieved when the cable has free access to the ambient air.

Table 21.5 *Unit circuit resistance and current rating of some common wire sizes*

Stranded wire size	Area mm²	Resistance mΩ/m	Current rating amps	Approx. size SWG	AWG
7/0.12	0.08	230	1.0	30	28
16/0.2	0.5	38	4.0	22	20
24/0.2	0.75	25	6.0	20	18
32/0.2	1.0	19	10.0	18	16
48/0.2	1.5	12	16.0	17	15
64/0.2	2.0	9.5	20.0	16	14
78/0.2	2.5	7.6	25.0	14	12
192/0.2	6.0	3.1	48.0	11	9
256/0.2	8.0	2.3	80.0	10	8
†24*12/0.2	9.3	2.0	90.0		
196/0.4	25.0	0.76	130.0		

† braided cable

Single unscreened wires are used to make connections with mains, power supplies and power transducers. Note that the send and return wires should be kept as close as possible in order to avoid being sensitive to, or give rise to, electromagnetic compatibility (EMC) problems. The larger the separation between the cables the larger the inductance. It is often advantageous to twist wire pairs together.

Multicore unscreened wires are found in a variety of guises and they can be used where signals are not much affected by, nor give rise to, an EMC problem. Examples include power cables and, very commonly, multicores used in telephone systems. Ribbon cables are often used within equipments and for digital interconnections. Note that EMC considerations may force the use of screened ribbon cables external to equipment. Unscreened aerial feeder is usually arranged to have a characteristic impedance of 300 ohms and although versions intended for audio loudspeaker cable look similar they can be manufactured to have a characteristic impedance of around 8 ohms.

Screened cables

Of the screened types of cable the most common form of screening uses a single layer of copper wire braid. The screening offered is usually more than adequate for most audio purposes. Two other ways of forming the screening can be met commonly. In the first the screening braid covers a thin layer of highly conductive plastic. This construction is often used in cables which need to show a low amount of triboelectric behaviour. The second replaces the braid by an aluminium foil wrapper which is usually accompanied by a 'drain' wire which should be used to assist connection. This format can offer improved screening since the electrical cover is clearly more complete. Where the impedance performance is required to be very consistent a double braided screen cable may be used but its use is, in general restricted to video circuits. Note that where precision coaxial cables are required that the performance can be prejudiced if the cable is kinked or bent through too tight a radius. This is because this mechanical deformation causes a local change in the characteristic impedance which may lead to distortion of high frequency pulse shapes.

For most modern cable types the outer insulation will usually be one of the PVC family. However greater flexibility over a wider temperature range is offered by silicone rubbers. Single core screened cable is available in a wide range of designs. In order to determine the correct selection you'll need to consider a list of characteristics which include the screening performance, the characteristic impedance, the mechanical strength, environmental limitations such as oil, water, radiation, dust and temperature. The insulation between the inner core and the outer braided sheath depends on the signals the cable is designed to carry. Most commonly this will be simply a PVC insulation similar to that used in conventional single unscreened cables. For applications where the centre conductor must be accurately held at the centre of the braided tube and the characteristic impedance must be uniform along the conductor length a core of polythene or teflon is used. The control of impedance is important for the digital audio signals since a cable with a uniform characteristic is much less likely to suffer from intersymbol interference.

Cables in this group of single core screened cables will be commonly used in interconnecting domestic audio systems where the cost of operating a properly balanced port is not justified.

Twin core screened cable is a very important category since low level signals are very commonly handled using balanced inputs and outputs in professional audio systems. Although mineral insulated cable (MICC) was not intended for audio work it is sometimes required. Since copper has a melting point around 1083° C such cables

can find use in installations where robustness in the face of fire is required. Typical examples include installations of voice evacuation systems where the sound and signalling system integrity requirements are set by BS6527 and BS7443.

It seems obvious that any installation of a cable system should be planned and that a consistent cable marking philosophy should be a part of the installation. There is a wide selection of cable marking and numbering systems available currently and, although their use during an original installation may seem a great way of absorbing time, this time is more than readily saved on the first occasion when the cabling must be modified. The wise engineer learns to make sure that the wiring schedules, charts and diagrams are kept in step with any modification.

For external installations a suitable alternative may be the use of steel reinforced cables. The outer steel braid is, however, not intended to be used primarily as a coaxial shield. Multicore screened cables come in a variety of forms from multiple single conductors and twisted pairs with an overall braided screen to complex mixtures of screened twin pairs, single conductors and multiple coaxial cables intended for RF or video signals such as might be used in a SCART connector.

The usage of cables

There are two ways in which an audio or video signal can be carried by a cable. It can either be single-ended (also known as unbalanced) or it can be balanced. A single-ended signal can be carried by a simple coax or single screened type of cable whilst the balanced signal requires twin conductors which are usually within an overall screen. The simpler scheme is most commonly used in domestic audio and semi-professional equipments. It becomes important to ensure a uniform earthing policy with such equipment. Balanced circuitry is more expensive to implement. In its simplest form it requires the use of transformers with their attendant limitations. More commonly transformerless electronic balanced circuitry has been implemented. The advantage of using transformers is that there can be complete electrical isolation between the sending and receiving equipments. Electronic balancing loses this advantage but substitutes lighter weight and potentially greater signal performance.

The use of the term 'balanced lines' should not be confused with the term '600 ohm line'. The concept of the characteristic impedance of a cable pair has been introduced above. The main point about using a cable over a sufficiently long distance such that its characteristic impedance dominates the performance is that the cable will always appear resistive irrespective of the length of cable used provided that it is correctly terminated. In addition the arrangement provides for the maximum transfer of power. For historical reasons the impedance which has been targeted is 600 Ω and you may recall that 1 mW developed in a 600 Ω load requires 0.7745 V rms. Thus this voltage has become fixed as the reference of the dBm (used when the sending and receiving impedances are 600 Ω) or as the reference for dBu (used whenever voltages are measured in audio systems without reference to 600 Ω).

These days you will find 600 Ω working being used in large broadcast studios, for example, where the lengths of cable between source and destination may be both long and variable. In smaller installations and systems the practice of sending balanced signals from a low impedance (between 30 Ω and 100 Ω) and receiving the signal at high impedance (typically between 10 kΩ and 100 kΩ) is far more common. The disadvantage of 600 Ω working is that the signal undergoes a minimum attenuation of 6 dB and this loss must be made up at the receiving end with all of the consequent problems such as noise. Note that the lines provided by the telephone companies are seldom 600 Ω and are around 140 Ω and this requires the use of matching transformers. In particularly noisy environments you will come across 'quad' cable. This is a cable which has two sets of twin cores within a screen jacket and the arrangement offers an improved rejection of interference over that obtainable with the more common twin screened cable.

For digital audio (and video) signals it is important to use cables and terminating impedances which are precise and well behaved over the signal spectrum since this avoids undue frequency sensitive losses and waveform distortion due to non-linear phase performance. Digital audio signals have a bandwidth with fundamental components up to 2.4 MHz, some 100 times that of the audio baseband signal.

We have shown above that the common cable constructions are ineluctably associated with capacitance and inductance. The significance of these associations depends much on the length of cable and the nature of the sending or receiving impedance. Since the combinations of cable and system impedance can be so varied a few examples will help to highlight some typical problem areas. A classic problem is the connection of a moving coil pick-up cartridge to a preamplifier, Fig. 21.12. The cartridge has an essentially inductive output whilst the usual coaxial cable has a significant capacitance. Thus the two will resonate at a frequency usually just beyond the top end of the audible bandwidth.

Most domestic equipment output impedances lie in the range of 100 Ω to 2 kΩ, Fig. 21.13. At the higher impedance a 1 metre length of interconnecting cable with a net capacitance of, say, 300 pF would cause the audio signal to be rolled off by 3 dB by 265 kHz. Conventional wisdom concerning the acuity of human hearing indicates that this error will not produce any audible artifact. Although this error may be small the wise audio system designer will note that there may be many such places where the audio bandwidth is being limited and will usually take steps to ensure that the sum of all of this kind of error will not exceed the sensitivity of trained ears. It would also be prudent to keep the lengths of interconnection as short as is needed. For example, a 20 metre length of the same cable in the same application will produce a response which is 3 dB down at 13 kHz and this is very likely to be audible.

Similarly the cable which connects an amplifier to a loudspeaker system can be selected by bearing in mind the maximum amount of power which can be lost in the cable. The power can be lost in the basic resistance of the wire used or in the interaction between the cable's inductance and capacitance and the reactance of the loudspeaker load. This topic deserves closer attention since a number of manufacturers have made some fairly outrageous claims for various specially devised cables. Most of these claims can be considered as pure fairy tale surrounded by plausibly used technical jargon. Much of the text above can be used by the astute reader to reveal the underlying fallacy of these claims.

Loudspeaker cables

The key parameters which describe a loudspeaker, or indeed any power cable, are its series resistance, series

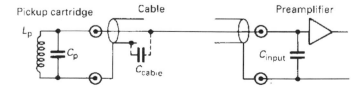

Figure 21.12 The simplified arrangement of a moving coil cartridge, the cable and the preamplifier. The cartridge has an inductance of L_p and an associated winding capacitance of C_p and it feeds a signal to the preamplifier and its own associated stray input capacitance C_{input} via the cable which has a self capacitance of C_{cable}.

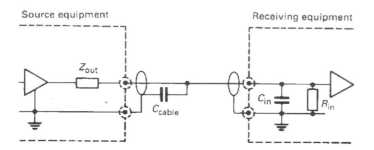

Source equipment   Receiving equipment

Figure 21.13   This diagram shows, in simple form, the problem which occurs for real signal sources when practical values are considered. The sending output impedance, shown here as Z_{out}, has to drive the input impedance of the following equipment and the cable capacitance C_{cable}. The reactance of the series inductance element of the cable will usually be a very small fraction of the input impedance and is neglected here. The input impedance Z_{in} is shown here as comprised of an input resistance R_{in} paralleled by an input capacitance C_{in}. Typical values for R_{in} and C_{in} are in the region of 1 MΩ and 50 pF for oscilloscopes and laboratory instruments, 100 kΩ and 100 pF for musical instrument amplifiers, 20 kW and 50 pF for much audio equipment. Although much professional audio equipment is required to work in a balanced 600 Ω environment it is usually preferable to operate equipment to send from a low impedance and receive the signal using a high input impedance. In this way the effects of the cable and the input capacitance are minimised. For high frequency signals, such as digital audio and video, the connectors and cable should be considered as components of a transmission line.

inductance and shunt capacitance and shunt conductance. The lengths of cable normally encountered are of the order of one thousandth of a wavelength even at 20 kHz thus a treatment of a loudspeaker cable on the basis of transmission line theory is essentially an academic exercise. Such a treatment will consider the characteristic impedance, the need for termination and matching, the effects of reflections and frequency dispersion. A cable of 10 m is approximately one wavelength long at 30 MHz and it will be at this frequency that the effects of poor termination would be manifest as reflections. Commonly, well behaved audio amplifiers do not perform at this high frequency and, indeed, there are thus good reasons why an audio power amplifier should have its high frequency response well controlled by 50 kHz. Audio power amplifiers are arranged to have as low an output impedance as practicable and the input impedance of a loudspeaker system is both varying and complex over the audio frequency band. Thus neither end of a loudspeaker cable will be matched to the characteristic impedance of the cable and one must conclude that, for cables of reasonable length, there are no transmission line effects of any audible consequence.

The matter of frequency dispersion arises since the series inductance of the cable is too small compared to the resistance of the cable. In telecommunications where very long distances are involved this is remedied by adding loading coils to long lines. The magnitude of this dispersion for a typical loudspeaker cable is minute however and amounts, typically, to around 20 ns per metre at 10 kHz. In principle this can be adjusted for acoustically by moving the tweeter 17 µm further from the listener and it should be immediately clear that this is not worth doing.

It is more fruitful to model the loudspeaker cable as lumped components. In addition we should consider a competently designed amplifier as having an output impedance which, over the full range of audio signal frequencies, will have an output impedance between 0.05 Ω and 0.42 Ω. It is usual for the amplifier to incorporate an output inductance of around 2 µH and this will increase the output impedance at 20 kHz by some 0.25 Ω. Amplifiers which cannot tolerate even slightly capacitive loading without becoming unstable require redesign to resolve their problem.

A simple diagram, Fig. 21.14, can be drawn to show the resulting simplified equivalent circuit. It is important to realise that the loudspeaker load represents a complicated impedance in which the effects of cabinet, drivers and crossover networks are incorporated. It is common to use the term damping factor to indicate the value of the amplifier's output impedance at a given frequency. The damping factor is the ratio of the intended load treated as a pure resistance to the output impedance. In practical terms this measurement is almost valueless since any practical loudspeaker system will incorporate the effects of the acoustic loading, the effect of the crossover unit and the loudspeaker driver coil resistance as elements in series with the cable and amplifier output impedance. This provides one reason for performing the crossover function at a low signal level and driving each loudspeaker element with its own appropriate amplifier. The major benefit of this is that it is possible to produce a more complex crossover filter more cheaply without incurring the costs associated with wasting power by using high power capacitors, inductors and resistors in a passive crossover filter.

A common technique involves splitting the high frequency and low frequency crossover elements and

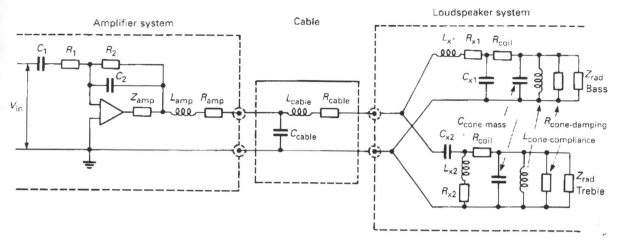

Figure 21.14 A simplified circuit for the overall interface between an amplifier and the loudspeaker load. Although this is simplified it is necessary to show the many reactive and resistive components which are in the path between the amplifier input and the required sound wave in the air. On its own the amplifier exhibits a frequency response which is limited at the bass end by the time constant of R_1C_1 and at the treble end by R_2C_2. The amplifier's effective output impedance Z_{amp} will change as a function of the amplifier's response. Most amplifiers will have an output inductance, shown here as L_{amp}, and an output resistance R_{amp}; which includes the wiring and terminal resistances. The loudspeaker cable is shown simply in terms of its lumped component values L_{cable}, R_{cable} and C_{cable}. It would be too much of a simplification to show the loudspeaker as simply the voice coil resistance when, in reality, many loudspeaker systems comprise a minimum of a bass and treble driver each handling a frequency band determined by the components used in a crossover filter. Here the bass and treble drivers have been shown as comprising R_{coil}, $C_{cone-mass}$, $L_{cone-compliance}$, $R_{cone-damping}$ and the radiation impedance Z_{rad}. The crossover filter (L_x, R_x and C_x) has been much simplified. We have shown that the values of the cable's lumped components are almost negligible when compared to those of the amplifier or the loudspeaker system. By a similar token the magnitude of the difference in performance which can be expected when suitable different cables are substituted can be clearly appreciated to be negligible.

arranging that the amplifier channel drives each segment via a separate cable. The approach, known as bi-wiring, does require that the loudspeaker crossover is designed to be split in this way. It is argued that there is an audible improvement in performance as 'the high frequency components do not travel down the same wire as the low frequency components and thus do not become confused as to which driver they should emerge from'. It is not an argument which is particularly convincing as there seems to be no reason why the principle of superposition should break down for certain loudspeaker loads but still be applicable to other engineering activities such as high power generation and space flights.

Note also that many amplifiers and loudspeakers are protected by a fuse. A fuse works by becoming so hot that it melts and breaks the circuit. Thus a fuse must have a resistance and this will change with the temperature caused by increasing the current handled by it. In turn this modifies the maximum power which will be delivered to the loudspeaker voice coil. Whilst it comprises yet another component between the amplifier and the loudspeaker driver it should be remembered that the loudspeaker voice coil itself will become warmer as higher powers are dissipated from it. This increase the voice coil temperature and thus its resistance and

thereby compresses the movement of the voice coil. It is worth commenting that this change in voice coil resistance (and possibly the change in the resistance of the crossover coils themselves) will temporarily shift the alignment of a passive crossover filter and this may produce an audible result. All of these effects are orders of magnitude greater than those caused by loudspeaker cables.

In summary, the performance of some typical loudspeaker cables, Figs. 21.15(a) and 21.15(b), can be shown. Once more the breath of reality in audible perception suggests that any performance which guarantees the transmission of an audio signal with no non-linear distortion and a frequency response error of less than 0.25 dB at 20 kHz will provide a performance which is essentially that of a perfect lossless wire. Indeed there is nothing much in the audibility of different loudspeaker cables save for those which are too short to reach the loudspeaker.

Much the same set of considerations apply when the choice of power cable (and connectors) is being considered. The prime requirement for a DC power supply has to be the minimisation of voltage drop along the supply cable. Provided that the power supply has been competently designed the length and type of cable connected to

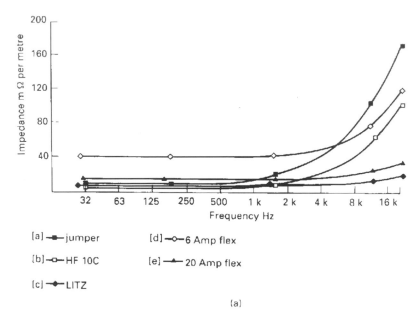

(a)

Figure 21.15   These curves show the performance of various types of cable which have been used to connect power amplifiers and loudspeakers. (a) Shows the impedance in ohms per metre length for a selection of types of cable. Type [a] is simply a set of car jumper leads around 8 mm² area; [b] is a special cable with a greater area whose conductive strands are not individually insulated, whilst [c] has been made as a Litz cable with an overall conductive area of 4 mm². Type [d] is a standard 6 ampere rated flex with a cross section of 0.75 mm² and type [e] is a similarly common 20 ampere rated flex of 2.5 mm². Note that the jumper lead, type [a], impedance rises with frequency as does that of type [b] due to the relatively wide separation between the two cores of the cable. Although Litz cable with an area of 4 mm² is an expensive cable it does not exhibit any particularly outstanding property when compared to a conventional 20 ampere rated flexible cable with an area of only 2.5 mm². The performance of these cables can be accurately described by the simple models developed above.

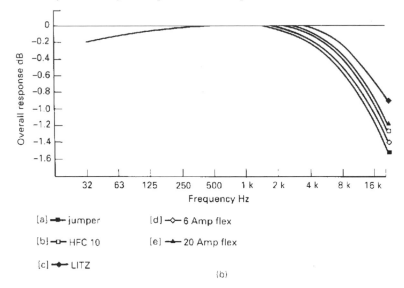

(b)

Figure 21.15   (b) When such cables are connected between practical amplifiers and practical loads the differences in performance become only marginally measurable. The power amplifier is typical of the majority available and it provides an output voltage response which has fallen 0.5 dB by 20 kHz and the curves shown should be appreciated accordingly. The loudspeaker was a three way design nominally 8 Ω but whose impedance falls to about 4.8 Ω above 8 kHz. Though the Litz cable has lost slightly less response than the 2.5 mm² flexible copper cable it is a matter of only 0.25 dB. The thin 0.75 mm² cable and the jumper leads are a further 0.25 dB worse than this. The differences between performance depend on the loudspeaker impedance and a loudspeaker system showing a more normal impedance response which rises with frequency will also show much reduced differences.

it should not affect its stability. Power supplies which are located some distance from the circuitry which they supply will often be able to sense the delivered voltage and adjust the output voltage in order to counteract for any voltage drop. Some thought should be given to the placing and type of the sensing cables since they could pick up interference. In some cases it may prove possible to distribute unregulated DC power and to carry out the regulation locally. Where a large amount of power is required from the mains supply an appropriate cable installation should be made and, in addition, some care should be taken to ensure that all of the relevant connectors are suited to the task of handling the high currents. Such installations benefit greatly from preventative maintenance.

The term '100 V line' is sometimes encountered in professional audio. The reader is referred to the relevant section for greater details. Note that the idea is to reduce the losses in the loudspeaker cabling due to carrying high currents by transforming the power amplifier output voltage up to a higher level. At locations where a loudspeaker is to be placed a step down transformer is placed. These are normally supplied with their secondary ratios marked in terms of the nominal output power into an 8 ohm load. An advantage is the freedom to use a thinner gauge of wiring. A natural choice of cable for installations where emergency voice announcements are not required to be made is to use a standard mains type of cable. It is thus important that this cable is clearly and frequently marked and identified as an audio loudspeaker feed cable during the installation. The importance of a consistent and properly effected cable identification plan cannot be better emphasised.

Summing up about cables

This outlines some of the factors which need to be taken into account when selecting a cable for audio use. The effect on the frequency response (and its concomitant the phase response) and the magnitude of any attenuative loss which may require compensation should be borne in mind. Where low level signals are concerned the choice will be made after considering the impedance of the source and destination and the degree of screening which is required. These considerations are fairly straightforward though they have been the subject of outrageous unsupportable claims. Thus the reader should find it easy to recognise that the main benefit in using exotic cables (read this as often exotically priced) is for the supplier of the cables. The audible performance of the cable can be calculated and verified by measurements to a resolution and an accuracy which is well beyond that of the human ear's ability to assess. The

situation has been made a bit murkier where loudspeaker (and to a much lesser extent power) cables are considered due to the over-zealous claims for performance which have been made in recent years.

Connectors

It is difficult to discuss the various types of cable used in audio without including a brief review of the connectors which are commonly used. Connectors for audio purposes can be neatly divided into two groups, those commonly used in domestic environments and those more associated with professional use. The domestic group includes a number of readily recognised types. In addition to the connector types it is necessary to recognise the wide variety of materials which are used in making connectors. A casual review of the pages of most electronic components stockists catalogue will reveal the very wide range of connectors which are available. The intention here is to cover some of the more common ones which an audio engineer might encounter.

Connector materials

There are two parts to a connector, the metallic conducting elements and the insulating and protective elements. The commonest metallic materials used for electrical contacts include brass and beryllium copper alloy. Often some environmental protection is given to the raw metal by plating or coating the contacts. The range of plating materials includes chromium, tin, nickel, silver and gold. The advantage of brass is that, when clean, it is readily solderable. A coating of tin increases the ease of soldering to brass without adding significantly to the cost. Silver is also readily solderable and makes a good contact which cleans easily on mating. Gold plating is amongst the more expensive treatments and for that reason it is often applied selectively to specific contact mating areas. Gold is readily solderable but is almost immune to common corrosive environmental attack. Beryllium copper offers the advantage that it will act as a spring without undergoing fatigue. Thus contacts can be formed to provide an acceptable contact pressure.

The insulating and protective elements use a much wider range of materials. Insulation is provided by materials such as ABS, teflon, bakelite, nylon, acetal resin, polycarbonate, neoprene, silicones and rubber. Each is used after the necessary compromises between material and manufacturing cost, thermal performance, electrical performance, mechanical strength and environmental requirements have been considered. ABS is an easily moulded thermoplastic with fair temperature performance (it has a melting point in the range 140° C to 180° C and this can be compared to the tip temperature of typical soldering irons which is around 340° C). Nylon and acetal resin are also thermoplastic and the mechanical performance of nylon can be significantly enhanced by incorporating fillers such as chopped glass fibre. Teflon has a very high melting point but it is an expensive material. Bakelite is a thermosetting plastic which since it is very cheap and withstands heat reasonably well continues to be commonly used for mains power and lighting fittings. Polycarbonate shows good dimensional stability as well as a good temperature performance and is readily mouldable. In addition to their use as supports for mating contacts the rubbers are often used to help seal the internal parts of the connector from the external world.

The connector shell materials include the plastics referred to above and a range of metals which include anodised and bare aluminium, die-cast zinc, brass, deep drawn steel, stainless steel and bronze. Again the choice of material is made after due consideration of the salient compromises. Die-cast zinc, usually nickel plated, is used for XLR connectors and provides a very strong shell for the connector but one which does not run the risk of rusting. The use of steel shells (usually plated with zinc) provides some strength and also some protection from the hazards of electromagnetic radiation whilst being very cheap to produce. Brass is almost always plated either with nickel or chromium.

Making the connections

An important role which the connector shell plays is in supporting the cable. The better connectors provide a means of securing the cable to the shell in order that any mechanical strain on the cable is not directly taken by the contact pins. Connectors which must not become accidentally disconnected under strain or vibration will incorporate some form of latching fastening. Connectors which must perform in adverse environments will need to be able to accommodate some form of seal between the cable and shell whilst the matching connector mounted in a panel will need to incorporate some form of seal to the panel.

There are four main methods of attaching wires to contact terminals and these are crimping, soldering, insulation displacement and mechanical fixing. Crimping can produce very reliable joints and you'll note that examples of this abound in every motor vehicle though the technique is not limited to high current connections. It is important to use the correct combination of tool, wire gauge and contact in order to form a reliable crimp. The technique of producing a reliable soldered joint is one

which must be learnt through practice. Part of the skill is in being able to control the time for which the connector pin is being heated whilst the joint is made. Too long a time may result in a molten or deformed connector housing! Insulation displacement (IDC) techniques are widespread and often require the use of proper tools. When properly carried out IDC connections can be very reliable. Simple mechanical connections means screw terminals and other simple methods of securing wiring.

IEC mains connector

This is the most common mains power connector in use with modern equipment. In its plain form it is rated at 5 A capacity and it should be fused at no more than this value. Higher current versions of this style are available and the chassis mounted socket carries an identifying lug which prevents the use of lower powered mains cables being used. The quality of the mains connection is important not for any audible reason but principally from the safety point of view. A good quality earth path needs to be provided in the wiring installation. Recording, radio and television stations are amongst users who make a very special point of achieving the very best technical earth installation with the additional target of ensuring the lowest possible risk from interference.

Alternative connectors have been used for mains power especially for professional audio and video equipment. Such connectors as the Cannon XLR-LNE and EP-LNE styles exist but the connectors do not necessarily satisfy the safety requirements for domestic environments and are thus decreasingly common.

Phono connectors

The phono (also known as the RCA jack) is intended for use with a single core screened coaxial cable, Fig. 21.16, of typically around 3 mm to 5 mm in diameter. This connector can be exceedingly cheap to manufacture and is in very common use for low power signals. Although so-called 'professional' styles of this connector are now available you should bear in mind that their cost can be more than forty times that of a well made economy model and there will be no measurable audible difference between the two types. Due to its ease of use and low cost the humble phono connector has found its way into semi-professional audio equipment. However, it does seem to be quite a dumb choice to use an overpriced professional version of the phono connector when a proper professional style connector can be had for less outlay. Both will offer a gold flash finish that will look expensive and hold back any atmospheric corrosion. The usual means of termination involves soldering and some dexterity with the soldering iron and hand tools will be involved. The convention with the connectors is that the cable carries the male plug and the chassis the socket. Thus the usual cable with phono connectors can be connected either way though cable mounted sockets can be purchased. The common form of phono connector suffers one minor disadvantage in practical use because its design means that the protective and signal earth connection is made after the active signal connection. This is one of the causes of loud crackles and bursts of hum which accompany the insertion of such cables into working equipment. You'll soon learn, however, to turn down the audio level of any channel of audio equipment into which a new connection is being made! It is worth pointing out that the most expensive phono connector currently available features a spring loaded shielding ring which does ensure that the earth connection is made before the signal connection.

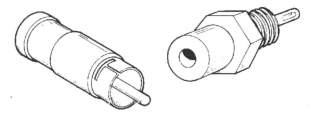

Figure 21.16 Phono connectors come in a variety of styles and their cost, robustness and suitability varies accordingly. In the USA the phono connector is sometimes referred to as an RCA connector. The phono connector is often used for video signals and it is the specified connector for the SPDIF or consumer digital audio interface. In these applications the connector should be attached to cable of nominally 75 Ω impedance.

Coaxial connectors

Aerial down leads for radio and television commonly use a very simple coaxial connector. Termination can be made without soldering. The connector is very cheap and is not suited for use outside the domestic environment. The professional coaxial connector in greatest use is the BNC, Fig. 21.17. This connector can be found in use for test equipment connections, in professional video systems and in radio frequency (RF) work. It is important to note that there are two impedence conventions in use and the connectors are mechanically interchangeable. The world of RF and data networks uses cables and connections based on the characteristic impedance of 50 Ω whilst the world of video follows a convention based on 75 Ω cables and terminations. Using connectors and cables of the incorrect characteristic impedance will give rise to phase and frequency response errors due to the mismatch of impedance.

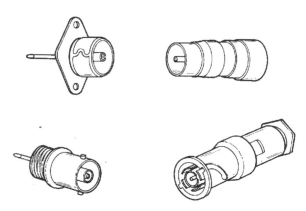

Figure 21.17 Coaxial cable connectors vary enormously. Those which the audio engineer might encounter include the common aerial feeder connector and the more robust and expensive BNC style of connector. Note that as for the phono connector the usual arrangement of the connector results in the outer part of the chassis mounted connector is directly attached to the chassis. An insulating bush can be readily purchased for BNC connectors.

DIN connectors

The DIN connector was similarly designed to be a cheap and reliable connector principally for domestic equipment. However, unlike the simple coaxial styled phono connector the DIN connector is essentially a multipole connector. The two-way version is intended to be used for loudspeaker circuits whilst the three, four, five, six and seven-way styles are intended for low power signal circuits, Fig. 21.18. Once again it is intended that the cable carries a plug connector at each end whilst the chassis of the audio equipment bears the mating socket. For the purposes of domestic equipment interconnection the signal circuits are assumed to be unbalanced single ended and this means that even the three pin DIN connector can carry both left and right stereo channels whilst the five pin connector is usually used to carry left and right signals to and from an equipment such as a tape recorder. When wiring up the five-pin connector bear in mind that since pin 1 on a preamplifier socket is the left audio output the audio connection at the cable's other end must be made on pin 4, the tape recorder's left channel input pin.

This swapping over of pin functions has caused confusion for many users in the past. The DIN connector can be a bit fiddly to assemble and to wire especially if individually screened signal leads are used since all of the screen connections will need to be soldered to the one central earth pin 2. The DIN connector also suffers from the failure to ensure that the earth connection is made before the signal pins make contact with their respective socket contacts. Although the DIN connector is small it can be manufactured to be robust and there are also several styles of the DIN connector which offer a latching or locking feature. The DIN connector can also be designed to provide strain relief for the signal cables, something that is not commonly possible in the phono connector.

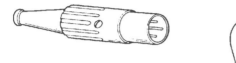

Figure 21.18 DIN type connectors can vary in price and quality in the same way as phono connectors. Note that there is a provision for a cable strain relief and that the shell of the connector does not have to carry signal currents.

The DIN connector is also specified in the MIDI (Musical Instrument Digital Interface) standard. The fixed equipment carries the socket whilst the cables are fitted with plugs. In the case of the MIDI equipment some freedom from earth and hum loops is achieved through the use of optical coupling of the incoming MIDI control signal.

The jack plug

The domestic headphone connector is the ¼ inch jack plug and its companion the jill socket, Fig. 21.19. This style connector is also widely used for low level signals

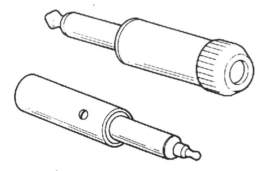

Figure 21.19 The 6.35 mm or 0.25 inch jack plugs come in mono and stereo versions. Note that the professional user will use the type B jack and that this has a smaller tip and ring diameter than the stereo jack. This style of jack has been made in smaller versions for particular purposes. For domestic headphones it has been reduced to 3.5 mm diameter and you may find 2.5 mm diameter occasionally. For the professional user a bantam jack is available in which the barrel diameter is 4.5 mm. It is common for the jills to have switch contacts which are actuated when the jack is inserted and these can be arranged to perform a range of useful functions. In professional installations there is often a very comprehensive switching arrangement available when there are a large number of jills and there are three common ways (some times referred to as normalling) used for wiring the sockets.

and it is very common in musical instruments and their associated amplifiers. The barrel, or sleeve, is always connected to the signal screen or earth and the tip carries the audio signal. There are three main variants. The commonest in domestic environments is the stereo headphone jack in which the left and right headphone signals are carried on the tip and ring thus leaving the barrel to carry the earth signal return. Note that the tip, ring and barrel all have the same diameter of 6.3 mm. The mono version of this connector simply carries the signal on the tip. Broadcasters and telecommunications users (in the UK) have a modified form of the stereo jack, the Type B or Type 316, in which the ring and tip are of progressively smaller diameters than the barrel. Depending on the details of a particular jill the two variants may not be mechanically interchangeable. The Type B is intended to be used in systems where the audio signal is balanced and the tip will carry the 'hot' phase of the signal while the ring will carry the cold.

The 4 mm plug and post

Possibly the most common form of loudspeaker connection is provided by the 4 mm socket and binding post, its associated plug, Fig. 21.20, and its variants. By long-standing international convention the spacing between pairs of such sockets is fixed at 19 mm and access, particularly when large wire sizes are involved, can be very restricted. Although examples of such sockets can be supplied with current ratings up to 30 A you will find it quite tedious to make a neat reliable connection as a consequence and it is more usual to terminate cables in a 4 mm plug. Stranded wire can cause shorting unless all of the strands are carefully twisted and properly incorporated into the connection. Plugs are commonly rated at 10 A but usually provide no protection against shorting to each other and adjacent metalwork. A range of self-shrouding plugs is available which meet the VDE 0100 and 0110 standards. The high current rating, low cost and ease of use have helped this style of connection to become commonplace not only for audio power amplifier outputs but also, for example, as an output connector for low voltage laboratory power supplies.

The problems with this style of connector stem from the very convenience with which connections can be made. Firstly, because it is so easy to make a poor connection in the relatively crowded space at the rear of an amplifier, and secondly, because the very simplicity of the connection means that it is very easy to come into contact with live metal parts. The issue of electrical safety is prescribed in the BS415 standard. Fortunately suitable high power connectors are available and an example is the NL4 Speakon from Neutrik, Fig. 21.21.

Figure 21.20   The 4 mm binding post has been in existence for a long time and it is very commonly found as the output terminal on low voltage bench power supplies. At the equipment side the termination will be made either by soldering directly to the threaded part of the post or, preferably, to a suitably rated solder tag. In use there are three main ways of making a connection to the terminal. The first is probably the most common and simply involves wrapping bared wire around the post, or threading it through the small central hole in the post, and then tightening the insulated boss. A second method requires that the cable ends are first terminated in a 'spade' terminal. This is an ideal method when a reliable connection is required and when the risk of whiskers of wire and poor connection needs to be minimised. The third method is often used for loudspeakers and it requires that the cable end is terminated in a 4 mm 'banana' plug. This can be inserted in the centre of the post. However the quality of the contact is a little more difficult to control because of the need to ensure that there is adequate contact pressure to clear the surface contamination and to ensure a low contact resistance. The two virtues of the 4 mm binding post system are its low cost and its relative ease of use. The user of this method should be aware of the safety aspects of using this connector. Under the current low voltage safety standards (BS415, IEC65, EN60065) its use may be questionable when voltages equivalent to exceeding 70 W into 8 Ω may be present. Thus for higher powers a better safer connector should be installed.

The use of members of the XLR family for power connections should be avoided because there are well established conventions for their use in low-power circuits.

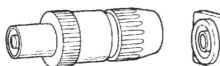

Figure 21.21   There are a few connectors which are intended solely for use in connecting loudspeakers and the Neutrik NL4 is the most commonly used. Alternatives (now shown) include the Utilux and ITT Cannon AXR PDN design. The NL4 design offers a fully shrouded set of contacts which meet fully the safety requirements in a shell outline which is consistent with the Cannon XLR series. Each contact is rated at 30 amps continuously and the tough plastic shell and connector body can provide the airtight seal which is required for loudspeaker connections. Although more expensive than the 4 mm binding post it is much cheaper than the alternative dedicated loudspeaker connectors. Termination within the equipment may be by soldering or, preferably, by the use of crimped connections. Connections to the cable may be made either by soldering or by the screw clamp provided and the contacts can accommodate conductors up to 10 mm². The connector shell is able to handle easily the large diameter cables which are needed for large powers. Note, however, that unlike the Cannon XLR series, and indeed the DIN and phono series, that the genders of the cable and chassis mounted connectors cannot be

changed and thus a fully compatible cable will have a cable-mounted plug at each end whilst the amplifier and loudspeaker will always carry a chassis-mounted socket. A simple double socketed barrel is available in order to join successive lengths of cable.

The XLR connector

For professional audio use the main connector for single screened cables has been the XLR-3 series of three-pin connectors, Fig. 21.22. There is an internationally agreed standard for the wiring of this connector series and this can be found in IEC 268 or in a standard published by the Audio Engineering Society (AES-14-1992). The essential element is that pin 1 of the male connector is arranged to mate with pin 1 of the female before electrical contact with the other pins is established. This allows the safety earth contact to be made first. The IEC 268 standard lays down the wiring convention that pin 2 carries the in phase signal (also known as the hot pin) and pin 3 carries the signal current return (often referred to as the cold pin). Note that the numbering of the pins is not obvious.

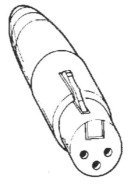

Figure 21.22   The Cannon XLR series of connector is, perhaps, the most common connector in use in professional audio systems. Its shell and connector body are high pressure die cast zinc and are mechanically very sound. The cable shell has proper provision for a cable clamp. The connections are made by soldering and good, tidy soldering skills are required to make a reliable durable connection. Higher numbers of contacts, with appropriately reduced contact current rating are available and may be used to provide extra signalling or power. The use of the XLR series for loudspeakers is not ideal since the contacts are not fully shrouded when the connector is used in the preferred IEC-268 convention and the maximum current rating of the three-pin version is 16 amps. Some manufacturers and users have modified their power amplifiers to use a gender convention opposite to that specified in the standard. This creates the risk that a low power signal source can be connected directly to a high power amplifier output. Many terminally damaged microphones and other equipment have resulted. This abuse also requires the use of gender changing connectors since the loudspeakers are often fitted with the XLR chassis mounted female connector.

Versions of the XLR connector bearing up to 7 pins can be found but they are not in common use in audio applications. The IEC 268 standard also supports the convention that the male XLR carries the signal whilst the female XLR receives it. Thus, in those cases where a male XLR has been used to carry a power amplifier output, you should remember that the pins are live and they are readily touchable. At low powers this may not cause a problem especially for healthy individuals, but at power levels above 100 W into 8 Ω there is an increasing risk of shock. Users of XLRs should be aware that for some time audio equipment originating from the USA followed neither the convention for the gender of connector nor the convention for the wiring polarity but instead followed the NAB convention.

It is worth pointing out here that the XLR-3 pin connector is specified as the connector for use with the AES/EBU digital audio interface (AES-3-1992) and with a cable with $Z_{ch} = 110\,\Omega$.

The D-type connector

A very common form of multi-pole connector is the D-type, Fig. 21.23. Some dexterity is required when soldering to the connector pins but the connector series does provide a reliable connection. A range of styles which reflects the various environmental and performance compromises is available. Forms of this connector are available which use insulation displacement connection (IDC) techniques and these are ideally suited to making a connection to ribbon cables.

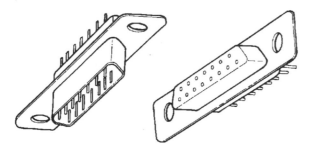

Figure 21.23   The D-type connector is available in a variety of styles. The connector series provides for multiway contacts from 9 way through to 50 way. The cable mounted connector (which may be either male or female) is usually associated with a shroud or shell, some forms of which are fitted with jack screws which allow the mated connectors to be mechanically secured. The quality of contact ranges from the cheap formed tin plated variety to the turned gold plated pin. Connecting the cable to the connector contacts is often achieved by soldering and, again, skill is needed. However, two further important

methods are available. The first allows the D-type connector to be directly attached to a ribbon style cable through the use of insulation displacement connection. This is a simple way of terminating a large number of wires in one go using nothing more than a modified vice. The second method recognises that in many cases not all of the contacts in a multiway connector are going to be used and thus only the intended ones need to be connected. This can be achieved by using individually crimped contacts which are then inserted into the connector shell in the required location.

The FCC 68 style

IDC terminations are also used to make up cables with the FCC68 connectors, Fig. 21.24. These are very similar to the current UK telephone jack socket and in clement environments they provide a reliable connection for the low power audio signals. Making IDC terminations does require that the proper tool is used.

In the future it is increasingly likely that audio signals will be distributed digitally along simple unshielded twisted pair (USP) cabling. The signal format will follow the requirements for ISDN. Although such systems will be inherently more reliable if the cabling has the required 110 Ω impedance the increasing processing intelligence which is being incorporated into the transmitting and receiving ends will make this a less critical element in the overall performance.

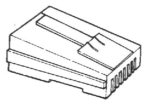

Figure 21.24   The use of insulation displacement connection is fairly widespread. The technique can produced durable reliable connections which can be made speedily either manually or by machine. The ones shown here are an example of the style of connector which is used to connect telephone systems. Traditionally these systems would be considered as analogue only but increasingly the cables will carry signalling information and digitally encoded voice signals.

Optical connectors

A mention should be made here for the role which optical connectors will be playing in the future in bearing digital audio and video signals between equipments. The discussion of optical fibre techniques warrants its own chapter. Though it is worth commenting that there has been much misrepresentation of the engineering issues, particularly within the hi-fi industry, concerning the benefits of optical coupling of digital audio signals between equipment such as CD players and their associated analogue to digital converters (ADCs). Between competently designed implementations of equipment there is no reason why simple electrical connection should not suffice. Very occasionally it is admitted that the improvement in performance can be traced to poor electrical grounding or some other simple curable fault. All too often it seems that pride and prejudice play ponderously upon the purse as the purveyor and purchaser of such over-complicated technological implementations fail to engage the faculty of rational thought.

22 NICAM Stereo and Satellite Radio Systems

Geoff Lewis

We take stereo radio for granted nowadays, but the FM system that we have used for so many years is now outdated and takes up a large amount of precious bandwidth. In this chapter, Geoff Lewis reviews the NICAM system that is used for stereo sound transmission on TV and notes the various proposals that exist for digital sound broadcasting.

It has long been recognised that the inclusion of high quality stereo audio with television significantly adds to the viewing experience and consequently, the analogue Zenith-GE Pilot tone system has been included in some parts of the world. The Zweiton analogue dual carrier system was developed in Germany and is used with some success. The monophonic signal for the single channel receiver is transmitted as the sum of left and right hand signals, but in the form of $(L + R)/2$, whilst the R signal only is transmitted on the second sub-carrier, both using frequency modulation (FM). This system, which considerably simplifies both the transmission and the reception circuitry, has been very successfully used on the PAL systems B&G, where the sound and vision carrier spacing is 5.5 MHz. When tested on the UK PAL I system with 6 MHz carrier spacing, the Zweiton modulation created unacceptable interference with either the vision or primary sound carriers. At about this time, the compact audio disc was available which, by using digital processing, was capable of producing very much better audio quality than was available by analogue FM broadcasting. The teletext system had demonstrated that analogue and digital signals could be included in the same multiplex without creating mutual interference. Taken together, this probably created the necessary impetus for the engineering departments of the BBC and what was formerly the IBA to develop jointly the system that is now known as NICAM-728 (Near Instantaneous Companded Audio Multiplex 728 kbit/s). NICAM which can be adapted to suit all PAL or NTSC systems, allows almost CD quality stereo audio to be delivered direct to the home alongside a television signal that also includes teletext.

The signal structure of the NICAM-728 system

In the interests of compatibility, NICAM transmissions retain the standard 6 MHz monophonic FM audio channel and place the digital signal on a second sub-carrier located at 6.552 MHz or 9×728 kHz above the vision carrier. For the PAL systems that use 5.5 MHz sound carriers, the NICAM sub-carrier is located at about 5.85 MHz above the vision carrier. The sub-carrier which is maintained at a level of −20dB and −10dB relative to the peak vision and sound carriers respectively, is differentially encoded with the digital signals for both channels of the stereo pair using quadrature (four phase) PSK modulation. Using this scheme, each carrier resting phase represents two data bits and so halves the bandwidth requirements. Because the data is differentially encoded (DQPSK) it is only the phase changes that have to be detected at the receiver decoder. The bits to phase change relationships are as follows:

$$00 = -0° \text{ phase change,}$$
$$01 = -90° \text{ phase change}$$
$$11 = -180° \text{ phase change}$$
$$10 = -270° \text{ phase change}$$

By using this Gray code scheme, only 1 bit change is needed for each step phase shift making the signal more robust in noisy conditions. Pre-emphasis/de-emphasis to CCITT Recommendation J.17 which provides 6.5 dB boost or cut at 800 Hz is applied to the sound signal either when in the analogue state, or by using digital filters in the digital domain.

The left and right hand audio channels are simultaneously sampled at 32 kHz, coded and quantised separately to 14 bits resolution and transmitted alternately at a frame rate of 728 kbit/s. Using the formula of $(1.76 + 6.02n)$ dB for bipolar signals shows that where $n = 14$ bits per sample this yields a quantisation noise of about 86 dB.

The NICAM compander processes the 14 bit samples in the manner shown in Figure 22.1 and the protocol for discarding bits can be summarised as follows:

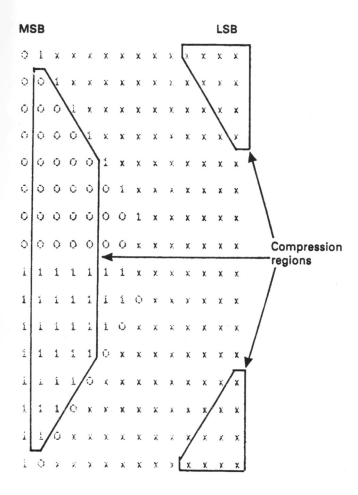

X = Don't care

Figure 22.1 NICAM-728 Data Compression scheme.

- The most significant bit (MSB) is retained and the four following bits are dropped if and only if they are of the same consecutive value as the MSB. If this leaves a word longer than 10 bits, then the excess bits are dropped from the least significant bit (LSB) region.

A single even parity bit is added to check the 6 MSB bits in each word. The data-stream is then organised into blocks of thirty-two 11-bit words in the 2's compliment format. This is used because it provides for both positive and negative signal values. A 3-bit compression scaling factor is calculated from the magnitude of the largest sample in each block and this is then encoded into the parity bits for that block. At the receiver, the scale factor can be extracted using a majority voting logic circuit. At the same time, this process restores the original parity bit pattern.

Two blocks of data are then interleaved in a 16×44 (704 bits) matrix to minimise the effects of burst data errors. Adjacent bits in the original data-stream are now 16 bits apart. A transmission frame multiplex is then organised in the manner shown in Figure 22.2 with additional bits being used as follows:

- 8 bits (0100 1110) are used as a frame alignment word (FAW);
- 5 control bits are used to select the mode of operation (C_0–C_4) either:
  - stereo signal composed of alternate channel A and B samples;
  - two independent mono signals, transmitted in alternate frames;
  - one mono signal plus one 352 kbit/s data channel on alternate frames;

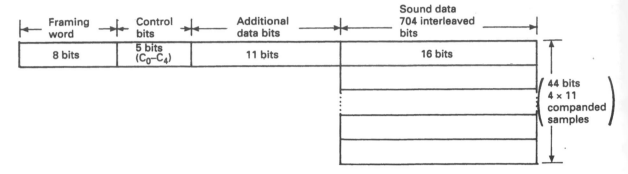

Figure 22.2   NICAM-728 frame multiplex.

– one 704 kb/s data channel, plus other concepts so far undefined;
• 11 data bits are entirely reserved for future developments.

Following the interleaving of the 704 sound data bits (64×11 bit samples), the complete frame with the exception of FAW is scrambled for energy dispersal by adding XOR to a pseudo random binary sequence (PRBS) of length $2^9 - 1$ (511). This 10-bit sequence (1000010001) is created by the generator polynomial, $X^9 + X^4 + 1$. At the decoder, the PRBS generator is reset on receipt of every FAW code.

To limit the signal bandwidth, the data-stream is passed through a spectrum shaping filter that removes much of the harmonic content of the data pulses. This combined with the effect of a similar filter in the receiver produces an overall response that is described as having a full or 100 per cent cosine roll-off.

The data-stream is finally divided into bit pairs to drive the DQPSK modulator of the 6.552 MHz sub-carrier.

The NICAM-728 receiver

The NICAM sub-carrier appears at either 32.948 MHz (39.5 – 6.552 MHz) or 6.552 MHz depending upon the adopted method for sound IF channel processing. As indicated by Figure 22.3, this component is amplified and gain controlled within the normal receiver IF stages. The spectrum shaping filter forms part of the system overall

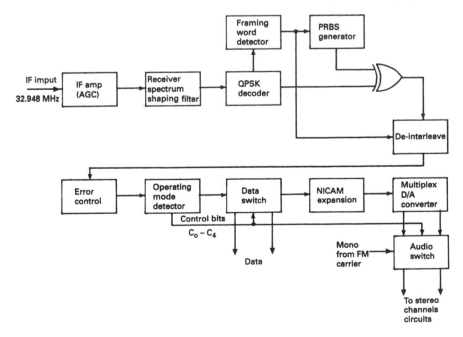

Figure 22.3   Decoding the NICAM-728 stereo signal.

pulse shaping and has an important effect upon the noise immunity. The overall filtering effect ensures that most of the pulse energy lies below a frequency of 364 kHz (half bit rate).

The QPSK decoder recovers the data-stream and the framing word detector scans this to locate the start of each frame and reset the PRBS generator. This sequence is then added XOR to the data for descrambling. The de-interleaving circuit is also synchronised by the arrival of the FAW code word.

Error control follows standard practice, but since this is buried within an IC, the process is transparent to outsiders.

The operating mode detector searches for the control bits C_0–C_4, to set up automatically the data and audio stage switches, the data outputs being those for the 352 or 704 kbit/s options.

The NICAM expansion circuit functions in a complimentary manner to the compressor, but using the scaling factor to expand the 10-bit data words into 14 bit samples.

Finally, the data-stream is converted back into analogue format for delivery to the audio amplifier stages. These should be designed to a very high standard, as NICAM-728 has an audio quality approaching that of compact disc systems.

The DQPSK decoder

This complex stage is commonly embedded within an Application Specific IC (ASIC) NICAM processor device. Figure 22.4 which represents this is very much simplified. The two main sections are associated with the recovery of the carrier and bit rate clock. The first section relies on a voltage controlled crystal oscillator running at 6.552 MHz and two phase detectors to regenerate the parallel bit pairs, referred to as the I and Q signals (in-phase and quadrature). A second similar circuit which is locked to the bit rate of 728 kHz is used to synchronise and recover the data-stream. Parallel adaptive data slicers and differential logic circuits are used to square up the data pulses and decode the DQPSK signals. The bit pairs are then converted into serial format. The practical decoder incorporates a third phase detector circuit driven from the Q chain. This is used as an amplitude detector to generate a muting signal if the 6.552 MHz sub-carrier is absent or fails. This mute signal is then used to switch the audio system over to the normal 6 MHz mono FM sound signal.

Satellite-delivered digital radio (ASTRA digital radio ADR)

Multiple carrier sound channels

Several years ago, Wegener Corporation showed that it was possible to incorporate up to 10 FM sub-carriers spaced by 180 kHz, within the spectrum of the satellite delivered sound and vision channel. Provided that the FM modulation index was maintained below 0.18 there was no worsening of the normal S/N ratio or reduction of the demodulator threshold. Since the sub-carrier deviations were small compared with that of the main carrier system, these added components made a negligible contribution to the total transmission bandwidth. Each 180 kHz slot was allocated to 15 kHz of audio or sub-divided to provide either 7.5 kHz or 3.5 kHz of audio or data

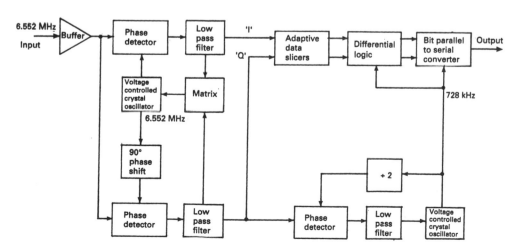

Figure 22.4 Decoding the DQPSK signal.

using FSK or QPSK. For the NTSC system these sub-carriers were placed between 5.2 and 8.5 MHz and for PAL, between 6.3 and 7.94 MHz.

In the Wegener 1600 stereo system, the L and R channels are separately used to frequency modulate adjacent sub-carriers spaced by 180 kHz. These are then used to frequency modulate the final RF carrier so that the system became known as FM^2. The two commonly used sub-carrier frequencies are 7.02 and 7.20 MHz. Total sub-carrier deviation of ±50 kHz is allowed and both channels are companded. There are two standards for companding, PANDA I and PANDA II. The former is a broadband 2:1 linear system that under certain conditions can suffer from system noise. PANDA II was developed to combat this by dividing each audio channel into two segments at 2.1 kHz and then applying companding with a ratio of 3:1 separately to each segment.

The ADR system

Figure 22.5(a) shows the frequency spectrum associated with the television baseband signal as applied to the ASTRA system before the introduction of ADR. The monophonic sound channel was placed at 6.5 MHz above the vision carrier and two Wegener type stereo channels positioned at 7.02 and 7.20 MHz. At further 180 kHz intervals a group of stereo radio service sub-carriers was included. The component at 8.64 MHz is used for network control. As the provision for stereo audio became a standard feature of satellite TV receivers, the mono channel virtually became redundant. The new frequency spectrum for ASTRA satellites 1A to 1D is

shown in Figure 22.5(b) with the original two stereo channels at 7.02 and 7.20 MHz being retained, leaving the spectrum to support 12 stereo radio channels each spaced by 180 kHz. Thus it becomes possible to provide 12 stereo (or 24 mono or any mix of the two) audio channels with each TV channel on the 16 transponders of the four satellites. The basic parameters of the ADR system are shown in Table 22.1.

Table 22.1   *ADR transmission parameters*

Audio frequency range	20 Hz to 20 kHz
Sampling frequency	48 kHz
Dynamic range	>90 dB
S/N ratio	>96 dB
Modulation	DQPSK
Audio coding	MPEG-Layer II (MUSICAM)
Data rate	192 kbit/s total
Error control	CRC for data and scale factor
Bandwidth	130 kHz (180 kHz carrier spacing)
Encryption	IDR/IBS implementation of CCITT v.35

Of the 192 kbit/s audio rate, 9.6 kbit/s is reserved for service use. This can provide on-screen display information such as channel number, station name, music title, artist etc., related to the programme being received. Encryption to the CCITT v.35 standard provides for pay channels which are switched on via a smart card system.

MPEG Layer II (Musicam) audio processing

The Musicam system exploits the masking effect whereby only those sounds above some threshold level of hearing

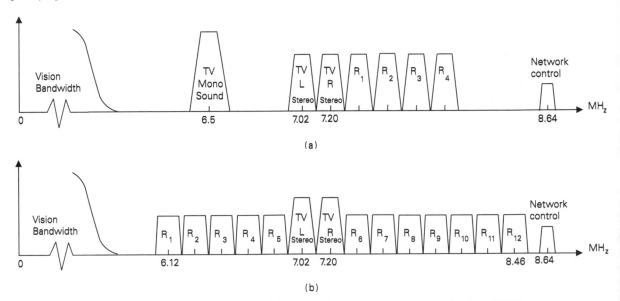

Figure 22.5   (a) Sound channel spectrum before ADR. (b) Reallocation of spectrum after introduction of ADR.

convey useful information (see Chapter 4). This threshold is sensitive to the level of nearby frequency components in the signal and also to the actual frequencies involved. A simplified block diagram of the processor is shown in Figure 22.6(a). The PCM audio samples are filtered into 32 sub-bands. A model based on the perception of human hearing is then used to generate a set of data that is used to control the quantising and coding process. Successive samples of each sub-band data are grouped into blocks in which the maximum level in each sub-band is used to calculate a scale factor. In this way, it is possible to code only those signal components above the individual threshold levels using just two or three bits to ensure that the quantisation noise is inaudible. The re-quantised and coded signal words together with scaling factor are then packed into MPEG frames in the manner shown in Figure 22.6(b). The formatter stage inserts the ancillary data bits and the error control checks. The signal is then passed to the digital modulator stage. At the receiver, the data from the set of code words is decoded to recover the different elements of the signal to reconstruct the quantised sub-band samples. An inverse filter bank then transforms the sub-bands back into digital PCM code at the original 48 kbit/s sampling rate.

The ADR receiver

Figure 22.7 shows the several ways that the receiver can be linked to a satellite TV system which allow for radio reception even when the TV receiver is switched off. In Figure 22.7(a), the ADR set is coupled via a loop-through to the normal satellite receiver, dividing the signal into its two constituent parts and avoiding the need for using a separate LNB as shown in Figure 22.7(b). Later developed receivers will most likely incorporate the ADR receiver module into the main system as implied by Figure 22.7(c). Whichever way the system is deployed, the receiver/decoder will rely heavily on microprocessor control and phase locked loops. Figure 22.8 shows how the processing may be achieved. The input to the receiver will be via an LNB which provides the first IF signal at 950 to 2050 MHz with the band switching, channel switching and polarisation selection under the control of the microprocessor. The satellite tuner section down-converts this to the second intermediate frequency typically at 480 MHz before being amplified under the control of the AGC system. The serial digital signal is formatted into 8-bit parallel words before being input to the QPSK demodulator. Note that this

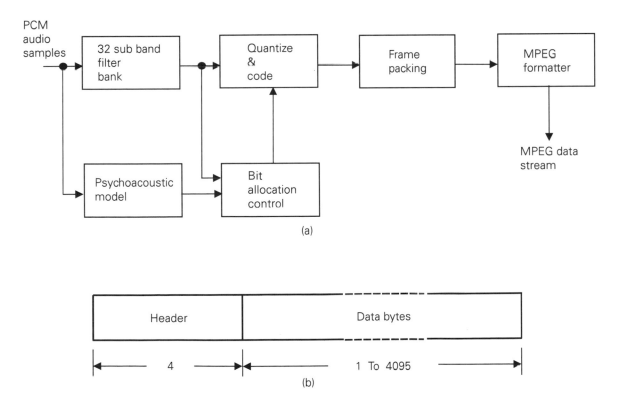

(a)

(b)

Figure 22.6 (a) MPEG audio processor/encoder. (b) MPEG audio bitstream frame format.

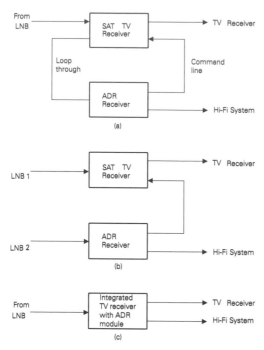

Figure 22.7   Linking a receiver to a satellite system: (a) Separate receivers with loop-through control. (b) TV/radio fed via satellite LNBs. (c) Combined TV/radio receiver.

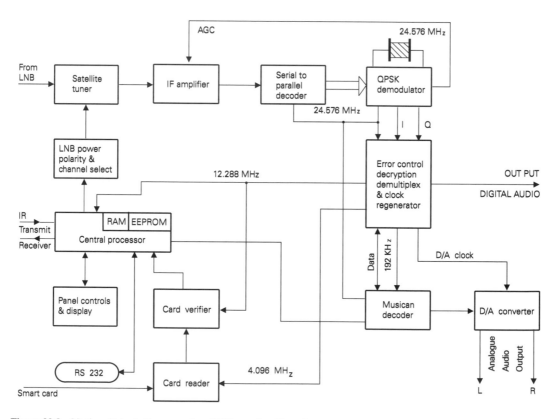

Figure 22.8   Notional block diagram of an ADR receiver/decoder.

stage is driven by a crystal controlled oscillator running at 24.576 MHz and that all the other frequencies that are derived from this are all multiples of the bit rate of 192 kbit/s or sub-multiples of 24.576 MHz. The demodulated I and Q signals are then input to an ASIC which carries out error control, decryption and demultiplexing as well as regenerating all the necessary clock signals. The data-stream is then decoded back into PCM format before being converted into its original analogue form.

A smart card is necessary to provide for pay per listen programmes and this function is controlled via the central processor and the card reader/verifier stages. The processor carries embedded RAM and EEPROM, with the former handling all the user type of data, whilst the EEPROM provides for the programmable memory necessary for channel tuning etc. Because the ancillary data (9.6 kbit/s) carries information about each programme, additional interfaces are provided which allow this to be displayed either on the front panel of the receiver or on the remote control handset.

The central processor adds another significant feature to the receiver: at power-up, the receiver automatically goes into a search and scan mode to identify all the ARD channels and their polarisation data. This is then stored in memory for future use. This system also allows the receiver to detect which channels are transmitting dual language mono signals so that the user can select the appropriate one.

The RS 232E interface provides for linking the receiver to a personal computer, not only does this provide for pre-programming but it is also valuable for service purposes.

Digital Audio Broadcasting (DAB)

This system which derived from the European Union (EU) Eureka 147 development project is accepted by the ITU-R organisation as one of the world standards for digital broadcasting. It is also covered by the European ETSI standards in document ETS 300/401. DAB is sometimes referred to internationally as **Digital System A**. The project developed from the perceived need to find a system that could deliver CD quality audio into moving vehicles with little or no multipath distortion effects. As the system has developed, it has been shown to be capable of delivering a much wider service to listeners, not only those that are mobile, but also as a possible future replacement for conventional sound broadcasting needs. The service can be delivered by terrestrial transmissions, satellites or even via cable networks. Bit rate reduction is achieved by using the MPEG-II (Musicam) algorithm and the data-stream is heavily error protected before transmission.

The system can provide a range of services as follows:

- audio, stereo, mono or bi-lingual;
- programme associated data, programme type, transmitter identification, date, time etc.;
- independent data for road maps, pictures and even differential GPS data.

Coded orthogonal frequency division multiplex (COFDM)

The OFDM concept consists of generating a large number of carrier frequencies with equal spacing. Each is digitally modulated with a sub-band of frequencies and then filtered to produce a $(\sin x)/x$ response as shown in Figure 22.9(a). The spectra of the individual neighbouring carriers thus overlap in the orthogonal manner shown. When these combine, the total spectrum becomes practically flat as indicated by Figure 22.9(b). The channel capacity approaches the Shannon limiting value and the spectrum of carriers thus behaves as a parallel transmission bus.

The allocated bandwidth is divided into N elementary frequencies and arranged to carry P programme channels. There are therefore N/P interleaved elementary carriers which carry sub-band modulation for a given programme in the manner shown. In practice, either 4 or 8 phase PSK or even 64-QAM modulation may be employed.

The system performance can be further improved by employing COFDM and it is this form that is actually in use. A fast Fourier transform (FFT) processor is used in conjunction with convolution error coding at the modulator stage. When used with a complimentary FFT processor and Viterbi decoding at the demodulator, the overall bit error rate is very low. Since the COFDM spectrum has noise-like properties and the signal can be transmitted at relatively low power, it produces very little adjacent channel interference. In addition, co-channel interference from transmitters radiating exactly the same data will actually reinforce the primary signal provided that the delay is within some specific limits.

The transmission system

The multiplex of bit rate reduced audio together with its service information, programme associated data and independent data provides a 2.048 Mbit/s data-stream which is distributed to all the transmitters in the network. The multiplex configuration information is transmitted within the programme associated data group and this is used to ensure correct decoding. This concept can also

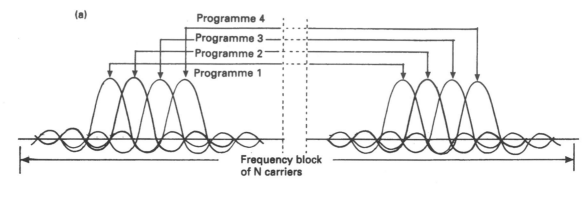

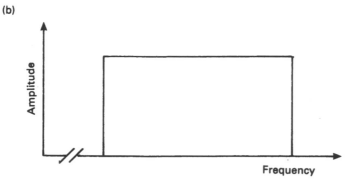

Figure 22.9   OFDM spectra (a) $(\sin x)/x$ response curves, (b) spectrum of combined carriers.

give warning to the receiver about impending changes and so ensure a seamless changeover. The correct synchronisation of the signals for each transmitter is important, hence one possible delivery system would involve the SDH (synchronous digital hierarchy) system of the telecom companies. The data is time-stamped before delivery to the distribution network so that the precise time for the transmission of each packet is accurately known. This time reference may easily be obtained from an incorporated GPS (global positioning system) receiver.

Generating the transmission signal

The COFDM generator converts the 2.048 Mbit/s datastream into I and Q signals using a DSP (digital signal processing) chip. An adjustable delay is arranged via software control to ensure synchronism with other transmitters in the network. The data is modulated using DQPSK on to the individual carriers of the COFDM system. Although the gross bit rate is high, the actual rate per individual carrier is relatively low when shared between all the carriers. This feature provides much of

the multipath immunity. This output is then modulated on to a relatively low RF carrier which can then be up-converted to the final transmission band. Before final radiation, the RF signal is band pass filtered to remove any spurious components that might create interference with other DAB transmissions. The radiated power is relatively low for example, using Band III VHF, the total radiated power may be around 1 kW.

Four modes of operation are defined in the standards as shown by Table 22.2.

Table 22.2   *System operation modes*

System parameters	Transmission mode			
	1	2	3	4
Max separation of TX sites (km)	96	24	12	48
Max receiver frequency (mobile Ghz)	0.375	1.5	3.0	1.5
Number of carriers	1536	384	192	768

Because all the same transmitters within a network radiate the same group of programmes within the same multiplex bandwidth, typically 1.5 MHz, DAB makes much better use of the frequency spectrum. By comparison, four VHF/FM stereo transmissions occupy at least 2.2 MHz of bandwidth.

The signal parameters for Mode 3 operation are shown in Table 22.3.

Table 22.3 *Mode 3 signal parameters*

Frame duration	24 ms
Sync signal duration	250 µs
Total signal symbol duration	156.25 µs
Useful signal symbol duration	125 µs
Guard period	31.25 µs

Typically each carrier of the OFDM system is unaffected by delayed signals up to 1.2 times the guard period. For the 12 km path for Mode 3 operation, this represents a signal arriving from an extra path distance of around 10 km. In fact, such a multipath signal might well be beneficial to the system. With a mobile receiver moving around town, the main signal might well become blocked by a high building. In this case, the reflected signal will fill-in until the vehicle moves out of the shadow.

Design of a possible receiver

Among the many proposals that have been made that might create consumer confidence and enthusiasm for this new system, the following are the result of crystal gazing. It is thought that the listener is unlikely to accept the new system unless the extra cost is somewhat below £200 within the next five years or so. Figure 22.10 based on the superhet principle summarises one possible approach. The front end tuner needs only to be frequency switchable between channels and so could easily be produced today using surface mount technology and a dedicated ASIC chip for the lower frequency operations. Eventually, the whole of this stage could be integrated into a single chip that is software controlled. If a relatively low IF is chosen, then the IF signal could be conveniently converted into digital format at this stage, thus removing much of the filtering normally associated with

such analogue amplifiers. Since quadrature modulation is employed, it has been suggested that a DSP chip using a 50 MHz clock could easily handle the conversion of a 12.5 MHz IF signal. The second DSP handles all the data packet decoding, error control and demodulation. Finally depending upon the success of the Musicam algorithm to audio coding, then within the next few years a dedicated single chip could well be available to generate the audio outputs.

The JPL digital system

This system, which is often referred to internationally as **Digital System B**, has been developed by the Jet Propulsion Laboratory (JPL) of America, under the sponsorship of the Voice of America (VOA). It is capable of delivering full CD quality stereo audio at 384 kbit/s together with a range of lower qualities at slower bit rates right down to 32 kbit/s. The system uses convolutional encoding followed by interleaving for a high degree of error protection. Once again, DQPSK modulation is employed with the data pulses being filtered to ensure bandwidth efficiency. In order to synchronise the demodulation and de-interleaving process, a unique sync word is included within the data multiplex. The receivers for this system are designed to be resilient to fading, echoes and multipath effects by the inclusion of the following features.

- Convolutional encoding at either the 1/2 or 1/3 rate can be selected by the service provider, with the 1/3 rate being preferred for use under the most difficult conditions.
- Viterbi maximum likelihood decoding is well established for high integrity error control systems.
- Because of the multipath effects in built-up areas and within buildings, it is suggested that space diversity reception will be necessary. This involves using more than one aerial on each receiver.

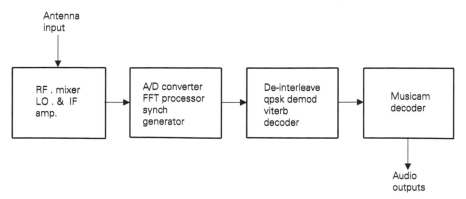

Figure 22.10   Notional block diagram of a 4-chip DAB receiver.

- The receiver circuitry will include an equaliser that automatically chooses between the direct and reflected signals.

Reality of digital sound broadcasting

There are many questions and problems that need to be resolved before such a service becomes widely accepted and it needs wide acceptance to ensure relatively low costs to the listener. Whilst Europe is well advanced with the deployment and testing of the Eureka-147 system, other countries are not quite so sure that this concept will suit their specific needs and applications. Another important consideration involves the availability of frequency spectrum and this varies around the world. Microwave frequency operation is more expensive than VHF. Within the VHF Band II, it might be possible to utilise the taboo channels that already exist between the present analogue FM transmission frequencies. Furthermore, the organisation of sound broadcasting differs widely between Europe and the USA. Whilst much of Europe has stations radiating four or five different programmes from the same transmitter site, the US tends to use single programmes per transmitter and this is not conducive to single frequency networks. The search for a world standard has a long way to go yet.

23 Modern Audio and Hi-fi Servicing

Nick Beer LCGI Mipre

It would be foolhardy in the extreme to suggest that in a chapter of a few thousand words we could cover all that there is to know about servicing modern audio and hi-fi equipment. It is thus the purpose of this chapter to précis the modern servicing scene whilst pondering upon some of the new techniques and problems brought about by modern equipment.

It is a fact (as indeed has been mentioned in other chapters of this book) that for many years audio equipment saw little development or revolution and thus in servicing environments similar stagnation occurred. There was greater interest for an engineer in, say, VCR servicing with its fast moving, innovative technology. This was of course largely paralleled with public perception of audio and hi-fi, and compact disc brought about large-scale changes. The impact on the market is widely appreciated but it introduced to the domestic engineer his first real digital signal processing system. Initial perceptions of great fear and trepidation have largely been replaced with the common experience that few faults, especially on newer generations of compact disc player, occur in the complex digital areas.

Now we have recordable digital formats and multimedia falling into the sphere of a hi-fi engineer. Thus the level of technology, breadth of knowledge and the abilities of a proficient hi-fi engineer are much higher and directly comparable with any other branch of consumer electronics – indeed the distinctions between TV, VCR, hi-fi, telecommunications and PCs are becoming extremely indistinct. Already home PCs can record audio from microphones or line inputs and store them on hard or floppy disks. The computerised musical instrument interface MIDI is present on virtually any PC with a sound card. We can look to the professional and broadcast audio markets to see where we are heading with computer and audio integration. Here audio is (as indeed is video) recorded on hard disks as a digital data and then edited on screen in a familiar Windows or Mac environment with the aid of signal traces and time scales. Sections can be cut and dropped and inserted at will. The audio engineer here needs to consider the massive implications of software set-up or design problems.

Let us then consider some key areas that influence the way in which modern audio and hi-fi is serviced.

Mechanism trends

Let us be under no misapprehension that the majority of servicing involves the mechanism components of a unit. The obvious cassette deck, compact disc player and turntable have now been supplemented by MD (mini-disc) players and recorders (see Figure 23.1) and theoretically at least, DCC (digital compact cassette). In the mainstream hi-fi market, sales of systems with a turntable have nose dived. This is reflected in the number seen for repair – indeed the majority are very old units being determinedly kept going by the vinyl stalwarts who cannot afford or justify the turntables of the enthusiast market and cannot tolerate such mainstream units that are still available!

As well as the conventional, mains-powered units of each format, there are portable and personal audio counterparts (with the exception of the turntable). To qualify my terms, 'portable' refers to battery/mains units with speakers and a handle, 'personal unit' refers to the smallest battery powered units using headphones – generically referred to as 'Walkmans' (a Sony trade mark). Such reduction of scale gives rise to its own problems. Personal stereo equipment, which very often also incorporates a radio tuner is, of course, by definition, very small. One factor that limits the reduction in size is the size of the software, be it cassette or disc. Figure 23.2 illustrates that by building the mechanism through the circuit board, one can achieve these space savings – one also introduces new servicing problems! Many designs today are externally only a fraction larger than the software.

When it comes to servicing, one has to contemplate tiny screws, high component density (although this is not peculiar to personal stereo of course), and often highly

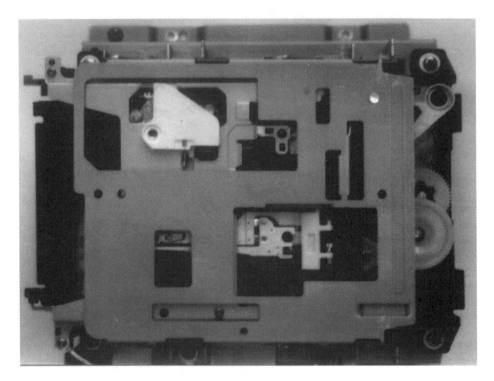

Figure 23.1   Plan view of a mini-disc mechanism showing the all-enclosing nature of the construction.

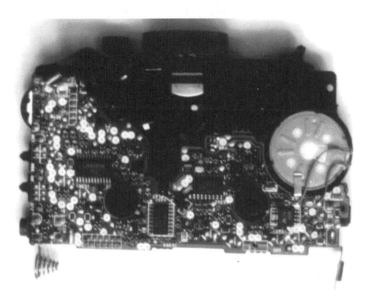

Figure 23.2   A cassette mechanism from a personal stereo illustrating the space saving technique of building the mechanism through the circuit board.

complex but tiny mechanisms. In order to achieve size (and doubtless cost) target, it is often the case that manufacturers make such mechanisms – or at least the base parts of them – in a 'once only' manner. By this I mean that certain parts cannot be removed or replaced –

riveted or melted rather than screwed. This results in the necessity to replace mechanism parts *en bloc* – a practice that is anathema to many engineers. This practice is typically employed on most items in the range, save the most expensive. Certain strategic parts will be available separately, pinch rollers, belts and heads being the obvious examples.

The question raised by this practice is of course viability of repair. A mechanism as a 'spare part' is often unfavourably compared with the cost of a replacement unit. The only converse argument is the smaller labour time involved in such a repair, which may or may not reduce the cost of repair. Under warranty, many manufacturers operate an exchange scheme for low cost items such as the cheaper personal stereos as paying labour claims to repair them is more expensive than throwing them away!

The one incontrovertible parallel that can be observed between personal and home hi-fi is the change in materials used for the components. Traditionally levers and plates were constructed of metals. Plastics found their niche in wheels that then had rubber tyres for drive purposes. This developed into clutches and idlers and then the replacement of rubber tyred wheels with all plastic gears. Today we see large-scale plastic mechanism construction. The extent of integration of plastic components varies but is greatest at the cheaper end of the market. All of the gears and pulleys, mechanism base and head plate, latching plates, key plates even the capstan flywheels are plastic but with a metal insert to retain some weight. The inclusion of plastic gears has meant that even in cheaper designs, rubber components, the traditional Achilles heel of mechanisms, are reduced to perhaps just the pinch roller and one drive belt.

The result is, despite what designers may wish to believe, an inferior mechanism. Life expectancy is generally poorer. To define this more precisely, rather than as was traditional, belts and pinch rollers would wear out and upon replacement, the mechanism would then be in good order again. Modern, plastic designs wear more generally and therefore it is often the case that a larger number of parts require replacement to return the unit to good order. Furthermore, plastic components break. One of the most annoying failures is when a clip or spring hook which is part of the moulding of the plastic mechanism base breaks necessitating replacement of the complete base. This can be at best a long job and at worst a nightmare as all of the mechanism components have to be rebuilt on to the new base.

The plastic-based construction of a mechanism gives rise to some specific problems which are not exactly service failures. Firstly, plastic gears meshing and running at relatively high speed as opposed to a rubber tyred idler on metal pulleys create a noisier mechanism. Complaints about this generally come from those who have just replaced an older unit. Furthermore, static can build up as plastics move across each other or metal components, and as this is discharged a noise will be generated via the audio path. In a cassette system this will often be a crackle on the audio which can be recorded on the tape even when in playback! There are various techniques used to try to counteract such problems: rubber belts can be made conductive as can the lubricating grease. It's another possibility in the list of causes of faults and it's not one you are likely to find if you are not aware that it can be a problem!

It is reassuring that higher specification mechanisms such as those used in commercial and industrial equipment still retain favourable qualities of earlier domestic units!

Circuit trends

Obviously there is a massive range of circuit techniques in use across all formats of audio and hi-fi. Here we will consider those that have been implemented into more recent units and that give rise to the need to reconsider how we tackle faults.

Compact disc

The compact disc really heralded the most major change ever in audio servicing practice. Digital techniques were used throughout a large proportion of the circuits. Lasers were present! When the dust settled for most engineers it became obvious that common sense played as much a part in successfully servicing the players as anything else. However, some engineers still simply cannot come to terms with them. The patterns of design and servicing have panned out much like cassette machines. Mechanical faults are more likely to be the cause of a symptom than electronic ones – this is the most important (and familiar) lesson to learn.

As with cassette decks, especially lower range models, across many different brands one will find similar or identical mechanisms used, often allied to bespoke electronics. With compact disc, two major manufacturers' mechanisms will be found to be used, extensively: Sony, with their KSS range of optical pick-ups and allied KSM mechanisms, and Philips with their CDM range of mechanisms. This brings a number of servicing advantages.

If an engineer is servicing any brand of player and encounters unfamiliar designs, the inclusion of one of these almost generic compact disc mechanisms means that a number of tests can be carried out without service information or a knowledge of the individual unit. This assumes experience of the mechanism concerned of course, which must surely follow any degree of involvement with compact disc servicing.

For example, one will be familiar enough with the unit to check the rf test point – usually obvious in any player – and find that there is 1.2 V p-p of signal with a Sony

KSS150 and be happy that this is adequate. One will also naturally be able to assess whether it is clean and undistorted. If one has a suspect motor, a DC resistance check on brush motors will often elicit a clue as to its fault – if one knows what to expect.

As with any design where there is familiarity there will be experience of failure patterns. Again using the Sony KSS150 as our example, it is commonplace in the event of a failure such as failure to register certain discs, to find that the rf level has dropped to say 400 mV p-p. Cleaning the objective lens (as should always be a first step) will often restore the level by up to 50 per cent but it will still be low and quite possibly 'dirty' i.e. indistinct. Diagnosis of the faulty optical pick-up unit can thus take minutes.

As we have discussed earlier with respect to cassette mechanisms in personal stereo units, the extent of the replacement part in certain compact disc mechanism designs can be quite large. With virtually all portable or personal unit mechanisms, one has to obtain the entire 'traverse deck' of optical pick-up unit, traverse and spindle motors and mounting base. The Philips CDM mechanisms also represent complete replacement units save the very earliest radial designs where the so-called 'light pin' (optical pick-up unit) could be replaced separately. This theoretically required the use of some fairly unlikely and expensive test equipment such as a glass disc to ensure that the laser was perpendicular to the disc.

Wherever a large assembly represents the replacement part, the emphasis shifts heavily on to correct diagnosis. The embarrassment of replacing an entire CD mechanism only to find that one has missed a dry joint on a tracking coil drive transistor can only be guessed at by anyone who hasn't been there!

As will have been gathered, different approaches have been taken by Philips to achieve the same ends – something characteristic of Philips! Whereas most other manufacturers opt for a separate tracking (of the lens) and traverse (of the optical pick-up unit complete) circuit, Philips have combined the two in their radial servo. Instead of the optical pick-up unit being moved across the radius of the disc in a linear plane, they move the optical pick-up unit, courtesy of a linear motor, in an arc across the radius. One has to accept that in terms of reliability of playing without skipping it is the best system. Such linear motor designs cannot be used in portable units as the optical pick-up unit arm is free to move around when not powered or when knocked. The alternative linear system invariably uses a worm drive which obviously has a permanent mechanical linkage.

The greatest incidence of compact disc player failure is where the unit sticks or skips whilst playing. A unit's ability to play without such error is called *playability*, a term considered insufficiently technical by the author. This problem, if not handled correctly, will lead to much wasted time and probably much customer antagonisation. Handling the problem correctly can only stem from adequate ability and experience. One has first to establish whether there is a problem with a player. We discuss towards the end of the chapter the implications and incidences of symptoms being caused by defective software, and poor compact disc playability is a classic case in point. Test discs are available which have designed-in defects to simulate the effects of damaged (interrupted signal layer) and dirty (simulated fingerprints) discs. The discs will have a series of such errors of increasing severity. The player should be easily able to play up to a certain level of error to be within specification. Invariably a working player will play all levels of error and a faulty one will fail at an early stage. If a player passes such tests and checks on the RF waveform indicate no defects (such as a dirty lens or low OPU) one has to look elsewhere, i.e. the software being used. The vast majority of problems will by now have been identified. Beyond this one needs to consider traverse problems, possibly motors or belts, or even dirt in the mechanism providing a physical encumbrance. Furthermore tracking or focus servo problems may hold the key to the problem.

Tuners

Tuners have seen modest development in recent years but the most significant in terms of aspects of servicing is the change from the use of tuning capacitors to varactor diodes and beyond this to electronic (PLL) tuning systems. Fault-finding is rather more straightforward due to the presence of a DC tuning voltage which can be easily checked and monitored. The development of this tuning voltage, however, may be rather more complex – a voltage or frequency synthesis system is employed enabling a microprocessor to interface between the user, a digital display, tuning sweep and preset buttons and the humble varactor diode. To those readers familiar with tuning systems in TV, VCR and satellite equipment, this is, of course, very familiar.

Power supplies

Conventionally anything other than a linear power supply in audio equipment would be a nonsense. All the noise generated by switched mode types would be intolerable to hi-fi users! Therefore power supplies tend to be very simple in principle, if sometimes elaborate in design, to achieve excellent regulation. The modern audio and hi-fi market place contains a massive amount of portable products and they typically are powered by rechargeable battery units; often these batteries are of bespoke design. The charging units can often be very complex units with oscillator based power supplies and all forms of charge detection and over-voltage and current protection.

System control

We cannot go far these days in domestic electronics before we encounter system control circuits in one form or another. They have gradually become commonplace in everything from turntables to compact disc players in audio and are also very prominent in modern TV and VCR designs. Their overall purpose is to interface the various sections of an appliance and ensure that they all operate together correctly. They also initiate appropriate measures, should a failure be detected, to protect the unit and its user. Their exact mode of operation depends, of course, on the type of unit which they control. In audio this inevitably involves an interface between electronics and mechanics as well as between electronic sections and the user.

The inclusion of a system controller circuit is not always a blessing for service engineers though! Whereas it may help the user by preventing misoperation and damage it tends to hinder and complicate fault finding due to the fact that it often prevents a unit from showing its true symptoms. If, for example, a cassette deck loses tacho pulses from one reel sensor it will probably be returned to stand-by by the system controller in exactly the same way as when the capstan stops turning. Without 'syscon' the symptoms would be quite different and the fault immediately obvious. On the other hand you would probably have a handful of ruined tapes.

Microprocessors

The heart of the whole circuit is a microprocessor, which receives information and instructs the other circuits. It has at least one power supply and one ground connection, a master clock oscillator and a reset line. Its inputs come from sensors of one form or another: rotation sensors, infrared remote signal detectors or manual switches on an operation keyboard. In line with these sensors are usually signal amplifiers/buffers, input expanders or A–D (analogue to digital) converters to provide suitable input signals for the processor. The sensors used in various mechanisms are covered in their own chapters.

Microprocessor outputs consist of data to other ICs, switching lines or pulses to operate other circuits, or 'on/off' control lines to turn on switching devices. Figure 23.3 demonstrates to what degree system control can be implemented in modern hi-fi.

A dead unit may well be caused by system control failure. There are three basic checks to make before getting deeply involved. Are the supply(ies), and ground lines, RESET and clock (oscillator) signals correct to the microprocessor? Once these have been established the possibility of a faulty IC can be considered!

Many problems within a unit can be attributed to status lines being incorrect – e.g. a permanently muted audio output will be invoked by system control. One needs to understand the operation of the unit to be able safely to disable such lines for further fault finding. It cannot be over-emphasised that incorrect outputs from a system control are usually caused by incorrect inputs to it rather than a fault within the system controller itself.

Amplifiers

Audio output or power amplifier stages have always tended to cause problems for service engineers. Today one has the two extremes of design. A completely discrete circuit employing many bi-polar transistors or FETs in a multi-stage amplifier with degrees of DC coupling between them brings with it a number of servicing problems. At the opposite end of the scale, many modern amplifiers, including higher range ones, include one or more ICs as the amplifying elements. The servicing of such circuits is generally easier than discrete designs but of course, the emphasis is again on getting diagnoses right!

Output stages are of course one of the highest power stages in any audio unit and thus one of the most prone to failure. This said, many designs are supremely reliable but others are not. One extra factor that puts them at a higher risk of failure is the interface with users, i.e. the speakers and their connections. Short circuits in cabling or connectors or indeed faulty speakers themselves give rise to many amplifier failures. Furthermore, excessive loading can be present where speakers (maybe more than one set) are connected such that their impedance is lower than that permitted by the amplifier. Over a period of such use (usually of running at fairly high volume as well) the output stage runs hot and then fails. An output stage failure should precipitate an investigation of the speakers and their connections to save further failures.

Naturally, one major factor in how amplifier circuits fail is the inclusion of a protection circuit. Indeed, the symptom presented may be that of a protection circuit having operated and one needs to find a masked problem

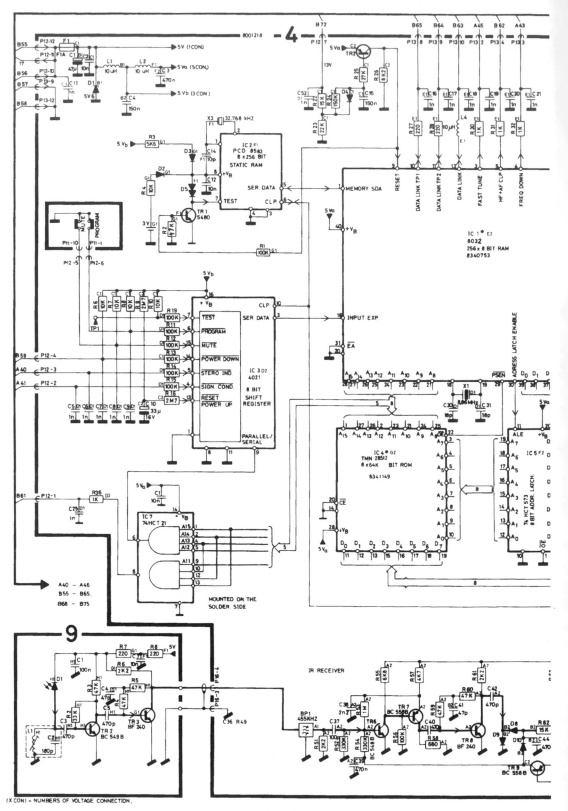

Figure 23.3 A system-control circuit from a sophisticated tuner/amplifier, illustrates ,most of the features discussed in the text: microprocessor control; memories; data buses; displays; remote control; and input expansion. It is referred to many times in the text. The circuit is also particularly relevant to the description, later in this chapter, of integrated control systems, specifically the Bang & Olufsen 'Master Control Link', for which it acts as a central 'master'. The extra features that this entails can be seen, including an infra-red remote control transmitter as well as receiver (Bang & Olufsen)

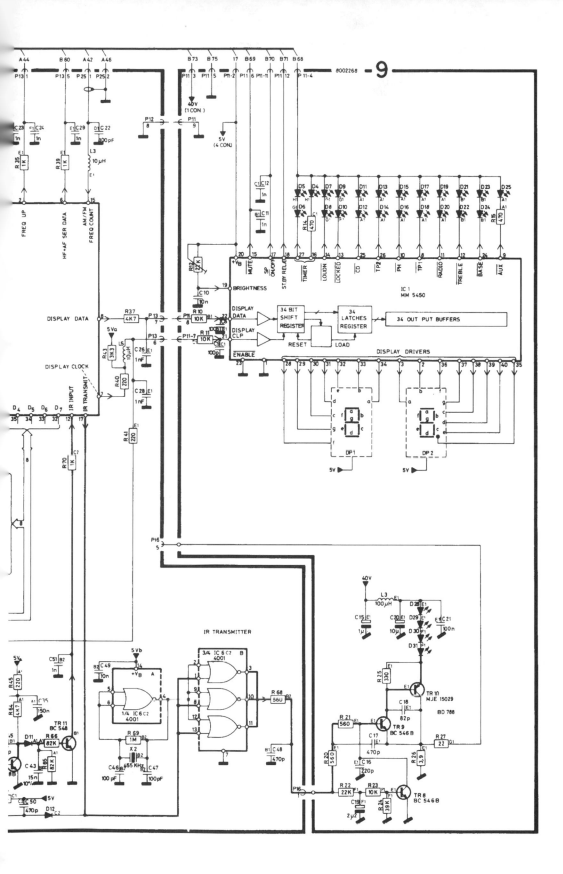

or be able safely to over-ride the protection for further investigation. This latter course is permissible only if severe failures have been ruled out by cold tests.

Some designs leave a lot to be desired in terms of their effectiveness and a lot to be replaced when a failure occurs; others are virulent and will cause muting or shut down under transient undesirable conditions. The result of the latter is fewer failed output stages but some intermittent and annoying symptoms.

It is quite common when pairing certain Japanese amplifiers with specialist British speakers, for example, to get this kind of problem. The amplifier's protection is very keen and despite the speakers being within specification for impedance (4–16 Ω) at usually 4 or 6 Ω, and well capable of handling the power delivered, the protection operates causing no sound after a variable period of time.

The interfacing problem seems to be that the speakers may have a nominal impedance of, say, 6 Ω, but at certain frequencies at the lower end of the spectrum, typically around 100 Hz, the impedance characteristic is such that impedance drops to say 2 Ω or less for the period of that frequency. It is not uncommon for high powers to be present for such IF audio content and so the worst possible set of circumstances are present.

It is interesting to note that in circumstances where such speakers have been used with amplifiers with much less 'fussy' protection, they have suffered rather than the amplifier. You may find open-circuit bass units and tweeters and crossover circuits that have got so hot as to melt the solder on the p.c.b.! There is a large-scale lack of appreciation of the need to get amplifiers and speakers correctly matched. The problem can be exacerbated by the introduction of sub-bass units (sub-woofers), purposely designed to increase disproportionately the bass content of the audio signal. If these are not used with speakers with sympathetic impedance characteristics, such over-running problems can be introduced.

All of this can be seen to increase when users purchase boxes from multiple stores with no installation or expert advice, so that no recourse is made to existing equipment etc.

Discrete output stage failures

Many have a fear of such faults. If one discovers that the output devices have failed then it is best service policy to replace all semiconductors in the area. Refer to Figure 23.4

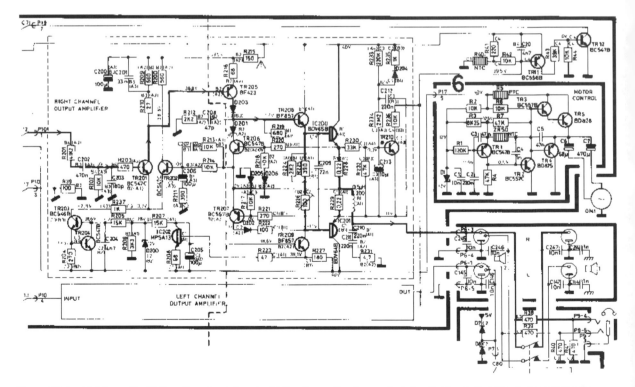

Figure 23.4 A high-quality hi-fi audio output stage including the cooling fan control circuit. The dotted line demonstrates the break point for failures. Semiconductors failing to the right of the line would probably result in protection circuits operating or more catastrophic failure. To the left, one is likely to end up with purely an audio fault- distortion or loss of audio (courtesy of Bang & Olufsen).

which illustrates the cut-off area for this work. Checks should also be made on very low-value resistors (usually less than 2.2 Ω) for even remotely increased values. It would also be sensible in units that are more than a few years old to replace all electrolytic capacitors.

Having rebuilt the stage, one should decrease the bias and power the whole unit up via a variac. Gradually increase the supply whilst monitoring the Iq (quiescent current). If all runs up without excitement, then carefully align the Iq to specification and run the unit for some hours. Ensure that the Iq does not vary outside the specification, if it does then there is another fault and it, of course, could be the underlying cause of the output stage failure. Check on bias stabilising and regulation circuitry – remember the advice to change all semiconductors.

There is always a question mark over the second channel in such circumstances. In most cases it will not be any more likely to fail than it was previous to the repair but in some cases one might find the unit returning to the workshop with the unrepaired channel down. If the primary failure has been found to be due to external causes such as over-running or shorts, then it may be worth rebuilding both stages together. If this involves significant extra cost it is unlikely to be worth it and more worth the risk – ultimately the decision is the owner's, of course, and one can only give informed guidance.

Digital signal processing

Reference here is really to the large-scale processing circuitry encountered in MD or DCC rather than the now well established methods in CD. We should also consider the imminent arrival of DAB (digital audio broadcasting). The problems are manifold. One needs firstly to understand what is to be achieved, then how it's done, at least in outline and finally to consider how one tackles failures (or indeed rules out the possibility of failures) in this area. Incumbent in the final factor is the application of test equipment.

The basic brief of the MD and DCC signal processing circuitry (aside from the obvious A–D and D–A conversions) is to provide data reduction during the record process and reconstruction during playback. The systems employed by MD and DCC are respectively ATRAC (Adaptive TRansform Coding) and PASC (Precision Adaptive Sub Coding). Both work on similar principles of data redundancy achieved through mathematical calculations (cosine transforms/ Fouriers etc.). The theory of all this is mind blowing at its deepest and much can be read elsewhere (a basic grounding may be achieved from *Servicing Audio and Hi-fi Equipment*). From our analysis of modern servicing we need consider these facts.

Firstly, the number of faults that occur in such circuitry are very small indeed. Witness the ratio of digital signal processing faults to mechanism faults in CD players. However, when working on any number of such units, the questions are bound to be there. There is a human tendency to assume that, because a particular circuit is very complex and carries out massive amounts of processes it is more likely to be at fault than a simple circuit. As realists will acknowledge, the inverse applies!

Secondly, the reduction/reconstruction processing is integrated to a large degree to within ICs. The PB and record processes are largely a simple case of direction reversal and so if a fault occurs on playback but not on record or vice-versa, then it is highly unlikely that the problem will be within this signal processing.

An oscilloscope is as useful and practical a tool as anything when tracing faults here. One can only really (at a servicing level) look for the presence or absence of data (especially where it is in serial form) and any obvious deformity in terms of level or shape. Beyond this we are given no information as to data form. If corruption is occurring in an area where parallel data processing is used, an oscilloscope may well show where there is no activity on a specific line. A further useful test may be to open-circuit or short-circuit (via a suitable capacitor) to ground the suspect line and see what effect it has on the fault symptom. If none, then you've hit the right spot, if the symptom varies, then you likely haven't.

Mini-disc

Specific to the mini-disc, we have the magneto-optical recording system. It is all fairly straightforward, especially to those who are experienced with CD but the laser power control is slightly more involved as it has to be able to switch powers between record and playback modes. Here, then, is a prime area to look for a one-mode only symptom. The magneto-optical system has also been seen now in back-up and storage media drives for PCs. Whereas MD has seen quite forcible marketing at a domestic level and fair inroads being made into broadcast radio, DCC has simply not impressed the market.

Alignment procedures

A number of factors have culminated in the greatly reduced requirement for alignment of much audio equipment. A traditional source of requirement is the radio receiver. Here, the use of ceramic filters and IC based circuitry has reduced typical alignment to a couple of adjustments. In turntables and cassette decks, speed adjustments can be simplified by the use of highly accurate servo systems. The fact of the matter is that the number of adjustments is generally reduced across all products but a whole new generation of adjustments and new ways in which adjustments are carried out generally are what we need to consider.

If we take DAT as an example, this is an audio tape recording system. There is no azimuth adjustment, however. The method of operation is helical scan as with most VCR formats. This means that there is tape path alignment of much greater complexity than for any other audio tape format. The alignment procedure will be by monitoring the off-tape FM with an oscilloscope and obtaining the correct (usually flat envelope) shape. Things are slightly more complex though. In a helical scan system there must be some form of tracking arrangement to ensure that the heads both read their correct tracks and run over the track completely rather than overlapping adjacent ones. In DAT (as with the Video 8 VCR system) an automatic tracking system called ATF is used which, to allow sensible alignment, has to be disabled or allowed for. Like a speed servo it will 'follow' adjustments.

In much current hi-fi, especially of digital format, electronic alignment is perpetrated not by a screwdriver but by some form of computer based interface. This may take several forms. An example may be a compact disc player. The player incorporates digital servo signal processing and so its adjustment values for gain and offset are written into non-volatile (NV) RAM within the servo signal processing device. However, the setting of these values has to be initiated. Some units may well have software to allow them to align themselves, others may need values adjusting and then storing. Either way the adjustment will be via a series of key press sequences from the front control panel or remote handset, or possibly from an external device such as a PC, connected via some form of (usually serial) interface. An alternative method of entering the electronic alignment mode is by enabling a link or switch within a unit. This may allow access to test modes useful for fault finding, as well as the alignment modes.

Test modes

With the complexity of modern designs and the inclusion of all-seeing system control, when a failure occurs it may be extremely difficult to fault find in any conventional sense. This is due to the nature of system control circuitry in that it will protect when anything it considers abnormal occurs. This protection usually involves making the unit largely inoperable – keeping it in stand-by or return-

ing it to stop mode within seconds of a problem. A compact disc player may fail to play a disc, it simply returns to stop mode after a couple of seconds of attempting to play. The sequence of events that leads to successful reading of the disc includes focusing, disc spinning on rough servo, tracking and disc speed locking. Dependent on the design of the player, any of these could lead to a disc spinning and then returning to stop having failed.

There is little or no time available to make meaningful tests. A service mode, allowing each stage of the process to be initiated by the engineer, is thus invaluable. This gives the vital indication of where a problem is occurring and by sticking at that stage of the sequence, tests and repairs can be carried out. Simple options on the test mode may allow for example, the laser to be turned on – one may wish to check its presence or indeed measure its power. Alternatively in any unit using a display, there may be a test to illuminate all segments or sections of the display. This helps to determine where a problem may occur by proving that the display and its drive device are clearly (or not) able to produce the display.

One final use of a PC or computerised interface with audio equipment is with in-car units that use a security code. The PC can be connected, via suitable hardware, to read existing codes and write in new ones. Naturally there are legitimate reasons for requiring to do this – flat batteries and lost codes etc. – and indeed when received for repair. There is equally though the obvious use of such technology for theft whereby stolen units are reprogrammed to make them useful again.

Surface mounted and VLSI devices

Virtually every piece of electronics from a remote handset to a mini-disc recorder contains surface mounted components (see Figure 23.5). Indeed we can also safely stipulate that they all contain VLSI (Very Large Scale Integration) devices also. The ideas behind their inclusion are the obvious ones of space saving and production cost reductions. From a servicing perspective they represent another new challenge – 'new' can be qualified as 'within the last few years' to some and 'in the future' to others! Much has been written on how to handle such devices and there are a plethora of soldering tools on the market all claiming to make the job easy. The basic facts are, however, that no method will be straightforward without practice. That practice should not come when one first needs to remove a device in anger. One should obtain scrap PCBs and put in some hours getting familiar with techniques. Similarly, simple and often cheaper methods (and tools) are as straightforward as those costing hundreds of pounds. One has to balance the costs against likely usage. Do not try to take the line that you won't need to work with surface mounted devices though! Even those engineers familiar with handling surface mount devices for some time have recently been facing new challenges with devices incorporating J-legs where the solder joint is under the body of the device. We are also beginning to consider what happens when one has to change VLSIs being used in current equipment where over 150 legs are present with an almost invisible pitch.

For a more detailed discussion of techniques see *Servicing Audio and Hi-fi Equipment* by the same author and publisher.

Obsolete formats

We have once again a situation where format wars have meant that certain units are no longer sold or have irrelevant market position despite their recent introduction. We have seen it previously with eight track and the Elcaset; today DCC and 3DO (the compact disc based multimedia/games format) are rapidly falling into this category. Whereas there may be little or no new software available for these formats, purchasers are not keen to

simply scrap them and so engineers are required to repair them wherever possible. DCC of course can be used to play conventional, analogue compact cassettes and will still perform as a digital tape recorder even if no pre-recorded software can be found.

Software problems

I refer to the tapes and discs that are used in the units that are accused of having failed. When a symptom (e.g. skipping) fails to show despite thorough testing of a compact disc player, it is time to inspect the owner's discs! If a symptom seems improbable ask for an example of the problem on tape or disc. This will often provide the answer. Cassette decks received with badly contaminated tape paths reveal that the user has not been taking care with their tapes – maybe they are damp or just plain worn out. Nip the problem in the bud by checking tapes and disposing of suspect ones – unless you wish to see the deck again soon!

The vast majority of the public fails to appreciate the delicacy of their equipment and the need for the software

Figure 23.5 The inclusion of SMDs is now commonplace – look here and see if a distinction can be easily made between solder joints and SM components!

to be in good condition. The obvious example of where this is most prevalent is with compact disc. Mainly due to press and media misinformation, the public believes that compact discs are indestructible and that a player should play them faultlessly whatever their condition. Nothing is further from the truth. The problem now exists of convincing someone that the problem is a software rather than hardware one.

Good servicing practice

There are a number of points to ponder if one is to consider oneself a service engineer. Many try to carry out repair and alignment without the requisite test equipment. Whereas this may pass unnoticed when repairing low grade, cheap audio units, it is totally unsatisfactory for those aiming to work on anything resembling hi-fi. In certain areas the safety factor comes into play. Can you be sure that laser power on a compact disc player is not at a dangerously high level? Most significantly there is the need to test all mains powered equipment, following repair, for electrical safety. The most convenient way to do this is with a Portable Appliance Tester (PAT) which tests for insulation resistance between the mains input and all exposed conductors and where appropriate the integrity of earth connections. For further information on safety and test equipment, refer to *Servicing Audio and Hi-fi Equipment* by the same author and publisher.

Test equipment

The level of test equipment that is necessary (or acceptable dependent on your perspective!) will vary but if true hi-fi servicing is envisaged then the following should be available to engineers.

- Analogue oscilloscope, minimum 30 MHz bandwidth, not necessarily because of encountering high bandwidth signals but because lower specification units have poor beam acceleration (dim displays) and

poorer triggering abilities. A delayed sweep will also be necessary for DAT amongst others.
- Digital multi-meter with at least 3.5 digit display, bar graph display and reasonable accuracy.
- A frequency counter for the ever increasing appearances of PLLs and VCOs.
- Audio and RF generators, sweep and FM stereo generators. The need for an RDS generation capability will increase, especially for those working extensively on in-car products but it is likely to remain useful rather than necessary for most.
- DC power supply to at least 30 V (to replace tuning voltages). A Variac for power supply and amplifier repairs.
- Wow and flutter/drift meter is essential as is an audio voltmeter (ideally two channel).
- A distortion meter may be dispensable but in the case of all of these items, combined test units at reasonable price (a few hundred pounds sterling) represent the ideal way of obtaining them.

A modern workshop must also have means to be able to work in a static-free environment to prevent damage to ICs, FETs and lasers. Conductive work surfaces and tools are becoming commonplace.

The final few bits and pieces would include a stylus pressure gauge and turntable strobe, a head demagnetiser and some form of infrared detector for remote handsets. The latter can be achieved by connecting an IR receive diode across an oscilloscope trace.

Never forget that test equipment is useless if used without the requisite knowledge. Similarly it is a burden if it only ever sits on a shelf looking pretty!

Conclusion

So a few thousand words to condense a lifetime's experience and a book's worth of advice and observation! This should have given the reader the feel for modern practice and conveyed the impression of the extremely satisfying job that it is. For further research refer to the much mentioned *Servicing Audio and Hi-fi Equipment* and then for the ultimate method of gaining knowledge, work with an experienced engineer.

24 Other Digital Audio Devices

Ian Sinclair

Digital radio, using Eureka-3 DAB, has been discussed in Chapter 23, and is now up and running, although receiver prices are high, and the emphasis so far has been on in-car units rather than home units. There are, however, several other digital options that have now opened out for audiophiles, particularly for those with computing interests. The problem might be that, with so many options either available or promised, no-one can really decide what to buy until the situation settles.

Video Recorders

At the time when Beta and VHS video recorders where competing for the UK market, several makes of Beta recorders, notably Sanyo, offered the option of sound recording by digitising an audio input and recording it as if it were a video signal. These recorders have become prized possessions of some audiophiles because of their good sound recording quality and low-cost media. That's assuming they can still get hold of Beta tapes which, although now rare in the UK, are still easily available in other parts of the world, and are still manufactured for the professional grade of Betamax camcorders.

Looking at more modern equipment, manufacturers such as Hitachi have incorporated audio facilities (including audio dubbing) into Nicam recorders. The input audio signals are converted to Nicam stereo digital format, which implies some compression (see Chapter 22), and recorded. This offers at least three hours of good-quality music on a standard E180 tape. One drawback is that automatic gain control settings often result in rather low-level recording, so that you need to adjust your volume control settings on replay.

HDCD

HDCD means High Definition Compatible Digital, and is one of several recent improvements in the cod-

ing of the familiar compact disc. HDCD discs, developed by Pacific Microsonics, are created using a faster sampling rate of 96 kHz, as compared to the conventional 44.1 kHz used on present CDs, and with 20-bit data units. If this were coded directly on to the CD it would not be compatible with existing CDs, so the data is compressed to 16-bit units and 44.1 kHz pulse rate.

The result is that the HDCD discs can be played on a normal CD deck, but on a player that uses HDCD decoding these discs will deliver more dynamic range and overall better sound. Discs prepared in this way can be recognised by the use of a distinctive HDCD logo.

The first firms to offer players with HDCD capability were Denon, Harman Kardon, Rotel, and Toshiba. Though more than 4000 titles are available in the USA at the time of writing, these are not easy to find in the UK, nor are the HDCD players (one of the recent offerings is from the respected firm of Linn). The players contain interpolation circuitry that can also enhance conventional CDs, and HDCD has, as you would expect, arrived mainly on the players in the £1000 upwards price bracket. Players indicate the presence of an HDCD disc by lighting an indicator. Note that the most recent Philips CD recorder caters for copying HDCD discs.

CD Writers

CD writer drives have become commonplace on PC computers where they are used primarily for making backup copies of valuable data and programs, but they are still a fairly rare sight on hi-fi installations. The CD writer drives as used in computers have several advantages:

1. They are considerably less expensive.
2. They allow you to store computer data, including still or moving images as well as sound.

3. They are compact, fitting into a 5¼ inch drive space.
4. You can decide for yourself what software to use with them.
5. They can use either write-once (CD-R) or read-rewrite (CD-RW) discs.
6. They can make recordings at a higher speed than music can be played.

If you do not have a suitable computer, of course, this option is not open to you, and you will need to look at one of the CD writer units that are intended to be used along with a hi-fi system. Such units are more expensive because they need to incorporate several items of circuitry and software that would be available within the computer. In this chapter, the term *drive* refers to a unit incorporated into, or connected to, a computer, and *deck* means a unit that is part of a hi-fi stack or assembly.

Until 1990, the idea of creating your own CDs would have been considered ridiculous, because the creation of a CD involved many processes that called for elaborate and expensive equipment. The availability of compact disc writing equipment that is well within a normal domestic budget is due to evolution of the CD technology, dispensing with the need to burn into the disk material. The system that is mainly used for home sound recording, or for computer use, is CD-R, meaning CD recordable. This system allows you to write once to a disc and read it as many times as you like. Early versions allowed this also, but the later technology allows you to add more tracks to a disc if you did not fill it on earlier sessions. A disc that permits this type of use is described as *multi-session*. At the time of writing, a blank CD-R disc costs around £0.75, making this the cheapest method of recording that has ever been devised. A CD-R disc will hold up to 74 minutes of full CD-quality music, or the equivalent in computer data, about 650 Mbytes.

Computers and some more recent hi-fi CD recording decks can also use a different form of technology, CD-RW, that allows a disc to be recorded, played, wiped, and recorded again, rather as you re-use a tape or a floppy disk. This technology is at present not so well suited to audio use, and although the blank discs that once cost around £10 each are down to less than £3 each, they are not so popular for computing use either. Many of the better computer CD writer drives can use either type of disc, and prices are remarkably low, typically £125 if you shop around. Most CD-R drives can write at 2× or 4×, or even 6×, depending on the model, which means that they can make a recording of existing digital files faster than a tape. A 2× recorder will record at twice the speed at which the music can be played. This is an advantage for the drive

in a computer, because data files of music can be processed as fast as the CD writer allows, but for the CD-writer drive in a hi-fi installation you cannot speed up the music at the input and high recording speed is pointless. A drive or deck that allows both CD-R and CD-RW discs to be recorded and replayed is known as a CD-R/RW drive or deck.

Unlike DAT, there are no copyright barriers to CD recording. DAT developed as a medium for sound recording, and the record industry worked overtime to make sure that the system was not released until it incorporated safeguards that prevented serial copying. This so hindered the acceptance of DAT that it never became widely used, certainly not in Europe. The writeable CD, by contrast, was developed as a computer peripheral, and the record industry did not realise what was happening until it was too late to stop it. Compared to DAT, CD recording is fast, cheap and easy, and with no hindrance to making copies. Copyright protection has, however, been developed for DVD (see later).

To understand how the change in technology has come about, think back to how the early CDs were manufactured, and are still manufactured. The CDs that you buy are made by burning indentations with a powerful laser into the track surfaces of a master disc, using the presence or absence of a pit to indicate a 1 or 0 digital bit. The player also uses a laser, operating at a much lower power level and aimed at the track. The amount of light that is reflected from the laser beam depends on whether the beam hits a pit or an unpitted piece of track. As the disk spins, these changes in intensity are detected and converted into electrical signals, duplicating perfectly the digital signals that were used to create the original. The advantage of this system is that it permits record pressing analogous to the old vinyl disc method. The CD that is burned by the recording laser is used as a master to make copies that can be used for stamping out plastic discs with the information intact.

This process, incidentally, is much cheaper than the method of recording tapes, which need to be recorded from one end to the other, albeit at a faster speed than they are played. It follows then that a CD is much cheaper to produce than a tape, and some bargain CDs, even in the UK, are sold at prices that reflect the lower cost. The majority of issues, however, maintain the 'CD premium' in prices, in the belief that buyers will pay more for them even if they have cost less to make. You may have noticed the low prices of magazines that have CDs attached, pointing out the low price that the magazine has paid for the CD.

The more modern CD-R drives are recorded using a low-power laser that does not burn pits into the plastic of the disc. Instead, the discs are coated with a dye that is affected by the intense light from the laser. The effect is to change the dye colour, and though the change is not a vast one it can be seen by the eye. If you look at a partly recorded CD-R disc you can see that the recorded portion (the inner part) is a quite distinctly lighter shade of blue (usually) than the outer un-recorded portion. This change is irreversible, so that the disc tracks can be written once only. This type of process is called *dye sublimation*, and the surface appearance of the disc is also due to a thin metallic coating, silver or gold, to make the surface more reflective. CDs created with CD-Rs are compatible with all other computer CD-ROM drives, and with audio CD players.

Oddly enough, a 'premier price' situation has developed with blank CD-R discs. The requirements for recording computer data are more onerous than for sound recordings; after all, your sound system does not shut down if there is a mistake in a tiny fraction of a musical note. This should mean that any blank CD good enough for data recording should certainly be good enough for audio. Some shops, however, will try to sell audio-grade CD-R blanks at a **very** substantial premium.

The rapidly growing use of CD-R/RW has spawned a whole set of new terms that are probably better known to computer users than to audio enthusiasts. Some of the more important terms that have not been explained so far are summarised below.

disc at once (DAO) A CD-R/RW writing mode that requires the whole of the data to be written in one uninterrupted session. Compare **track at once**, **incremental writing**.

finalised disc A CD-R disc that has had its overall lead-in and lead-out information written, so that no further sessions can be recorded.

fixation The set of actions used at the end of a writing session on a CD-R drive. Fixation writes lead-in and lead-out information and creates a table of contents for the disc, so that the disc can be read on a normal CD-ROM drive or audio CD player. If the option of *fixation for append* is used, further sessions can be added to the disc until it is full. See also **finalized disc**.

incremental writing or **packet writing** A method of writing data to a CD-R or CD-RW disc in which several sets of data can be written in each track. This reduces the effect of the overhead of 150 recorded blocks that are used for *run-in*, *run-out* and linking.

lead-in A section of all CD ROM or music discs, pre-recorded, CD-R or CD-RW, that contains information on the data or music contents. The lead-in area immediately precedes the recorded area. For a fully-recorded disc, the lead-in contains the **table of contents**.

lead-out A section of all pre-recorded compact discs that follows the recorded area (on the outer rim of a fully-recorded disc). On the CD-R or CD-RW discs, the lead-out is not created until the disc is declared as fully-recorded (preventing further recording). With no lead-out, the disc cannot be replayed on music players, and some older CD-ROM drives on computers may not accept it.

multi-session Refers to a CD-ROM that can be recorded more than once, adding new material on the subsequent recording, until the disc is full. All computer CD-ROM drives and most hi-fi CD decks should be capable of playing CDs recorded in this way.

PCA Program Calibration Area, the portion of a CD-R disc that is used for making a trial recording to calibrate the laser intensity needed for the disc that is being used. This allows for differences in disc materials, particularly between CD-R and CD-RW discs.

PMA Program Memory Area, the portion of a CD-R or CD-RW disc that contains a table of track numbers along with start and stop data positions for each track.

session A recording made on CD-R or CD-RW that can consist of between 1 and 99 tracks. A session is preceded by a lead-in and ended by a lead-out, and a multisession disc is one that can be recorded at different times, writing a complete session on each occasion, with all the data readable.

table of contents (TOC) A table of track locations and extents prepared by the CD-R/RW software so that the player can locate each track and the data it contains.

track at once A system for writing a CD-R or CD-RW disc that writes the session as a set of complete tracks, compare **disc at once**.

Uses

The hi-fi version of the CD recorder is used much as you would use a cassette recorder, to record music from any other sections of the equipment, such as tuner, cassette deck, vinyl-disc deck, DAT deck, etc. You may also, subject to the restrictions of equipment and copyright, be able to record from an existing CD player, and this type of transfer is much better if the CD player allows a direct digital output that can be connected to the recorder.

The computer type of CD-R/RW drive must be used

along with a sound card that allows line and microphone analogue audio inputs. The quality of recording that you can obtain depends very much on the quality of the analogue–digital conversion in the sound card, and few provide anything like what we accept as CD sound quality. If you are using the system to copy sound tracks from a cassette recorder to a CD, however, the quality level of most cards is acceptable. The line input level of most sound cards is lower that we are accustomed to in hi-fi equipment, and you may need to use the microphone input. This, on the other hand, may be too sensitive, causing distortion at high sound levels, and an attenuator may be needed. Some cassette decks allow you to vary the output, and this is an ideal way of tackling the problem.

The computer type of drive is well suited to CD copying, and to making compilations from a variety of discs. This is not to say that these actions cannot be carried out on the hi-fi type of deck, but you can be certain that the computer type is using direct digital transfers, not converting the CD output into analogue and then converting back to digital in the recorder. The main advantages of using the computer drive are that you can add images, text and other data into the same CD if you wish (and if you can cope with the mixture). This is particularly useful if you want to make multimedia shows of sound and images.

We shall look at audio and other file transfers for the computer CD-R/RW drives in more detail in the following sections, since many of the steps are almost identical. For the moment, a description of a popular hi-fi CD-recording deck will give you an idea of what is currently 'state of the art' in this field.

The Philips CDR 770, Figure 24.1, was launched in September 1999, and was initially marketed mainly in Germany, where the main demand for CD recorders seems to be at present (as an Internet search will confirm). The initial price in Germany was DM 699, roughly £233, which compares well with earlier models from other manufacturers. The CDR 770 uses the 43.5 cm width that is now standard for hi-fi components. Like any other recorders, the CDR 770 allows consumers to make their own recordings from digital sources, as well as from any analogue sources connected to their audio system.

The CDR 770 performs analogue to digital conversion using the Philips system called DLR (Direct Line Recording). This uses the normal CD 44.1 kHz sampling frequency for bit-by-bit conversion, and for CD copying actions it ensures highly accurate recordings by matching the speed of the recording disc to that of the playing (source) disc. For work with other digital sources, different sampling rates are automatically detected, allowing the CDR 770 to deal with any sampling rate from 11 to 56 kHz.

DLR also allows full bit-by-bit recordings to be made of the new HDCD-encoded discs (see earlier, this chapter). The entire encoding of these discs is therefore reproduced on the copy, and is available for playback on compatible CD players with HDCD decoder.

Audio conversions from other analogue sources are also of high quality using the analogue inputs provided. This makes it easy for the user to transfer LP or older disc collections to CD, as well as to record from other analogue sources like tapes, radio or even live music (given suitable microphones).

The CDR 770 also incorporates a CD text function, allowing the consumer to put in text information such as album, artist or track name. When you make a CD recording using the CDR 770 you can enter your own personal text for each disc, for each track or for each artist. Each of these text items can contain up to 60 characters. The text is then shown on the display during playback. CD text that is present on pre-recorded discs will also be displayed when playing the disc back in a Philips Audio CD-Recorder.

Conscious that the hi-fi user is less accustomed to setting up digital equipment than a computer user, Philips have redesigned the user interface of the CDR 770 so as to make the recording action easier and more intuitive. This uses clear messages at every stage to prevent errors and show the user exactly what to do next. For example, the new *Make CD* function allows discs to be recorded and finalised quickly and conveniently, using a single command rather than a set of operations in sequence. Another useful feature is multi-track erase, allowing multiple tracks to be selected and erased at the same time. In addition, the CDR 770 features 99-track programming, easy recording start, an FTD display which gives a clear, at-a-glance indication of the set status, and a music calendar with track bar.

One common problem that users have with hi-fi CD recording is mistaken starts, starting the wrong piece for recording or starting in the wrong place. Because CD-R is a medium that does not allow erasing, this action either makes a set of tracks that you do not want to play, or makes the whole disc unusable. This is no problem for the computer user who makes use of a CD-R/RW drive, because the digital files are stored and can be edited before recording, but this is not the

Figure 24.1 The Philips CDR 770 CD recorder deck.

CDR 770 technical specifications

Number of channels: 2 (stereo)	Applicable supply: AC 230 volt (50/60 Hz)
Power consumption: 15 w	Operating temperature: 5–35ºC
Weight: 4 kg	Dimensions: 435 × 305 × 88 mm (w × d × h)

Audio, general

Frequency response (digital in): 20 Hz–22.05 kHz	Playback S/N: 100 dB
Playback dynamic range: 95 dB	Playback total harmonic distortion: 85 dB (0.0056%)
Recording S/N (analogue): 90 dB	Recording S/N (digital): recording quality equal to source
Recording dynamic range: 92 dB	Recording total harmonic distortion: 85 dB (0.0056%)
Headphones 0–5 V rms / 8-2000Ω	

Recording values for line input/output

Digital coaxial input (direct recording): 12–56 kHz ±100 ppm
Digital optical input (direct recording): 12–56 kHz ±100 ppm
Analogue input (level potentiometer): 700 mV rms/50 kΩ = 0 dB
Line output voltage: 2 V rms ±2 dB
Digital coaxial output: 0.5 Vpp/75 Ω

Recording functions, CD-R & CD-RW discs

Auto start recording per disc	Erase last track (CD-RW disc)
Erase disc (CD-RW disc)	Erase table of contents (for re-recording on finalised RW-disc)
Manual/auto track increment	Remaining recording time display
Auto-finalise (make disc compatible to CD-player)	SCMS (serial copy management system)
RID code (recorder unique identifier)	

Playback functions

Play	Pause	Stop	Direct track selection	Next/Previous track selection

Accessories

Search forward/reverse	Remote control (+batteries)
Repeat (all/ per track)	Audio cable (×2)
Program play (30 tracks)	Digital coaxial cable (×1)
Time display switching	AC mains cord

way that the hi-fi type of CD recording deck works. Philips has included a buffer memory into the CDR 770, allowing storage of up to three seconds of music.

The buffer allows a mistakenly started recording to be stopped within the first 3 seconds, before the start of any disc writing actions, and it also ensures that recordings can be made without loss of music at the start of a track when you are using synchronised CD recording. Buffering also permits the use of synchronised starting from analogue sources. This is done by monitoring the incoming audio signal for the rise in level that indicates the start of play, after which the first few seconds of music are recovered from the buffer for recording.

The recorder incorporates a digital recording level and balance control that allows manual adjustments. This can be used to correct variations between individual discs, a valuable feature if you are making compilations, since it allow you to adjust the volume levels of all tracks. Digital recording level adjustments can easily be made by using the *Easy Jog* control.

The CDR 770 has three sets of input sockets, allowing easy connection to a wide variety of audio sources. Both optical and coaxial digital inputs are provided for the highest-quality connection to digital sources, along with standard stereo sockets for connection to virtually all analogue sources, as well as coaxial digital and analogue outputs.

Like its computer drive counterpart, the CDR 770 can make use of both CD-R and CD-RW blanks, and the CD-R discs can be played on any other CD player, either on audio equipment or in a computer system. The CD-RW discs can be replayed on the CDR 770, and on many of the most recent CD playing decks or drives. If your CD player is not of recent design, however, it will not be able to read the CD-RW discs.

MPEG Systems

When the CD system was launched, following the commercial failure (in the UK, but not elsewhere) of the earlier laser-disc moving picture system, there was no form of compression of data used. The whole system was designed with a view to recording an hour of music on a disc of reasonable size, and the laser scanning system that was developed from the earlier 'silver-disc' was quite capable of achieving tight packing of data, sufficient for the needs of audio.

Data compression was, by that time, fairly well developed, but only for computer data, and by the start of the 1980s several systems were in use. Any form of compression for audio use had to be standardised so that it would be as universal as the compact cassette

and the CD, and in 1987 the standardizing institutes started to work on a project known as EUREKA, with the aim of developing an algorithm (a procedure for manipulating data) for video and audio compression. This has become the standard known as ISO MPEG Audio Layer-3. The letters MPEG means Motion Picture Expert Group, because the main aim of the project was to find a way of tightly compressing digital that data could eventually allow a moving picture to be contained in a compact disc, even though the CD as used for audio was not of adequate capacity (see DVD, later).

As far as audio signals are concerned, the standard CD system uses 16-bit samples that are recorded at a sampling rate of more than twice the actual audio bandwidth, typically 44 kHz. Without any compression, this requires about 8.8 Mbytes of data per minute of playing time. The MPEG coding system for audio allows this to be compressed by a factor of 12, without losing perceptible sound quality. If a small reduction in quality is allowable then factors of 24 or more can be used. Even with such high compression ratios the sound quality is still better than can be achieved by reducing either the sampling rate or the number of bits per sample. This is because MPEG operates by what are termed perceptual coding techniques, meaning that the system is based on how the human ear perceives sound.

The MPEG-1 Layer III algorithm is based on removing data relating to frequencies that the human ear cannot cope with. Taking away sounds that you cannot hear will greatly reduce the amount of data required, but the system is lossy, in the sense that the removed data cannot be reinstated. The compression systems that are used for computer programs, by contrast, cannot be lossy because every data bit is important; there is no unperceived data. Compressing other computer data, notably pictures, can be very lossy, so that the JPEG (Joint Photographic Expert Group) form of compression can achieve even higher compression ratios.

The two features of human hearing that MPEG exploits are its non-linearity and the adaptive threshold of hearing. The threshold of hearing is defined as the level below which a sound is not heard. This is not a fixed level; it depends on the frequency of the sound and varies even more from one person to another. Maximum sensitivity occurs in the frequency range 2–5 kHz. Whether or not you hear a sound therefore depends on the frequency of the sound and the amplitude of the sound relative to the threshold level for that frequency.

The threshold of hearing adapts to the sounds that are heard, so that the threshold greatly increases, for example, when loud noises accompany soft music.

The louder sound masks the softer, and the term masking is used of this effect. Note that this is in direct contradiction of an earlier theory, the 'cocktail-party effect' that postulated the ability of the ear to focus on a wanted sound in the presence of a louder unwanted sound.

The masking effect is particularly important in orchestral music recording. When an full orchestra plays *fortissimo* then the instruments that contribute least to the sound are, according to many sources, not heard. A CD recording will contain all of this information, even if a large part of it is redundant because it cannot be perceived. By recording only what can be perceived, the amount of music that can be recorded on a medium such as a CD is greatly increased, and this can be done without any perceptible loss of audio quality.

Musicians will feel uneasy about this argument, because they and many others feel that every instrument makes a contribution. Can you imagine what an orchestra would sound like if the softer instruments were not played in any fortissimo passage? Would it still be fortissimo? Would we end up with a brass band, without strings or woodwinds? My own view is that the masking theory is not applicable to live music, but it may well apply to sound that we head through the restricted channels of loudspeakers. In addition, how will a compressed recording sound when compared to a version using HCDC technology?

MPEG coding starts with circuitry described as a *perceptual subband audio encoder*. The action of this section is to analyse continually the input audio signal and from this information prepare data (the masking curve) that defines the threshold level below which nothing will be heard. The input is then divided off in frequency bands, called subbands. Each subband is separately quantised, controlling the quantisation so that the quantisation noise will be below the masking curve level for that subband. Data on the quantisation used for a subband is held along with the coded audio for that subband, so that the decoder can reverse the process. Figure 24.2 shows the block diagram for the encoding process.

Layers

MPEG1, as applied to audio signals, can be used in three modes, called Layer I, II and III. Ascending layer number means more compression and more complex encoding. Layer I is used in home recording systems and for solid state audio (sound that has been recorded on chip memory, used for automated voices etc.).

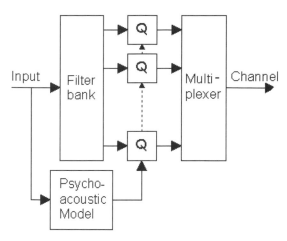

Figure 24.2 Block diagram for MPEG encoding

Layer II offers more compression than Layer I and is used for digital audio broadcasting, television, telecommunications, and multimedia work. The bit rates that can be used range from 32 to 192 kb/s for mono, and from 64 to 384 kb/s for stereo. The highest quality, approaching CD levels, is obtained using about 192–256 kb/s per stereo pair of channels. The precise figure depends on how complex an encoder is used. In general, the encoder is from 2 to 4 times more complex than the Level I encoder, but the decoder need be only about 25% more complex. MPEG Level II is used in applications such as CD-i full motion video, video CD, solid state audio, disk storage and editing, digital audio broadcast (DAB), DVD, cable and satellite radio, cable and satellite TV, ISDN links and in film sound tracks.

Layer III offers even more compression, and is used for the most demanding applications for narrow band telecommunications and other specialised professional audio areas of audio work. It has found much more use as a compression system for MP3 files (see later).

MPEG-1 is intended to be flexible in use, so that a wide range of bit rates from 32 kbit/s to 320 kbit/s can be used, with a low sampling frequency (LSF) of 8 kbit/s later added. Layer III allows the use of a variable bit rate, with the figure in the header taken as the average. Decoders for Layers I and II need not support this feature, but most do.

Table 24.1 shows relative complexity of encoding and decoding for the three levels of MPEG-1. The encoding process is always more complex, but the relative complexity of the decoder is less.

MPEG-1 coding can be applied to mono or stereo signals, and the stereo system makes use of joint stereo coding, a system that achieve further compression by seeking out redundancy between the two

Table 24.1   Comparing complexity of circuitry for MPEG-1 levels.

	Complexity	
Layer	Encoder	Decoder
I	1.5 to 3	1.0
II	2 to 4	1.25
III	7.5 or more	2.5

channels of a stereo signal. The system supports four modes:

mono stereo	joint stereo
(intensity stereo or mid/side stereo)	dual channel (two independent channels e.g. for two languages)

When the digital signal has been encoded, it is divided into blocks of 384 samples (Layer I) or 1152 samples (Layers II and III) to form the unit MPEG-1 frame. A complete MPEG-1 audio stream consists of a set of consecutive frames, with each frame consisting of a header and the encoded sound data. The header of a frame contains general information such as the MPEG Layer, the sampling frequency, the number of channels, whether the frame is CRC protected, whether the sound is an original, and so on. Each audio frame uses a separate header so as to simplify synchronisation and bitstream editing, even if much of the information is repeated and hence redundant. A Layer III frame can achieve further compression by distributing its encoded sound data over several consecutive other frames if those frames do not require all of their bits.

One important point about all digital audio systems is that the analogue concept of S/N ratio is no longer relevant, and so far no replacement has been found. If we try to measure S/N in any of the ways that work perfectly well for analogue signals, the results are widely variable and have no correspondence with the signal as heard by the listener.

- Note that the MUSICAM algorithm is no longer used, it was developed into MPEG-1 Audio Layers I and II. The name MUSICAM is a trademark used by several companies.
- MPEG-1 is one of several (seven at the last count) MPEG standards, and we seem to be in danger of being buried under the weight of standards at a time when development is so rapid that each standard becomes out of date almost as soon as it has been

adopted. Think, for example, how soon NICAM has become upstaged by digital TV sound.

MP3

MP3 is a high-compression coding and decoding system that is now used for transmitting audio signals over Internet links and for storing audio signals in compact computer file form. MP3 allows the construction of small players that store, typically, 40 minutes of music, but contain no moving parts. Because MP3 is a lossy form of compression the MP3 deck for hi-fi systems has not emerged so far, but we should remember that the compact cassette was also considered unfit for hi-fi uses in its initial days. The Minidisk™ uses similar compression methods.

The name MP3 began as an extension to a filename, devised to distinguish sound files created using MPEG-1 Layer III encoding and decoding software. The PC type of computer makes use of these extension letters, up to three of them, placed following a dot and used to distinguish file types. For example, *thoughts.txt* would be a file called thoughts, consisting purely of text, and *thoughts.doc* would be a document called thoughts which could contain illustrations and formatted text, even sounds. A file called *thoughts.jpg* would be a compressed image file and *thoughts.bmp* would be an uncompressed image file. There are many such extensions, each used to identify a specific type of file.

The same MP3 extension is used for sound files that have used MPEG-2 Layer III with a reduced sampling rate, but there is no connection between MP3 and MPEG-3. MP3 files use a compression ratio of around 12:1, so that MP3 files stored on a recordable CD will provide about 12 hours of sound. See later for a discription of DAM-CD.

The main use of MP3, however, has been the portable MP3 player which allows MP3 files to be recorded from downloads over the Internet. This has made MP3 very much of an audio system for the computer buff, but like all matters pertaining to computing, this use is likely to spread. MP3 is unlikely to appeal to those who seek perfection in orchestral music (let's face it, what system does?) but for many other applications it offers a sound quality that is at least as good as anything that can be transmitted by FM radio or obtained from a high-quality cassette.

- The advantages are many. You can load the memory up with music that you like, deleting anything you don't want to hear again. You can play tracks in any order, select tracks at random, and store other music on your PC until you want it on your MP3 player.

One other attraction is rather an illusion, that music is free. Many Internet sites offer MP3 files at no direct cost, but you have to pay for the large amount of telephone time you need for downloading them. Unless you want only fairly small-scale works you will need a fast Internet connection, such as you get with cable TV firms. The alternative is to spend a lot of your income paying for BT phone calls, although alternatives are appearing almost daily. If you really want to download a lot of music it makes sense to take out a fast line or use one of the offers of a fixed charge for unlimited Internet use. Either way, your music is not exactly free.

In addition, 'free' music is often made by unknown artists in strange places. Sometimes you will find an excellent recording made by a Russian orchestra that is unable to raise the money to make CDs or to go on tour. Other recordings may be a quite awful quality, and there is always the suspicion about some of the worst recordings that some tracks may even be illegally acquired, using miniature recorders taken to live concerts. Some may even be copied from existing CDs. On the other hand, the MP3 system is an excellent way for any person or group to make and record their own music and distribute it world-wide without the costs involved in making a CD.

- There is nothing illegal about possessing MP3 files, no matter how they were obtained originally, on your computer. If you make copies and distribute them, that's another matter, and the usual laws of copyright apply. It is certainly illegal to convert CD tracks and distribute the music in MP3 format without the permission of the copyright holder.

Transcribing a Recording by Computer

The standard hi-fi methods of copying music for your own use include cassette recording, DAT and now CD-R or CD-RW. With the help of a computer you can go considerably further by editing the music (cutting out scratches, for example, in old vinyl-disc recordings) or by recording to MP3. The following paragraphs summarise the methods used for computer manipulation of sound for any digital form of recording, mainly CD-R and MP3. The computer must contain a sound card with a A-D converter that is up to CD quality standards, and if you want to record your own CDs you will also need a CD writing drive with appropriate software such as Adaptec *Easy CD Creator*. For MP3 files you will need software such as *Winamp*.

This is not intended to be an exhaustive guide to using a computer for manipulating audio files, because space does not permit thorough treatment of such a large topic. If you are an experienced user of a PC computer, this is a guide to its use for audio work, and if you do not use a computer, it is a guide to what you are missing.

The first step to either the creation of an MP3 file or of a CD-R disc is to extract music tracks and digitise them in an uncompressed format, using a type of file distinguished by the extension letters WAV, hence called a WAV-file. Some software will carry out this action automatically, reading in the audio tracks and converting to MP3 or to CD-R without leaving a WAV file behind on the computer's hard disk. As applied to a CD as source, this action is often termed *CD ripping*. Whether you are aware of it or not, the WAV files are always created as an intermediary, and it's an advantage if you can store them in the computer, check them and possibly edit them, before you save them in MP3 or CR-R format and delete the WAV versions.

You are not obliged to use a CD as source, though, and many users of MP3 or CD-R are more concerned with taking tracks from old 78s, from LPs, or from cassettes, even from radio or private recordings. Remember, however, that no matter what source you use, working at CD quality will require disk space on your computer of around 700 Mbytes for a full CD.

If you are using a CD as your source, you must use the digital output from the CD drive or deck. It is certainly possible to connect the audio output of a CD deck to the line input of the sound card on your computer and to create WAV files in this way, but this sacrifices quality. Most computers that are fitted with a CD writer will also have a fast CD reader, allowing you to read digital data at 36 times (or more) the normal recording speed. This also ensures that the digital output of the CD is used.

- The normal setting on most CD copying software gives you a 2-second gap between tracks when you are working in 'normal mode', which is *track-at-once*. If you specify *disc-at-once*, you will not get any added gaps between the tracks, so if you want extra time between the tracks you have to edit the WAV files so as to include silent intervals. If your list of tracks shows separate files, the recording will always place track markers so that you can move to any track in the usual way.

You can also create WAV files using any other audio source, such as 78s, LPs, cassettes, DAT tapes, etc. The conversion quality will be lower, because these sources all provide analogue signals of varying quality and signal level. You will need to do a few experiments with connections and signal levels, and this is why it is such an advantage to make a separate conversion to WAV, because you can play back a WAV file that is stored in the computer without the need to waste space on CD-R with an unsatisfactory recording, or to make a useless MP3 file. You must, incidentally, use a modern 16-bit sound card – do not try to work with analogue to digital conversions using the older 8-bit type of card. Any computer that is fast enough to cope with audio work will almost certainly be fitted with a 16-bit card. Remember, however, that the quality of A-D conversion may not be as good as you would like.

The usual advice is to connect the audio output from the source device to the line-in connector, usually a 3.5 mm stereo jack socket, on the sound card of the computer. Depending on the sound card that you are using, you may find that the line-in is much too insensitive, and that you hear virtually nothing when you replay the WAV file. The only option, unless you have a spare preamp to connect between the signal source and the sound card, is to use the MIC input. This, by contrast, may be too sensitive, leading to overloading.

The important thing is to try this out with a short piece of music before you start making any recordings to CD-R or MP3. The typical software that you will be using for creating the WAV file is Creative *Wave Studio*, and this permits you to make a short recording to test sound levels. You can adjust the level of the signal on its way to the WAV file, using the software control panel illustrated in Figure 24.3, but this will not help if the input stage of the sound card is overloading.

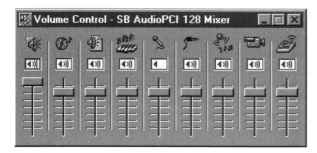

Figure 24.3   A typical software mixer panel as seen on the computer screen

You can then play this back, either through the loudspeakers of the computer or by way of a connection to a hi-fi system, using the output jack of the sound card. The same software, incidentally, allows you to edit a WAV file to remove gaps and, after some practice, unwanted sounds such as scratches and thumps.

You may need to make some setting-up steps, and though some software will do this almost automatically you should check the following:

- Record options must be set to stereo, 8 or 16 bits, 44.1 kHz sampling rate. Use 16-bit data for CD or other high-quality sources.
- Type in the name that you want to use for the WAV file and a folder (directory) on the computer's hard disk where you want to store the file. The usual pattern is to select a name that will describe the music, such as Beethoven 5 Symphony, and store it where you can accommodate a large file of up to 700 Mbytes.
- Check that you have set the recording levels correctly.
- Click the Record icon on the screen and start the source playing. The screen display will probably show the progress of the recording.

You can make your recording one track at a time, making a separate WAV file from each track, or you can make a single file of the whole input. A single file uses less space (because it eliminates the 'overhead' involved in making a separate file for each track), but it makes finding individual tracks (if you need to) more difficult. There are software products such as *LP Ripper* or Adaptek *Spin Doctor* that will work on the WAV file and separate out the tracks for you, and others that will edit the WAV files manually.

The topic of editing a WAV file is too specialised for this book, but the principles are not difficult, and some practice with a short file is more useful than any amount of text instruction.

WAV Onwards

The WAV file is, however, an intermediary. It takes up a large amount of space on your hard disk because it is totally uncompressed, and its main purposes are to allow you to edit the sound and to provide a source for conversion to MP3 or CD-R. Conversion to CD-R is a critical process, because the recording will be ruined if the digital data is not available when the CD writer software needs it. By using the WAV intermediate you are assured of this, because this is a file that is already in digital form, and there is no need to wait for audio signals to be converted at the risk of not keeping up with the demand from the CD writer.

To record the WAV files in CD form, place a blank CD-R disc into the recorder and use a good piece of software such as Adaptek's *Easy CD Creator*. This allows you to choose to make a Data or a Music CD, to select WAV files to record, and put them into the order you want. Once you have files ready, the software will test the files and then make the recording. The disc will be ejected once the recording has been made. You can use another option of the software to print front and back covers for the CD jewelcase. A CD made in this way will play in any reasonably modern CD player. On test, a CD I prepared in this way worked even on a very old Philips CD desk (the first model sold in the UK) with no problems.

If you are preparing MP3 files you need no hardware, only software. All MP3 software is not equal, and some are concerned much more with tricks than with quality or speed of conversion. The software will usually allow a choice of bitrate, and the usual rates are 128 Kbp/s for files sent over the Internet, and 198 Kbp/s for files to be stored on CD-R. At 128 Kbp/s, a four minute piece of sound will need 3.8 Mbytes of storage space, as distinct from closer to 40 Mbytes for an uncompressed CD-quality file.

Recommended software includes MusicMatch *JukeBox 4.1*, AudioSoft *Virtuosa Gold 3.1, and XingMP3.*

DAM CD

DAM is an acronym for Digital Automatic Music, and a DAM CD is one that can contain music both in MP3 format and in normal uncompressed CD format. The MP3 tracks can be copied into any MP3 player, and the normal CD tracks can be played using any normal CD player. It is equally possible to make a DAM CD that contains only MP3 tracks, so packing about ten times more music on to the CD than would be possible using uncompressed CD methods. If you use the computer extensively as a music player, you can transfer your favourite music into this format for easy access and compact storage.

You can also buy DAM CDs over the Internet. They often feature unknown artists, and are priced accordingly, though you should not expect a large selection of classical music to be available. The CD is usually offered on the Web page for the artist, and costs are kept low by using CD-R, so that the music you want is transcribed to a CD when you place your order.

• Some older CD decks cannot cope with DAM CDs that inevitably use multi-session methods.

DVD and Audio

The CD format was standardised at a time when digital recording of sound on disc was still an uncharted realm, full of possibilities and surprised, and CD technology strained at the limits of what was possible, particularly A-D and D-A conversion methods. The use of lasers to write the master discs, though not new because of the Philips 'Silver Discs' used for video recordings, was unfamiliar to many recording companies, and the extent of the packing of bits on the CD stretched the pressing capabilities of all but a few users. Now, at least two decades on, we can see that the potential of the little CD is much greater than we could have hoped for.

DVD, originally the acronym of Digital Video Disc, is now taken to mean *Digital Versatile Disc*, and it refers to a more recent development of CD technology. This was originally directed to recording full-length films on CD, hence the 'Video' in the original title, but the idea has been extended to a universal type of disc that can be used for films, audio, or computer data interchangeably. The main difference, at present is that there are very few DVD writing drives available at present, and these few are expensive by computing (though not by hi-fi) standards.

• An important feature of a modern DVD computer or TV drive unit is that it will accept conventional CDs as well as the DVD discs.

The DVD holds much more data, can transfer it faster, but is as easy to reproduce by stamping processes as the older CDs (which, alas, does not mean that it will be sold at reasonable prices in the UK, even if a DVD costs so much less to produce than a videotape). Eventually, DVD will be the one uniform recording format, replacing cassettes, DAT, videotape, and CD-ROM. A DVD drive is already virtually a standard item on computers, and the manufacturers claim that in a time of three years it has become the most successful electronics product of all time for home use.

Computers are the main end-use of DVD at present, but DVD drives to replace videocassette players are already widely available. The spread of DVD as a replacement for VCR, however, is not likely to spread widely until the recording version reach an acceptable price level. Surveys have shown repeatedly that the most common use for VCRs in the UK is to record TV programmes either when the viewer is not at home, or when two interesting programs are being broadcast at the same time. Use of DVD simply to play pre-recorded discs is very restrictive – I cannot think of more than a handful of films I would ever want to see again, and some of my own videotapes have not been played since the day I recorded them. This is mainly a UK attitude, and the laser disc that was rejected in the UK has survived up until now in other countries.

Figure 24.4   The rear of an APEX AD 600A DVD player which plays also CDs and MP3 files, priced in the USA at about $150.

- With the primary markets of computers and film viewing now being supplied, we are waiting for a standardised DVD format for audio that reportedly will allow up to 17 hours of CD-quality to be stored on a single disc. DVD Audio is already available in Japan, but Europe is unlikely to see these decks/drives before the end of the year 2000.

DVD offers so much more storage space than CD that the options it allows are more than most users can cope with at first. A single-layer disc can store just over 2 hours of digital video signals at a higher quality than is possible using VCR (which relies on considerable bandwidth reduction). More than one layer of CD recording can be placed on a disc, however, because the layers are transparent, and by altering the focus of the reading laser, it is not technically difficult to read either of two superimposed layers that are only a fraction of a millimetre apart.

By making two-layer DVDs the recording time can be doubled, and by adding double-sided recording it can be doubled again to 8 hours of video. The discs can contain up to 8 audio tracks, each using up to 8 channels, so that films can contain soundtracks in more than one language, and cater for surround sound systems.

The DVD can also end the concept of a film as a single story, because unlike tape it can switch from one set of tracks to another very quickly, allowing films to be recorded with several options endings, for example. Different camera angles can also be selected by the viewer from the set recorded on the disc, and displays of text, in more than one language can be used for audio and video tracks. Like CD and so unlike VCR, winding and rewinding are obsolete concepts, and a DVD can be searched at a very high speed that seems instantaneous compared to VCR. The disc is also smaller than a videocassette, does not wear out from being played many times, and resists damage from magnets or heat.

- DVD for video uses MPEG-2 coding and decoding, but there is nothing to prevent cut-price producers from coding with MPEG-1, producing the same

video quality as a VCR. Even MPEG-2, however, is a lossy compression method, and this sometimes shows in video quality as shimmering, fuzzy detail and other effects.
- DVD audio quality, by contrast, is excellent. DVD audio can optionally use CD methods (PCM) with higher sampling rates for even better quality than CD. Other options, used mainly in connections with films, are Dolby Digital or DTS compressed audio.

Regionalisation

Unlike audio CD, DVD is more regionalised than we would wish. Taking a cynical view, this is done to prevent European users from flying over to the USA to stock up with DVDs at bargain prices. Film studios have taken the same attitude to DVD as the record companies did to DAT – if you can't ban it, cripple it. The official reasons are that regionalisation prevents premature release of a file in another country and protects the distribution rights of suppliers in different countries. Regionalisation does **not** apply to DVDs that consist purely of audio signals.

Apart from regional codes, DVDs must be designed to work with the type of colour TV coding that different countries use, so that DVDs have to be manufactured in NTSC, PAL and SECAM versions.

The DVD standard includes regional codes and each DVD drive or deck is allocated a code for the region in which it is marketed. A disc bought in one region will not play on a deck/drive bought in another region, because the codes will not match. Several DVD users in the UK have countered this by buying their DVD equipment in the USA and then buying the discs also in the USA.

The established regions are:

1   US, Canada, US Territories;
2   Japan, Europe, South Africa, Middle East (including Egypt);
3   Southeast Asia, East Asia (including Hong Kong);

4 Australia, New Zealand, Pacific Islands, Central America, Mexico, South America, Caribbean;

5 Eastern Europe (Former Soviet Union), Indian Subcontinent, Africa (also North Korea, Mongolia);

6 China;

7 Reserved;

8 Special international venues such as airliners and cruise ships.

The manufacturers of discs are not obliged to use these codes, and if they do not do so the discs can be used on any drive/deck anywhere in the world. Some types of drives/decks can be modified so that they will play DVDs irrespective of regional coding.

- DVD-ROM discs that are used for computer software are not subject to region codes, nor are audio DVDs.

Copy protection

DVDs can use four different methods of copy protection systems. The *Macrovision* system includes signals that will cause a VCR to record incorrectly by feeding incorrect information to the synchronisation and automatic level control circuits. *CGMS* is designed to prevent serial copying (making copies of copies). *CSS* (Content Scrambling System) is a form of data coding supported by film studios, but the coding algorithm has been cracked and posted in the Internet (along with methods for defeating other protection systems), casting doubt on the future of this method. Finally, the *DCPS* (Digital Copy Protection System) is designed to prevent perfect digital copying between devices that incorporate this coding system.

DVD-Audio

The first DVD drives started to appear around 1996, but at that time there was no agreed format for DVD-Audio, despite the obvious advantages of DVD for audio recording. Considerable effort has gone into defining standards, but the final specification was approved only in February 1999, and the first consumer products are expected in 2000. As usual, the delay has been caused by the introduction of copy-protection codes as demanded by the music industry.

The situation now is that it is possible to design universal DVD players that will deal with both DVD-Video and DVD-Audio, but decks intended for DVD-Audio only will not play DVD-Video. As a further complication, because DVD-Audio is a rather different format, some DVD-Audio discs will not be fully usable in any DVD-Video player other than a universal type which at the moment is not in production or even planned. With some co-operation from manufacturers it would be possible to turn out DVD-Audio discs that would operate on all DVD decks or drives. As usual, it is unwise to be a pioneer consumer, just as it was in the Beta/VHS days.

The protection system that has been adopted uses what the manufacturers call a digital watermark. This adds signals that appear as low-level noise, and the recording companies claim that this is completely inaudible. If enough audiophiles can hear the difference, then it is a distinct possibility that two separate audio markets could develop, one using the older CD format for music acceptable to enthusiasts, with DVD used for all other recordings. One the other hand, we may feel that the golden-eared brigade can always detect the inaudible, even on discs that have not had the coding added, but most users will not be affected.

- Sony and Philips, who developed the CD standards have joined forces again to make their Super Audio CD format that competes directly with DVD-Audio. This takes us all back to the VCR battles of VHS and Beta, but manufacturers are likely to respond by making playing decks that will allow the use of either type of disc. At the moment, neither players nor discs are in plentiful supply.

Though DVD drives will read CDs, they will not in general read CD-R discs, and since recordable DVD is still rather distant this might be a stumbling block to anyone who is contemplating transferring a treasured collection of tracks to CD-R. ON the other hand, most DVD players can read CD-RW discs. This difference arises because DVDs use a laser whose light colour is not the same as is used for CD players, and this light does not match that used for CD-R, though it is better adapted to CD-RW.

At present, I would not urge anyone to rush out and buy DVD-Audio, even if equipment becomes available on the UK market. The list of incompatibilities already suffered by DVD-Video (films that will not play on specific players) is very long, and we simply can't guess how many problems of the same type we might see with DVD-Audio. For the long term, however, the medium must be the future of audio and video distribution. In the UK, much depends on making recordable DVD at a price that is not too much out of line with VCR, and that's a tall order even when all TV is digital. As to DVD-Audio, its day will come when all record companies start to distribute on DVD rather than on CD.

Index

0 VU, 215
100 V line, 320
13-bit PCM, 135
14-bit symbols, 76
4 mm socket, 400
600 ohm load, 215
600 ohm line, 387
70 V line, 320
78 rpm speed, 163
8/14 code, 104
8/10 code, 104

Absolute temperature, 1
Absorbers:
   performance, 37
   studio, 34
   types, 37
Absorption, 4
   dielectric, 384
AC parameters, valves, 174
Academy curve, 335
Acceleration, 159
Accelerometers, 24
Access time, 97
Accuracy, 41
Acetate, 158
Acicular crystals, 113
Acoustic characteristics, 4
Acoustic energy, 276
Acoustic environment, 28
Acoustic gain, 333
Acoustic loading, 290
Acoustic radiator, 285
Acoustic response, 33
Acoustic stiffness, 18
Acoustic transformer, 300
Acoustically dead rooms, 39
Acoustics:
   car, 350
   listening room, 4
   studio, 33
Action, ESL, 307
Action substitution, 153
Active cross-over, 348

Active loads, 228, 267
Active noise absorbers, 12
Active sensing, MIDI, 376
A-D conversion quality, 436
Adaptive bit allocation (ABA), 138
Adaptive digital signal processor (ADSP), 352
Adaptive filter, 140
Adaptive threshold, 432
ADC, 59, 97
Adding circuit, binary, 51
Addition and analogue mixing, 51
Additive effects, speech, 329
Adjacent channel interference, 188
ADR receiver, 407, 409
Advanced designs, power amplifier, 263
Advantages, digital form, 43
Aerials:
   car radio, 353
   feeder, 391
   noise, 201
   signal strength, 211
   testing, 354
AFC, 204
AGC effects, 204
Age, frequency response, 1
Air conditioning noise, 28
Air ionisation, 314
Airtight seal, 31
Alcons equation, 328
Aliasing noise, 68
Aliasing, PCM, 193
Alignment, crystal, 113
Alignment procedures, 423
Alignment protractor, 165
Alignment, tape domains, 113
Alternative transmission methods, 195
Alternator interference, 356
AM car radio, 337
AM linearity, 211
AM qualities, 189
AM rejection ratio, 205
Ambisonics, 27, 340
Ampere turns unit, 120
Amplification factor (μ), 174

Amplifier noise, tape replay, 126
Amplifier, voltage controlled, 362
Amplifiers, 419
Amplitude compensation, 160
Analogue and digital, 41
Analogue audio signal, 47
Analogue delay lines, 364
Analogue for loudspeaker, 287
Analogue, infinite baffle, 290
Analogue, labyrinth enclosure, 295
Analogue model, LS, 287
Analogue oscilloscope, 426
Analogue-digital conversion, 430
Analogue to digital converter (ADC), 59, 97
Analogue waveform, 47
ANC system, 338
AND gate, 43
Angel zoom, 172
Anode, 173
Anode-bend demodulator, AM, 212
Anode (slope) resistance (r_a), 174
Anti-aliasing, 135
Anti-aliasing filters, 193
Anti-resonance, 294
Anti-skating force, 165
Anvil ossicle, 5
Application-specific ICs (ASIC), 44, 217, 407
Arithmetic, binary, 51
Armstrong, Major Edwin, 199
Artificial reverberation enhancement, 336
Artificial waveforms, 366
Assumed remanent flux, tape, 116
ASTRA, 408
Astra digital radio (ADR), 407, 409
AT-cut crystal, 202
ATF system, 424
ATRAC, 423
Attaching wire, 398
Attack, AGC, 204
Attack time, compressor, 145
Attenuation rate, filter, 250
Audibility, bit error, 100
Audible range, 1
Audio conversions, 430
Audio edit, digital, 110
Audio generator, 426
Audio mixing, 53
Audio power amplifier, 252
Audio transducers, 390
Audio witch-doctor, 56
Auditory canal, 5
Auditory masking, 110
Auto-changer deck, 163
Auto-reverse cassette player, 342
Auto selection, 337
Automated bias setting, 132
Automatic frequency control (AFC), 204

Automatic gain control (AGC), 204
Automatic level control, 334
Automatic noise level sensing, 329
Automatic noise reduction. FM, 338
Automatic power control (APC), 80
Average sound insulation, 31
Averaging, 57
AWG, 391
Axial modes, 35
Azimuth recording, 97, 108

B-H curve, 113
Backboxes, 321
Background noise level, 236
Background noise requirement, 28
Back-polarised construction, 21
Back radiation, 295
Backwards compatible cassette, 137
Bailey clamp design, 259
Balance controls, 248
Balanced connections, 392
Balanced line inputs, 133
Balanced operation, 335
Balanced use, cable, 392
Balanced working, 177
Balcony infill LS, 331
Ballistic characteristics, 131
Band splitting, Dolby A, 147
Band-limited impulse, 58
Band-limited noise, 2
Bandpass coupled circuits, 198
Bandwidth allocations, 189
Bandwidth, digital signals, 63
Bandwidth, FM, 190
Barlow-Wadley loop, 202
Barrel, jack plug, 400
Base channel, MIDI, 370
Base of ten, 42
Basic digital recording, 94
Basic waveforms, 367
Basilar membrane, 6
Bass and room size, 3
Bass frequencies, 3
Bass response and AFC, 204
Bass response, maximising, 294
Bass traps, 38
Bass units in car, 347
Batteries, microphone, 20
Baxandall:
    explanation of current dumping, 270
    output stage, 255
    tone control, 241
BBC criteria, noise, 28
BBC FM/VHF parameters, 214
BBC NICAM, 195
BBC PCM distribution, 192
Beam tetrode, 173, 252

Beast, the, 180
Bell modes, 282
Belt drive, 164
Berliner, 157
Beta recorders, 427
Bi-linear companders, 144
Bi-wiring, 395
Bias compensation, 165
Bias frequency, 120
Bias field, 94
Bias level setting, 122
Bias noise, 124
Bias oscillator, 129
Bias setting, 122
Bias, tape, 114
Biaxial orientation, 112
Bidirectional microphone, 15
Bimorph, 21
Binary, 42
Binary adding circuit, 51
Binary arithmetic, 43, 51
Binary division, 58
Binary logic, 43
Binary multiplication, 51
Binary point, 52
Binaural hearing, 8
Binding forces in metals, 173
Binding post, 400
Biphonics, 311
Bipolar transistors, 226
Bit, 43
Bit errors, 100
Bit rate reduction, 411
Bitstream, 91, 136
Blank CD-R, 428
Block address, 105
Block diagram, tape recorder, 125
Blocked impedance, 309
  values, 288
Blocks, data, 98
Blomley non-switching circuit, 269
Blumlein, 157, 249, 310
  arrangement, microphone, 25
  work on stereo, 25
BNC connector, 399
Bode plot, 261
Boltzmann's constant, 219
Bone-conducted sound, 6
Boolean algebra, 43
Boomy speech, 6
Bootstrapped load, 233
Bose, 102 unit, 321
Boundary layer, 94
Boundary microphones, 21
Boundary (PZM) microphone, 21
Box within a box, 33
Braided cable, 391
Brake light interference, 359

Braking systems, 123
Breakage, plastic components, 417
Breakdown, MOSFET, 229
Break in/out, 31
Breakthrough, image, 200
Breathing, 145, 146
Brick wall filters, 136
Bridging, 338
Broadband noise, 125
Broadcast studios, 34
Brushless AC motor, 131
Buffer amplifier, 59
Buffer areas, 29
Buffer memory, 104
Buffer use, 432
Building materials, 33
Bumps, 68
Burst errors, 100
Byte, 44

Cabinet loudspeakers, 322
Cabinet volume, 291
Cable connections, 378
Cable losses, 322
Cable power loss factor, 320
Cable support, 398
Caddy, CD, 344
Calrec Soundfield microphone, 27
Capacitive audio transducers, 390
Capacitive effects, 383
Capacitor microphone, 19
Capacitor model, 389
Capacitor, stabilising, 262
Capacitors, 273
Capacitors, power amplifier, 273
Capstan, 123
Capstan motor, 164
Capture ratio, 206
Car and information (CARIN) system, 360
Car audio, 337
Car loudspeaker, 347
Carbon microphones, 21
Card reader/verifier, 411
Cardioid microphone, 15
Cardioid pattern, 18
Care of CDs, 426
Care of LPs, 170
Care of tapes, 425
Carriage servo, 84
Cartridge, 165
Cascode connection, 267
Casings and interference, 354
Cassette units, car, 342
Cat's whisker, 310
Cathode, 172
Cathode follower, 175
Cavity absorber, 38

CCD echo unit, 365
CCIT J17 pre-emphasis, 143
CD copying, 430
CD premium, 428
CD recorder drives, 429
CD ripping, 435
CD text, 430
CD writer, 427
CD-R, 428
CD-RW, 428
Ceiling loudspeakers, 321
Central cluster system, 326
Ceramic cartridge, 216
Channel allocation, 188
Channel balance controls, 248
Channel bits, 104
Channel coding, 103
Channel note on, MIDI, 371
Channel separation controls, 248
Channels, MIDI, 370
Characteristic impedance, 277, 387
Characteristics:
    cartridge, 165
    distortion, 254
    Dolby B, 150
    Dolby C, 151
    microphone, 14
Choke input filter, 184
Chorus, 364
Chrome tape, 341
Chromium dioxide, 112
Chromium dioxide tapes, 113
Cinema systems, 319, 335
Circuit capacitances, 241
Circuit design, tuners, 212
Circuit layout, 132
Circuit layout, tape, 125
Circuit magnification, Q, 197
Circuit trends, 417
Circular memory, 98
Circumaural headphone, 316
Clamp transistors, 259
Clapham Junction tone control, 246
Clarity, 186, 206
Clarity, FM receiver, 205
Clarity, radio, 186
Class A, 269
Class A designs, 269
Class A mode, 254
Class B mode, 254
Class S amplifier, 271
Class S system, 271
Cleaning, capstan/rollers, 123
Clipping, 263
Clock frequency, 70
Clock, MIDI, 376
Clock rate, final, 77
Closed headphone, 316

Clumping, 113
CMOS bilateral switches, 224
CMOS gate, 45
CMOS logic, 45
Co-channel interference, 411
Coaxial cable, 383
Coaxial connector, 399
Cobalt coating, tape, 112
Cochlea, 5
Coded orthogonal frequency division multiplex
    (COFDM), 411
Coding, stereo, 190
Coercivity, 119
    of tape, 93
Coincident pair, microphone, 25
Cold pin, 401
Collector-base leakage noise, 126
Collimation lens, 81
Collisions, 381
Column loudspeakers, 322
Comb filter effect, 22
Combination, code, 76
Combination microphone, 17
Common cathode amplifier, 174
Common code, system, MIDI, 372
Compact cassette, 341
Compact disc, 67, 231, 343, 417
    care, 426
    construction, 74
    juke box, 344
    player, block diagram, player, servicing, 79
Compact storage, 97
Compander, NICAM, 404
Companding, 143
Comparison of loudness, 6
Compensation, reflection, 352
Complementary noise reduction, 142
Complementary output transistors, 255
Complementary processing, 160
Complementary system, NR, 139
Complex waveform, 2
Compliance, 168
    LS, 280
Components, network, 298
Compresser characteristic, 144
Compression drivers, 319, 325
Compression, time, 98
Compressor/limiters, 171, 334
Compromises:
    design, 274
    disc cutting, 160
Computer-aided editing, 135
Computer based interface, 424
Computers and synthesizers, 368
Concealment, 100
Condenser microphone, 19
Conduction band, 378
Conduction, electrical, 378

Conductors, 378
Cone drive units, 322
Cone flexure, 282
Confidence replay, 101
Conical drive, 164
Conical horn, 300
Connection cables, MIDI, 370
Connector shell, 398
Connectors, 397
Consonants, intelligibility, 328
Constant angular velocity (CAV), 67
Constant charge principle, 306
Constant current sources, 232, 267
Constant current sink, 177
Constant directivity horn, 324
Constant linear velocity (CLV), 67
Constant time delay, 48
Construction, CD, 74
Contact noise, 124
Contact potential, 382
Continuous sine wave, 48
Control layout, 371
Control link, companding, 143
Control room acoustics, 211
Control word, 73
Controller values, MIDI, 371
Controls, 226, 240
Conversion efficiency, microphone, 15
Convolutional cross interleave, 101
Convolutional encoding, 413
Cook-books, 55
Copper wire braid, 392
Copy protection, DVD, 439
Copyright barriers, CD, 428
Copyright infringement, DAT, 344
Corona discharge, 112, 304
Correction, errors, 71
Corruption, 423
    CD, 71
Cosine roll-off, 406
Cosine wave, 48
Counters, 44
Counting base, digital, 42
Coupling bits, 77
Coverage, LS, 325
Creating CD, 428
Crimping, 398
Critical angle prism, 81
Critical distance, 327
Critical frequency, 187
Croft Series III, 178
Cross-blending, 311
Cross-fader, 111
Cross interleave Reed-Solomon code, 135
Crossover, 348
Crossover distortion, 254
Crossover/equaliser, 325
Crosstalk, 118, 168

Crosstalk attenuators, 30
Crystal control, motor speed, 132
Crystal control, tape speed, 132
Crystal form, tape coating, 113
Crystal ovens, 202
Curie point, 96
Current dumping circuit, 270
Current mirrors, 233, 267
Current switching circuit, 60
Current transfer characteristic, 226
Curvature of Earth, 186
Cut and splice editing, 135
Cutter head, 158
Cylinder address, 104
Cylindrical lens, 81

D layer, 187
D MOSFET, 256
D-type connector, 402
D-type flip-flop, 44
DAM CD, 437
Damping factor, 394
Damping, LS, 291
DAO, 429
Darlington push-pull, 267
DASH, 94
DAT, 137, 424, 428
DAT, in car, 345
Data compression, 432
Data presence/absence, 423
Data reduction, 49, 105, 110
Data separator, 103
dbx, 148
DC power supply, 426
DC to DC converter, 339
DCC *see* Digital compact cassette
Dead rooms, 4
Decay, AGC, 204
Decibel unit, 3
Decimal base, 42
Deck, 428
Decoder, CD, 86
Decoding, stereo, 190
Decoding PCM, 194
De-essers, 334
Deformation of cable, 392
Degree of insulation, 30
De-interleaving, 86
Delay, 49, 365
Delay line, 336
Delayed echo, 8
Delays, MIDI, 373
Delta-sigma modulators, 64
Demodulator misalignment, 205
Demodulator systems, FM, 207
De Morgan's theorem, 44
Design compromises, 274
Design requirements, output, 260

Destructive interference, 95
Detectability, 201
Detectable frequency range, 1
Detectors:
    AM diode, 211
    anode-bend AM, 212
    Foster-Seeley, 208
    grid leak AM, 212
    phase coincidence, 208
    pulse counter, 210
    phase locked loop, 209
    ratio, 208
    Round-Travis, 207
    slope, 207
    synchronous AM, 212
Diagnosis, 420
Diaphragms, 280
    materials, 346
    microphone, 16
    passive, 294
    size, ESL, 306
Dielectric absorption, 384, 388
Dielectric loss, 384
Difference signal, stereo, 190
Differential logic circuits, 407
Differential network, 144
Differential non-linearity, 62
Differential pair, 177
Differentially encoded data (DQPSK), 404
Diffraction grating, CD player, 80
Diffraction limitation, 95
Digital audio, 41
Digital audio broadcasting (DAB), 411
Digital audio disc, 67
Digital audio interlink, 387
Digital audio recording, 93
Digital audio tape, car, 344
Digital compact cassette (DCC), 65, 92, 94, 110, 137
Digital compression, 63
Digital data, 68, 415
Digital delay lines, 332
Digital filtering, 54, 87
Digital mastering, 28
Digital methods, synthesis, 365
Digital multimeter, 426
Digital processes, 97
Digital pseudo-random shift register, 363
Digital record head, 93
Digital recording, 161, 343
    tape, 125
Digital reverb, 366
Digital signal processing, 43, 54, 415, 423
Digital system A, 411
Digital to analogue converter (DAC), 60, 90
Digitised audio signals, 387
Digitising waveform, 42
DIN, 215
    connector, 399

Diode detector, AM, 211
Diode ring balanced modulator, 201
Diode switching, 224
Dipole radiator, 299
Direct broadcast by satellite (DBS), 361
Direct-coupled output, 273
Direct-drive turntables, 164
Directional microphone, 22
Directivity, 277
Directivity effect, 280
Directivity, ESL, 307
Directivity factor, 328
Directivity, microphone, 14
Direct line recording, 430
Disc at once, 429, 435
Disc identification information, 73
Disc pressing, 162
Disc servo, 84
Discrete cosine transform (DCT), 65
Discrete fast-Fourier transform (DFFT), 62
Discrete output stage faults, 422
Discrimination, capture, 206
Discrimination, ear, 7
Displaced heads, 111
Display information, 408
Distortion, 9, 237
    horn, 301
    meter, 426
    microphone, 16
Distributed system, 325
Dither, 52, 136
Dither noise, 61
Diversity reception, 337
Dividers, 44
DLR system, 430
Dolby, 144
Dolby A system, 147
Dolby B, 149, 341
Dolby C, 150
Dolby HX Pro, 133
Dolby laboratories, 65
Dolby S, 155
Dolby SR, 152
Dolby stereo, cinema, 335
Dolby surround sound, 27
Dolby system, 130
Domains, 113
Domestic tape recording, 136
Doors, studio, 32
Dopants, tape, 112
Double-leaf construction, 31
Double superhet, 200
Doublet source, ESL, 307
Drain wire, 392
Drift, electron, 379
Drive, 428
Drive amplifiers, 158
Drive servo, 158

Drive systems, 163
Drive unit, 285
Drive unit, horn, 302
Dropouts, 100, 112
Dual-capstan, 123
Dual-diaphragm microphone, 18
Dual gap head, 119
Dual-gate MOSFETs, 201
Ducts, air, 30
Dummy aerial, 354
Dummy head, 16, 311
DVD, 437
DVD-Audio, 439
DVMs, 60
Dword, 45
Dye sublimation, 429
Dynamic element balancing, 61
Dynamic impedance, 228
Dynamic microphone, 18
Dynamic noise filter, 140
Dynamic range, 28, 168
    CD, 231

E layer, 187
Ear, 5
Eardrum, 5
Early high fidelity amplifiers, 10
Earmax, 177
Earth path, 398
Easy jog control, 432
Echo, 365
Eddy currents, 386
    losses, 115
Edge triggering, 44
Edgy sound, 254
Edit gaps, 106
Editing digital tape, 110
Effective mass, 290
Effective series resistance (ESR), 239
Effects units, 334
Efficiency:
    horn, 302
    LS, 292
EFM comparator, 83
Eigentones, 34
Eight bit set, 44
Eight to fourteen modulation (EFM), 74, 77
Elapsed time displays, 132
Electret formation, 273
Electret headphone, 314
Electret microphone, 20
Electrical analogue, LS, 279
Electrical breakdown, air, 304
Electro-acoustic efficiency, 259
Electro-acoustic transducer, 303
Electrodynamic headphone, 313
Electro-forming, 157

Electrolytic capacitors, 184, 423
Electromagnetic pickups, 165
Electromagnetic wave, 384
Electron charge, 379
Electron cloud, 172
Electron movement, 379
Electronic crossover, 334
Electronic enhancement, 134
Electronic noise absorbers, 12
Electronic (PLL) tuning, 418
Electronic signal delay units, 330
Electronic sound sources, 362
Electrostatic headphone, 313
Electrostatic loudspeaker (ESL), 303
Electrostatic microphone, 19
Electrostatic stress, 303
Electrostatics, 383
Elliptical stylii, 168
Emergency override, 329
Emitter follower, output, 266
Emphasis, 142
Enclosures, 290
End correction, 288
Energy dispersal, 406
Energy gap, 378
Enhancement MOSFET, 256
Envelope generator, 363
Environmental protection, 397
Equal loudness contours, 6
Equalisation, 95, 116, 140, 217, 332
Equivalent circuit, 394
    LS cable, 395
Equivalent SPL, microphone, 15
Erase head, 119, 129
Error, azimuth, 108
Error control, NICAM, 407
Error correction, 71, 100, 343
Error, group delay, 49
Errors, quantisation, 69
Excess noise, 126
Excessive loading, 420
Exclusive code, system, MIDI, 372
EXOR gate, 44
Expanded Polystyrene, 283
Expander, 141
Expansion, time, 98
Exponential horn, 301
Extension letters, 434
External airborne noise, 29
External processing, 364
External reference locking, 99
External RF noise, 206
Eye pattern, 103
    waveform, 82
Eyring formula, 39

F layer, 187

Fader, 53
Fairy dust, 171
Fall time, 47
Fast Fourier transform (FFT), 411
Faulty optical pickup, 418
FCC, 68, 402
Feed-forward circuits, 271
Feedback equalisation, 217
Feedback loop stability, 261
Female XLR, 402
Fermi-Dirac statistics, 378
Ferrite materials, tapehead, 118
Ferro-chrome tapes, 113
Field effect transistors, 228
Figure of eight response, 15, 17
Filters, 249
    car audio, 359
    circuits, 249
    noise, 2
    preceding NR circuit, 130
    structures, digital, 56
    voltage controlled, 362
Filtering and oversampling, 89
Finalised disc, 429
Fine speed adjustment, 163
Fingerprint, system, 142
Finite impulse response, 55
FIR filters, 54
First data word, 70
First detector, 199
Fixation, 429
Fixed pattern noise, 172
Flags, MIDI, 371
Flangers, 365
Flash converters, 63
Flat disc source, 281
Flat frequency response, 116
Flat honeycomb sandwich, 347
Flat response, 332
Flip-flop (FF), 44
Floating construction, 33
Floating point converter, 63
Flutter, 163
Flux reversals, 94
Flying comma converter, 63
FM², 408
FM, car reception, 337
FM carrier wavelength, 337
FM qualities, 190
FM reception, car, 337
Focus error amplifier, 82
Focus error signal, 81
Focus OK (FOK) amplifier, 83
Focus servo, 84
    problems, 418
Focusing, 296
Format wars, 425
Foster-Seeley detector, 208

Four-bit coding, 43
Fourier transform, 48
Fractions, 43
Frame, 45
Frame alignment, 405
Frame, MPEG, 434
Frame multiplex, NICAM, 405
Free-air conditions, 291
Free-air headphone, 314
Free-air resonance, 290
Free electrons, 172
Free music, 435
Free-space sound, 3
Frequency, 1, 6
    and insulation, 31
    changer, 199
    choice, 189
    counter, 426
    deviation, 190
    discrimination, 7
    dispersion, 394
    distortion, 9
    maximum absorption, 37
    meter display, 205
    modulated systems, 142
    modulation distortion, 9, 11
    modulation (FM), 103
    multiplier, PLL, 203
    response curve, 198
    response, LS, 292
    response, optical disc, 96
    shift keying (FSK), 106
    synthesis, 418
    synthesiser techniques, 203
Fringe reception areas, 338
Front suspension, 284
Fudging, 134
Fuel pump interference, 359
Full adder, 51
Fundamental resonance, 290
Fuse, 395
Future, headphone, 317

Gain controls, 240
Gamma ferric oxide, 112
Gap, record head, 119
Gates, 43
Gaussian distribution, 53
Gears, plastic, 417
Germanium power transistors, 253
Gibbs phenomenon, 49
Glass, window, 33
Glide, 370
Glitches, 72, 91
Global positioning system (GPS), 412
Golden ratios, 35
Gold plating, 223

Graceful degradation, 100
Granularity, 135, 193
Graphic equaliser, 246, 332, 365
  car, 351
Gray code, 404
Grid, 173
Grid-leak detector, AM, 212
Groove packing density, 157
Groove pitch, 161
Groove radius, 167
Groove walls, 159
Ground attenuation, 186
Ground wave, 188
Grounded grid amplifier, 175
Grounding and shielding, car, 359
Group codes, 104
Group delay, 49
Guard bands, 97
Guitar pickups, 24

Haas effect, 8, 331
Hair cells, ear, 6
Half-track, 132
Hall effect, 382
Hammer ossicle, 5
Hammond organ, 362
Hang-over, 11
Hang-up, 263
Hard clipping, 254
Hard disc recorder, 104
Hard magnetic material, 119
Hard peak clipping, 193
Hardware, MIDI, 370
Harmonic distortion, 237
HDCD, 427
High-definition CD, 427
Head demagnetiser, 426
Head design, 117
Head losses, 116
Headphones, 310
Headroom, 142, 230, 320
Heater motor interference, 356
Heating effect, 382
Helical scan, 424
Helical tape path, 107
Helmholtz absorber, 38
Helmholtz radiator, 293
Helmholtz resonator, 280
Hemispherical tip, 167
Hertz, H., 186
Hertz (unit), 21
Hexadecimal base, 42
HF bias, 114, 120
HF bias rejection, 130
HF saturation, 152
HF stabilisation, 263
High dynamic impedance load, 228

High fidelity designs, 10
High-frequency beaming, 324
High-frequency parasitic oscillation, 258
High-impedance distribution, 320
High level monitoring, 296
High permeability core, 118
High polymer headphone, 314
High-power connectors, 401
High recording densities, 102
Hill and dale recording, 157
Hiss, stereo background, 192
Hole in the middle, 26
Holes, 380
Hole storage, 229
Home keyboards, 364
Homodyne, 201
Homodyne demodulator, 213
Honeycomb sandwich construction, 347
Horn, 300
Horn loading, ribbon, 299
Hot pin, 401
Hot vertical signal, 159
Hum, micrbphone, 16
Hum pickup, 221
Human auditory system, 5
Hyperbolic horn, 301
Hysteresis, 93, 385

IC current mirror, 233
IC opamps, 234
IC solutions, constant current, 233
IC use, voltage amplifiers, 235
IDC connectors, 398
Idler wheel drive, 164
IEC mains connector, 398
Ignition interference, 355
Illusion, stereo, 24
Image breakthrough, 200
Image frequencies, 200
Impact noise, 30
Impedance, 388
  LS, 286
impedance matching in ear, 5
Improvements, cassette system, 341
Impulse amplitude limiter, 206
Impulse noise, 205
Impulse signal, 50
Impurities, in crystal, 381
In-car audio, 337
In car DAT, 345
In car hi-fi, 352
In-room frequency imbalance, 330
Inaccuracy, 41
Incompatibility, car audio, 353
Incorrect inputs, 419
Incus, 5
Index, directivity, 328

Indicator, recording level, 131
Induced hum, microphone, 16
Inductance of electrolytics, 273
Inductive sources, 390
Inductor model, 390
Inductors, 298
Inertance, LS, 279
Infill, 331
Infinite baffle, 290
Infinite baffle systems, 259
Infinite impedance demodulator, AM, 212
Infinite impulse response (IIR), 56
Infrared detector, 426
Infrared laser, 80
Inhomogeneous medium, 124
Injection moulding, 158, 163
Innercode, 101
Inner ear, 6
Input capacitance, 165
Input circuitry, 217
Input connections, 223
Input impedance models, 390
Input push-pull layout, 232
Inputs, 215
Installation, car audio, 352
Instantaneous sound level, 290
Insulation, connector, 397
Insulation displacement, 398
Insulation displacement connection (IDC), 402
Insulation materials, cable, 391
Insulators, 379
Intelligibility of speech, 4, 330
Intensity, 1, 276
Intensity level (IL), 276
Intensity, sound, 6
Intensity stereo, 24
Interconnections, 335, 378, 390
Interdigital transducers (IDT), 199
Interfacing problems, 422
Interference, car, 353
Interleaving, 72, 101, 406, 413
Intermediate stage accuracy, 52
Intermediate frequency (IF), 199
Intermodulation, 201
    testing, 10
Internally generated interference, 355
Interpolation, 57, 88
Inter-symbol interference, 95
Intra-aural headphone, 316
Intrinsic noise level, 230
Inverse square law, 1, 326
Ionisation, 306
Ionosphere, 187
Incremental writing, 429
Input level, 430
ISDN, 402
Isolated transient, 48

J-legs, 425
Jack plug, 400
Jill socket, 400
Jitter, 92, 103
Jitter performance, ADC, 60
JK FF, 44
Johnson noise, 126
JPEG, 432
JPL digital system, 413
Juke box, CD, 344

Kerr effect, 96

Labyrinth enclosure, 295
Lacquer, 157
Laser diode, 80
Latching fastening, 398
Late-arriving sounds, 330
Lateral devices, 229
Lateral PNP, 236
Lateral recording, 157
Lathe, 158
Layering, MIDI, 368
Layers, ionosphere, 187
Layers, MPEG, 433
Layout, stereo listening 24
Lead-in, 429
Lead-out, 429
Lead screw, 158
Left data word, 70
Legality, MP3, 435
Level control, 133
Level of performance, 236
Leyden jar, 303
LF notch filter, tape replay, 128
Life expectancy, 417
Light-pin unit, 418
Light spot size, 95
Limiting temperature, voice coil, 285
Limiting function, 334
Lin circuit, 253
Linear drive motor, 85
Linear gain stages, 239
Linear integrated circuits, 226
Linear interpolation, 72
Linear motor, 418
Linear phase system, 48
Line-matching transformer, 320
Line microphone, 23
Line source, 322
Linearity, 62, 186, 226
    FM, 207
Linking limiters, 193
Linn HDCD, 427
Linn-Sondek turntable, 164
Lip-ribbon microphone, 23

Listener fatigue, 254
Litz wire, 386
Live end dead end (LEDE), 40
Load impedance, 168, 231
Lobe, aerial, 186
Local keyboard on/off, 376
Local oscillator, 199, 201
Locked loop 203
Logarithmic characteristic, 240
Logic, binary, 43
Logic circuits, 43
Logical processes, 43
Long throw, 280
Long-tailed pair, 227, 231, 267
Look-up table, 76
Loop filter, 203
Loss, dielectric, 384
Loss of consonants method, 328
Loss of signal, 11
Loudness, 6, 276
Loudness level, 6
Loudspeaker cables, 393
Loudspeaker enclosures, 290
Loudspeaker pickup, 239
Loudspeakers, 276
    car, 345
    commercial system, 322
Low frequencies, ESL, 306
Low frequency isolation, 30
Low frequency resonance, 183
    LS, 291
Low-impedance distribution, 321
Low noise, bias signal, 121
Low-noise cables, 384
Low-noise levels, 29
Low-noise opamps, 126, 222
Low-pass filter, 296
Low standards, AM, 189
Low-value resistors, 423
LP record, 157
LS protection circuitry, 265
LSB, 43
Lumped components model, 394

M-S stereo, 26
Machinery, noise, 30
Magnetic effects, 384
Magnetic medium, 93
Magnetic tape, 112
Magnetic tape recording, 363
Magnetising force, 120
Magneto-optical discs, 93
Magneto-optical system, 423
Magneto-resistive (MR) head, 94
Mains-borne interference, 239
Mains ripple, 265
Maintaining old recordings, 169

Majority voting logic, 405
Male XLR, 402
Malleus, 5
Manufacture, vinyl disc, 157
Manufacturer's ID, 372
MASH circuit, 91
Masked problem, 419
Masking, 7, 146
Masking effect, 408
Mass law of insulation, 31
Master tape, 157
Matching transformer quality, 321
Material, cone, 283
Materials, 417
    connector, 397
    disc, 162
Matrix circuit, 191
Maximising bass, LS, 294
Maximum output level (MOL), 121
Maximum usable frequency (MUF), 188
Maxwell's equations, 186
Maxwellian distribution, 378
MCC, Philips system, 338
MDS/MDP, 86
Measurement methods, 169
Measurement systems, 274
Measuring headphones, 316
Mechanical connectors, 398
Mechanical faults, 417
Mechanical fixing, 398
Mechanical impedance, 278
Mechanical resonance, 160
Mechanical services, 30
Mechanism components, 415
Mechanisms, CD, 417
Media, digital, 93
Melinex, 112
Membrane absorbers, 37
Memory log, tape, 132
Metallisation, 162
Metal tapes, 113, 341
Metals, connector shell, 398
Methods, MIDI, 369
Microcomputer control (MCC), 338
Microcomputer wave creation, 362
Microphone channels, 171
Microphones, 14
Microphone signal, 335
Microphonic effects, 239
Microprocessor control, 409
Microprocessor technology, 213
Middle ear, 5
Middle-side (M-S) arrangement, 25
MIDI, 415
MIDI layering, 368
MIDI sequencer, 366
MIDI specification, 368
MIDI time code, 376

Miller effect, 175
Mineral insulated cable (MICC), 392
Mini-disc, 96, 423, 434
MiniMoog, 363
Minimum phase behaviour, 49
Mirror amplifier, 83
Mismatching units, 422
Mistaken starts, 430
Missing codes, 61
Mitsubishi noise reduction, 338
Mixers, 199, 335
Mixer noise, 201
Mode messages, MIDI, 373
Mode, reverberation, 3
Modern standards, 236
Modes, MIDI, 370
Modular absorbers, 34
Modular synthesisers, 364
Modulation control, 154
Modulation noise, 124
Modulation systems, 189
Mono mode, MIDI, 370
Mono sound, 24
Moog, Dr. R., 362
MOS devices, 228
MOSFETs, 228
Mother, 158
Motional impedance, 286
Motor unit, 285
Mouth size, horn, 300
Moving car, FM, 337
Moving-coil actuator, 104
Moving-coil cartridge, 217
Moving-coil head amplifier, 219
Moving-coil headphone, 312
Moving-coil loudspeaker, 279, 285, 346
Moving-coil microphone, 18
Moving-coil pickup, 165
Moving-iron headphone, 311
Moving-iron principle, 310
Moving magnet cartridge, 216
Moving magnet pick-up, 165
MP3, 434
MPEG 2, 411, 432
MPG Layer 2, 408
MSB, 43
Mullard, 5–20, 180
Multicore unscreened wires, 391
Multimedia, 430
Multi-miking, 134
Multipath immunity, 412
Multipath reception, 337
Multiple-platter drives, 97
Multiple reflections, 330
Multiplexer, channel, 70
Multiplication, binary, 51
Multi-session, 429
Multi-session CD, 428

Multi-speaker replay, 340
Multi-stage noise shaping (MASH), 90, 91
Multi-track machines, 133
Music centre, 272
Musical telegraph, 362
Musicam, 65, 408, 411
Musician's amplifier, 171
Musique concrète, 363
Muting, 72
Mutual conductance (g_m), 174
Mylar, 112

NAB convention, 402
Nakamichi noise reduction, 338
NAND gate, 44
Narrow-band noise, 2
Narrow-bandwidth FM, 195
Near-coincident arrays, 26
Near-instantaneous companding, 136
Negative feedback, 11, 254
Negative numbers, 43
Networks, 296
Nibble, 45
NICAM, 63, 404, 728, 143
NICAM recorders, 427
NICAM sub-carrier, 404
Nodal drive, 347
Noise, 41, 239
Noise, aliasing, 68
Noise behind the signal, 124
Noise-cancelling microphone, 23
Noise control, 28
Noise criterion (NC), 28
Noise, digital generator, 363
Noise floor, 63
Noise gate, 141
Noise generator, 363
Noise killer, 338
Noise level, tape, 122
Noise margin, 45
Noise, microphone, 16
Noise rating, 28
Noise reduction, 139
Noise reduction circuitry, 133
Noise shaping, 64, 90
Noise spectrum, 142
Noise waveform, 2
Non-complementary system, NR, 139
Non-ideal behaviour, 388
Non-linear mixer, 200
Non-linearity, 9, 41, 432
    errors, 91
Non-magnetic gap, head, 93
Non-polar dielectric, 298
Non-uniform frequency response, 114
NOR gate, 44

Norton model, 174
Notch filter, 128
NOT gate, 43
Number damage, 52
Nyquist, 69
Nyquist criterion, 261
Nyquist's sampling theorem, 58

Objective lens, 81
  CD, 418
Octal base, 42
Odd/even shuffle, 100
Off-axis radiation, 330
Off-axis response, 15
Offset angle, 165
Off-tape FM, 424
Ohm's law, 378
Old recordings, 168
Omnidirectional microphone, 15
Omni mode, MIDI, 370
On-axis response, 14
Once-only assembly, 415
One-bit system, 91
One-mode symptom, 423
Opamps, IC, 234
Open headphone, 314
Open reel digital recorder, 106
Operating mode detector, NICAM, 407
Operational amplifier designs, 219
Optical assembly, 80
Optical connectors, 402
Optical discs, 95
Optical drives, 169
Optical inputs, 432
Optical isolator, MIDI, 370
Optical medium, 93
Optical pickup fault, 418
Optimum bias, 122
Optimum diaphragm size, 292
Opto-isolators MIDI, 373
OR gate, 43
Orthodynamic headphone, 313
Oscillation cone, 283
Oscilloscope use, 422
Outer code, 101
Outer ear, 5
Output fuse, 265
Output impedance, audio, 393
Output level, 168
Output stage faults, 423
Output stages, 419
Output transformer, 253
Output transformerless (OTL) design, 178
Output transistor protection, 258
Output voltage swing, 227
Oval stylii, 168
Overflow, arithmetic, 51

Overload characteristic, 265
Override, power level, 329
Overriding protection, 422
Oversampling, 64, 88, 136
Overshoot, 238

PA cone drive unit, 319
Packet writing, 429
Packing density, 107
Paging use, 324
PAL systems, 404
Panel absorbers, 37
Panpot, 25
Parabolic reflector microphone, 22
Parallel adaptive data slicer, 407
Parallel plate capacitor, 383
Parallel transmission, 46
Parallel-tuned circuit, 196
Parallel wire capacitance, 383
Paralleled input transistors, 219
Parametric controls, 246
Parametric equalisers, 334, 365
Parasitic oscillation, 258
Parity bits, 71, 405
PASC, 138, 423
Pascal unit, 2
Passive components, 388
Passive cross-over, 348
Passive diaphragm, 294
Passive equalisation, 217
Passive radiator, 294
Passive tone control, 241
Patchable design, synthesiser, 363
Pause control, tape, 119
P-bit, 74
PCA, 429
PCM adaptor, 105
PCM encoder, 193
PD array outputs, 82
Peak amplitude, 281
Peak recorded velocity, 159
Peak-shift distortion, 95
Peltier effect, 382
Pen cell operation, 221
Penetration depth, 386
Pentode, 171, 173
Perceived volume level, 240
Perceptual subband audio encoder, 433
Perfect piston, 346
Performance characteristics, 235
Permanently muted audio, 419
Permanently polarised material, 21
Permeability tuning, 337
Personal unit, 415
Phantom powering, 20
Phantom sound images, 24
Phase coincidence detector (PCD), 208

Phase comparator, 203
Phase detector, 208
Phase difference, 388
Phase distortion, 11
Phase/frequency shifter, 334
Phase-linear SAW, 199
Phase lock loop, 337
Phase locked loop, 76, 103, 202, 203, 409
    demodulator, 209
Phase modulation, 195
Phase shifter, 364
Phasers, 364
Philips CDR, 770, 430
Philips DCC process, 137
Phon unit, 6
Phono connector, 398
Phono connectors, 215
Phono inputs, 223
Photo detector (PD) array, 81
Photodiode (PD), 80
Photodiode assembly, CD player, 80
Photo-resist, 75
Physics of conduction, 378
Pickup arms, 165
Pickup inputs, 216
Piezoelectric behaviour, quartz, 202
Piezoelectric microphone, 21
Pilot tone, 130, 190
Pink noise, 332
Pinna, 5
Piston radiation, 277
Pit lengths, 77
Pitch, 1, 6
Pitch bender, 371
Pitch perception, 7
Pits, 68, 75
Place theory, 7
Plane wave, 1
Plastic gears, 417
Plasticiser effect, 284
Plastic mechanisms, 417
Plate reverb, 366
Plating, 391
Platter, 158, 163
Playability, 418
Playing weight, 165
Plug, 4 mm, 400
PMA, 429
PNP silicon transistor use, 126
Point-source, 308
Point-source system, 326
Polar graph paper, 15
Polar response, piston, 281
Polarised prism, 81
Polarising filter, 81
Polarising voltage, 20, 306
Pole-pieces, tapehead, 115
Poly mode, MIDI, 370

Polyphonic glissando, 370
Polyphony, 366
Polypropylene film capacitors, 273
Pollywogs, 112
Poor earthing, 354
Poor quality pressing, 83
Poor quality sound, PA, 321
Porous absorbers, 37
Port, 293
Portable appliance tester (PAT), 426
Portable unit, 415
Portamento, 370
Potentiometer law, 241
Power amplifiers, 335
    car, 338
    output stages, 267
Power factor, 384
Power handling, ribbon, 299
Power levels, 260
Power MOSFET, 255
Power output stages, 252
Power supply, amplifier, 264
Power supply design, 183
Power supply leads, car, 354
Power supply problems, 418
PRBS generator, 406
Preamplifier, 215, 226
    moving-coil cartridge, 165
    with gain, 176
Pre-compensation, 95
Pre-emphasis, 190
Precision adaptive sub-band coding (PASC), 1
Precision resistor chain, 241
Predictability, 186, 205
Preferred ratios, room dimensions, 35
Pressure doubling, 16
Pressure drive units, 300
Pressure fluctuation, 1
Pressure gradient microphone, 17
Pressure headphone, 316
Pressure operation, microphone, 16
Pressure pads, 123
Pressure, sound wave, 2
Pressure zone microphone (PZM), 21
Prestimulatory masking, 8
Previous word hold, 72
Prices, CD, 428
Principle of superposition, 395
Print through, 113
Probability, noise, 53
Problems, disc cutting, 161
Product code, 101
Professional RDAT, 109
Program calibration area, 429
Programmability, 366
Programmable delay, 97
Programme modulated noise, 145, 146
Program memory area, 429

Propagation delay, 46
Propagation velocity, 46
Prophet 5 synthesiser, 368
Protection, output transistor, 258
Protective diodes, 230
Proximity effect, 17, 387
Pseudo-chrome, 341
Pseudo random binary sequence (PRBS), 406
Public address, 319, 322
Public service broadcasting, 186
Pulse code modulation (PCM), 134
Pulse counting demodulator, 210
Pulse density modulation, 92
Push-pull, 227
Push-pull Darlington, 267
Push-pull ESL, 305
Push-pull oscillator, 129
Push-pull symmetry, 231

Q-bit, 74
Q factor, 197
QPSK decoder, 407
Quad cable, 393
Quad current dumping circuit, 270
Quad design, 252
Quad, 44
    disc input, 223
Quadratic residue diffuser, 38
Quality, 160
Quality factor, Q, 197
Quantisation, 58, 68
Quantisation noise, 87, 89, 135, 193, 409
Quarter-track, 132
Quarter wave plate, 81
Quartz crystal, 337
Quartz crystal oscillator, 202
Quasi-complementary circuit, 253
Quiescent current, 273

Radial carriageway tracking, 166
Radial magnetic field, 285
Radial modes, 282
Radial servo, 418
Radiation area, ribbon, 299
Radiation from a piston, 277
Radiation impedance, 276, 277
Radiation of sound, 276
Radiation patterns, 277
Radiator mass, 293
Radio, car, 337
Radio data system (RDS), 360
Radio interference, car, 353
Radio microphone, 23
Radiophonics, 363
Radio receivers, 186
Random access memory (RAM), 73, 96, 97

as recording medium, 96
Random errors, 41, 100
Random noise, 2
Random noise generator, 58
Random sound field, 3
Rapid access, 97
Rate of attenuation, 250
Rate of flare, horn, 300
Ratio detector, 208
Rayl unit, 277
RDAT, 94, 104, 137
RDS specification, 214
Re-alignment, tape head, 342
Re-entrant horn LS, 324
Re-entrant overload characteristic, 265
Reactance, acoustic, 278
Reactive components, 388
Reactively coupled noise, 45
Readout, CD, 95
Real sounds, waveform, 47
Real-time code, system, MIDI, 372
Reargap, 118
Rear suspension 284
Receiver design, 196
Receiver, digital, 413
Receiver, NICAM-728, 406
Recordable compact disc (CDR), 137
Recordable CD, 95
Record amplifier, 130
Record head gap, 119
Recorded wavelength effects, 115
Recording level indication, 131
Recording, magnetic, 93
Recording media compared, 96
Recording speed, CD, 428
Recording system, CD-R, 428
Recovery time, compressor, 145
Rectangular probability density function, 53
Recording WAV file, 436
Rectangular windowing, 54
Recursive filter, 64
Red noise, 363
Redundant bits, 100
Reference bias, 122
Reflected sound, 3
Reflections, 34
    in car, 351
    signal, 337
Reflex cabinets, 291
Refraction in ionosphere, 187
Regionalisation, DVD, 438
Regulator interference, 356
Rejection filter, 130
Relative permeability, 385
Relay circuits, 266
Relays, switching, 223
Release time, compressor, 145
Remanent flux characteristics, 116

Remanent magnetic flux, 114
Repeat, 365
Replay amplifier, 125
  design, 116
Replay equalisation, 127
Replay head gap, 115, 119
Reproduction, vinyl disc, 163
Requirements, equalisation, 128
Requirements, radio reception, 186
Requirements, tape, 112
Reset code, MIDI, 372
Resistance, metallic conductor, 380
Resistive cable loss, 320
Resistor behaviour, 388
Resistor model, 389
Resistor noise, 217
Resolution, 41
Resolution and word length, 51
Resonance setting, 365
Resonances, 34, 169, 324
Response contouring, 332
Response, inner ear, 6
Response shaping components, 217
Response smoothing, 332
Reverberation, 4
Reverberant component, 327
Reverberant environment, 328
Reverberation time, 38, 319
RF condenser microphone, 23
RF generator, 426
RF test point, CD, 417
RIAA, 216
RIAA curve, 160
Ribbon cables, 391, 402
Ribbon LS, 298
Ribbon microphone, 19
Rice and Kellog, 279
Rigid piston source, 281
Ringing, 49, 238, 291
Ringing on transient, 123
Ripple rejection, 2654
Rise time, 47
RMS detection, 149
Room dimensions, preferred, 36
Room modes, 34
Rotary head digital audio tape (RDAT), 107
Rotary head recorder, 97, 107
Rotation point, 148
Rotation speed, CD, 67
Round-Travis detector, 207
RPDF, 53
Rumble, 163, 248
Run-length limits, 104

Sabine formula, 38
Safe operating area (SOA), 258
Safety factor, 426

Sample and hold, 59
Sampler, 97
Sampling, 58
Sampling clock, 100
Sampling frequency, 68, 88, 135
Sampling rates, 366
Sandman, A.S., 271
Satellite delivered digital radio, 407
Satellite tuner section, 409
Saturation output level (SOL), 122
Saturation recording, 93
SAW filter, 198
Scale factor, 405
Scan and search mode, 411
Schottky diodes, 201
SCMS, 109, 137
Scratch filter, 249
Scratched CDs, 344
Screen grid, 173
Screened cables, 392
Screening, car audio, 354
SDAT, 137
Sealed box woofer, 347
Sectors, 104
Security code, 424
Seebeck potential, 382
Selectivity, 186, 196
Self-clocking, 103
Self-demagnetisation, 115
Self-noise, conductors, 380
Self-noise, microphone, 15
Self-shrouding plugs, 400
Semiconductors, 379
Semi-distributed system, 326
Sendust, 119
Sermheiser unit, 312
Sensitivity, 122, 186, 201
  ear, 5
  headphone, 312
  microphone, 15
Separate power amplifiers, 339
Separate power supplies, 265
Separation controls, 248
Sequential copies, 137
Serial copying, 428
Serial copying management system (SCMS), 109, 13
Serial data recording, 102
Serial transmission, 46
Series of 0s, 76
Series of 1s, 75
Series feedback network, 217
Service mode, 424
Servicing, 415
Servo speed control, 131
Servo systems, 80, 343
Session, 429
Set-reset latch, 44
Shapcd windows, 54

Shaw/Baxandall circuit, 255
Shell, connector, 398
Shift and add, 51
Shift registers, 44
Short circuits, 420
Shot noise, 126
Shotgun microphone, 23
Shuffling network, 25
Shunt regulators, 183
Shuttle action, 132
Sideband, 145
Sign bit, 43
Signal amplitude range, 390
Signal characteristics, 230
Signal delay line, 331
Signal distribution, 319
Signal inputs, 215
Signal interconnections, car, 354
Signal intrusion, 265
Signal path length, 46
Signal processing circuitry, 423
Signal rise time, 46
Signal strength and height, 186
Signal to noise ratio, 201
    microphone, 16
Signal to noise, 169
    ultimate, 211
Silencer, duct, 30
Silicon power transistor, 253
Silicone rubber, 392
Simple ramp, 61
Sine response, ear, 6
Single crystal model, 381
Single ended use, cable, 392
Single-function fault, 423
Single sideband (SSB), 195
Single unscreened wires, 391
Sinusoidal waveform, 1
Size of conductors, cable, 320
Sizes, CD, 67
Skin depth, 386
Skip distance, 187
Skipping, CD, 343
Skying effect, 365
Sleeve, jack plug, 400
Slew-rate distortion, 238
Slew-rate limiting, 123
Slew-rate limiting distortion, 262
Sliding stereo separation, 213
Slitting machines, 113
Slope controls, 245
Slope detection, 207
Slow time constants, 147
Small power transistors, 221
Smart card, 411
S/N equivalent, 434
Soft domes, 284
Soft magnetic alloy, 312

Soft magnetic material, 119
Software, 415
Software problems, 425
Soldering, 398
    tools, 424
Solid-state linear circuit, 226
Sone unit, 7
Song pointer, MIDI, 376
Sound absorbing materials, 31
Sound card, 430
Sound editing, 436
Sound field, 3
Sound field system, 134
Sound insulation, 30
Sound localisation, 330
Sound lock, 33
Sound masking systems, 336
Sound pressure, 276
Sound pressure level (SPL), 7, 276
Sound quality, 4, 274
Sound reinforcement, 319, 322
Sound sampling, 362
Sound synthesis, 362
Sound workshops, 364
Source of interference, car, 354
Source impedances, typical, 390
Sources, background noise, 29
Space charge, 172
Space wave, 188
Spaced microphones, 26
Spare part, 417
Specifications, CDR, 770, 431
Specific resistivity, 381
Spectral density characteristics, noise, 53
Spectral shaping, 53
Spectral skewing, 151
Spectrum shaping, 103
Spectrum shaping filter, 406
Speech intelligibility, 328
Speech signal to noise, 329
Speed of sound, 1
Speed servo control, 164
Spherical wave, 1
Spindle servo, 84, 85
Sporadic E, 189
Spring line reverb, 364, 366
Spurious signals, 239
Square law characteristic, 200
Square wave, 1, 49
Square wave performance, 123
SRC system, 338
SSB, 195
St. Paul's Cathedral system, 322
Stabilised power supplies, 265
Stability, 186, 201
Stage gain, 227
Staircase waveform, 134
Stamper, 158

Standards, FM, 90
Standing waves, 3
Staples, 5
State machine analysis, 44
Static build-up, 417
Station identifying codes, 338
Station PA system, 319
Stationary head recorder, 97
Status bytes, MIDI, 371
Step function, 11
Step-up transformer, 19, 219
Stepped pulleys, 164
Stepping motor, 104
Stereo, 190, 310
Stereo effect, 5
Stereo headphone jack, 400
Stereo hiss, 213
Stereo microphones, 24
Stereophones, 310
Stereo radio, 404
Stereo signal, 190
Stirrup ossicle, 5
Stored charge effects, 258
Straight sided cones, 282
Strain gauges, 24
Strain relief, 400
Stranded copper, 391
Stray capacitance, 228, 241
Structural noise, 30
Structure, waveform, 47
Studio acoustics, 28
Studio design, 5
Studio monitor, 296
Stylii, 167
Stylus preheating, 159
Stylus pressure, 165
Stylus pressure gauge, 426
Stylus wear, 168
Sub-bass units, 422
Subcarrier, NICAM, 406
Subcarrier phase, 192
Sub-code detector, 86
Subcode, RDAT, 109
Sub-woofers, 319, 336, 339, 422
Subjective testing, 172, 237
Successive approximation register (SAR), 60
Summary, cables, 397
Summation amplifier, 82
Sun spot activity, 188
Super-Permalloy, 119
Superhet system, 199
Supermatch pair, 230
Supermatched devices, 221
Superposition of waveforms, 47
Superposition principle, 395
Supplementary broadcast signals, 195
Suppression, interference, 355
Surface acoustic wave (SAW), 198

Surface mounted components, 424
Surface recombination noise, 126
Surround sound, 27
Suspended sub-chassis, 164
Suspension, 280
Suspensions, LS, 284
Swarf, 159
SWG, 391
Swishing, 146
Switch-on current demand, 335
Switched mode supplies, 4i8
Switching systems, 223
Syllabic compander, 63
Symmetrical speaker arrangement, 5
Symmetry, bias waveform, 121
Sync. pattern, 103
Sync. word, 73, 78, 413
Synchrodyne, 201
Synchronisation, 99
Synchronous demodulation, AM 212
Synchronous digital hierarchy (SDH), 412
Synchronous switching decoder, 191
Synthesised tuning, 337
Synthesiser, frequency, 203
System equalisation, 332
System exclusive code, MIDI, 372
System common, MIDI, 372
System control, 419
System control circuitry, 424
System frequency response, 329
System real time, MIDI, 372
System reset code, MIDI, 372
System reset, MIDI, 372

Table, 105
Table of contents (TOC), 74
    CD, 74, 429
Tape, 96
Tape backing, 112
Tape care, 425
Tape drive control, 131
Tape heads, 341
Tape path alignment, 424
Tape recording, 112
Tape speed effects, 115
Tape thickness tables, 113
Tape transport, 123
Tapered motor shaft, 164
Telecom C4 system, 148
Telharmonium, 362
Temperature coefficient of resistance, 382
Temperature inversion layers, 188
Temporal masking, 8, 146
Terphan, 112
Test discs, 418
Test equipment, 426
Test modes, 424